Methods in Plant Biochemistry

Volume 3
Enzymes of Primary Metabolism

METHODS IN PLANT BIOCHEMISTRY

Series Editors

P. M. DEY
Department of Biochemistry, Royal Holloway and Bedford New College, UK

J. B. HARBORNE
Plant Science Laboratories, University of Reading, UK

1 Plant Phenolics: J. B. HARBORNE

2 Carbohydrates: P. M. DEY

3 Enzymes of Primary Metabolism: P. J. LEA

4 Lipids, Membranes and Aspects of Photobiology: J. L. HARWOOD and J. R. BOWYER

Methods in Plant Biochemistry

Series editors
P. M. DEY and J. B. HARBORNE

Volume 3
Enzymes of Primary Metabolism

Edited by

P. J. LEA

*Institute of Environmental and Biological Sciences
University of Lancaster, UK*

ACADEMIC PRESS
Harcourt Brace Jovanovich, Publishers
London San Diego New York Berkeley
Boston Sydney Tokyo Toronto

ACADEMIC PRESS LIMITED
24–28 Oval Road
London NW1 7DX

US edition published by
ACADEMIC PRESS INC
San Diego, CA 92101

Copyright © 1990, by
ACADEMIC PRESS LIMITED

All Rights Reserved
No part of this book may be reproduced in any form, by photostat, microfilm or any other means, without written permission from the publishers

This book is printed on acid-free paper. ∞

British Library Cataloguing in Publication Data is available

ISBN 0-12-461013-7

Filmset by Bath Typesetting Limited, Bath, Avon
Printed by Galliard (Printers) Ltd, Great Yarmouth, Norfolk

Contents

Contributors	ix
Series preface	xiii
Preface	xv
Glossary	xvii

1	Ribulose Bisphosphate Carboxylase/Oxygenase and Carbonic Anhydrase *A. J. Keys and M. A. J. Parry*	1
2	Enzymes of the Calvin Cycle *R. C. Leegood*	15
3	Enzymes of C_4 Photosynthesis *A. R. Ashton, J. N. Burnell, R. T. Furbank, C. L. D. Jenkins and M. D. Hatch*	39
4	Enzymes of Sucrose Metabolism *L. Copeland*	73
5	Fructose 2,6-Bisphosphate *M. Stitt*	87
6	Enzymes of Starch Synthesis *A. M. Smith*	93
7	Starch Degrading Enzymes *M. Steup*	103
8	Enzymes of the Photorespiratory Carbon Pathway *R. D. Blackwell, A. J. S. Murray and P. J. Lea*	129
9	Glycolysis *W. C. Plaxton*	145
10	The Mitochondrial Pyruvate Dehydrogenase Complex *D. D. Randall and J. A. Miernyk*	175

11	Enzymes of Fatty Acid Synthesis *J. L. Harwood, M C. Walsh and K. A. Walker*	193
12	Enzymes of Lipid Degradation *A. H. C. Huang*	219
13	Enzymes of Phospholipid Synthesis *T. S. Moore Jr.*	229
14	Nitrate Reductase and Nitrite Reductase *J. L. Wray and R. J. Fido*	241
15	Enzymes of Ammonia Assimilation *P. J. Lea, R. D. Blackwell, F.-L. Chen and U. Hecht*	257
16	Aminotransferases *R. J. Ireland and K. W. Joy*	277
17	Enzymes of Asparagine Metabolism *K. W. Joy and R. J. Ireland*	287
18	Enzymes of Lysine Synthesis *P. L. R. Bonner and P. J. Lea*	297
19	Threonine Biosynthesis *S. E. Rognes*	315
20	Enzymes of Leucine, Valine and Isoleucine Biosynthesis *R. M. Wallsgrove*	325
21	Sulphur Metabolism	
	A ATP-Sulphurylase *D. Schmutz*	335
	B Adenosine 5'-phosphosulphate sulphotransferase *C. Brunold and M. Suter*	339
	C Sulphite Reductase *C. Von Arb*	345
	D. Cysteine Synthase *A. Schmidt*	349
	E Synthesis of Glutathione *S. Klapheck and H. Rennenberg*	355
	F Enzymes Involved in the Synthesis of Methionine *J. T. Madison*	361
22	Protein Kinase *A. M. Hetherington, N. H. Battey and P. A. Millner*	371

23 Tonoplast Adenosine Triphosphatase and Inorganic Pyrophosphatase   385
P. A. Rea and J. C. Turner

Index   407

Contributors

Anthony R. Ashton, Division of Plant Industry, CSIRO, GPO Box 1600, Canberra City ACT 2601, Australia
Nicholas H. Battey, Department of Horticulture, University of Reading, Reading RG6 2AS, UK
Ray D. Blackwell, Division of Biological Sciences, University of Lancaster, Lancaster LA1 4YQ, UK
Phillip L. R. Bonner, Division of Biological Sciences, University of Lancaster, Lancaster LA1 4YQ, UK
Christian Brunold, Pflanzenphysiologisches Institut der Universität Bern, Altenbergrain 21, CH-3013 Bern, Switzerland
James N. Burnell, Division of Plant Industry, CSIRO, GPO Box 1600, Canberra City ACT 2601, Australia
Feng-Lin Chen, Department of Biological Sciences, University of Warwick, Coventry CV4 7AL, UK
Les Copeland, Department of Agricultural Chemistry, University of Sydney, NSW 2006, Australia
Roger J. Fido, Department of Agricultural Sciences, University of Bristol, AFRC Institute of Arable Crops Research, Long Ashton Research Station, Long Ashton, Bristol BS18 9AF, UK
Robert T. Furbank, Division of Plant Industry, CSIRO, GPO Box 1600, Canberra City ACT 2601, Australia
John L. Harwood, Department of Biochemistry, University of Wales College of Cardiff, PO Box 903, Cardiff CF1 1ST, UK
Marshall D. Hatch, Division of Plant Industry, CSIRO, GPO Box 1600, Canberra City ACT 2601, Australia
Ursula Hecht, Biologisches Institut II der Universität, Schanzlestrasse 1, D-7800 Freiburg i. Br., FRG
Alistair M. Hetherington, Department of Biological Sciences, University of Lancaster, Lancaster LA1 4YQ, UK
Anthony H. C. Huang, Department of Botany and Plant Sciences, University of California, Riverside, CA 92521, USA
Robert J. Ireland, Department of Biology, Mount Allison University, Sackville, NB, E0A 3C0, Canada

CONTRIBUTORS

Colin L. D. Jenkins, Division of Plant Industry, CSIRO, GPO Box 1600, Canberra City ACT 2601, Australia

Kenneth W. Joy, Department of Biology, Carleton University, Ottawa, Ontario K1S 5B6, Canada

Alfred J. Keys, Institute of Arable Crops Research, Rothamsted Experimental Station, Harpenden, Hertfordshire, AL5 2JQ, UK

Sigrid Klapheck, Botanisches Institut der Universität zu Köln, Gyrhofstrasse 15, D-5000 Köln 41, FRG

Peter J. Lea, Division of Biological Sciences, University of Lancaster, Lancaster LA1 4YQ, UK

Richard C. Leegood, Robert Hill Institute and Department of Animal and Plant Sciences, University of Sheffield, Sheffield S10 2TN, UK

James T. Madison, Agricultural Research Service, United States Department of Agriculture, United States Plant, Soil and Nutrition Laboratory, Tower Road, Ithaca, New York 14853, USA

Jan A. Miernyk, Department of Biochemistry, University of Missouri-Columbia, Columbia, MO 65211, USA

Paul A. Millner, Department of Biochemistry, University of Leeds, Leeds LS2 9JT, UK

Thomas S. Moore, Jr., Louisiana State University, Baton Rouge, LA 70803-1705, USA

Alan J. S. Murray, William Grant and Sons, The Distillery, Girvan, Ayrshire, KA26 9PT, UK

Martin A. J. Parry, Institute of Arable Crops Research, Rothamsted Experimental Station, Harpenden, Hertfordshire, AL5 2JQ UK

William C. Plaxton, Department of Biology, Queen's University, Kingston, Ontario, K7L 3N6, Canada

Douglas D. Randall, Department of Biochemistry, University of Missouri-Columbia, Columbia, MO 65211, USA

Philip A. Rea, Department of Biochemistry, AFRC-IACR, Rothamsted Experimental Station, Harpenden, Hertfordshire AL5 2JQ, UK

Heinz Rennenberg, Fraunhofer Institut für Atmosphärische Umweltforschung, Kreuzeckbahnstrasse 19, D-8100 Garmisch-Partenkirchen, FRG

Sven E. Rognes, Department of Biology, Botany Division, University of Oslo, PO Box 1045, Blindern, 0316 Oslo 3, Norway

Ahlert Schmidt, Botanisches Institut, Tierarztliche Hochschule Hannover, Bunteweg 17d, D-3000 Hannover 71, FRG

Daniel Schmutz, Pflanzenphysiologisches Institut der Universität Bern, Altenbergrain 21, CH-3013 Bern, Switzerland

Alison M. Smith, Department of Applied Genetics, AFRC-IPSR, John Innes Institute, Colney Lane, Norwich, NR4 7UH, UK

Martin Steup, Botanisches Institut der Westf., Wilhelms-Universität Münster, Schloßgarten 3, D-4400 Münster, FRG

Mark Stitt, Lehrstuhl für Pflanzenphysiologie, Universität Bayreuth, 8580 Bayreuth, FRG

Marianne Suter, Pflanzenphysiologisches Institut der Universität Bern, Altenbergrain 21, CH-3013 Bern, Switzerland

Janice C. Turner, Department of Biochemistry, AFRC-IACR, Rothamsted Experimental Station, Harpenden, Hertfordshire AL5 2JQ, UK

Christoph von Arb, Pflanzenphysiologisches Institut der Universität Bern, Altenbergrain 21, CH-3013 Bern, Switzerland

Kevin A. Walker, Schering Agrochemicals Ltd, Chesterford Park Research Station, Saffron Walden, Essex CB10 1XL, UK

Roger M. Wallsgrove, Biochemistry Department, AFRC Institute of Arable Crops Research, Rothamsted Experimental Station, Harpenden AL5 2JQ, UK

M. C. Walsh, Department of Biochemistry, University of Wales College of Cardiff, PO Box 903, Cardiff CF1 1ST, UK

John L. Wray, Plant Molecular Genetics Unit, Sir Harold Mitchell Building, University of St. Andrews, St. Andrews, Fife KY16 9TH, UK

Preface to the Series

Scientific progress hinges on the continual discovery and extension of new laboratory methods and nowhere is this more evident than in the subject of biochemistry. The application in recent decades of novel techniques for fractionating cellular constituents, for isolating enzymes, for electrophoretically separating nucleic acids and proteins and for chromatographically identifying the intermediates and products of cellular metabolism has revolutionised our knowledge of the biochemical processes of life.

While there are many books and series of books on biochemical methods, volumes specifically catering for the plant biochemist have been few and far between. This is particularly unfortunate in that the isolation of DNA, enzymes or metabolites from plant tissues can often pose special problems not encountered by the animal biochemist. For a long time, the Springer series *Modern Methods in Plant Analysis*, which first appeared in the 1950s, provided the only comprehensive guide to experimental techniques for the investigation of plant metabolism and plant enzymology. This series, however, has never been completely updated; a second series has recently appeared but this is organised on a techniques basis and thus does not provide the comprehensive coverage of the first series. One of us (JBH) wrote a short guide to modern techniques of plant analysis *Phytochemical Methods* in 1976 (second edition, 1984) which showed the need for an expanded comprehensive treatment, but which by its very nature could only provide an outline of available methodology.

The time therefore seemed ripe to us to produce an entirely new multi-volume series on methods of plant biochemical analysis, which would be both thoroughly up-to-date and comprehensive. The success of *The Biochemistry of Plants*, edited by P. K. Stumpf and E. E. Conn and published by Academic Press, was an added stimulus to produce a complementary series on the methodology of the subject. With these thoughts in mind, we planned individual volumes covering: phenolics, carbohydrates, amino acids, proteins and nucleic acids, terpenoids, nitrogen and sulphur compounds, lipids, membranes and light receptors, enzymes of primary and secondary metabolism, plant molecular biology and biological techniques in plant biochemistry. Thus we have tried to cover all the major areas of current endeavour in phytochemistry and plant biochemistry.

The main aim of the series is to introduce to the scientist current knowledge of techniques in various fields of biochemically-related topics in plant research. It is also intended to present the historical background to each topic, to give experimental details of methods and analyses and appraisal of them, pointing out those methods that are most suitable for immediate application. Wherever possible illustrations and structures have been used and one or more case treatments presented. The compilation of known data and properties, where appropriate, is included in many chapters. In addition, the reader is directed to relevant references for further details. However, for the sake of clarity and completeness of individual reviews, some overlap between chapters of volumes has been allowed.

Finally, we extend our warmest thanks to our volume editors for undertaking the important task of organising each volume and cooperating in preparing the contents lists. Our special thanks go to the staff of Academic Press and to the many colleagues who have made this project a success.

P. M. DEY
J. B. HARBORNE

Preface to Volume 3

The original aim of this book was to cover all the enzymes of primary and secondary plant metabolism in ten chapters of 25 pages each! It soon became clear that this was not possible and I limited the initial volume to primary metabolism and extended the number of chapters to 24. As a result of this, the size of the book is considerably larger than initially intended and for this I owe my apologies to Academic Press.

It has, however, not proved possible to cover all enzymes of primary metabolism, but I think the subjects covered are a reasonable attempt. I apologise in advance to those workers who feel that their 'pet' enzyme has been left out. The chapters start at reputedly the 'most important protein on earth' and work through carbohydrate synthesis and breakdown to lipid metabolism. There follows what some would say is a considerable emphasis on nitrogen metabolism, but in this area I have called upon my editor's prerogative. Chapter 21 covers all aspects of sulphur metabolism and I am extremely grateful to Christian Brunold for his help in arranging this section. Finally, I have included two enzymes, protein kinases and ATPases, both of which are becoming more important in studies on plant metabolism.

In my original letter to the authors I included the following statement: 'The chapter should include a brief historical background to the topic, full experimental details of enzyme assays and methods of purification. It should also include an appraisal of methods and modern techniques, pointing out those most suitable for specific purposes. Details of the properties of the enzymes should also be included, wherever possible illustrations and structures should be used'. Taking into account the confines of space, I feel that these aims have been achieved. I am particularly pleased that with some well characterised enzymes it has been possible to include work using specific antisera raised against purified proteins. Unfortunately, from some enzymes in my own field, it has been extremely difficult to confirm their presence, let alone purify them to homogeneity!

With a volume of this size it would be impossible to finish without expressing my thanks to a large number of people. Initially I would like to thank Jeffrey Harborne who suggested the idea to me while he was acting as an external examiner at the University of Lancaster in the summer of 1987. I greatly appreciate the advice and encouragment received from Andrew Richford and Carol Parr of Academic Press. I am of course totally indebted to all of the 45 authors who produced their manuscripts within three months of the original submission date. Unfortunately one author was unable to submit by the final closing date. During the two year gestation period, I have received valiant service from three secretaries Connie Atkinson, Carol Barlow and Terry Bowden. They all deserve considerable thanks for not only coping with the large volume of correspondence, but also for typing the chapters that carry my own name. Finally, I would like to express my sincere gratitude to Janice Turner for carrying out a large amount of the editorial work on what is now her fourth book. Her patience and perseverance in checking the authors' manuscripts for errors, idiosyncrasies in reference citation and in the final production of the index have been far beyond the call of duty.

PETER J. LEA

Glossary

BSA, bovine serum albumin.
DTT, dithiothreitol.
EDTA, ethylenediamine tetraacetic acid.
EGTA, ethyleneglycol-bis (β-aminoethyl ether)N,N,N',N' tetraacetic acid.
FPLC, fast protein liquid chromatography.
GLC, gas–liquid chromatography.
Hepes, N-(2-hydroxyethyl)piperazine-N'-(2-ethane sulphonic acid).
HPLC, high pressure liquid chromatography.
PAGE, polyacrylamide gel electrophoresis.
PMSF, phenyl methyl sulphonyl fluoride.
PVP, polyvinyl pyrrolidone.
SDS, sodium dodecyl sulphate.
Tris, tris(hydroxymethyl)aminomethane.

1 Ribulose Bisphosphate Carboxylase/Oxygenase and Carbonic Anhydrase

ALFRED J. KEYS and MARTIN A. J. PARRY

Institute of Arable Crops Research, Rothamsted Experimental Station, Harpenden, Hertfordshire, AL5 2JQ, UK

I.	Introduction	1
II.	Ribulose 1,5-bisphosphate carboxylase/oxygenase (Rubisco)	3
	A. Historical background	3
	B. Structure	3
	C. Essential knowledge about activation	4
	D. Extraction from leaves: initial and total activity of extracts	5
	E. Purification	6
	F. Measurement of carboxylase activity	8
	G. Measurement of oxygenase activity	9
	H. Measurement of the specificity factor	10
	I. Measurement of amount of Rubisco protein	11
III.	Carbonic anhydrase	11
	A. Historical background	11
	B. Preparation of extracts from leaves	12
	C. Measurement of carbonic anhydrase activity	12
	D. Purification	12
	E. Properties	13
	References	13

I. INTRODUCTION

Ribulose 1,5-bisphosphate carboxylase/oxygenase (Rubisco) catalyses two known reactions, the carboxylation of ribulose 1,5-bisphosphate (RuBP) to form two molecules of

3-phosphoglycerate (PGA) and, alternatively, the oxygenation of the same substrate to form one molecule each of PGA and 2-phosphoglycollate.

1. $RuBP + CO_2 \longrightarrow 2\,PGA$
2. $RuBP + O_2 \longrightarrow 1\,PGA + 1\,phosphoglycollate$

Carbonic anhydrase catalyses the reversible hydration of CO_2.

3. $CO_2 + H_2O \longleftrightarrow HCO_3^- + H^+$

The order in which the two enzymes are described is arbitrary; the sequence in which they are involved in the assimilation of inorganic carbon by plants is almost certainly the reverse of the order of treatment here. The precise role of carbonic anhydrase in aquatic autotrophs is currently the subject of intensive research. In C_4 plants the initial assimilation of inorganic carbon in the mesophyll cells involves bicarbonate ion (Cooper et al., 1968), so carbonic anhydrase must intervene in the pathway from CO_2 in the air. The situation in the bundle sheath is somewhat different and carbonic anhydrase may not be essential immediately before the operation of Rubisco (Burnell and Hatch, 1988).

There are problems in defining the concentration of CO_2 (one of the substrates for both enzymes) in solutions. In a buffered solution in equilibrium with air, or with an artificial atmosphere containing a known partial pressure of CO_2, the concentration of CO_2 in solution is specified by the solubility coefficient (the Bunsen coefficient). As the CO_2 dissolves, bicarbonate ion and carbonate ion will be formed to an extent that depends on the value finally attained by the pH of the solution once the equilibria involved have all been reached. The uncatalysed hydration of CO_2 is rapid enough for equilibrium to be established in a few minutes ($t_{0.5}$ at 25°C = 18 s). The H_2CO_3 is dissociated into HCO_3^- and CO_3^{2-} to an extent determined by the effective dissociation constants for the protons.

It is usually more convenient to produce CO_2 in solution by adding a known amount of sodium or potassium bicarbonate to a buffered mixture, previously depleted of inorganic carbon, in a vessel that precludes significant gaseous exchange with the ambient air. The CO_2 in solution then depends on the pH and on the effective dissociation constants for carbonic acid; if there is also a gas phase (head space) above the solution, some of the inorganic carbon added as bicarbonate will diffuse from the solution as CO_2 to an equilibrium related to the solubility of CO_2 in the solution.

The effective solubility coefficient for CO_2 and the pK' values for the protons of carbonic acid will depend on the composition of the solution and on the temperature. Arbitrary values can be extrapolated from the values given by Harned and Davis (1943) and Harned and Bonner (1945) against the ionic strength of the buffered solution to be used (Yokota and Kitaoka, 1985; McNeil et al., 1981; Bird et al., 1982). Such extrapolated values strictly refer to the situation when the ions in solution are Na^+ and Cl^- rather than the buffer ions normally used in enzymology. Therefore it is essential to give full details of the composition of reaction mixtures used to determine kinetic constants.

II. RIBULOSE 1,5-BISPHOSPHATE CARBOXYLASE/OXYGENASE (RUBISCO)

A. Historical Background

The existence of an enzyme to catalyse the reaction of CO_2 with RuBP was predicted from radioactive tracer studies of the path of carbon in photosynthesis. The enzyme was subsequently detected in extracts of algae and leaves (Quayle et al., 1954; Weissbach et al., 1954, 1956); its more general occurrence in photoautotrophs is now fully established. The additional competitive oxygenase activity of the enzyme became evident from the work of Bowes et al. (1971). Subsequent studies by Andrews et al. (1973), Laing et al. (1974) and Jordan and Ogren (1981) show that O_2 and CO_2 are competing alternative substrates for the enzyme in reactions with ribulose 1,5-bisphosphate involving the oxygenase and carboxylase catalytic functions. Lorimer (1981) has provided a detailed review of the reactions. A transition state intermediate of the carboxylation reaction has been isolated (Lorimer et al., 1986) but the intermediate stages of the oxygenation reaction are still mainly subject to speculation. The RuBP binds to the enzyme first and undergoes tautomerisation to enolate anions (Pierce, 1988). Most probably there is no formal binding site for either CO_2 or O_2, so differences in discrimination between the two competing substrates with enzymes from different photoautotrophic species must depend on the reactivity of the bound RuBP. Increasing knowledge of the structure of Rubisco, together with the use of genetic manipulation to change the primary structural elements of the protein, form the basis of many contemporary investigations.

B. Structure

The enzyme varies in its structural complexity from a dimer of identical polypeptide subunits in the photosynthetic bacterium, *Rhodospirillum rubrum*, to a hexadecameric protein with eight subunit polypeptides (L) of between 50 and 60 kDa and eight of 12–20 kDa (S) found in most other photosynthetic organisms. It is convenient to refer to the form of the enzyme in *R. rubrum* as L_2; L_4, L_6 and L_8 forms have also been reported in other photosynthetic bacteria as well as the L_8S_8 form (Akazawa, 1979; McFadden et al., 1986). Amino acid sequences have been deduced for the L and S subunits of Rubisco of many species from studies of cloned DNA. Models showing the secondary, tertiary and quaternary structures are being perfected for the enzymes from *R. rubrum*, *Nicotiana tabacum* and *Spinacea oleracea* from X-ray diffraction data for crystals of the protein (Andersson et al., 1989). A common feature is probably that the L subunits are intimately associated in pairs with an N-terminal domain of each partner interacting with the C-terminal domain of the other. The catalytic site is associated with an α/β-barrel in the C-terminal domain of each subunit. The models show that the N-terminal domains are each close to the mouth of the α/β-barrel in the partnering L subunit. Experiments have shown that the N-termini do influence the catalytic properties. In the L_8S_8 enzymes, the L_2 pairs form a cylinder with four small S subunits round the axis at either end.

C. Essential Knowledge about Activation

Firstly, the concentrations of CO_2 and Mg^{2+} present with the enzyme immediately before assay determine the activity measured (Lorimer et al., 1976, 1977). This is because an activation process involves a reversible reaction of CO_2 with a lysine residue in the catalytic sites of the large subunit of the enzyme to form a carbamate group (Andersson et al., 1989). The carbamate is stabilised by divalent metal ions; Mg^{2+} is probably the only metal involved significantly in vivo and is normally used in routine work with the enzyme. In CO_2-free and Mg^{2+}-free buffer the active enzyme loses its activity in about 5 min at 25°C; the activity is fully restored in a similar time once the cofactors are added back to the mixture at concentrations saturating for activation, 10 mM HCO_3^- and 20 mM Mg^{2+} at pH 8 for the enzyme from higher plants. There are important practical consequences. Thus, the activity of Rubisco extracted from leaves in CO_2-free buffer usually decreases at room temperature while extracts made in buffers containing saturating concentrations of the cofactors often increase in activity. Difficulties also arise in the determination of $K_m(CO_2)$ because of spontaneous decrease in activity during measurements of rate at low concentrations of the substrate. All measurements of the oxygenase activity, where the concentration of competing CO_2 is decreased, are subject to error because of loss of activity of the enzyme through decarbamylation.

Secondly, where extracts are subjected to low temperatures, especially in the absence of CO_2, Rubisco is liable to change into a form that can only be reactivated quickly if it is heated. This slow activating (E_s) form is best known in the crystalline tobacco enzyme: this requires some 10 min at 50°C for activation (Singh and Wildman, 1974). The activation by heat can be conducted in the absence of CO_2 and Mg^{2+} or in their presence. In the former situation the rapidly activating form of the enzyme is obtained; in the latter situation the active carbamylated form is made directly. The extent to which Rubisco is converted to the E_s form depends on the species of origin and the conditions used during purification (Schmidt et al., 1984; Parry et al., 1987).

Thirdly, there are complicated effects of various oxyanions, including sulphate, phosphate, NADPH, fructose 1,6-bisphosphate and 6-phosphogluconate, which increase the activation state of the enzyme in the presence of low concentrations of CO_2 and Mg^{2+} but are competitive inhibitors of the enzyme with respect to RuBP. The mechanism of action of these oxyanions is not fully understood; it is best to avoid them in routine measurements of Rubisco activity but to remember that such anions will be present in crude extracts of leaves. For further discussion of the topic see Jordan et al. (1983) and Parry et al. (1985). In the absence of such effectors, the concentration of CO_2 required to saturate the activating carbamylation reaction are at least 10 times higher than the concentrations in aqueous solutions in equilibration with the ambient air. The concentrations of Mg^{2+} required to stabilise the carbamate in the absence of the oxyanions are also above those thought likely in the chloroplast stroma. An enzyme called Rubisco activase has been detected in plants and catalyses the activation of Rubisco in the presence of ATP when the CO_2 and Mg^{2+} are suboptimal (Robinson et al., 1988). This enzyme may also affect the activity of Rubisco in crude extracts as the ATP is hydrolysed by phosphatases.

D. Extraction from Leaves: Initial and Total Activity of Extracts

In attempting to relate the activity of Rubisco in leaves to the rate of photosynthesis observed, many experiments have been described in which the enzyme is extracted and its activity measured immediately after extraction and after subsequent incubation with saturating concentrations of CO_2 and Mg^{2+} for periods of 5–10 min. The resulting measurements of activity have been called 'Initial' and 'Total' activities. Leaf tissue may be frozen quickly and where necessary stored in liquid nitrogen until activities can be measured (Vu et al., 1983; Servaites et al., 1984). The tissue is ground to a fine powder in a mortar cooled with liquid nitrogen. Subsamples are taken, with a spatula cooled in liquid nitrogen, after evaporation of the liquid nitrogen but while the sample is still very cold; the subsamples are then ground in a mortar with 5 volumes of extraction buffer until thawing is complete. To avoid changes in activity, the temperature of the homogenate must not rise much above 2°C before the sample is taken for measurement of initial activity. Where it is practicable, it is equally effective to drop fresh leaf samples into 5 volumes of buffer at 0°C, in a mortar standing on ice, and to grind the sample immediately for 2 min, or until an appropriate degree of homogenisation has been achieved (Perchorowicz et al., 1981; Boyle and Keys, 1982). Provided that the time is short and the temperature kept low, the homogenates can be centrifuged for clarification. Alternatively, rapid filtration through nylon mesh of 25 μm apertures can provide extracts that can be sampled using microsyringes.

The composition of buffers used for making extracts for the determination of initial Rubisco activity affects the result obtained. Machler and Nosberger (1980) and Boyle and Keys (1982) favour the inclusion of Mg^{2+} ions to prevent changes in activity of the cold extracts. However, the changes are slow even in the absence of Mg^{2+} provided the extract is kept cold. In other studies the inclusion of both CO_2 and Mg^{2+} has been favoured (Servaites et al., 1986); we have noticed that the activity of extracts of wheat leaves harvested in dull light does increase significantly even if the temperature is close to 0°C if both Mg^{2+} and CO_2 are present. The inclusion of both cofactors is important in studies of the occurrence of the tight-binding (nocturnal) inhibitor of Rubisco, 2-carboxy-D-arabinitol-1-phosphate (CA1P) (Gutteridge et al., 1986; Berry et al., 1987); it has a much greater affinity for the catalytic site in the presence of the cofactors (Servaites et al., 1986). The buffers suggested for extraction to measure initial activity are as follows: (1) 20 mM Tris (or Bicine) (pH 8.2) containing 10 mM 2-mercaptoethanol and 20 mM $MgCl_2$ for studies with young wheat leaves (Machler and Nosberger, 1980). (2) 100 mM Hepes (pH 7.7) containing 20 mM KCl (Perchorowicz et al., 1981) for studies of wheat and maize leaves. For leaves with high concentrations of polyphenols, a higher concentration of 2-mercaptoethanol may help but a change to more acid pH and addition of polyvinyl pyrollidone and other substances may be necessary (Vu et al., 1983; Hurewitz and Janes, 1987).

To measure Rubisco activity in crude extracts, we recommend use of the carboxylase assay with $^{14}CO_2$ as described in Section II.F.1; measurement of oxygenase activity in crude extracts is subject to many interfering factors. The initial carboxylase activity is determined by adding some 20 μl of fresh extract to a reaction mixture that is otherwise complete; total activity is measured by adding the same volume of extract to a reaction mixture containing all constituents except RuBP, leaving the mixture to incubate at 25°C for 5 min, then adding the RuBP. The reaction time is best restricted to 30 s.

Extending the time of pre-activation beyond 5 min to distinguish initial and final activities is not advised because unexplained decreases in activity have been observed (Hurewitz and Janes, 1987). Even where neither Mg^{2+} nor CO_2 is present in the extraction buffer, the measured value of total activity does not include catalytic sites containing the nocturnal inhibitor, CA1P.

E. Purification

A basic method for the purification of Rubisco from wheat leaves is presented together with a survey of alternative methods and comments on storage.

1. Rubisco from wheat leaves

Wheat seed is sown thickly in trays of soil-based compost in a glasshouse. Natural daylight over the trays is supplemented where neccessary with artificial lighting to give a 16 h photoperiod with a photon flux density $>250\,\mu mol$ quanta $m^{-2}s^{-1}$ of photosynthetically active radiation. Additional nitrogen fertiliser is added after 2 weeks (50 ml Hoagland's solution, containing 4 times the normal nitrogen, per tray every third day). The tops of the plants, mainly leaves 2 and 3, are harvested about 3 weeks after sowing and after at least 3 h of illumination. The leaf material, approximately 100 g from four trays, is cut into pieces about 1 cm long and homogenised in 600 ml of ice-cold buffer, containing 20 mM Tris (pH 8.0), 10 mM $NaHCO_3$, 10 mM $MgCl_2$, 1 mM EDTA, 5 mM dithiothreitol, 0.002% chlorohexidine diacetate (Hibitane) and 1% acid-washed (Loomis, 1974) insoluble polyvinylpyrrolidone, using a Waring blender at maximum speed for four periods of 15 s with 15 s intervals. This and subsequent steps are carried out at between 0 and 4°C. The homogenate is filtered through four layers of muslin. After the froth has dispersed (approximately 10 min), solid ammonium sulphate is added (19.4 g per 100 ml) to the filtrate to give 35% saturation. After 20 min the suspension is centrifuged at $20\,000 \times g$ for 15 min and the pellet obtained is discarded. The supernatant liquid is raised to 55% saturation by the further addition of 11.8 g per 100 ml of solid ammonium sulphate. After 15 min centrifugation at $20\,000 \times g$ the pellet is carefully drained and dissolved in 30 ml of 20 mM Tris containing 1 mM dithiothreitol, 1 mM $MgCl_2$ and 0.002% Hibitane at pH 8.0. The solution is centrifuged at $100\,000 \times g$ for 30 min to remove fine particles and the supernatant liquid is layered above linear gradients of from 8 to 25% sucrose (w/v) in 32 ml of the buffer used to dissolve the protein. A total of eight tubes is required for the sample with the tubes being finally filled to the top with more buffer. After centrifugation in a fixed angle rotor (Beckman type 60 Ti) for 2.5 h at $215\,000 \times g$ at 4°C (alternatively for 1.5 h at $361\,000 \times g$ using a Beckman Type 70 Ti rotor) the contents of the tube are fractionated by pumping the contents from the bottom. Essentially the first 4 ml of the liquid is discarded and the next 15 ml collected; this liquid can be shown by absorbance at 280 nm to contain the main band of protein (the Rubisco) that has entered into the sucrose gradient. The solutions containing the Rubisco are bulked together and loaded onto a column 25×360 mm containing Sepharose Q (fast flow) (Pharmacia). The column is washed with 100 ml of buffer with the same composition as that used to resuspend the final ammonium sulphate pellet and then with a linear gradient of 0 to 0.5 M NaCl in 700 ml of the same buffer. Carboxylase activity is associated with the major protein component detected by absorbance at 280 nm. The fractions containing the carboxylase are bulked together and the solution desalted by passage through a column of Sephadex G-25 (fine), 50 mm × 310 mm, equilibrated and operated with 5 mM Bicine buffer, pH 8.0, at a flow rate of 200 ml h^{-1}. Fractions containing the protein (usually at least 250 mg in 120 ml solution) are combined, and the solution frozen as a shell in a flask or a bottle using a bath of acetone and solid CO_2. The container is connected to a commercial freeze-drier for one or two days. Secondary drying is over P_2O_5 at room temperature in a vacuum desiccator in vials of about 20 ml capacity. The dried product is stored at 0 to 4°C over dry silica gel. Vials are warmed to room temperature before opening to remove samples.

Freezing of the protein solution prior to freeze-drying at temperatures below that of the CO_2/acetone mixture, e.g. in liquid nitrogen, or the inclusion with the protein of dithiothreitol or 2-mercaptoethanol, causes the dried product to be almost totally insoluble in buffers at pH values near to 8. Solutions containing too low a concentration of protein give hygroscopic powders; the best results are obtained by drying from solutions containing between 2 and 3 mg protein per ml. Protease activity is a particular problem in the extraction of Rubisco from older leaves; the addition of casein during homogenisation has proved to be more effective than specific protease inhibitors.

2. *Purification from other leaves: alternative methods*

A rapid method for purifying Rubisco from spinach leaves has been described by Salvucci *et al.* (1986). Chloroplasts prepared from the leaves are lysed, and the enzymes released are separated on a Mono Q column (Pharmacia) by fast protein liquid chromatography (FPLC). The freshly prepared protein has a specific activity close to the maximum that has been observed. Another rapid preparation of Rubisco from spinach leaves involves fractionation of whole leaf extracts by the use of polyethylene glycol and $MgCl_2$ (Hall and Tolbert, 1978). This procedure is not successful with wheat leaves.

Extraction of Rubisco from leaves of maize and tobacco plants requires extra precautions to prevent interference from the oxidation of polyphenols. Extracts are made at pH 7.0 in the presence of pre-soaked 5% insoluble polyvinyl pyrrolidone and 10 mM sodium diethyldithiocarbamate. Leaves from trees pose special problems. Protein is often difficult to extract and very liable to browning due to the presence of polyphenols. In addition to the steps taken for maize and tobacco it is necessary to add 1% Tween 80 (Gezelius, 1975). The addition of the Tween 80 causes some difficulties in the fractionation of extracts; ammonium sulphate additions may produce precipitates containing air expelled from the solution. Special problems have also been encountered in the extraction and purification of Rubisco produced in cultures of *Escherichia coli* by the expression of genes for the large and small subunit polypeptides (Kettleborough *et al.*, 1987).

3. *Storage*

The freeze-dried enzyme stores well provided that it is exposed to the air infrequently. The specific activity depends heavily on the suitability of the leaves used to provide the protein. Much of the enzyme may be in the slowly activating form and it is recommended that the powder is dissolved in buffer, at pH near to 8, containing 10 mM $NaHCO_3$ and 20 mM $MgCl_2$ and heated at 40°C for 40 min for activation. The advantage of freeze-drying is the convenience of the product for use in experiments. In solution at 0 to 5°C, Rubisco loses activity due to the formation of E_s and also because of its tendency to aggregate to the extent that material comes out of solution. Initially at least, the aggregates retain catalytic activity (Kleinkopf et al., 1970). Aggregation and inactivation are also observed in Rubisco stored under ammonium sulphate solutions. Activity of Rubisco stored in this manner can be restored by treatment with dithiothreitol (Hall et al., 1981a). Storage at −80°C of enzyme made into a slurry in 50% saturated ammonium sulphate and frozen as pellets in liquid nitrogen has also been recommended (Hall et al., 1981a). In this form the spinach enzyme loses less than 30% of its activity in six months, and most of this activity can be recovered by incubating the protein with dithiothreitol after dialysis or gel filtration to remove the ammonium sulphate (Hall et al., 1981a). Where the enzyme from spinach was frozen in liquid nitrogen in the absence of ammonium sulphate and stored at −80°C, more activity was lost but reversal by dithiothreitol was virtually complete.

F. Measurement of Carboxylase Activity

1. By use of $^{14}CO_2$ ($H^{14}CO_3^-$)

To ensure the highest level of accuracy, all the solutions used, except the solution of $NaH^{14}CO_3$, should be CO_2-free. This is most difficult to achieve with the buffer component because of the pH. The buffer should be prepared using CO_2-free water (boiled out distilled water) to dissolve the acid component (Bicine in the method described below). The solution should then be gassed with CO_2-free air or N_2 for some 10 min before adjusting the pH to 8.2 with CO_2-free NaOH (e.g. 4 M solution from BDH, Poole, UK). Any contaminating inorganic carbon will cause an underestimate of activity. To optimise the conditions, the oxygen in solution should be decreased; for many purposes it is satisfactory to use solutions equilibrated with air. The buffer and distilled water needed to adjust the volume can be gassed with N_2. Also the vessel in which the reaction mixture is being prepared can be gassed with N_2 while the buffer, the $MgCl_2$ and RuBP are added; if gassing is continued during the addition of the $NaH^{14}CO_3$ a loss of CO_2 will result and some small shift in pH will take place. Typical reaction mixtures are prepared by mixing the following reagent solutions: 0.2 M Bicine containing 40 mM $MgCl_2$ (0.5 ml), CO_2-free H_2O (0.38 ml), 33 mM RuBP (0.01 ml), 0.1 M $NaH^{14}CO_3$ (0.10 ml) and solution containing enzyme (0.01 ml). The $NaH^{14}CO_3$ is made by mixing solutions supplied by Amersham International plc with the appropriate amount of AR-grade $NaHCO_3$ and diluting to 0.1 M. The solution of RuBP is made by dissolving the tetra sodium salt (Sigma Chemical Co. Ltd.) in water; this solution is normally close to pH 6.5 and can be stored indefinitely at −25°C provided that the pH is near to this value and always below pH 7. Where many assays are to be made it is convenient to prepare RuBP from 5'AMP following the methods described by Wong et al. (1980), but without immobilising the enzymes. The barium salts obtained following treatment with charcoal are converted to sodium salts by ion-exchange and purified by ion-exchange chromatography as described by Pierce et al. (1980).

For the routine accurate measurement of carboxylase activity in extracts of leaves, or in fractions obtained during the purification of the enzyme, reaction mixtures should contain 0.1 M Bicine (pH 8.2), 20 mM $MgCl_2$, 10 mM $NaH^{14}CO_3$ (5 µCi cm^{-3}; 300 mBq cm^{-3} of reaction mixture), 10–50 µg Rubisco protein and 330 µM RuBP. Reaction can be started either by the addition of the enzyme or of RuBP; the RuBP should not be added too much in advance when starting the reaction with enzyme because of its instability above pH 7. Reaction is stopped after 30 or 60 s by the addition of 100 µl of 10 M formic acid per ml of reaction mixture. Alternatively samples of 100 µl of reaction mixture are withdrawn by a syringe and delivered into 100 µl of 10 M formic acid. The acidified samples are evaporated to dryness, in vials suitable for measurements using a scintillation spectrometer, either in an oven at 60°C or *in vacuo* over anhydrous $CaCl_2$ and KOH pellets. Alternative methods for evaporation have also been described, in which a stream of air is passed over the liquids in vials standing in a fume hood. Where measurements of activity in crude extracts are involved it is essential to keep the temperature low during drying to prevent caramelisation and consequently colour quenching during scintillation counting. Residues are dissolved in 1 ml of distilled water and the solution mixed with 10 ml of Cocktail T Scintran (BDH, Poole, UK) or the reagent described by Patterson and Greene (1965).

The short reaction times require special precautions to prevent timing errors. The reaction mixtures should be prepared in small vials containing a stirring bar turned by an appropriate magnetic stirrer submerged in a water bath. The vials should have screw caps containing a silicone rubber septum to allow the addition of the bicarbonate and enzyme after the solutions of the other components and the head space have been adequately gassed with N_2. An alternative is to use the vessel of an oxygen electrode unit (e.g. Type DW 1 from Hansatech Ltd., King's Lynn, UK). In this vessel there is no head space for escape of CO_2 from the solution, the oxygen concentration can be monitored and a stirring mechanism is a component part together with good temperature control.

2. Spectrophotometric assay

An important spectrophotometric method for carboxylase activity has been improved by Lilley and Walker (1974). Reaction mixtures, final volume 1 ml, contain 50 mM Bicine or Hepes (pH 7.8), 10 mM $NaHCO_3$, 20 mM $MgCl_2$, 660 µM RuBP, 200 µM NADPH, 5 mM ATP, 5 mM creatine phosphate and 80 nkat each of creatine phosphokinase, glyceraldehyde phosphate dehydrogenase and phosphoglycerate kinase with 40–100 µg of protein in crude leaf extracts or 20–50 µg purified Rubisco [1 kat = 1 mol substrate converted per s]. Reaction is started, after establishing a steady base rate, by the addition of the RuBP. Reaction is measured by the decrease in absorbance at 340 nm due to the oxidation of the NADH. This method is best used in routine measurements of Rubisco activity; it should be used with caution where studies are being made of the activation status of Rubisco.

G. Measurement of Oxygenase Activity

The simplest measurement of oxygenase involves use of the oxygen electrode and reaction mixtures as described in Section II.F.1, but without the $NaH^{14}CO_3$ and with the buffer and any distilled water used having been previously equilibrated with air or oxygen at 25°C. It is impossible to conduct oxygenase activity measurements at saturating O_2 in the ordinary laboratory because this would require a partial pressure of O_2 greater than 1 atmosphere. The enzyme needs to be at least partly purified and RuBP-dependent O_2 oxygen uptake is measured a short while after the addition of a concentrated solution of activated enzyme. The enzyme is activated at 10–20 mg ml^{-1} so that little CO_2 is carried over into the reaction mixture. The accuracy can be improved if the sensitivity of the oxygen electrode is increased and some of the signal is 'backed off'.

H. Measurement of the Specificity Factor

Because CO_2 and O_2 are alternative competing substrates for reaction with RuBP, $v_c/v_o = (V_c K_o/V_o K_c)$ (C/O) where v_c and v_o are the rates of oxygenation and carboxylation, respectively, measured in solutions containing the concentrations of O_2 and CO_2 of O and C, respectively, and V_c, V_o, K_c and K_o are the V_{max} and Michaelis constants referring to the carboxylation and oxygenation reactions. This reduces to the simple relation that $v_c/v_o = \tau$ (C/O) where τ = the specificity factor. The values of v_c and v_o can be measured for several relative concentrations of CO_2 and O_2 and the value of τ determined from measurements made in the oxygen electrode. The constitutents of the reaction mixture are equilibrated with CO_2-free air at 25°C giving a concentration of 260 μM O_2 at 760 mm air pressure. The reaction mixtures are as described in Section II.F.1 except that the Rubisco protein is increased to near to 250 μg per ml, 25 units of carbonic anhydrase (Sigma Chemical Co., from erythrocytes) per ml are added, and the NaH^{14}CO$_3$ is decreased to a range 0.5–3.0 mM. The reaction is conducted in the oxygen electrode with continuous record of the oxygen concentration amplified 10 times from the normal signal and with 90% of the signal suppressed by a back-off potential. Reaction is initiated by adding the activated protein in 10 μl of solution and is stopped after 40 s by the addition of 50 μl of 10 M formic acid. To make accurate measurements in this way the oxygen electrode needs to be set up carefully with efficient stirring; the oxygen uptake rate is deduced from the recorded trace of oxygen concentration in the solution and the carboxylation from the non-volatile ^{14}C in samples of the acidified reaction mixtures. The inclusion of carbonic anhydrase is essential in this test system to maintain a sufficient rate of conversion of HCO$_3^-$ to CO_2; the rate of uptake of CO_2 exceeds the rate of uncatalysed conversion of HCO$_3^-$ to CO_2 when high concentrations of enzyme are used with low concentrations of bicarbonate (Bird et al., 1980). The value of τ is calculated from the relationship discussed above, i.e. $\tau = (v_c/v_o)$ (O/C). The inclusion of several amounts of added H^{14}CO$_3^-$ in each set of measurements allows a check for contamination with non-radioactive bicarbonate. When comparing the specificity factors of several samples of Rubisco it is wise to make the comparison using the same buffer and reagent solutions and preferably on the same day.

A simplified measurement of the specificity factor requires similar reaction mixtures but the HCO$_3^-$ need not be radioactive. Reaction is allowed to proceed at the chosen concentration of HCO$_3^-$ until oxygen uptake due to added RuBP ceases. Knowing the total amount of RuBP added, and the amount used to produce the observed uptake of oxygen, the amount of carboxylation and hence v_c/v_o can be deduced. Another set of methods for determining oxygenation and carboxylation simultaneously in the same reaction mixtures employs a dual radioisotopic approach (Jordan and Ogren, 1980; Kent and Young, 1980).

I. Measurement of Amount of Rubisco Protein

Because the activity of Rubisco is affected by many factors, it is not an effective measure of the amount of Rubisco protein present in extracts. A method that is independent of activation status, and involves only equipment and reagents found in most biochemistry departments, is used to measure the amount of Rubisco protein in extracts of leaves (Servaites et al., 1984; Rintamaki et al., 1988). The soluble proteins are separated by electrophoresis on polyacrylamide gels and the Rubisco measured by the amount of Coomassie Brilliant Blue R-250 stain bound by the protein bands. Samples of extract containing between 2 and 12 µg of Rubisco are separated in 1.5 mm thick native 6% polyacrylamide slab gels using slots 6 mm wide. Samples containing measured amounts of purified Rubisco are separated in adjacent tracks with some tracks left without protein to provide appropriate standards and blanks. Alternatively, the samples can be heated for 2 min with 2-mercaptoethanol and SDS and the polypeptides separated by electrophoresis in an SDS 4–20% polyacrylamide gradient gel using the discontinuous electrophoresis buffer system of Laemmli (1970). The gels are stained for 16 h in darkness in 0.1% Coomassie Brilliant Blue R-250 dissolved in water–propan-2-ol–glacial acetic acid (65:25:10). Gels are destained overnight in the same solvent and then in several changes of 7% acetic acid over a period of a further 24 h. Pieces of the gels containing the stained holoenzyme or the large and small subunit polypeptides are cut from the slab using a scalpel and extracted with 1.5 or 2 ml of 1% aqueous SDS in separate glass vials at 40°C during 18 h. Pieces of gel from adjacent tracks without protein are treated similarly. Protein content was calculated by reference to the standards from the absorbance of extracts of the gel pieces at 585 nm.

Rubisco in crude extracts has also been estimated by rocket immunoelectrophoresis (Pyke and Leech, 1985), by enzyme linked immunosorption assay (Catt and Millard, 1988), by radial immunodiffusion (Tingey and Andersen, 1986) and by radioimmunoassay based on the binding of [^{14}C]2-carboxyarabinitol 1,5-bisphosphate, an analogue of the transition state intermediate of the carboxylation catalysed by Rubisco (Collatz et al., 1979). This latter method can be modified so that antibody to Rubisco is not involved but the enzyme complex precipitated instead by the addition of polyethylene glycol (PEG) in the presence of 20 mM $MgCl_2$ (Hall et al., 1981b). The precipitate is washed with 20% PEG containing 20 mM $MgCl_2$ and then the bound radioactivity determined by scintillation spectrometry. Because of non-specific binding of the [^{14}C] analogue, precipitation by antibody is preferred.

III. CARBONIC ANHYDRASE

A. Historical Background

The extensive review by Reed and Graham (1981) gives the background information on the plant carbonic anhydrases from the first report of detection by Neish (1939) to the current view of the enzyme activity. The occurrence of the enzyme in plants was confirmed by Bradfield (1947) when the importance of including sulphydryl reagents in the extraction buffers was demonstrated. Carbonic anhydrase in plants has a special association with photosynthetic activity. The enzyme contains zinc very tightly bound into the catalytic site. The catalytic mechanism of the hydration of CO_2 involves the formation of an enzyme-bound hydroxyl ion from water followed by the reaction of this with the CO_2 to form a bicarbonate ion. The complete hydration to carbonic acid (H_2CO_3) depends then on the addition of a proton from the medium and the extent to which this happens depends on the pH of the solution. Carbonic acid itself is a minor constituent of the equilibrium mixture and readily dissociates into water and CO_2.

B. Preparation of Extracts from Leaves

The following procedure has been used for young leaves of wheat and barley. Fresh leaf samples (0.2–1.0 g) are ground in an ice-cold mortar with 3 volumes of 50 mM Imidazole (HCl) buffer, pH 7.5, containing 1 mM dithiothreitol, 0.5 mM ethylenediaminetetraacetic acid (EDTA) and 0.1% Triton X-100. The homogenate is centrifuged for 3 min at $9000 \times g$ in a microfuge at 0–4°C. The supernatant liquid is desalted using small columns of Sephadex G-25 that are operated either by gravity or by the centrifugal method of Helmerhorst and Stokes (1980). Desalted extracts are stored on ice and the measurements for carbonic anhydrase conducted at the earliest possible moment.

C. Measurement of Carbonic Anhydrase Activity

The method recommended is based on the description by Everson and Slack (1968) of a decreased scale version of the indicator method of Wilber and Anderson (1948). The use of small glass beads to facilitate mixing is avoided by using tubes of 1 cm diameter. Take 0.06 ml of the crude extract prepared as in Section III.B and mix with 0.54 ml of 1.5 mM NaOH. Prepare a control mixture of 0.06 ml of the 50 mM Imidazole buffer used to prepare the extracts mixed with 0.54 ml of 1.5 mM NaOH. Use clear plastic or pyrex glass tubes 10×75 mm to make the following test at 0°C. Mix 0.1 ml of diluted extract with 0.2 ml of 25 mM Veronal/H_2SO_4 buffer made up in boiled out distilled water (pH 8.20 at 0°C or pH 7.85 at 25°C) and containing 2 mg Bromothymol Blue per 100 ml. Cool the tube carefully in a mixture of ice and water then add from a syringe water saturated with CO_2 at 0°C. Mix gently while observing the change in colour. Record the time in seconds for the blue colour to change to a distinct yellow (T_c). Repeat the test using the control solution and record the time for the uncatalysed change (T_b). The activity is given by the relationship:

$$\text{Activity} = 10 \times (T_b/T_c - 1)$$

If necessary the initial dilution of the extract should be adjusted so that T_c is in the range 20 to 40 s. T_b should be in excess of 90 to 100 s. A more precise end-point can be established by making a solution of 10 mM Veronal at pH 6.3 (H_2SO_4) containing 0.8 mg of Bromothymol Blue per 100 ml.

Similar methods of assay use a glass electrode instead of the indicator to detect the release of a proton resulting during the reaction (Wilbur and Anderson, 1948).

D. Purification

The methods described by Atkins et al. (1972a) for partial purification of the enzyme from tradescantia and pea leaves can be adapted for leaves from wheat. The tissue is homogenised in 0.3 M Tris(hydroxymethyl)aminomethane (Tris)/sulphate (pH 8.3) containing 1 mM EDTA and 0.1 M 2-mercaptoethanol at 0–5°C. The homogenate is clarified by filtration through nylon mesh (60 μm) and by centrifugation. Protein precipitating between 45 and 70% saturation of the solution by ammonium sulphate (reached by the addition of solid $(NH_4)_2SO_4$) is dissolved in a minimum volume of 10 mM Tris/sulphate (pH 8.3) containing 1 mM EDTA and 0.1 M 2-mercaptoethanol. This solution is desalted by gel filtration using Sephadex G-25 (Pharmacia) and the protein loaded onto a column of diethylaminoethyl cellulose (DEAE)–Sephacel equilibrated with the same buffer. The protein is eluted by a linear gradient of Na_2SO_4 of 0 to 0.2 M. Further purification is achieved by gel filtration on a column of Sephadex G-100 using the 10 mM Tris/sulphate pH 8.3 buffer containing 1 mM EDTA and 0.1 M 2-mercaptoethanol. The partly purified enzyme is stored as a precipitate in 70% saturated $(NH_4)_2SO_4$ at pH 8.3 with 0.1 M 2-mercaptoethanol.

E. Properties

Atkins *et al.* (1972b) show that many plants contain isozymes of carbonic anhydrase. The carbonic anhydrase isozymes in monocotyledons have native molecular weights in the region 40–50 kDa, whilst in dicotyledons the native molecular weights are in the range 180–205 kDa. Activity is inhibited by sulphonamide drugs including Acetozolamide (Diamox), ethoxzolamide and suphonilamide. Azide is a potent inhibitor and many monovalent anions including nitrate, acetate and chloride are inhibitors.

REFERENCES

Akazawa, T. (1979). *In* "Encyclopedia of Plant Physiology," New Series. (M. Gibbs and E. Latzko, eds) Vol. 6, pp. 209–229. Springer, New York.
Andersson, I., Knight, S., Schneider, G., Lindqvist, T., Branden, C.-I. and Lorimer, G. H. (1989). *Nature* **337**, 229–234.
Andrews, T. J., Lorimer, G. H. and Tolbert, N. E. (1973). *Biochemistry* **12**, 11–18.
Atkins, C. A., Patterson, B. D. and Graham, D. (1972a). *Plant Physiol.* **50**, 214–217.
Atkins, C. A., Patterson, B. D. and Graham, D. (1972b). *Plant Physiol.* **50**, 218–223.
Berry, J. A., Lorimer, G. H., Pierce, J., Seemann, J. R., Meek, J. and Freas, S. (1987). *Proc. Natl. Acad. Sci. USA* **84**, 734–738.
Bird, I. F., Cornelius, M. J. and Keys, A. J. (1980). *J. Exp. Bot.* **121**, 365–369.
Bird, I. F., Cornelius, M. J. and Keys, A. J. (1982). *J. Exp. Bot.* **33**, 1004–1013.
Bowes, G., Ogren, W. L. and Hageman, R. H. (1971). *Biochem. Biophys. Res. Commun.* **45**, 716–722.
Boyle, F. A. and Keys, A. J. (1982). *Photosynth. Res.* **3**, 105–111.
Bradfield, J. R. G. (1947). *Nature* **159**, 467–468.
Burnell, J. N. and Hatch, M. D. (1988). *Plant Physiol.* **86**, 1252–1256.
Catt, J. W. and Millard, P. (1988). *J. Exp. Bot.* **39**, 157–164.
Collatz, G. J., Badger, M. R., Smith, C. and Berry, J. A. (1979). *Carnegie Inst. Yr. Bk.* **78**, 171–175.
Cooper, T. C., Tchen, T. T., Wood, H. G. and Benedict, C. R. (1968). *J. Biochem.* **243**, 3857–3863.
Everson, R. G. and Slack, C. R. (1968). *Phytochemistry* **7**, 581–584.
Gezelius, K. (1975). *Photosynthetica* **9**, 192–200.
Gutteridge, S., Parry, M. A. J., Burton, S., Keys, A. J., Mudd, A., Feeney, J., Servaites, J. and Pierce, J. (1986). *Nature* **324**, 274–276.
Hall, N. P. and Tolbert, N. E. (1978). *FEBS Lett.* **96**, 167–169.
Hall, N. P., McCurry, S. D. and Tolbert, N. E. (1981a). *Plant Physiol.* **67**, 1220–1223.

Hall, N. P., Pierce, J. and Tolbert, N. E. (1981b). *Arch. Biochem. Biophys.* **212**, 115–119.
Harned, H. S. and Bonner, F. T. (1945). *J. Am. Chem. Soc.* **67**, 1026–1031.
Harned, H. S. and Davies, R. (1943). *J. Am. Chem. Soc.* **65**, 2030–2037.
Helmerhorst, E. and Stokes, G. B. (1980). *Anal. Biochem.* **104**, 130–135.
Hurewitz, J. and Janes, H. W., (1987) *Photosynth. Res.* **12**, 105–117.
Jordan, D. B. and Ogren, W. L. (1980). *Plant Physiol.* **67**, 237–245.
Jordan, D. B. and Ogren, W. L. (1981). *Nature* **291**, 513–515.
Jordan, D. B., Chollet, R. and Ogren, W. L. (1983). *Biochemistry* **22**, 3410–3418.
Kent, S. S. and Young, J. D. (1980). *Plant Physiol.* **65**, 465–468.
Kettleborough, C. A., Parry, M. A. J., Burton, S., Gutteridge, S., Keys, A. J. and Phillips, A. L. (1987). *Eur. J. Biochem.* **170**, 335–342.
Kleinkopf, G. E., Huffaker, R. C. and Matheson, A. (1970). *Plant Physiol.* **46**, 204–207.
Laemmli, U. K. (1970). *Nature* **227**, 680–685.
Laing, W. A., Ogren, W. L. and Hageman, R. H. (1974). *Plant Physiol.* **54**, 678–685.
Lilley, R. McC. and Walker, D. A. (1974). *Biochim. Biophys. Acta* **358**, 226–229.
Loomis, W. D. (1974). *Meth. Enzymol.* **31**, 524–544.
Lorimer, G. H. (1981). *Ann. Rev. Pl. Physiol.* **32**, 349–383.
Lorimer, G. H., Badger, M. R. and Andrews, T. J. (1976). *Biochemistry* **15**, 529–536.
Lorimer, G. H., Badger, M. R. and Andrews, T. J. (1977). *Anal. Biochem.* **78**, 66–75.
Lorimer, G. H., Andrews, T. J., Pierce, J. and Schloss, J. V. (1986). *Phil. Trans. Roy. Soc. Lond. B* **313**, 397–407.
Machler, F. and Nosberger, J. (1980). *J. Exp. Bot.* **31**, 1485–1491.
McFadden, B. A., Torres-Ruiz, J., Daniell, H. and Sarojini, G. (1986) *Phil. Trans. Roy. Soc. Lond. B* **313**, 347–358.
McNeil, P. H., Foyer, C. H., Walker, D. A., Bird, I. F., Cornelius, M. J. and Keys, A. J. (1981). *Plant Physiol.* **67**, 530–534.
Neish, A. D. (1939). *Biochem. J.* **33**, 300–308.
Parry, M. A. J., Schmidt, C. N. G., Cornelius, M. J., Keys, A. J., Millard, B. N. and Gutteridge, S. (1985). *J. Exp. Bot.* **36**, 1396–1404.
Parry, M. A. J., Schmidt, C. N. G., Cornelius, M. J., Millard, B. N., Burton, S., Gutteridge, S., Dyer, T. A. and Keys, A. J. (1987). *J. Exp. Bot.* **38**, 1260–1271.
Patterson, M. S. and Greene, R. C. (1965). *Anal. Chem.* **37**, 854–857.
Perchorowicz, J. T., Raynes, D. A. and Jensen, R. G. (1981). *Proc. Natl. Acad. Sci. USA*, **78**, 2985–2989.
Pierce, J. (1988). *Physiol. Plant.* **72**, 690–698.
Pierce, J., Tolbert, N. E. and Barker, R. (1980) *Biochemistry* **19**, 934–942.
Pyke, K. A. and Leech, R. M. (1985). *J. Exp. Bot.* **36**, 1523–1529.
Quayle, J. R., Fuller, R. C., Benson, A. A. and Calvin, M. (1954). *J. Am. Chem. Soc.* **76**, 3601–3611.
Reed, M. L. and Graham, D. (1981). *Prog. Phytochem.* **7**, 47–94.
Rintamaki, E., Keys, A. J. and Parry, M. A. J. (1988). *Physiol. Plant.* **74**, 326–331.
Robinson, S. P., Streusand, V. J., Chatfield, J. M. and Portis, A. R. (1988). *Plant Physiol.* **88**, 1008–1014.
Salvucci, M. E., Portis, A. R. and Ogren, W. L. (1986). *Anal. Biochem.* **153**, 97–101.
Schmidt, C. N. G., Cornelius, M. J., Burton, S., Parry, M. A. J., Millard, B. N., Keys, A. J. and Gutteridge, S. (1984). *Photosynth. Res.* **5**, 47–62.
Servaites, J. C., Torisky, R. S. and Chao, S. F. (1984). *Plant Sci. Lett.* **35**, 115–121.
Servaites, J. C., Parry, M. A. J., Gutteridge, S. and Keys, A. J. (1986). *Plant Physiol.* **82**, 1161–1163.
Singh, S. and Wildman, S. G. (1974). *Plant Cell Physiol.* **15**, 373–379.
Tingey, S. V. and Andersen, W. R. (1986). *J. Exp. Bot.* **37**, 625–632.
Vu, C. V., Allen, L. H. and Bowes, G. (1983). *Plant Physiol.* **73**, 729–734.
Weissbach, A., Smyrniotis, P. Z. and Horrecker, B. L. (1954). *J. Am. Chem. Soc.* **76**, 3611–3615.
Weissbach, A., Horrecker, B. L. and Hurwitz, J. (1956). *J. Biol. Chem.* **218**, 795–810.
Wilbur, K. M. and Anderson, N. G. (1948). *J. Biol. Chem.* **176**, 147–155.
Wong, C.-H., McCurry, S. D. and Whitesides, G. M. (1980). *J. Am. Chem. Soc.* **102**, 7938–7939.
Yokota, A. and Kitaoka, S. (1985). *Biochem. Biophys. Res. Commun.* **131**, 1075–1079.

2 Enzymes of the Calvin Cycle

RICHARD C. LEEGOOD

Robert Hill Institute and Department of Animal and Plant Sciences, University of Sheffield, Sheffield S10 2TN, UK

I.	Introduction	15
	A. The Calvin cycle	15
	B. Light activation of Calvin cycle enzymes	17
	C. Measurement of enzyme activity in leaf or chloroplast extracts	18
II.	Enzymes of the Calvin cycle	20
	A. 3-Phosphoglycerate kinase	20
	B. NADP-glyceraldehyde phosphate dehydrogenase	21
	C. Aldolase	23
	D. Triose phosphate isomerase	23
	E. Fructose 1,6-bisphosphatase	24
	F. Transketolase	28
	G. Sedoheptulose 1,7-bisphosphatase	28
	H. Ribose 5-phosphate isomerase	30
	I. Ribulose 5-phosphate 3-epimerase	30
	J. Ribulose 5-phosphate kinase	30
	References	34

I. INTRODUCTION

A. The Calvin Cycle

The Calvin cycle, or Reductive Pentose Phosphate pathway, comprises thirteen reactions catalysed by eleven enzymes (Fig. 2.1). The cycle acts as an autocatalytic 'breeder' reaction and, for every three turns of the cycle, one molecule of triose phosphate is generated from three molecules of CO_2. Triose phosphate may be utilised either to regenerate the CO_2 acceptor, RuBP, or may be utilised in the synthesis of starch or sucrose. The Calvin cycle lies at the interface between electron transport and product synthesis and it must therefore be very responsive to changes in the relationship between supply of, and demand for, the products of electron transport, both in the short term (e.g. during light transients) and during longer-term changes in environmental conditions (e.g. acclimation to sun and shade). The activities of Calvin cycle enzymes are therefore under several forms of control. Light-activation and metabolite modulation allow rapid short-term modulation of enzyme activity, while protein synthesis (which, with the exception of the large subunit of RuBP carboxylase, is wholly under the control of the nuclear genome) allows long-term modulation of the amounts of Calvin cycle enzymes.

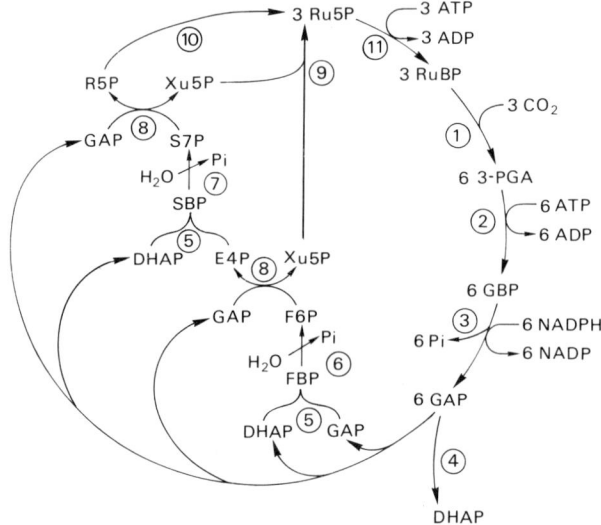

FIG. 2.1. The Calvin cycle. The Calvin cycle can be divided into three phases. The first phase is carboxylation catalysed by RuBP carboxylase (1). The second phase is the reductive phase by which PGA is converted to triose-P, catalysed by phosphoglycerate kinase (2) and $NADP^+$-dependent glyceraldehyde-P dehydrogenase (3). The third phase is regeneration of the CO_2 acceptor in the sugar phosphate shuffle, in which 5 C_3 units are rearranged to form 3 C_5 units (Xu5P and R5P) by the actions of triose-P isomerase (4), aldolase (5), FBPase (6), SBPase (7) and transketolase (8). Xu5P and R5P are then converted to RuBP by ribulose-5-P 3-epimerase (9) and ribulose-5-P isomerase (10). Glyceraldehyde-P dehydrogenase, the bisphosphatases and product synthesis recycle the Pi required for continuous ATP synthesis.

Recently interest has focused on the organisation of the Calvin cycle within the chloroplast. There are a large number of reports of associations between enzymes of the Calvin cycle and thylakoids (e.g. Miller and Staehelin, 1976; Fischer and Latzko, 1979; McNeil and Walker, 1981; Furbank, et al., 1986) and increasing interest in multi-enzyme complexes of the Calvin cycle isolated from chloroplasts in associations which may be similar to other multi-enzyme complexes such as the Krebs cycle 'metabolon' isolated from mitochondria (Srere, 1987). Ribulose 5-phosphate (Ru5P) kinase and the chloroplastic glyceraldehyde-P dehydrogenase have been extracted from leaves as large complexes, which upon activation release smaller active forms. The role of these two complexes has yet to be established and they may be linked as they tend to copurify (Nicholson et al., 1986a,b; Wolosiuk and Buchanan, 1976, 1978b). Complexes between ribose 5-phosphate (R5P) isomerase, Ru5P kinase and ribulose 1,5-bisphosphate (RuBP) carboxylase have been reported by Sainis et al. (1989) and evidence of metabolite channelling between 3-phosphoglycerate (PGA) kinase and glyceraldehyde-P dehydrogenase by Macioscek and Anderson (1987) and between ribose 5-P isomerase and phosphoribulokinase by Anderson (1987). A functional association between all five of these enzymes (which catalyse the complete sequence from R5P to triose-P), plus an unidentified 65 kDa polypeptide, has been reported by Gontero et al. (1988), although the molecular mass of the complex is considerably less than that expected from the usual subunit composition of these enzymes. Further evidence for weak interactions between six Calvin cycle enzymes has been obtained by Persson and Johansson (1989).

B. Light Activation of Calvin Cycle Enzymes

Many enzymes function only in the reduced form, but in photosynthetic organisms a virtue has been made out of a necessity by linkage of the catalytic activity of Calvin cycle and other enzymes to the availability of photosynthetically-generated reductants (Anderson, 1979; Buchanan, 1980). Five enzymes of the Calvin cycle (RuBP carboxylase (see Chapter 1, this volume), fructose 1,6-bisphosphatase (FBPase), sedoheptulose 1,7-bisphosphatase (SBPase), Ru5P kinase and NADP-glyceraldehyde 3-phosphate (GAP) dehydrogenase) show light-induced transitions to active forms. The last four of these undergo reductive activation. In Buchanan's scheme (shown below) ferredoxin reduces thioredoxin in a reaction catalysed by ferredoxin–thioredoxin reductase (FTR) and thioredoxin, in its turn, reduces the enzyme.

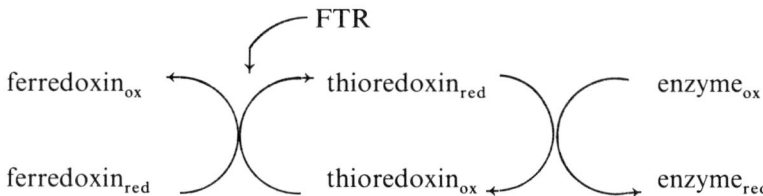

Anderson's protein modulase (the soluble part of the light effect mediator or LEM: Ashton and Anderson, 1981) has also been shown to be a complex of ferredoxin, thioredoxin and ferredoxin–thioredoxin reductase (Ford et al., 1987). *In vitro*, activation of enzymes can be mimicked by the addition of thiol-reducing agents such as dithiothreitol (DTT), although this is also facilitated by thioredoxin (e.g. Wolosiuk and Buchanan, 1978a,b; Wolosiuk et al., 1980).

All the components of the ferredoxin–thioredoxin system have been purified and characterised. Ferredoxin–thioredoxin reductase is an iron–sulphur protein (M_r 30 kDa) comprising two dissimilar subunits and contains a reducible disulphide bridge (Miginiac-Maslow et al., 1987; Droux et al., 1987). It catalyses the reduction of both types of chloroplast thioredoxin: thioredoxin *f*, which activates FBPase preferentially, and thioredoxins *mb* and *mc*, which activate NADP-malate dehydrogenase preferentially. Thioredoxins *mb* and *mc* differ only in that *mb* has an additional lysine residue at its N-terminus (Schürmann et al., 1981). Thioredoxin *f* exists in monomeric (*fA*) and dimeric (*fB*) forms (Buc et al., 1984). All three thioredoxins have M_r of 12 kDa (Schürmann et al., 1981) and possess a single reducible disulphide bridge per monomer.

The mechanism of light modulation is a dynamic one (Leegood et al., 1985) which means that enzyme activity can potentially respond to changes in irradiance. In purified systems, Soulié et al. (1981, 1987) and Clancey and Gilbert (1987) have demonstrated the extreme thermodynamic sensitivity of the spinach thioredoxin–FBPase system to dithiol oxidation, resulting from the fact that the system has a redox potential comparable to the lowest yet observed for protein disulphide bond formation. In leaves, activities of enzymes such as FBPase and NADP-malate dehydrogenase readily respond to electron availability, and electron acceptors such as oxaloacetate inhibit light-activation of enzymes, especially in low light (Leegood and Walker, 1980b, 1983). In the dark, during which time enzymes revert to the oxidised state, it is evident from studies both of purified systems (Shürmann and Kobayashi, 1984) and of leaves (Leegood and Walker, 1982) that the natural terminal electron acceptor is molecular O_2.

C. Measurement of Enzyme Activity in Leaf or Chloroplast Extracts

Measurement of the activity of the enzymes of the Calvin cycle may be attempted for a variety of reasons. We may wish to know the maximum catalytic activity of each component in order to assess, for example, whether coarse control is operating in response to changes in environmental conditions. On the other hand, we may wish to measure how the activity of an enzyme is modulated in the short-term by measuring changes in its activation state. These different measurements require quite different assay conditions and quite different precautions.

Measurement of the maximum catalytic activity requires optimisation of the pH and all of the components of the extraction and assay media for each tissue studied. Evidence should also be sought that all of the enzyme has been recovered from the tissue. This can be done by checking for complete cell breakage, and by recovery experiments using comparable amounts of the commercially available enzyme or by mixing known quantities of the tissue under study with known quantities of other tissues (e.g. spinach) in which it is known that the enzyme of interest is not readily inactivated.

Measurement of the activation state of an enzyme requires that the activation state of the enzyme be maintained during extraction and assay. Modulation of enzyme activity following exposure of leaves or chloroplasts to light can be very rapid. There are a number of methods to ensure maintenance of the activation state of enzymes. First, the leaf may be freeze-clamped and stored in liquid nitrogen. The activation states of all the enzymes are maintained in tissue frozen in liquid nitrogen for several days or weeks. Rapid extraction can then be achieved in less than 10 s using a small pestle and mortar or a glass-in-glass homogeniser. Removal of solids can be accomplished by a short centrifugation (10 s) in an Eppendorf centrifuge. Second, sap may be expressed directly from leaves after placing them in a syringe which can be illuminated (Wirtz et al., 1982) and the sap used immediately in non-spectrophotometric assays.

A wide range of extraction media may be employed, but the following features are desirable:

(1) pH between 7.5 and 8.5;

(2) a relatively high concentration of Mg^{2+} (10 mM or more);
(3) inclusion of a thiol protecting agent such a mercaptoethanol which protects the enzymes but which does not itself lead to appreciable activation;
(4) inclusion of a detergent such as Triton in the extraction medium is useful because it solubilises the thylakoid membrane, thereby preventing binding of enzymes to the membrane which is promoted by Mg^{2+}, and it facilitates chlorophyll estimation;
(5) in the case of the bisphosphatases inclusion of the substrate stabilises the enzyme against oxidative inactivation;
(6) if the maximum activity is to be measured then extraction and pre-incubation with DTT is necessary;

and

(7) protective agents such as glycerol, bovine serum albumin (BSA) or polyvinyl pyrrolidone (PVP) can be added, but are not normally necessary with young leaves, for which the following medium has been found suitable:

1 mM ethylenediamine tetraacetic acid (EDTA)
10 mM $MgCl_2$
15 mM 2-mercaptoethanol
0.05% Triton X-100
100 mM Tris(hydroxymethyl) aminomethane (Tris)-HCl (pH 7.8)

In order to detect light-activation, it is usually necessary to assay the enzyme under sub-optimal conditions. For example, the V_{max} of FBPase is unaffected by light activation, so that when measured under optimal conditions no difference in the activity is detected between illuminated and darkened samples.

All the light-activated enzymes of the Calvin cycle exhibit hysteretic behaviour (Frieden, 1970) (i.e. catalysis during activation by ligands may be slower than the subsequent rate of catalysis). Two aspects of an enzyme's behaviour in relation to its substrate are therefore open to investigation. A rapid assay (lasting a few seconds) will reveal the instantaneous effect of changed conditions of substrate upon the rate of catalysis. However, prolonged incubation of an enzyme under assay conditions which differ from those of its previous environment will often lead to a slow, but reversible, change in activation state (Fig. 2.2). Rapid assays of hysteretic enzymes have the advantage that they will give an estimate of the activation state of the enzyme under *in vivo* conditions rather than under the conditions of the assay (see, e.g., Fig. 2.2). They also allow accurate determination of enzyme activities at low concentrations of substrate and are thus useful for the determination of kinetic constants. Laing *et al.* (1981) report that for the rapid assays of FBPase, SBPase and Ru5P kinase rates are linear for about 30 s. However, care must be taken to ensure linearity under all assay conditions. The rapid assays are two-stage assays which involve measurement of the reaction products in a separate assay from the catalytic step. Two-stage assays have the advantage that large numbers of samples may be handled in parallel, but they are more laborious. Another advantage is that the influence of the product upon the reaction can be measured in a two-stage assay (Gardemann *et al.*, 1986). Direct spectrophotometric assays have the advantage of speed and simplicity and, if samples are stored in liquid nitrogen, their number presents no problem. Longer-term two-stage assays (e.g. Pi release) are also useful, but the linearity must be checked carefully and care must be taken in crude extracts as side reactions may occur (e.g. FBP formation from SBP, see Section II.G.1.a). A lag will also be present in coupled spectrophotometric assays due solely to the build-up of pools of intermediates. In any preliminary investigation it is generally best to employ both types of assay. The measurement of Calvin cycle enzymes in C_4 plants is also discussed by Ashton et al. (Chapter 3, this volume).

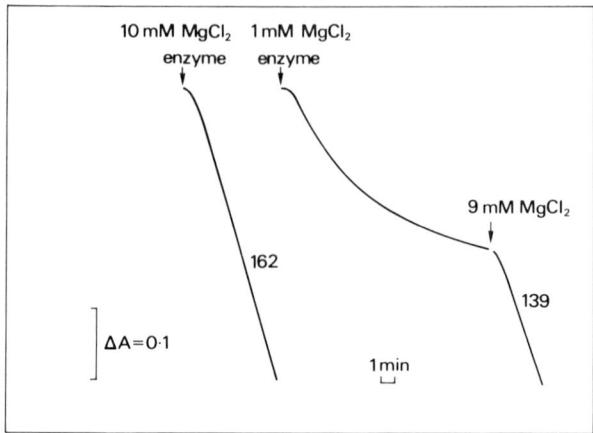

FIG. 2.2. Example of hysteretic behaviour as shown by the effect of Mg^{2+} on the activity of FBPase. Illuminated chloroplasts (equivalent to 5 µg chlorophyll) were broken into an assay medium (1 ml) containing 1 mM EDTA, 1 mM FBP, 0.05% Triton X-100, 100 mM Tris-HCl (pH 7.8). Addition of the enzyme is indicated by the arrows. The initial lag is largely due to the build-up of pools within the coupled assay. Addition of 9 mM $MgCl_2$ to the cuvette containing 1 mM $MgCl_2$ largely restored the rate to that observed with 10 mM $MgCl_2$ from the outset.

II. ENZYMES OF THE CALVIN CYCLE

A. 3-Phosphoglycerate Kinase (EC 2.7.2.3)

$$PGA^{3-} + ATP^{4-} \longrightarrow ADP^{3-} + GBP^{4-}$$

PGA kinase is a monomer with M_r of 48 kDa. The kinetic constants are: $K_m(PGA)$ 0.6 mM, $K_m(ATP)$, 0.7 mM, (Pacold and Anderson, 1975; Cavell and Scopes, 1976). The relatively large positive $\Delta F'$ value (+4.5 kcal) makes this reaction very sensitive to the [ATP]/[ADP] ratio. It has also been reported that the enzyme is allosterically regulated by ADP (Pacold and Anderson, 1975), and McMorrow and Bradbeer (1987a) have reported that the chloroplastic enzyme is inhibited by both AMP and ADP to a greater degree than is the cytosolic enzyme. These effectors will therefore reinforce the influence of mass-action (Robinson and Walker, 1979).

The chloroplastic enzyme accounts for about 90% of the total leaf activity (McMorrow and Bradbeer, 1987b). The activity of PGA kinase is between 2000 and 3000 µmol h^{-1} mg^{-1} chlorophyll in isolated chloroplasts (Latzko and Gibbs, 1968; Leegood and Walker, 1980a) which is considerably in excess of the rate of photosynthesis. There is little or no change in activity upon illumination of chloroplasts or leaves (Leegood and Walker, 1980a; Wirtz et al., 1982).

Purification of the enzyme from spinach by affinity chromatography has been achieved by Kuntz et al. (1978) and separation of the two isoforms by McMorrow and Bradbeer (1987b). Persson and Olde (1989) have described a rapid procedure for purification of the enzyme from chloroplasts by affinity partitioning on ATP-dextran, yielding a preparation which is 80% pure.

B. NADP-glyceraldehyde Phosphate Dehydrogenase (EC 1.2.1.13)

$$GBP^{4-} + H^+ + NADPH \longrightarrow GAP^{2-} + Pi^{2-} + NADP^+$$

1. Properties

Regulation of the activity of glyceraldehyde-P dehydrogenase is probably the most complex of any of the Calvin cycle enzymes. The enzyme has been isolated in a number of aggregated forms. Pupillo and Giuliani-Piccari (1973) showed that spinach $NADP^+$-linked glyceraldehyde-P dehydrogenase aggregates in the presence of NAD^+ during gel filtration of partially purified extracts, while $NADP^+$ causes dissociation of the enzyme to the form most active with NADP(H). In the dissociated form the enzyme comprises a tetramer of 145 kDa with dissimilar subunits of 37–38 kDa (A) and 39–42 kDa (B) in white mustard, spinach, peas and barley (Ferri et al., 1978; Pupillo and Giuliani-Piccari, 1973, 1975; Pawlizki and Latzko, 1974; Cerff, 1978b). Subunit A is the polypeptide containing the catalytic site, whose structure is distinct from that of its cytosolic counterpart and is related to bacterial glyceraldehyde-P dehydrogenases (Martin and Cerff, 1986). The enzyme can be separated into two active forms, A_2B_2 (isoenzyme 2) and A_4 (isoenzyme 1) (Cerff, 1979). There is a single active site per AB protomer (Zapponi et al., 1983). Subunit B is highly homologous to, though smaller than, β-tubulin of *Chlamydomonas* (Cerff et al., 1986). Cerff et al. (1986) have suggested that the tendency for aggregation and disaggregation may be related to the nature of this subunit.

The first demonstration of light activation of a Calvin cycle enzyme was made with glyceraldehyde-P dehydrogenase. Light activation results in an increase in the $NADP^+$, but not the NAD^+-linked activity (Ziegler and Ziegler, 1965). Schulman and Gibbs (1968) showed that the NAD^+-linked activity was not separable from the $NADP^+$-linked activity. The enzyme has Michaelis–Menten kinetics and has a ten-fold higher affinity for NADP(H) than it does for NAD(H) with: K_m(GBP) 1 μM, K_m(NADPH) 23–39 μM, K_m(NADH) 170–300 μM, K_m($NADP^+$) 41 μM, K_m(NAD^+) 200 μM (Ferri et al., 1978; Cerff, 1978b). The enzyme shows hysteretic properties and its activation is sensitive to NADPH, ATP and Pi (Wolosiuk and Buchanan, 1978a; Wara-Aswapati et al., 1980; Müller et al., 1969; Pupillo and Giuliani-Piccari, 1973, 1975) as well as the ferredoxin–thioredoxin system (Wolosiuk and Buchanan, 1978a). There is no difference in K_m between oxidised and reduced forms (Ferri et al., 1978). Wolosiuk and Buchanan (1978a) have shown that pre-incubation with effectors causes a rapid increase in activity and that selected pairs of effectors act synergistically. Thus addition of 10 mM Pi, which itself causes no change in activity, decreases the K_a for ATP from 4 to 0.8 mM and the K_a for NADPH from 0.8 to 0.3 mM (FBP and SBP are also effective in this regard). Similarly, addition of ATP or NADPH lowered K_a for Pi. The mechanism remains unknown, but enhancement of the $NADP^+$-linked activity by metabolites and by DTT is a concerted process (Wolosiuk et al., 1986). The enzyme is strongly inhibited in isolated chloroplasts by low concentrations of iodoacetol phosphate, an analogue of dihydroxyacetone phosphate (DHAP) (Usuda and Edwards, 1981).

In contrast to the enzyme from higher plants and algae, the enzyme from *Euglena gracilis* has no tendency to associate and exhibits little activity with NAD^+ as substrate (Theiss-Seuberling, 1984).

2. *Assay of glyceraldehyde-P dehydrogenase activity*

In chloroplasts the activity of the enzyme is between 150 and 1500 µmol h^{-1} mg^{-1} chlorophyll (Leegood and Walker, 1980a; Huber, 1978). Light activation in chloroplasts and leaves is relatively slight (1.5- to 5-fold) (Steiger et al., 1971; Leegood and Walker, 1980a; Akamba and Anderson, 1981; Wirtz et al., 1982; Heber et al., 1982). The pH optimum is between pH 7.5 and 8.8 (Ferri et al., 1978; Cerff, 1982). No satisfactory rapid assay has been developed for this enzyme.

(a) *Continuous spectrophotometric assay.* The enzyme can be assayed in the direction of PGA formation, but this assay is likely to be complicated in crude extracts by side reactions of the substrate, GAP (which is also expensive), and by the presence of non-phosphorylating glyceraldehyde-P dehydrogenase from the cytosol (Kelly and Gibbs, 1973). The assay can be made non-reversible by the inclusion of arsenate as an alternative substrate to phosphate. When assayed by $NADP^+$ formation (in the direction of photosynthesis), sufficient PGA kinase must be added to ensure that the generation of glycerate-1,3-bisphosphate is not limiting.

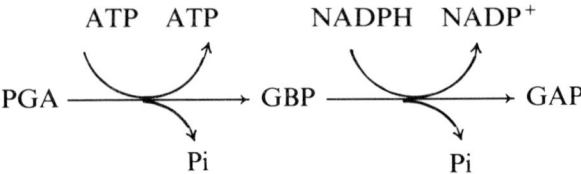

4 mM 3-PGA
5 mM ATP
10 mM $MgCl_2$
0.2 mM NADPH
20 units phosphoglycerate kinase
100 mM Tricine-KOH (pH 8.0)

Initiate the reaction (total volume 1 ml) with extract containing 1 to 5 µg chlorophyll.

3. *Purification of glyceraldehyde-P dehydrogenase*

Cerff (1978a, 1979, 1982) has described parallel purification of the cytosolic and the two chloroplastic isoenzymes from white mustard, yielding a specific activity for the $NADP^+$-linked activity of 100–150 U mg^{-1} protein. The separation is based on the sequential gel filtration of the extracts on Sephadex G-200, first in the presence of

NAD$^+$ (leading to aggregation of the cytosolic, but not of the chloroplastic enzyme), followed by gel filtration in the presence of NAD$^+$. The applicability of this technique to other plant tissues is discussed by Cerff (1982).

C. Aldolase (EC 4.1.2.13)

$$GAP^{2-} + DHAP^{2-} \longrightarrow FBP^{4-}$$

$$E4P^{2-} + DHAP^{2-} \longrightarrow SBP^{4-}$$

Aldolase from chloroplasts exists as a tetramer of 148 kDa, comprising subunits of 38 kDa (slightly smaller than the subunits of the cytosolic enzyme) (Krüger and Schnarrenberger, 1983; Lebherz et al., 1984). The same protein catalyses the synthesis of FBP and SBP. The chloroplastic enzyme is of the Class I type, like aldolases from other eukaryotic cells, and in contrast to the Class II metalloaldolases of prokaryotic cells (Krüger and Schnarrenberger, 1983).

Kinetic constants from a range of plants are similar, with K_m(FBP) 20 μM, K_m(SBP) 8–20 μM, K_m(GAP) 0.3 mM, K_m(DHAP) 0.4 mM (Brooks and Criddle, 1966; Anderson and Pacold, 1972; Murphy and Walker, 1981). Anderson and Pacold (1972) reported inhibition by RuBP (K_i 0.14 mM), ADP (K_i 2 mM) and PGA (K_i 2.9 mM). An important regulatory aspect of the aldolase reaction is that the FBP concentration changes in proportion to the square of the concentration of triose-P (Leegood et al., 1985).

The activity of the enzyme in chloroplasts is about 300 μmol h^{-1} mg^{-1} chlorophyll (Latzko and Gibbs, 1968) and there is no evidence for light activation. In leaves of spinach and wheat, the chloroplastic enzyme constitutes about 85% of the total aldolase activity (Murphy and Walker, 1981; Lebherz et al., 1984; Krüger and Schnarrenberger, 1983). The pH optimum is about pH 8 (Murphy and Walker, 1981).

Lebherz et al. (1984) and Krüger and Schnarrenberger (1983) have both described procedures for the simultaneous purification to homogeneity of cytosolic and plastidic enzymes from spinach and maize, attaining specific activities of about 8 U mg^{-1} protein. Persson (1988) has also described an aqueous two-phase system for aldolase purification from spinach chloroplasts.

D. Triose Phosphate Isomerase (EC 5.3.1.1)

$$GAP^{2-} \longrightarrow DHAP^{2-}$$

Triose-P isomerase from spinach and lettuce contains two identical subunits of 27 kDa which have a distinct primary structure from, and are slightly larger than, the subunits of the cytosolic enzyme (Pichersky and Gottlieb, 1984; Kurzok and Feierabend, 1984b). The kinetic constants are: K_m(DHAP) 1.1 to 2.5 mM, K_m(GAP) 0.42 to 0.68 mM (Anderson, 1971a; Kurzok and Feierabend, 1984a). The enzyme is competitively inhibited by a wide range of intermediates, e.g. RuBP (K_i 0.56 mM), Pi (K_i 6.3 mM), FBP (K_i 2.3 mM), PGA (K_i 2.9 mM) and strongly by 2-phosphoglycollate; K_i 15 μM in peas (Anderson, 1971a), K_i 0.2 mM in rye (Kurzok and Feierabend, 1984a).

The activity of the enzyme in isolated chloroplasts is about 10 000 μmol h^{-1} mg^{-1} chlorophyll (Latzko and Gibbs, 1968), which is vastly in excess of the rate of photosynthesis. In leaves of rye, the activities of the chloroplastic and cytosolic enzymes are approximately equal. The pH optimum of the chloroplastic enzyme is about pH 8.4, compared with a pH optimum for the cytosolic enzyme of pH 7.2–7.5, but the pH optimum of the chloroplastic enzyme is lowered to about pH 7.5 in the presence of phosphate (Kurzok and Feierabend, 1984a). The pI is 4.4.

Pichersky and Gottlieb (1984) and Kurzok and Feierabend (1984a) have both described procedures for the simultaneous purification to homogeneity of cytosolic and plastidic enzymes, attaining specific activities of up to 9200 U mg^{-1} protein (for the plastidic enzyme from lettuce).

E. Fructose 1,6-Bisphosphatase (EC 3.1.3.11)

$$FBP^{4-} + H_2O \longrightarrow F6P^{2-} + Pi^{2-}$$

1. Properties

The enzyme is a tetramer (M_r 160 kDa), but spinach FBPase dissociates into a dimer at alkaline pH (Zimmermann et al., 1976). FBPase from *Synechococcus leopiliensis* undergoes a transition from an inactive dimer to the active tetramer which is promoted by FBP and Mg^{2+} (Gerbling et al., 1984, 1985). The enzyme is largely specific for FBP, although it is also capable of a low rate of SBP hydrolysis (Cadet and Meunier, 1988b), particularly in *S. leopiliensis* (Gerbling et al., 1986). Gontero et al. (1984a,b) and Cadet et al. (1987) have also purified a chloroplastic phosphatase (M_r 100 kDa) which is distinct from FBPase, and which is activated by thioredoxin, although its role is not yet clear.

The chloroplastic FBPase differs from the cytosolic enzyme (and mammalian and yeast enzymes) in not being inhibited by AMP and in being light-activated. Raines et al. (1988) have deduced the amino acid composition of the mature wheat chloroplast enzyme from the cDNA sequence. About 45% of the sequences of the chloroplast, mammalian and yeast enzymes are homologous and a further 20% of substitutions are conservative changes. The chloroplastic enzyme from wheat and spinach possesses a unique sequence of twelve amino acids in the variable region of the protein (residues 165–190), which contains three cysteine residues which may be involved in light-activation (see also Marcus et al., 1988). The wheat FBPase neither possesses the N-terminal extension found in the yeast enzyme, which is believed to be the site of action of a protein kinase (Rittenhouse et al., 1986), nor does it possess the lysine residue at position 174, which is believed to be involved in AMP inhibition of the mammalian enzyme (Marcus et al., 1982).

The enzyme is regulated by the ferredoxin–thioredoxin system (Wolosiuk et al., 1980; Schürmann and Kobayashi, 1984). The native oxidised enzyme contains six disulphide bridges, two of which are cleaved upon activation (Pradel et al., 1981; Aragnol et al., 1985). The thioredoxin binding site is different from the catalytic site, and thioredoxin fB binds tightly and interacts with the two disulphide bridges of the tetrameric FBPase (Soulié et al., 1985). FBP does not influence the binding of thioredoxin. Reduction of the enzyme results in a decrease in the K_m for FBP from 800 μM to 33 μM and in the K_m for Mg^{2+} from 9 mM to 2 mM (Charles and Halliwell, 1980) and the K_m for the

[FBP^{4-}.Mg^{2+}]$^{2-}$ complex from 130 to 6 μM (Cadet and Meunier, 1988b), although the maximal velocity of the enzyme is unaltered. The enzyme also shows a change from hyperbolic to sigmoidal kinetics upon reduction.

A complex interplay of electron transfer, ligand binding and the stromal pH determines the activation state of the enzyme (Leegood et al., 1982). FBPase activity is particularly sensitive to pH and the concentration of Mg^{2+}. In its oxidised form the enzyme has a pH optimum of pH 8.8, whereas the reduced form is optimally active at pH 7.5 to 8.5 (Baier and Latzko, 1975; Kelley et al., 1976; Zimmermann et al., 1976). This pH sensitivity does not result from changes in V_{max} but is due to the fact that the substrate for the enzyme is the [FBP^{4-}.Mg^{2+}]$^{2-}$ complex (Gardemann et al., 1986). A pH shift from pH 7 to pH 8 also induces a slow conformational change in the enzyme (Minot et al., 1982; Gontero et al., 1984c; Pradel and Aragnol, 1986), resulting in exposure of disulphide bridges (Aragnol et al., 1985) and enhancing the binding of thioredoxin fB (Soulié et al., 1985).

FBP is reported to enhance enzyme activity in the absence of a reductant (K_a 0.25 mM), although together with a reductant it is effective at a much lower concentration (Chehebar and Wolosiuk, 1980; Hertig and Wolosiuk, 1983; Wolosiuk et al., 1980). SBP, RuBP and glucose 1,6-bisphosphate are also effective activators (Hertig and Wolosiuk, 1983). Similarly, the activity of oxidised FBPase is enhanced reversibly by Mg^{2+}, independently of a reductant (Schürmann and Wolosiuk, 1978). However, Gontero et al. (1984b) have suggested that FBP and Mg^{2+} have little or no activating effect either on oxidised FBPase or on its reductive activation. FBP stabilises the enzyme against oxidative inactivation (Leegood, 1981; Leegood et al., 1982; Schürmann and Kobayashi, 1984), although such observations have yet to be reconciled with the properties of the purified enzyme (see Soulié et al., 1985; Gontero et al., 1984b). Ca^{2+} has been found to facilitate the activation (as distinct from catalysis) of FBPase (Hertig and Wolosiuk, 1980, 1983; Charles and Halliwell, 1981; Kreimer et al., 1988).

The enzyme has been shown to be inhibited by F6P^{2-} at millimolar concentrations (Gardemann et al., 1986). The oxidised form of the enzyme is inhibited by Pi ($I_{0.5}$ 10 mM) and the reduced enzyme is relatively insensitive to Pi (Charles and Halliwell, 1980), although Cadet et al. (1987) claim that only the alternate phosphatase is inhibited by Pi.

2. Assay of FBPase activity

The activity is between 50 and 500 μmol h^{-1} mg^{-1} chlorophyll in leaves and is predominantly stromal FBPase (Leegood and Walker, 1982; Wirtz et al., 1982). The activity of the enzyme increases markedly on illumination of chloroplasts or leaves (Leegood and Walker, 1982; Laing et al., 1981; Wirtz et al., 1982) but the activity of the cytosolic form of the enzyme is not influenced directly by light (Foyer et al., 1982; Stitt and Grosse, 1988). Extraction is best done in the presence of 1 mM FBP as this stabilises the enzyme against oxidative inactivation. Desalting and storage on ice causes a loss of activity (Kobza and Edwards, 1986). The magnitude of the contribution of the alternate phosphatase (Gontero et al., 1984a) to the assay of FBPase and SBPase in crude extracts is unknown. An immunological method for the quantitative determination of FBPase protein, without interference from the cytosolic FBPase or the non-specific phosphatase, has been described by Hermoso et al. (1987).

(a) *Continuous spectrophotometric assay.* A range of concentrations of FBP and pH values may be used in the assay of light activation (Leegood, 1981; Gardemann et al., 1986). Assays are less linear at low pH (less than pH 7.8), and at low concentrations of Mg^{2+} and FBP. Excessive FBP (greater than 1 mM) or too high a pH (greater than pH 8.2) will tend to minimise apparent light activation.

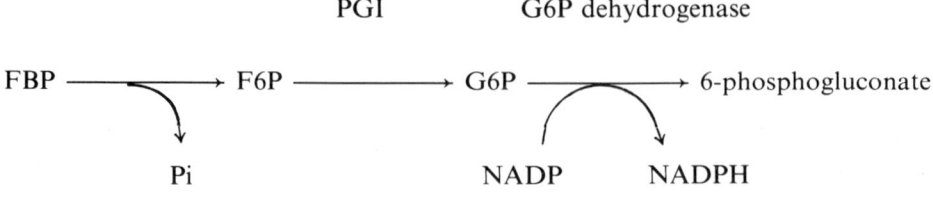

0.1–1 mM FBP
1 mM EDTA
0.4 mM $NADP^+$
10 mM $MgCl_2$
0.05% Triton X-100
100 mM Tris-HCl (pH 8.2)
4 units phosphoglucose isomerase (PGI)
2 units glucose 6-P dehydrogenase

Initiate the reaction (total volume 1 ml) with extract containing 1–5 µg chlorophyll.

(b) *Rapid assay.* The rapid assay can be employed to measure release of F6P or Pi. In the case of Pi release, malachite green is employed to increase the sensitivity of the standard acid molybdate assay for Pi (Gardemann et al., 1986; Itaya and Ui, 1966). Alternatively, the release of ^{32}Pi from F-1,6-[1-^{32}P]BP can be employed (Laing et al., 1981). For the assay of Pi release, care must be taken to ensure that all glassware, etc. is scrupulously clean.

The assay is the same as for the continuous spectrophotometric assay (omitting coupling enzymes), but in a total volume of 500 µl. Terminate the reaction after 20–30 s with 200 µl 1 M $HClO_4$.

(c) *Assay of F6P release (Gardemann et al., 1986).* Neutralise the sample with an ice-cold mixture of 5M KOH–1 M triethanolamine (to pH 7, using pH paper) and centrifuge to remove the precipitated $KClO_4$. The supernatant can be assayed for F6P (Lowry and Passonneau, 1972).

100 µl neutralised supernatant
1 mM $NADP^+$
10 mM $MgCl_2$
100 mM Tris-HCl (pH 7.8)
2 units glucose 6-P dehydrogenase

Wait until the zero reading is stable after gentle stirring and initiate the reaction by the addition of 4 U phosphoglucose isomerase.

(d) *Assay of Pi release.* Centrifuge the acidified sample and use the supernatant for the determination of Pi.

500 µl supernatant
1 ml 1M HCl
1 ml reagent A

(Reagent A: 1 vol 4.2% $(NH_4)_6Mo_7O_{24}$ in 5 M HCl mixed with 3 vol 0.2% malachite green. Stand for 30 min before filtering.)

After 15 min add 100 µl 1.5% Tween-20. Read the absorbance at 660 nm and estimate Pi reference to a standard curve using between 0.1 and 2 µg Pi.

3. Purification of FBPase

The following is the procedure for purification of FBPase from spinach developed by Zimmermann *et al.* (1976), which achieves a specific activity of 109 U mg^{-1} protein with a yield of 80%. A number of other procedures have been developed by Nishizawa *et al.* (1982), Schürmann and Wolosiuk (1978), Marcus *et al.* (1987) and affinity chromatography by Plá *et al.* (1981). The p*I* is 4.5.

(a) *Homogenisation.* Two kilograms of de-ribbed and washed spinach leaves are homogenised in a blender. To each 100 ml of the extract, 10 ml 100 mM Na phosphate buffer (pH 7.5), 2 mM EDTA is added and the pH adjusted to pH 7.5 with 5 M NaOH. The mixture is centrifuged at 20 000 × *g* for 20 min.

(b) *$(NH_4)_2SO_4$ fractionation.* The protein which precipitates between 42% and 66% saturation by $(NH_4)_2SO_4$ is collected by centrifugation at 20 000 × *g* for 20 min. The precipitated protein is dissolved in 50 ml 100 mM Na phosphate buffer (pH 7.5), 2 mM EDTA and $(NH_4)_2SO_4$ added (18 g per 100 ml). The precipitated protein is removed and $(NH_4)_2SO_4$ added (14 g per 100 ml) and the mixture centrifuged at 20 000 × *g* for 20 min. The precipitate is dissolved in 10 ml 20 mM Na phosphate buffer (pH 7.5), 0.4 mM EDTA, 20 mM NaCl.

(c) *DEAE-Sephadex chromatography.* The solution from Step (b) is placed on a column (40 cm × 8 cm) of DEAE-Sephadex equilibrated with 20 mM Na phosphate buffer (pH 7.5), 0.4 mM EDTA, 20 mM NaCl. The column is first washed with 360 ml of the same buffer, and the protein then eluted with 1.2 l of a linear gradient of NaCl (50 to 700 mM) in the same buffer. The bulk of the FBPase elutes at around 500 mM NaCl in about 90 ml. Solid $(NH_4)_2SO_4$ is added to 80% saturation (56 g/100 ml) and the precipitate collected by centrifugation and dissolved in 5 ml 100 mM Na phosphate buffer (pH 7.5), 2 mM EDTA, 100 mM NaCl.

(d) *Sephadex G-200.* The solution from Step (c) is passed through a column of Sephadex G-200 (90 × 5.3 cm^2) equilibrated with 100 mM Na phosphate buffer (pH 7.5), 2 mM EDTA, 100 mM NaCl. Active fractions (*c.* 30 ml) are concentrated by ultrafiltration under N$_2$ over a Diaflo XM-10 membrane to a volume of about 3 ml.

(e) *Storage.* The activity is stable for at least 6 months at −20°C.

F. Transketolase (EC 2.2.1.1)

$$F6P^{2-} + GAP^{2-} \longrightarrow E4P^{2-} + Xu5P^{2-}$$

$$S7P^{2-} + GAP^{2-} \longrightarrow R5P^{2-} + Xu5P^{2-}$$

The enzyme is a tetramer with M_r 150 kDa with K_m(xylulose 5-phosphate (Xu5P), erythrose 4-phosphate (E4P), R5P) of 100 to 130 μM (Horecker et al., 1956; Murphy and Walker, 1982). The activity is about 300 μmol h^{-1} mg^{-1} chlorophyll (Latzko and Gibbs, 1968) and a high proportion (90%) of the activity in leaves of peas and rye is found in the plastids (Feierabend and Gringel, 1983). No effectors have been reported, but thiamine pyrophosphate is an essential cofactor; in its absence the enzyme dissociates into a dimer. The chloroplastic enzyme is stimulated by Mg^{2+} and has a pH optimum between pH 7.5 and 8.5 (Murphy and Walker, 1982). Murphy and Walker (1982) have reported a 400-fold purification of the enzyme from spinach.

G. Sedoheptulose 1,7-Bisphosphatase (EC 3.1.3.37)

$$SBP^{4-} + H_2O \longrightarrow S7P^{2-} + Pi^{2-}$$

1. Properties

The enzyme from spinach is a dimer (M_r 66 kDa) comprising two identical 35 kDa subunits (Cadet and Meunier, 1988a). The enzyme is quite distinct from FBPase and is immunologically unrelated (Cadet et al., 1987). The true substrate of the reaction is probably $[SBP^{4-}.Mg^{2+}]^{2-}$ (Woodrow et al., 1984; Cadet and Meunier, 1988b). The reduced enzyme has a K_m(SBP) of 13.5 μM and K_m(Mg^{2+}) of 1.6 mM (Woodrow et al., 1983). The specificity of the spinach enzyme for SBP is high but is not absolute (Cadet and Meunier, 1988b). Like FBPase, the enzyme is inhibited by Fru-2,6-bisP (Cadet and Meunier, 1988b).

The enzyme is regulated by the ferredoxin–thioredoxin system (Schürmann and Buchanan, 1975) and interacts with thioredoxin *f*B. Unlike FBPase, the oxidised enzyme has virtually no activity (Woodrow and Walker, 1983; Cadet et al., 1987). Reduction of the enzyme results in a decrease in the K_m for the $[SBP^{4-}.Mg^{2+}]^{2-}$ complex from 0.18 to 0.05 mM (Cadet and Meunier, 1988b). The enzyme undergoes spontaneous oxidative inactivation in the absence of a reductant (Cadet and Meunier, 1988a) The affinity of the enzyme for SBP is diminished in the presence of Pi. At 5 mM Pi, the K_m(SBP) is raised from 13.5 μM to 17.5 μM and at 10 mM Pi to 33.5 μM (Woodrow et al., 1983), although Cadet et al. (1987) claim that only the alternate phosphatase is sensitive to Pi.

2. Assay of SBPase

Like FBPase, the activity of SBPase is increased markedly upon illumination of chloroplasts or leaves and SBP is required for reductive activation (Woodrow and Walker, 1980; Laing et al., 1981; Wirtz et al., 1982). The activity is between 50 and 200 μmol h^{-1} mg^{-1} chlorophyll in leaves (Wirtz et al., 1982). As with FBPase, the enzyme is stabilised by the presence of its substrate; 1 mM SBP should be included in the extraction medium.

(a) *Continuous spectrophotometric assay.* In crude extracts, the rapid assay of Pi release or the release of ^{32}Pi from Sed-1,7-[1-^{32}P]bisP can be employed (Laing *et al.*, 1981). Although longer assays of Pi release (over minutes) have been employed (and are suitable for the purified enzyme) the action of aldolase, triose-P isomerase and FBPase in crude extracts or in chloroplast extracts ensures that a considerable portion of the SBP is converted to F6P with concomitant release of Pi (Woodrow and Walker, 1980). Although a correction can be made for F6P generated, this assay is cumbersome and less reliable in crude extracts. The continuous spectrophotometric assay is suitable only for the purified enzyme and is based upon the ability of phosphofructokinase to convert sedoheptulose 7-phosphate (S7P) to SBP. This has the advantage that the concentration of substrate is conserved during the assay.

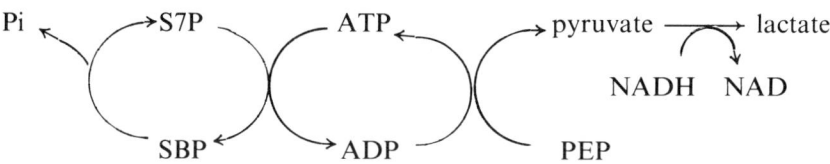

0.1 mM SBP
10 mM MgCl$_2$
0.2 mM NADH
1 mM phosphoenolpyruvate (PEP)
20 mM KCl
0.1 mM ATP
100 mM Tricine-NaOH (pH 8.2)
2 units pyruvate kinase
2 units lactate dehydrogenase
0.5 unit phosphofructokinase

Initiate the reaction (total volume 1 ml) with the enzyme.

(b) *Rapid assay of Pi release.* See Section II.E.2(b) replacing FBP by SBP as the substrate. In principle S7P release could also be measured using the above assay.

3. *Purification of SBPase*

Woodrow and Walker (1982) devised the following purification procedure for wheat SBPase, achieving about 30% purity and a specific activity of 9 U mg^{-1} protein (Woodrow *et al.*, 1983). Cadet *et al.* (1987) have reported a procedure from spinach which gives a clear separation of the alternate phosphatase and in which affinity chromatography on Blue-B dye matrix gel (to which SBPase does not bind) is included giving a final specific activity of 70 U mg^{-1} protein. Nishizawa and Buchanan (1981) have purified the enzyme from maize leaves.

(a) *Homogenisation.* One hundred grams of twelve-day-old wheat leaves are homogenised for 60 s in a blender containing 150 ml 50 mM Tris-HCl (pH 7.8), 0.5 mM EDTA, 10 mM mercaptoethanol. The homogenate is filtered through four layers of muslin and centrifuged at 30 000 × g for 15 min.

(b) *$(NH_4)_2SO_4$ fractionation.* The protein which precipitates between 50% and 90% saturation by $(NH_4)_2SO_4$ is collected by centrifugation at 30 000 × g for 15 min. The precipitated protein is dissolved in a minimum volume of 50 mM Tris-HCl (pH 8.2), 0.5 mM EDTA, 10mM NaCl. The solution is desalted on a column of Sephadex G-25M (PD-10) equilibrated with the same buffer.

(c) *DEAE-Sephadex chromatography.* The solution from Step (b) is placed on a column (3.6 cm × 20 cm) of DEAE-Sephadex A-50 equilibrated with 50 mM Tris-HCl (pH 8.2), 0.5 mM EDTA, 10 mM NaCl. The column is first washed with two bed volumes of the same buffer, and the SBPase then eluted with a linear gradient of NaCl between 10 and 300 mM. The bulk of the SBPase elutes at around 150 mM NaCl. The active fractions are pooled and concentrated to a volume of 3 ml by ultrafiltration.

(d) *Sephadex G-100.* The solution from Step (c) is passed through a column of Sephadex G-100 (2 cm × 90 cm) equilibrated with 50 mM Tris-HCl (pH 8.2), 0.5 mM EDTA, 10 mM mercaptoethanol, 200 mM NaCl. Active fractions, eluting after the main protein peak, are concentrated by ultrafiltration in an Amicon cell with a Diaflo PM-10 membrane to a volume of about 3 ml.

(e) *Storage.* The enzyme is desalted on a column of Sephadex G-25M equilibrated with 20 mM Tricine-NaOH (pH 8.2) and stored in 25% glycerol at $-20°C$.

H. Ribose 5-Phosphate Isomerase (EC 5.1.3.4)

$$R5P^{2-} \longrightarrow Ru5P^{2-}$$

The tobacco enzyme has M_r 54 kDa and pI 5.1. The K_m(R5P) is 1.6 mM (Rutner, 1970; Kawashima and Tanabe, 1976). The activity is about 600 µmol h^{-1} mg^{-1} chlorophyll (Latzko and Gibbs, 1968). The enzyme is competitively inhibited by AMP (K_i 1.3 mM) and activated by RuBP (K_a 1.2 mM) (Anderson, 1971b). The pH optimum is between pH 7.5 and 8.2. An assay is described by Wood (1970).

I. Ribulose 5-Phosphate 3-Epimerase (EC 5.3.1.16)

$$Xu5P^{2-} \longrightarrow Ru5P^{2-}$$

The enzyme has M_r 46 kDa, K_m(Xu5P) 0.5 mM and an activity of about 1600 µmol h^{-1} mg^{-1} chlorophyll (Latzko and Gibbs, 1968). An assay is described by Wood (1970).

J. Ribulose 5-Phosphate Kinase (EC 2.7.1.19)

$$Ru5P^{2-} + ATP^{4-} \longrightarrow RuBP^{4-} + ADP^{3-} + H^+$$

1. Properties

The purified enzyme from wheat and spinach is a dimer of 83 kDa comprising subunits of 42 to 44 kDa which are probably identical (Surek et al., 1985; Porter et al., 1986). The dimer is the smallest active form of the enzyme. However, the enzyme can also be isolated from crude extracts in a number of oligomeric forms (245 kDa from spinach (Surek et al., 1985), higher M_r from other sources (Wara-Aswapati et al., 1980; Lazaro et al., 1986)).

The kinetics are Michaelis–Menten with respect to Ru5P and ATP (K_m(Ru5P) 25–70 μM, K_m(ATP) 35–65 μM), but are sigmoidal with respect to Mg^{2+} ($S_{0.5}$ (Mg^{2+}) less than 0.5 mM) (Gardemann et al., 1982, 1983; Surek et al., 1985). This enzyme shows hysteretic properties and is regulated by the ferredoxin–thioredoxin system (Wolosiuk and Buchanan, 1978b). It has been shown that thiol regulation involves intra-subunit reduction of two cysteine residues (Cys-16 and Cys-55), one of which (Cys-16) is part of the nucleotide binding domain of the active site (Porter et al., 1988). The purified enzyme is very sensitive to pH and redox control, and has an activity at pH 6.8 in the absence of DTT (conditions approximating to the darkened stroma) which is only 2% of the activity at pH 7.8 in the presence of DTT (an approximation of the illuminated stroma) (Surek et al., 1985). The enzyme undergoes spontaneous slow oxidation in the absence of a reductant (Porter et al., 1986).

The enzyme shows complex isosteric control by metabolites (Anderson, 1973; Lavergne et al., 1974; Laing et al., 1981; Gardemann et al., 1982, 1983). In studies with the enzyme activated by light in spinach chloroplasts, several compounds were found to inhibit competitively with respect to Ru5P; 6-phosphogluconate (K_i 0.07 mM), RuBP (K_i 0.7 mM), PGA (K_i 2 mM) and Pi (K_i 4 mM) (Gardemann et al., 1983). PGA inhibits only at low concentrations of R5P (Surek et al., 1985) and only the PGA^{2-} form inhibits the enzyme. Inhibition by PGA in vivo is therefore enhanced by a decrease in stromal pH (Gardemann et al., 1982). ADP inhibits the dark (oxidised) form of the enzyme, acting competitively (K_i 0.04 mM) with ATP (Gardemann et al., 1983). Although the K_m of the light-activated form of the enzyme for ATP alone is about 30 μM, addition of ADP and AMP to adenylate kinase equilibrium concentrations increases the ATP concentration required for half-saturation of activity to 220 μM (Laing et al., 1981), indicating that adenylate regulation of the enzyme is probably important. FBP shows a mixed-type inhibition (Gardemann et al., 1983).

2. Assay of Ru5P kinase

The activity is between 1100 and 1500 μmol h^{-1} mg^{-1} chlorophyll in chloroplasts or leaves (Leegood and Walker, 1980a; Wirtz et al., 1982). Measurements of in vitro light-activation of the enzyme show considerable variation, from two- to three-fold (Latzko and Gibbs, 1969; Leegood and Walker, 1980a) to 10-fold or more (Laing et al., 1981; Fischer and Latzko, 1979). One reason may be that the enzyme is activated even at very low irradiance (Fischer and Latzko, 1979). Light-activation of the enzyme results in a dramatic increase in V_{max}, whereas the affinity for substrates is not appreciably altered (Gardemann et al., 1983). The pH optimum is between pH 7.8 and pH 8.4.

(a) *Continuous spectrophotometric assay.* The continuous spectrophotometric assay is reliable in crude extracts, although there is often an extended activation lag. However, the complexity of the assay means that great care must be taken with blanks. Potential contributors to the blank are: (1) the combined action of PEP carboxylase and malate dehydrogenase; (2) ATPase (which in the thylakoids is also light-activated); (3) PEP phosphatase activity; and (4) NADH oxidase. R5P is frequently employed as the substrate in crude extracts, with reliance upon endogenous R5P isomerase to convert the R5P to Ru5P. However, a check should be made that an excess of this coupling enzyme is present, and if necessary, more should be added. Alternatively, an equilibrium mixture of R5P and Ru5P can be generated in advance of the assay and employed as the substrate (see below). This is clearly a prerequisite for the rapid assay and for assays of the activity of the purified enzyme. Rapid assays have been based both on ADP formation and upon RuBP formation (Gardemann, 1983; Gardemann et al., 1983).

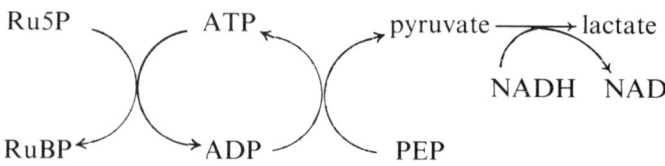

0.5 mM R5P (or Ru5P)
1 mM ATP
50 mM KCl
0.4 mM NADH
10 mM MgCl$_2$
5 mM PEP
100 mM Tricine-KOH (pH 8.0)
7 units pyruvate kinase
10 units lactate dehydrogenase
R5P isomerase if necessary

Initiate the reaction (total volume 1 ml) with extract containing 1–5 µg chlorophyll.

(b) *Rapid assay.* The enzyme may be assayed by the release of ADP (Gardemann, 1983).

0.5 mM Ru5P
1 mM ATP
50 mM KCl
10 mM MgCl$_2$
0.05% Triton X-100
100 mM Tricine-KOH (pH 8.0)

Initiate the reaction (total volume 500 µl) with extract containing 1–5 µg chlorophyll. Terminate the reaction after 20–30 s with 200 µl 1 M HClO$_4$. Neutralise the sample with an ice-cold mixture of 5 M KOH/1 M triethanolamine (to pH 7, using pH paper) and centrifuge to remove the precipitated KClO$_4$. Blanks should take into account the activity and light-activation of the thylakoid ATPase.

ADP may be assayed by adding 300 µl of the neutralised supernatant to 300 µl of the following mixture (Lowry and Passonneau, 1972):

100 mM Tris-HCl (pH 7.8)
0.2 mM NADH
10 mM $MgCl_2$
0.5 mM PEP
10 U lactate dehydrogenase

Wait until the zero reading is stable after gentle stirring and initiate the reaction by the addition of 2 U pyruvate kinase.

(c) *Generation and assay of Ru5P.* D-Ribulose 5-phosphate may be prepared enzymically from R5P (Slabas and Walker, 1976). One gram R5P is dissolved in 8 ml 20 mM glycylglycine buffer (pH 7.4). One thousand units phosphoriboseisomerase (Sigma Chemical Co., St. Louis, MO) is added and the reaction mixture incubated for 30 min at 37°C. The reaction is terminated with 2 ml of 1 M HCl and centrifuged to remove precipitated protein. Ru5P may either be precipitated as the barium salt (Slabas and Walker, 1976) or freeze-dried. The yield is about 25%. Ribulose-5-P may be assayed by the carbazole method (Dische and Borenfreund, 1951): Add 6 ml Reagent B to 0.6 ml of the solution to be tested. After 30 min at 37°C read the absorbance at 540 nm. Fructose (up to 15 µM) may be employed as a standard. (Reagent B: 225 ml conc. H_2SO_4 plus 95 ml water followed by 0.2 ml of a 0.12% solution of carbazole in ethanol and 0.2 ml 1.5% cysteine-HCl; the last two components added and mixed within 30 s.)

3. Purification of Ru5P kinase

A number of purification procedures have been described. Surek *et al.* (1985) have reported the following purifications from wheat, giving a high specific activity (900 U mg^{-1} protein) and a yield of 41%. Other procedures have been reported by Porter *et al.* (1986), Krieger and Miziorko (1986) and Lavergne and Bismuth (1973) for spinach, by Kagawa (1982) for tobacco, by Satoh *et al.* (1985) for *Bryopsis maxima* and by Lazaro *et al.* (1986) for *Scenedesmus obliquus*.

(a) *Homogenisation.* Twenty grams of wheat leaves are homogenised for 3 × 15 s in a blender in 80 ml 50 mM Tris-HCl (pH 8.0), 30 mM EDTA, 10 mM mercaptoethanol, filtered through two layers of Miracloth and centrifuged at 27 000 × *g* for 20 min.

(b) *$(NH_4)_2SO_4$ fractionation and isoelectric precipitation.* The protein which precipitates following 35% saturation by $(NH_4)_2SO_4$ and stirring for 60 min is collected by centrifugation. The pellet is discarded and the pH of the supernatant brought to pH 4.5 with HCl. The precipitated protein is collected by centrifugation and dissolved in 10 ml 50 mM Tris-HCl (pH 8.0), 10 mM EDTA, 1 mM mercaptoethanol. Following solubilisation of the protein, $(NH_4)_2SO_4$ is added to 20% saturation to facilitate binding to the pentylagarose column.

(c) *Pentylagarose.* A column of pentylagarose (25 ml bed volume) is equilibrated with 50 mM Tris-HCl (pH 8.0), 10 mM EDTA, 1 mM mercaptoethanol, 35% $(NH_4)_2SO_4$. (Gels regenerated between one and four times with a sequence of water, ethanol, butanol, ethanol, water are reported to give the most reliable results.) The solution from Step (b) is applied to the column, which is first washed with the same buffer containing 20% $(NH_4)_2SO_4$, and the protein then eluted with 50 ml of a linear gradient of $(NH_4)_2SO_4$ between 15 and 5% saturation, followed by the same buffer containing 5% $(NH_4)_2SO_4$. The bulk of the Ru5P kinase elutes at around 5% $(NH_4)_2SO_4$. Active fractions are concentrated by addition of solid $(NH_4)_2SO_4$ to 60% saturation and the precipitate collected by centrifugation and dissolved in 100 mM Tris-HCl (pH 6.8), 0.5 mM EDTA, 5 mM DTT. The solution is desalted on a column of Sephadex G-25M (PD-10) equilibrated with the same buffer.

(d) *Blue-Sepharose CL-6B.* The solution from Step (c) is passed through a column of Blue-Sepharose (bed volume 20 ml) equilibrated with 100 mM Tris-HCl (pH 6.8), 0.5 mM EDTA, 5 mM DTT. Ru5P kinase activity is eluted with the same buffer containing 1 mM ATP. Active fractions are buffer exchanged on a column of Sephadex G-25M (PD-10) equilibrated with 50 mM Tris-HCl (pH 8), 10 mM EDTA, 1 mM mercaptoethanol.

(e) *Storage.* The preparation was stable for at least a month when stored at 0–5°C in the presence of DTT and EDTA. The enzyme was stable stored frozen in 50% glycerol.

REFERENCES

Akamba, L. M. and Anderson, L. E. (1981). *Plant Physiol.* **67**, 197–200.
Anderson, L. E. (1971a). *Biochim. Biophys. Acta* **235**, 237–244.
Anderson, L. E. (1971b). *Biochim. Biophys. Acta* **235**, 245–249.
Anderson, L. E. (1973). *Biochim. Biophys. Acta* **321**, 484–488.
Anderson, L. E. (1979). *In* "Encyclopedia of Plant Physiology New Series" (M. Gibbs and E. Latzko, eds), Vol. 6, pp. 271–281. Springer, Berlin.
Anderson, L. E. (1987). *FEBS Lett.* **212**, 45–48.
Anderson, L. E. and Pacold, I. (1972). *Plant Physiol.* **49**, 393–397.
Aragnol, D., Pradel, J. and Cecchini, J.-P. (1985). *Biochim. Biophys. Acta* **829**, 275–281.
Ashton, A. R. and Anderson, L. E. (1981). *Biochim. Biophys. Acta* **638**, 242–249.
Baier, D. and Latzko, E. (1975). *Biochim. Biophys. Acta* **396**, 141–148.
Brooks, K. and Criddle, R. S. (1966). *Arch. Biochem. Biophys.* **117**, 650–659.
Buc, J., Rivière, M., Gontero, B., Suave, P., Meunier, J.-C. and Ricard, J. (1984). *Eur. J. Biochem.* **140**, 199–202.
Buchanan, B. B. (1980). *Ann. Rev. Plant Physiol.* **31**, 341–374.
Cadet, F. and Meunier, J.-C. (1988a). *Biochem. J.* **253**, 243–248.
Cadet, F. and Meunier, J.-C. (1988b). *Biochem. J.* **253**, 249–254.
Cadet, F., Meunier, J.-C. and Ferté, N. (1987). *Biochem. J.* **241**, 71–74.
Cavell, S. and Scopes, R. K. (1976). *Eur. J. Biochem.* **63**, 483–490.
Cerff, R. (1978a). *Plant Physiol.* **61**, 369–372.
Cerff, R. (1978b). *Eur. J. Biochem.* **82**, 45–53.
Cerff, R. (1979). *Eur. J. Biochem.* **94**, 243–247.
Cerff, R. (1982). *In* "Methods in Chloroplast Molecular Biology" (M. Edelmann, R. B. Hallick and N.-H. Chua, eds), pp. 683–694. Elsevier, Amsterdam.
Cerff, R., Hundreiser, J. and Friedrich, R. (1986). *Mol. Gen. Genet.* **204**, 44–51.
Charles, S. A. and Halliwell, B. (1980). *Biochem. J.* **185**, 689–693.
Charles, S. A. and Halliwell, B. (1981). *Cell Calcium* **2**, 211–224.
Chehebar, C. and Wolosiuk, R. A. (1980). *Biochim. Biophys. Acta* **613**, 420–438.
Clancey, C. J. and Gilbert, H. F. (1987). *Biochim. Biophys. Acta* **262**, 13 545–13 549.

Dische, Z. and Borenfreund, E. (1951). *Biochim. Biophys. Acta* **192**, 583–587.
Droux, M., Miginiac-Maslow, M., Jacquot, J.-P., Gadal, P., Crawford, N. A., Kosower, N. S. and Buchanan, B. B. (1987). *Arch. Biochem. Biophys.* **256**, 372–380.
Feierabend, J. and Gringel, G. (1983). *Z. Pflanzenphysiol.* **110**, 247–258.
Ferri, G., Comerio, G., Iadorola, P., Zapponi, M. C. and Speranza, M. L. (1978). *Biochim. Biophys. Acta* **522**, 19–31.
Fischer, K. H. and Latzko, E. (1979). *Biochem. Biophys. Res. Commun.* **89**, 300–306.
Ford, D. M., Jablonski, P. P., Mohamed, A. H. and Anderson, L. E. (1987). *Plant Physiol.* **83**, 628–632.
Foyer, C. H., Latzko, E. and Walker, D. A. (1982). *Z. Pflanzenphysiol.* **107**, 457–465.
Frieden, C. (1970). *Biochim. Biophys. Acta* **245**, 5788–5799.
Furbank, R. T., Foyer, C. H. and Walker, D. A. (1986). *Biochim. Biophys. Acta* **852**, 46–54.
Gardemann, A. (1983). PhD thesis, University of Göttingen.
Gardemann, A., Stitt, M. and Heldt, H. W. (1982). *FEBS Lett.* **137**, 213–216.
Gardemann, A., Stitt, M. and Heldt, H. W. (1983). *Biochim. Biophys. Acta* **722**, 51–60.
Gardemann, A., Schimkat, D. and Heldt, H. W. (1986). *Planta* **168**, 536–545.
Gerbling, K.-P., Steup, M. and Latzko, E. (1984). *Arch. Microbiol.* **137**, 109–114.
Gerbling, K.-P., Steup, M. and Latzko, E. (1985). *Eur. J. Biochem.* **147**, 207–215.
Gerbling, K.-P., Steup, M. and Latzko, E. (1986). *Plant Physiol.* **80**, 716–720.
Gontero, B., Meunier, J.-C. and Ricard, J. (1984a). *Plant Sci. Lett.* **36**, 137–142.
Gontero, B., Meunier, J.-C. and Ricard, J. (1984b). *Plant Sci. Lett.* **36**, 195–199.
Gontero, B., Meunier, J.-C. and Ricard, J. (1984c). *Eur. J. Biochem.* **145**, 485–488.
Gontero, B., Cardenas, M. L. and Ricard, J. (1988). *Eur. J. Biochem.* **173**, 437–443.
Heber, U., Takahama, U., Neimanis, S. and Shimizu-Takahama, M. (1982). *Biochim. Biophys. Acta* **679**, 287–299.
Hermoso, R., Chueca, A., Lazaro, J. J. and Lopez-Gorge, J. (1987). *Photosyn. Res.* **14**, 269–278.
Hertig, C. and Wolosiuk, R. A. (1980). *Biochem. Biophys. Res. Commun.* **97**, 325–333.
Hertig, C. and Wolosiuk, R. A. (1983). *Biochim. Biophys. Acta* **258**, 984–989.
Horecker, B. L., Smyrniotis, P. Z. and Hurwitz, J. (1956). *J. Biol. Chem.* **223**, 1009–1019.
Huber, S. C. (1978). *FEBS Lett.* **92**, 12–16.
Itaya, K. and Ui, M. (1966). *Clin. Chim. Acta* **14**, 361–366.
Kagawa, T. (1982). In "Methods in Chloroplast Molecular Biology" (M. Edelmann, R. B. Hallick and N.-H. Chua, eds), pp. 695–705. Elsevier, Amsterdam.
Kawashima, N. and Tanabe, Y. (1976). *Plant Cell Physiol.* **17**, 757–764.
Kelly, G. J. and Gibbs, M. (1973). *Plant Physiol.* **52**, 111–118.
Kelly, G. J., Zimmermann, G. and Latzko, E. (1976). *Biochem. Biophys. Res. Commun.* **70**, 193–199.
Kobza, J. and Edwards, G. E. (1986). *Austral. J. Plant Physiol.* **13**, 627–636.
Kreimer, G., Melkonian, M., Holtum, J. A. M. and Latzko, E. (1988). *Plant Physiol.* **86**, 423–428.
Krieger, T. J. and Miziorko, H. M. (1986). *Biochemistry* **25**, 3496–3501.
Krüger, I. and Schnarrenberger, C. (1983). *Eur. J. Biochem.* **136**, 101–106.
Kuntz, G. W. K., Eber, S., Kessler, W., Krietsch, H. and Krietsch, W. K. G. (1978). *Eur. J. Biochem.* **85**, 493–501.
Kurzok, H.-G. and Feierabend, J. (1984a). *Biochim. Biophys. Acta* **788**, 214–221.
Kurzok, H.-G. and Feierabend, J. (1984b). *Biochim. Biophys. Acta* **788**, 222–233.
Laing, W. A., Stitt, M. and Heldt, H. W. (1981). *Biochim. Biophys. Acta* **637**, 348–359.
Latzko, E. and Gibbs, M. (1968). *Z. Pflanzenphysiol.* **59**, 184–194.
Latzko, E. and Gibbs, M. (1969). *Progress in Photosynthesis Research* **3**, 1624–1630.
Lavergne, D. and Bismuth, E. (1973). *Plant Sci. Lett.* **1**, 229–236.
Lavergne, D., Bismuth, E. and Champigny, M. L. (1974). *Plant Sci. Lett.* **3**, 391–397.
Lazaro, J. J., Sutton, C. W., Nicholson, S. and Powls, R. (1986). *Eur. J. Biochem.* **156**, 423–429.
Lebherz, H. G., Leadbetter, M. M. and Bradshaw, R. A. (1984). *Biochim. Biophys. Acta* **259**, 1011–1017.
Leegood, R. C. (1981). Annual Report of the ARC Research Group on Photosynthesis, University of Sheffield, pp. 26–32.
Leegood, R. C. and Walker, D. A. (1980a). *Arch. Biochem. Biophys.* **200**, 575–582.

Leegood, R. C. and Walker, D. A. (1980b). *FEBS Lett.* **116**, 21–24.
Leegood, R. C., Kobayashi, Y., Neimanis, S., Walker, D. A. and Heber, U. (1982). *Biochim. Biophys. Acta* **682**, 168–178.
Leegood, R. C. and Walker, D. A. (1982). *Planta* **156**, 449–456.
Leegood, R. C. and Walker, D. A. (1983). *Plant Physiol.* **71**, 513–518.
Leegood, R. C., Walker, D. A. and Foyer, C. H. (1985). In "Photosynthetic Mechanisms and the Environment" (J. Barber and N. R. Baker, eds), pp. 189–258. Elsevier, Amsterdam.
Lowry, O. H. and Passonneau, J. V. (1972). "A Flexible System of Enzymatic Analysis". Academic Press, New York.
Marcus, F., Edelstein, I., Reardan, I. and Heinrikson, R. L. (1982). *Proc. Natl. Acad. Sci. USA* **79**, 7161–7165.
Marcus, F., Harrsch, P. B., Moberly, L., Edelstein, I. and Latshaw, S. P. (1987). *Biochemistry* **26**, 7029–7035.
Marcus, F., Moberly, L. and Latshaw, S. P. (1988). *Proc. Natl. Acad. Sci. USA* **85**, 5379–5383.
Macioszek, J. and Anderson, L. E. (1987). *Biochim. Biophys. Acta* **892**, 185–190.
McMorrow, E. M. and Bradbeer, J. W. (1987a). In "Progress in Photosynthesis Research" (J. Biggins, ed.), Vol. 3, pp. 333–336. Martinus Nijhoff, Dordrecht, The Netherlands.
McMorrow, E. M. and Bradbeer, J. W. (1987b). In "Progress in Photosynthesis Research" (J. Biggins, ed.), Vol. 3, pp. 483–486. Martinus Nijhoff, Dordrecht, The Netherlands.
McNeil, P. H. and Walker, D. A. (1981). *Arch. Biochem. Biophys.* **208**, 184–188.
Martin, W. and Cerff, R. (1986). *Eur. J. Biochem.* **159**, 323–331.
Miginiac-Maslow, M., Droux, M., Jacquot, J.-P., Crawford, N. A., Yee, B. C. and Buchanan, B. B. (1987). In "Progress in Photosynthesis Research" (J. Biggins, ed.), Vol. 3, pp. 241–247. Martinus Nijhoff, Dordrecht, The Netherlands.
Miller, K. R. and Staehelin, L. A. (1976). *J. Cell Biol.* **68**, 30–47.
Minot, R., Meunier, J. C., Buc, J. and Ricard, J. (1982). *FEBS Lett.* **142**, 118–120.
Müller, B., Ziegler, I. and Ziegler, H. (1969). *Eur. J. Biochem.* **9**, 101–106.
Murphy, D. J. and Walker, D. A. (1981). *FEBS Lett.* **134**, 163–166.
Murphy, D. J. and Walker, D. A. (1982). *Planta* **155**, 316–320.
Nicholson, S., Easterby, J. S. and Powls, R. (1986a). *FEBS Lett.* **202**, 19–22.
Nicholson, S., Easterby, J. S. and Powls, R. (1986b). *Eur. J. Biochem.* **162**, 423–431.
Nishizawa, A. N. and Buchanan, B. B. (1981). *Biochim. Biophys. Acta* **256**, 6119–6125.
Nishizawa, A. N., Yee, B. C. and Buchanan, B. B. (1982). In "Methods in Chloroplast Molecular Biology" (M. Edelmann, R. B. Hallick and N.-H. Chua, eds), pp. 707–713. Elsevier, Amsterdam.
Pacold, I. and Anderson, L. E. (1975). *Plant Physiol.* **55**, 168–171.
Pawlizki, K. and Latzko, E. (1974). *FEBS Lett.* **42**, 285–288.
Persson, L.-O. (1988). *Photosyn. Res.* **15**, 57–65.
Persson, L.-O. and Johansson, G. (1989). *Biochem. J.* **259**, 863–870.
Persson, L.-O. and Olde, B. (1988). *J. Chromatogr.* **457**, 183–193.
Pichersky, E. and Gottlieb, L. D. (1984). *Plant Physiol.* **74**, 340–347.
Plá, A., Chueca, A. and López-Gorgé, J. (1981). *Photosyn. Res.* **2**, 291–296.
Porter, M. A., Milanez, S., Stringer, C. D. and Hartmann, F. C. (1986). *Arch. Biochem. Biophys.* **245**, 14–23.
Porter, M. A., Stringer, C. D. and Hartmann, F. C. (1988). *Biochim. Biophys. Acta* **263**, 123–129.
Pradel, J. and Aragnol, D. (1986). *Biochim. Biophys. Acta* **871**, 293–301.
Pradel, J., Soulié, J.-M., Buc, J., Meunier, J. C. and Ricard, J. (1981). *Eur. J. Biochem.* **113**, 507–511.
Pupillo, P. and Giuliani-Piccari, G. G. (1973). *Arch. Biochem. Biophys.* **154**, 324–331.
Pupillo, P. and Giuliani-Piccari, G. G. (1975). *Eur. J. Biochem.* **51**, 475–482.
Raines, C. A., Lloyd, J. C., Longstaff, M., Bradley, D. and Dyer, T. (1988). *Nucleic Acids Res.* **16**, 7931–7942.
Rittenhouse, J., Haarsch, P. B., Kim, J. M. and Marcus, F. (1986). *J. Biol. Chem.* **261**, 3939–3953.
Robinson, S. P. and Walker, D. A. (1979). *Biochim. Biophys. Acta* **545**, 528–536.
Robinson, S. P. and Walker, D. A. (1980). *Arch. Biochem. Biophys.* **202**, 617–623.
Rutner, A. C. (1970). *Biochemistry* **9**, 178–184.

Sainis, J. K., Merriam, K. and Harris, G. C. (1989). *Plant Physiol.* **89**, 368–374.
Satoh, H., Okada, M., Nakayama, K. and Murata, T. (1985). *Plant Cell Physiol.* **26**, 931–940.
Schulman, M. D. and Gibbs, M. (1968). *Plant Physiol.* **43**, 1805–1812.
Schürmann, P. and Buchanan, B. B. (1975). *Biochim. Biophys. Acta* **376**, 189–192.
Schürmann, P. and Kobayashi, Y. (1984). *In* "Advances in Photosynthesis Research" (C. Sybesma, ed.), Vol. III, pp. 629–632. Martinus Nijhoff/Dr W Junk, The Hague.
Schürmann, P. and Wolosiuk, R. A. (1978). *Biochim. Biophys. Acta* **522**, 130–138.
Schürmann, P., Maeda, K. and Tsugita, A. (1981). *Eur. J. Biochem.* **116**, 37–45.
Slabas, A. R. and Walker, D. A. (1976). *Biochem. J.* **153**, 613–619.
Soulié, J.-M., Buc, J., Meunier, J. C., Pradel, J. and Ricard, J. (1981). *Eur. J. Biochem.* **119**, 497–502.
Soulié, J.-M., Buc, J., Rivière, M. and Ricard, J. (1985). *Eur. J. Biochem.* **152**, 565–568.
Soulié, J.-M., Rivière, M., Buc, J., Gontero, B. and Ricard, J. (1987). *Eur. J. Biochem.* **162**, 271–274.
Srere, P. A. (1987). *Ann. Rev. Biochem.* **56**, 89–124.
Steiger, E., Ziegler, I. and Ziegler, H. (1971). *Planta* **96**, 109–118.
Stitt, M. and Grosse, H. (1988). *J. Plant Physiol.* **133**, 129–137.
Surek, B., Heilbronn, A., Austen, A. and Latzko, E. (1985). *Planta* **165**, 507–512.
Theiss-Seuberling, H.-D. (1984). *Plant Cell Physiol.* **25**, 601–609.
Usuda, H. and Edwards, G. E. (1981). *Plant Physiol.* **67**, 854–858.
Wara-Aswapati, O., Kemble, R. J. and Bradbeer, J. W. (1980). *Plant Physiol.* **66**, 34–39.
Wirtz, W., Stitt, M. and Heldt, H. W. (1982). *FEBS Lett.* **142**, 223–226.
Wolosiuk, R. A. and Buchanan, B. B. (1976). *Biochim. Biophys. Acta* **251**, 6456–6461.
Wolosiuk, R. A. and Buchanan, B. B. (1978a). *Plant Physiol.* **61**, 669–671.
Wolosiuk, R. A. and Buchanan, B. B. (1978b). *Arch. Biochem. Biophys.* **189**, 97–101.
Wolosiuk, R. A., Perelmuter, M. E. and Chehebar, C. (1980). *FEBS Lett.* **109**, 289–293.
Wolosiuk, R. A., Hertig, C. M. and Busconi, L. (1986). *Arch. Biochem. Biophys.* **246**, 1–8.
Wood, T. (1970). *Anal Biochem.* **33**, 297–306.
Woodrow, I. E. and Walker, D. A. (1980). *Biochem. J.* **191**, 845–949.
Woodrow, I. E. and Walker, D. A. (1982). *Arch. Biochem. Biophys.* **216**, 416–422.
Woodrow, I. E. and Walker, D. A. (1983). *Biochim. Biophys. Acta* **722**, 508–516.
Woodrow, I. E., Murphy, D. J. and Walker, D. A. (1983). *Eur. J. Biochem.* **132**, 121–123.
Woodrow, I. E., Murphy, D. J. and Latzko, E. (1984). *Biochim. Biophys. Acta* **259**, 3791–3795.
Zapponi, M. C., Berni, R., Iadorola, P. and Ferri, G. (1983). *Biochim. Biophys. Acta* **744**, 260–264.
Ziegler, H. and Ziegler, I. (1965). *Planta* **65**, 369–380.
Zimmermann, G., Kelly, G. J. and Latzko, E. (1976). *Eur. J. Biochem.* **70**, 361–367.
Zimmermann, G., Kelly, G. J. and Latzko, E. (1978). *Biochim. Biophys. Acta* **253**, 5952–5956.

3 Enzymes of C$_4$ Photosynthesis

ANTHONY R. ASHTON, JAMES N. BURNELL,
ROBERT T. FURBANK, COLIN L. D. JENKINS and
MARSHALL D. HATCH

Division of Plant Industry, CSIRO, GPO Box 1600, Canberra City ACT 2601, Australia

I.	Introductory comments		40
II.	Extraction of enzymes from leaves		43
III.	Enzymes of C$_4$ Photosynthesis		44
	A.	Phosphoenolpyruvate carboxylase	44
	B.	Pyruvate,Pi dikinase	46
	C.	NADP-malate dehydrogenase	49
	D.	NADP-malic enzyme	51
	E.	NAD-malic enzyme	53
	F.	Phosphoenolpyruvate carboxykinase	55
	G.	Aspartate aminotransferase	57
	H.	Alanine aminotransferase	59
	I.	Adenylate kinase	61
	J.	Pyrophosphatase	63
	K.	Pyruvate,Pi dikinase regulatory protein	64
IV.	Enzymes of the photosynthetic carbon reduction cycle in C$_4$ plants		66
	A.	Ribulose 1,5-bisphosphate carboxylase/oxygenase	66
	B.	3-Phosphoglycerate kinase and NADP-glyceraldehyde 3-phosphate dehydrogenase	67
	C.	Fructose 1,6-bisphosphate aldolase	67
	D.	Fructose 1,6-bisphosphatase and sedoheptulose 1,7-bisphosphatase	67
	E.	Ribulose 5-phosphate isomerase and ribulose 5-phosphate kinase	68
	References		68
	Addendum		71

I. INTRODUCTORY COMMENTS

The history of the resolution of C_4 photosynthesis follows a pattern of demonstrating the operation of unique photosynthetic biochemistry by various means and then identifying the enzymes necessary to support that biochemistry. Critical to the developing understanding of this process was the recognition of two types of photosynthetic cells in C_4 plants (mesophyll and bundle sheath) with quite different enzyme complements and distinct biochemical roles (see Fig. 3.1). As currently interpreted (see Edwards and Walker, 1983; Hatch, 1987) the reactions unique to the C_4 pathway serve, in association with some remarkable modifications of leaf anatomy and ultrastructure,

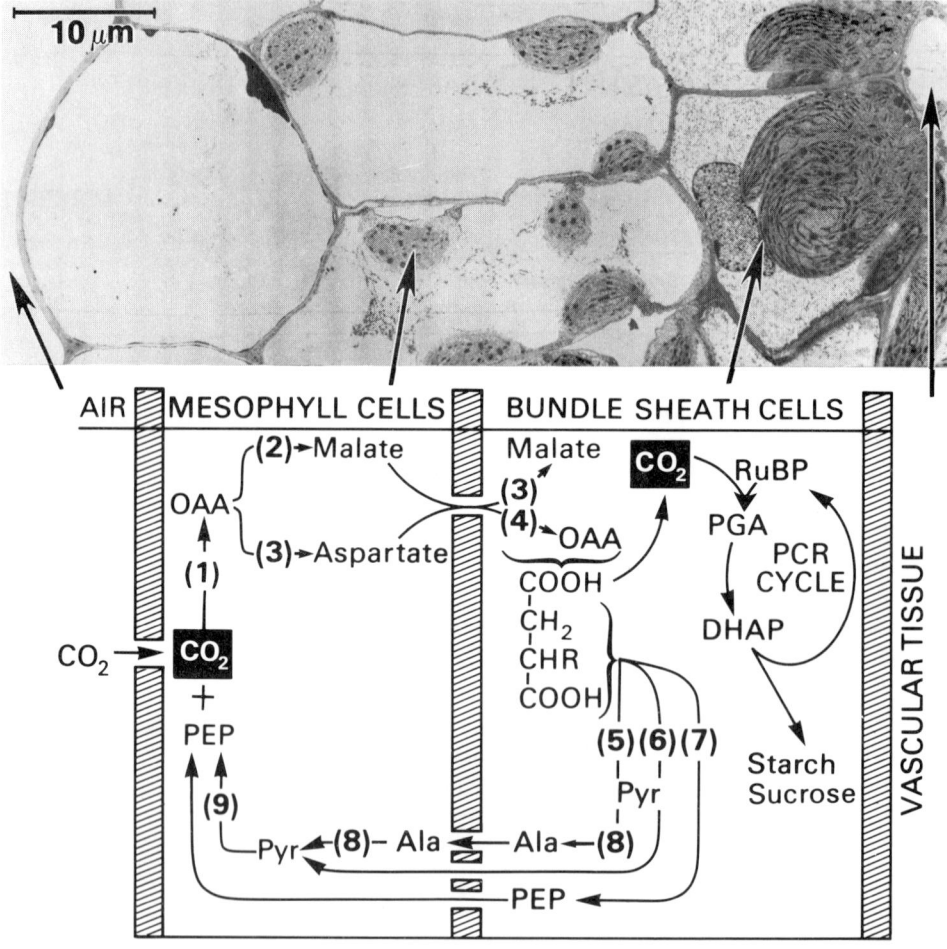

FIG. 3.1. The anatomy of a C_4 leaf and a simplified scheme outlining the options for the biochemical steps of C_4 photosynthesis, the enzymes involved, and their intercellular location. The enzymes indicated by the numbers in brackets are: 1, PEP carboxylase; 2, NADP-malate dehydrogenase; 3, aspartate aminotransferase; 4, NAD-malate dehydrogenase; 5, NADP-malic enzyme; 6, NAD-malic enzyme; 7, PEP carboxykinase; 8, alanine aminotransferase; 9, pyruvate, Pi dikinase. The subcellular location of enzymes is shown in Table 3.1. The electron micrograph of a *Panicum miliaceum* leaf was kindly provided by S. Craig. A more detailed account of the reactions involved in the different C_4 subgroups (NADP-ME-type, NAD-ME-type and PCK-type) has recently been provided elsewhere (Hatch, 1987).

to concentrate CO_2 in bundle sheath cells for utilisation by the photosynthetic carbon reduction cycle carboxylase, ribulose 1,5-bisphosphate carboxylase-oxygenase (Rubisco). The Rubisco-mediated oxygenase reaction and associated photorespiration are thereby eliminated.

The recognition in the mid-1960s of this modified form of photosynthesis and the role of C_4 acids as primary carboxylation products (Kortschak et al., 1965; Hatch and Slack, 1966; Hatch et al., 1967), led to the identification of phosphoenolpyruvate (PEP) carboxylase as the enzyme involved in the initial assimilation of CO_2 (Slack and Hatch, 1967). Following this, special roles were demonstrated for other enzymes involved in various aspects of C_4 acid metabolism including NADP malic enzyme, NADP malate dehydrogenase, and then pyruvate, Pi dikinase (PPDK) and associated high activities of adenylate kinase and pyrophosphatase (see Hatch and Slack, 1970).

By the early 1970s it became clear that not all C_4 plants decarboxylated C_4 acids via NADP malic enzyme; aspartate aminotransferase and alanine aminotransferase were implicated in those species lacking NADP malic enzyme although the exact role of these enzymes was unclear (Andrews et al., 1971). Later PEP carboxykinase (Edwards et al., 1971) and then a mitochondrial-located NAD malic enzyme (Hatch and Kagawa, 1974) were identified as alternative C_4 acid decarboxylating enzymes to NADP malic enzyme. This led to the classification of C_4 plants into three biochemically distinct subgroups, the NADP-malic enzyme (ME)-type, the NAD-ME-type, and the PEP carboxykinase (PCK)-type based on the operation of these decarboxylating enzymes (Gutierrez et al., 1974; Hatch and Kagawa, 1974; Hatch et al., 1975). There followed a period of consolidation which concentrated on the mechanism and function of the C_4 pathway, the properties and location of enzymes, and the mechanisms of regulation of the process (Hatch and Osmond, 1976; Edwards and Walker, 1983; Hatch, 1987).

A feature of all the enzymes implicated in the C_4 pathway is their high activity in leaves relative to the activity recorded for C_3 plants (15- to >100-fold; see Table 3.1). In addition, these activities are low or negligible in leaves developed on etiolated plants germinated in the dark. However, following illumination of these plants, normal activities rapidly develop together with other photosynthetic enzymes and in parallel with leaf greening and chloroplast biogenesis (see Hatch, 1978). Table 3.1 also provides information on the cellular and subcellular location of the photosynthetic enzymes specific to the C_4 pathway. Not shown in Table 3.1 is that C_4 leaves also contain high activities of adenylate kinase and pyrophosphatase (20–50 times C_3 activities: Hatch et al., 1969), largely located in mesophyll chloroplasts together with pyruvate, Pi dikinase (Slack et al., 1969).

Figure 3.1 provides a simplified scheme illustrating the elements of C_4 photosynthesis and the way individual reactions are partitioned between the mesophyll and bundle sheath cells in the C_4 leaf. The different options shown in the scheme operate in the different biochemical subtypes of C_4 plants mentioned above (NADP-ME-type, NAD-ME-type, PCK-type). The detailed pathways of metabolism operating in these three subgroups have recently been described elsewhere (Hatch, 1987). As Fig. 3.1 shows, in general terms C_4 photosynthesis involves the initial assimilation of CO_2 into C_4 acids via PEP carboxylase in mesophyll cells, the decarboxylation of these C_4 acids after transfer to bundle sheath cells, the assimilation of the released CO_2 via Rubisco and the PCR cycle, and the regeneration of the primary CO_2 acceptor PEP in the mesophyll cells from the C_3 product of C_4 acid decarboxylation. The subcellular location of the

TABLE 3.1. Activity and location of enzymes with a specific role in C_4 photosynthesis.

Photosynthetic group	PEP carboxylase	Pyr, Pi dikinase	NADP malic enzyme	NAD malic enzyme	PEP carboxykinase	NADP malate dehyd.	Asp, Ala aminotransferases
		Median value for range of activity[a] (μmol min^{-1} (mg Chl)$^{-1}$)					
C_3 Species	0.8	0.1	0.4	0.3	<0.15	0.8	2.1
		Range of activity relative to median value for C_3 spp. $\equiv 1$[a]					
C_4 Species							
NADP-ME-type	16–30	40–80	25–40	≈1	≈1	12–21	1–3
NAD-ME-type	15–31	40–90	≈1	18–60	≈1	1–2	12–27
PCK-type	21–34	15–40	≈1	3–10	40–110	2–6	15–35
Location (C_4)[b]	M, cyt	M, chl	BS, chl	BS, mit	BS, cyt	M, chl	Varies[c]

[a] For sources of data see Hatch (1987).
[b] Enzyme location. Abbreviations: M, mesophyll; BS, bundle sheath; cyt, cytoplasm; chl, chloroplast; mit., mitochondria.
[c] High aspartate and alanine aminotransferase activities occur in both M and BS cells. All are located in the cytosol except that BS aspartate aminotransferase in NAD-ME-type species is located in mitochondria.

enzymes listed in Fig. 3.1 is shown in Table 3.1. As inferred above, the reactions unique to C_4 photosynthesis combine to give a C_4–C_3 cycle which operates to concentrate CO_2 in bundle sheath cells where it is assimilated by the conventional photosynthetic carbon reduction (PCR) cycle.

II. EXTRACTION OF ENZYMES FROM LEAVES

It is commonly required to extract leaves to measure the total enzyme activity present in the tissue. For photosynthetic enzymes, activity is usually expressed per unit of extracted chlorophyll. If the enzyme and chlorophyll are equally extracted then it is not necessary to ensure complete extraction of the leaf sample. However, C_4 plants present a special problem in this respect because when leaves are macerated mesophyll cells are much more readily broken, and hence extracted, than bundle sheath cells. For instance blending leaves in a Waring blender, a Sorvall Omnimixer, or a Polytron homogeniser, will leave most of the bundle sheath cells unbroken (see Bjorkman and Gauhl, 1969; Andrews *et al.*, 1971). As a result, accurate determination of the activity of enzymes arising from bundle sheath cells in particular, but also from mesophyll cells, requires essentially complete extraction of the tissue. After addressing this problem over a period of many years, and considering various alternatives, our laboratory has been unable to develop any substitute for the procedure of mortar and pestle extraction of leaf tissue.

A commonly used procedure for extracting leaves is described below. Vigorously grind deribbed leaf tissue (preferably a small sample, 0.5–1.5 g) in a chilled mortar for at least 40 s with 1–2 vols (v/w) of extraction medium (usually containing 50 mM 4-(2-hydroxymethyl)-1-piperazine ethanesulphonic acid (Hepes)-KOH, N-tris-(hydroxymethyl) methylglycine (Tricine)-KOH or tris-(hydroxymethyl) aminomethane (Tris)-HCl buffer depending on the pH required, 10 mM dithiothreitol (DTT) and other special additions for individual enzymes indicated in the following sections). For effective extraction, a pasty consistency must be maintained during the grinding procedure. Add additional extraction buffer to the homogenate as required and then filter the mixture through Miracloth (Calbiochem) or an 80 μm nylon net, and remove a sample of the filtrate for the determination of chlorophyll content [measured in acetone (Arnon, 1949) or methanol (Wintermans and DeMots, 1965)]. Centrifuge the filtered extract (2 min 10 000 × g is usually sufficient) and either assay a sample of the supernatant directly or, if required, after processing on a small-volume column of Sephadex G-25 (Pharmacia) to remove small molecular weight compounds in the extract. Where this treatment is required columns of about 5 ml are prepared in small tapered tubes (for example Gilson Pipetman P5000 tips) plugged at the tip with glass wool. After equilibrating the column with two or three volumes of an appropriate buffered medium, a sample of about 1 ml of the leaf extract is added and the emerging protein band (indicated by the green colour of the chlorophyll–protein complexes) is collected. Generally the mid 0.5 ml of the emerging green band is collected and this is assumed to be undiluted.

A similar procedure can be used to extract enzymes from isolated strands of bundle sheath cells (see Hatch and Kagawa, 1974). A possible alternative for the extraction of leaves or isolated cell preparations would be to macerate frozen tissue in a mortar with liquid nitrogen and then extract the frozen powder with an appropriate buffered medium.

III. ENZYMES OF C_4 PHOTOSYNTHESIS

A. Phosphoenolpyruvate Carboxylase (EC 4.1.1.31)

$$PEP + HCO_3^- \longrightarrow OAA + Pi$$

PEP carboxylase catalyses the irreversible carboxylation of PEP to oxaloacetate (OAA) in the cytosol of mesophyll cells of C_4 plants. This is the primary carboxylation reaction of C_4 photosynthesis (Slack and Hatch, 1967; see Section I).

1. Extraction for assay

In desalted crude leaf extracts (see Section II) of many C_4 species the enzyme is stable at pH 7 and 25°C, and sometimes increases in activity; loss of activity can occur at a pH greater than 7.5, especially at 0°C (Hatch and Oliver, 1978). This loss can be alleviated by adding bovine serum albumin (BSA) (5 mg ml^{-1}) to extraction mixtures. Although under some conditions the enzyme is susceptible to inactivation by thiol binding reagents, the presence of thiol protecting agents is not essential for activity. Mg^{2+}, and DTT or 2-mercaptoethanol are usually included in extraction mixtures. The PEP carboxylase content of leaves is reduced several fold in plants grown in low light (Hatch et al., 1969).

2. Assay

PEP carboxylase is usually assayed spectrophotometrically at 340 nm by following the reduction of OAA by NADH in the presence of excess malate dehydrogenase. The reaction mixture (1 ml) contains enzyme, 25 mM Tris-HCl, pH 8, 5 mM $MgCl_2$, 2 mM DTT, 1 mM $KHCO_3$, 5 mM glucose 6-phosphate, 5 mM PEP, 0.2 mM NADH, and 2 units malate dehydrogenase. The reaction can be initiated with enzyme or PEP and remains linear for several minutes. For measuring maximum activities glucose 6-phosphate is usually included as an allosteric activator. This assay procedure has the potential to underestimate true activities due to non-enzymic OAA decarboxylation (Walker et al., 1986a) but lactate dehydrogenase can be included to catalyse the NADH-dependent reduction of any pyruvate formed (Meyer et al., 1988). With crude C_4 leaf extracts and excess malate dehydrogenase addition of lactate dehydrogenase has no effect on the rate. Typical activities in extracts of C_4 plants are in the range 10–25 µmol min^{-1} (mg chlorophyll)$^{-1}$ (Hatch and Oliver, 1978). Compared with plants grown largely in full sunlight, the PEP carboxylase content of leaves from plants growing at lower light intensities can be several fold lower (Hatch et al., 1969).

Alternative assay methods include: (1) determination of Pi production (Walker et al., 1986a); (2) measurement of incorporation of acid-stable radioactivity from $H^{14}CO_3^-$ after stopping reactions with acid (Slack and Hatch, 1967); or (3) following the increase in extinction at about 280 nm due to the formation of OAA. The latter method is certainly not suitable for assaying crude extracts (O'Leary, 1982) and is not recommended.

3. Purification

The following relatively simple procedure giving pure enzyme in good yields was developed in our laboratory. Early purification steps are carried out at 0–4°C; ion-exchange chromatography and subsequent steps may be carried out at room temperature. Homogenise chopped leaves from *Zea mays* in a Waring blender with about 3 vols of 50 mM Hepes-KOH, pH 7.1, 5 mM $MgCl_2$, 5 mM DTT, 1 mM ethylene diaminetetraacetic acid (EDTA), 0.5% (w/v) iso-ascorbate, 2% (w/v) Polyclar AT, and filter the homogenate through Miracloth. The filtrate is centrifuged and the supernatant fractionated by adding solid polyethylene glycol (PEG) 6000. The protein which precipitates between 8% and 15% (w/v) PEG is collected by centrifugation and resuspended in 25 mM Hepes-KOH, pH 7.1, containing 5mM $MgCl_2$, 5 mM DTT, 1 mM EDTA (termed column buffer below) and applied to a diethylamino ethyl cellulose (DEAE)-Sepharose CL-6B ion-exchange column (2.7 × 25 cm) equilibrated with the same buffer. After washing with about 40 ml of column buffer, PEP carboxylase is eluted from the column with a linear gradient of KCl (50–400 mM) in 400 ml column buffer. At this stage the enzyme has a specific activity up to about 20 µmol min^{-1} (mg protein)$^{-1}$, is free of pyruvate,Pi dikinase, and is suitable for use as a coupling enzyme for pyruvate,Pi dikinase assays (see Section II.B). Judicious selection of column fractions can give apparently homogeneous enzyme (Hatch and Heldt, 1985). The pooled active fractions from DEAE-Sepharose can be further purified by gel-filtration on a Sephacryl S-300 column equilibrated with the column buffer described above (yielding enzyme with specific activity up to about 30 units per mg protein) and then passing the active fractions through an Amicon Blue A column which removes remaining contaminants. This gives fractions with specific activities up to 60 µmol min^{-1} per mg protein which are homogeneous as judged by sodium dodecyl sulphate (SDS)-polyacrylamide gel electrophoresis (our unpublished observations). The procedure described gives a purification of PEP carboxylase of about 30-fold with good yield (about 50%). The enzyme can be stored without loss of activity as a precipitate in 65% saturated $(NH_4)_2SO_4$ solution at 4°C or as a solution containing 50% (v/v) glycerol at −20°C.

A variety of other methods have been used for purification of PEP carboxylase from a range of plant species, as discussed by O'Leary (1982). Size-exclusion HPLC methods have also been used as a final purification step for the *Zea mays* leaf enzyme (Walker *et al.*, 1986b). Recently, two regulatory forms of the enzyme, occurring in illuminated or darkened maize leaves, have been purified so that each retains its different enzymic properties (Huber *et al.*, 1986; Jiao and Chollet, 1988).

4. Properties

Maize leaf PEP carboxylase is a tetramer made up of four identical 100 kDa subunits and has a sedimentation coefficient ($s_{20,w}$) of 12.3 S (Uedan and Sugiyama, 1976). The primary structure from DNA sequence analysis (Izui *et al.*, 1986) and the amino acid composition (Stiborová and Leblová, 1986) have been determined, and some active site residues (histidine, cysteine, lysine, and arginine) have been identified (discussed by Andreo *et al.*, 1987). A tetrahedral arrangement of subunits has been suggested for the maize enzyme (Wagner *et al.*, 1988). Under various conditions the tetramer is dissociated into dimer and monomer forms (e.g. low temperature, pH 6 or 9, thiol reagents, histidine reagents, high salt: Walker *et al.*, 1986b; Stiborová and Leblová, 1986; Wagner *et al.*, 1987; Andreo *et al.*, 1987). The tetramer is the most active form.

The reaction catalysed by PEP carboxylase displays a sequential kinetic mechanism which occurs in two steps with transient formation of enzyme-bound carboxyphosphate and enolpyruvate; bicarbonate, not CO_2, is the substrate, and a divalent metal ion activator is essential for activity (see O'Leary, 1982). The K_m for Mg^{2+}, the assumed physiological cofactor, is about 1 mM at pH 8. At the pH optimum of 8.0–8.5 the reaction velocity responds to variation in substrate concentrations in a hyperbolic fashion with K_m values of about 0.8 mM for PEP and 30 μM for HCO_3^- in the presence of Mg^{2+} (Uedan and Sugiyama, 1976; O'Leary, 1982; Bauwe, 1986). At lower pH, the response to variation in PEP concentration is sigmoidal (K_m of 2.6 mM and Hill coefficient near 2 at pH 7: Uedan and Sugiyama, 1976).

5. Regulation

PEP carboxylase activity in C_4 leaves increases greatly on greening of etiolated tissue, due to new enzyme synthesis (Vidal and Gadal, 1983). The enzyme is activated by glucose 6-phosphate and glycine (see Hatch, 1978) and is inhibited by the product, oxaloacetate (Lowe and Slack, 1971; Coombs *et al.*, 1973a) and by aspartate and malate (Huber and Edwards, 1975).

The forms of the enzyme rapidly extracted from illuminated or darkened C_4 leaves display different activities when assayed under suboptimal conditions (pH 7–7.5, subsaturating PEP concentration). The dark form has lower activity, lower affinity for PEP and greater sensitivity to malate inhibition (Huber and Sugiyama, 1986; Doncaster and Leegood, 1987). Under these suboptimal conditions enzyme activity is increased by the metabolite effectors glucose 6-phosphate, fructose 6-phosphate, triose phosphate, and glycine, serine, and alanine, mainly due to increasing the affinity for PEP (Huber and Sugiyama, 1986; Doncaster and Leegood, 1987). Under conditions of pH and metabolite concentrations approximating the physiological situation, differences in activity between dark and light forms of the enzyme may be 10- to 20-fold (Huber and Sugiyama, 1986; Doncaster and Leegood, 1987). Conversion of the dark to light form is associated with increased phosphorylation on serine residues (Budde and Chollet, 1986; Nimmo *et al.*, 1987; Jiao and Chollet, 1988). The two forms of the enzyme have recently been purified (see above).

B. Pyruvate,Pi Dikinase (EC 2.7.9.1)

$$\text{Pyruvate} + \text{ATP} + \text{Pi} \rightleftharpoons \text{PEP} + \text{AMP} + \text{PPi} + \text{H}^+$$

Pyruvate,Pi dikinase (PPDK) catalyses the conversion of pyruvate to PEP in the chloroplasts of mesophyll cells of C_4 plants. This reaction regenerates the primary CO_2 acceptor PEP (see Section I).

1. Extraction for assay

Leaf tissue is extracted as described in Section II. To obtain maximum activity in leaf extracts the enzyme should be extracted from pre-illuminated leaves; the enzyme is inactivated in darkened leaves (Hatch and Slack, 1969). Dithiothreitol (5–10 mM) and $MgCl_2$ should be included in the extraction medium to prevent the irreversible inactivation of the extracted enzyme (Hatch and Slack, 1968). Further significant increases in PPDK activity in leaf extracts are often obtained by incubating with 2 mM Pi to activate inactive enzyme (see Section II.K). The enzyme is inactivated below 10°C (Hatch and Slack, 1968; Hatch, 1979; Sugiyama, 1973). The PPDK content of leaves is low in etiolated leaves and in leaves developed in low light intensities; the levels increase several-fold following the illumination of leaves at higher light for 2–3 days (Hatch *et al.*, 1969).

2. Assays

PPDK is usually assayed spectrophotometrically at 340 nm in the forward direction by coupling the production of PEP to NADH oxidation via PEP carboxylase and malate dehydrogenase. The following procedure follows that used by Jenkins and Hatch (1985). Reaction mixtures (1 ml, 25°C) contain 25 mM Hepes-KOH, pH 8.0, 8 mM $MgSO_4$, 10 mM DTT, 10 mM $NaHCO_3$, 2 mM pyruvate, 5 mM $(NH_4)_2SO_4$, 1 mM glucose 6-P, 1 mM ATP, 2.5 mM Pi, 0.2 mM NADH and 0.5 units of PEP carboxylase plus 2 units of malate dehydrogenase. The reaction is usually started by adding PPDK. The assay is linear using crude leaf extracts as a source of enzyme. It is non-linear with purified enzyme assayed in the absence of adenylate kinase and pyrophosphatase because of inhibition due to the build-up of the products AMP and PPi (see Sections II.I and II.J). Where the activity of only partially-activated PPDK is being examined in leaf extracts it may be necessary to include 10 μM Cibacron Blue F3GA to prevent activation of the enzyme during the course of the assay (Burnell and Hatch, 1986).

In the reverse direction the enzyme is assayed spectrophotometrically by measuring pyruvate production coupled to lactate dehydrogenase (Andrews and Hatch, 1969; Jenkins and Hatch, 1985). Using the forward direction assay, activities in leaf extracts fall within the range of 2 to 10 $\mu mol\ min^{-1}\ (mg\ Chl)^{-1}$ with the lower activities being recorded for PCK-type C_4 plants (Hatch *et al.*, 1975; Edwards *et al.*, 1982; Ku and Edwards, 1975).

3. Purification

The following procedure for the purification of PPDK from maize leaves is based on that described previously (Burnell and Hatch, 1985a). Deribbed and sliced maize leaf tissue is homogenised at 0°C in 4 vols (w/v) of 50 mM Hepes-KOH, pH 7.4, containing 10 mM $MgSO_4$, 5 mM DTT, 1 mM EDTA, 2 mM Pi, 2% (w/v) Polyclar AT, 0.5% (w/v) sodium ascorbate and 1 mM pyruvate. Following filtration through Miracloth and centrifugation the supernatant is incubated at 25°C for 30 min to activate the enzyme (see Section II.K) and then cooled to 0°C. Solid PEG-6000 is added to the supernatant at a rate of 15 g per 100 ml and the precipitated protein is removed by centrifugation. The pH of the supernatant is then slowly lowered to 5.0 by adding 1 M acetic acid. The protein precipitate, collected by centrifugation, is dissolved in medium containing 25 mM Hepes-KOH, pH 7.4, 5 mM $MgSO_4$, and 5 mM DTT (referred to below as the buffer medium) and brought to pH 7.4 by the addition of 1 M KOH. All subsequent steps are run at room temperature. The protein solution is applied to a column of DEAE-Sepharose CL-6B equilibrated with the buffer medium and PPDK is eluted with a linear KCl gradient (0.05 to 0.4 M in the buffer medium). The active fractions are pooled and concentrated by $(NH_4)_2SO_4$ precipitation (to 65% saturation) and applied to a column of Sephacryl S-300 equilibrated with the buffer medium. Peak fractions are at least 70% pure as judged by sodium dodecyl sulphate (SDS)-polyacrylamide gel electrophoresis (Burnell and Hatch, 1985a). This procedure purifies PPDK about 15-fold with a 20% yield and with an average specific activity between 5 and 9 μmol min^{-1} (mg protein)$^{-1}$. PPDK has been purified to homogeneity from chloroplast extracts by $(NH_4)_2SO_4$ fractionation, Sephacryl S-300 column chromatography and affinity chromatography on Amicon Blue A (Ashton et al., 1984), however the yields are low. The specific activity of this preparation of PPDK ranged between 10 and 18 μmol min^{-1} (mg protein)$^{-1}$. Active fractions may be stored at 2°C as a precipitate in a 65% saturated solution of $(NH_4)_2SO_4$ without significant loss of activity. Activity is also effectively retained by storing samples in 50% (v/v) glycerol at −20°C.

Sugiyama (1973) has also described a procedure for purifying PPDK from Zea mays. Dark-inactivated PPDK for use in activation studies (see Section II.K) can be purified by a procedure based upon the protocol described above (Burnell and Hatch, 1985a).

4. Properties

Purified PPDK from Z. mays leaves has a sedimentation coefficient of 8.86 S and a molecular weight of about 387 kDa with subunits of about 94 kDa (Sugiyama, 1973). Inactivation in the absence of Mg^{2+} is accompanied by dissociation of the tetramer to a dimer (Sugiyama, 1973) while cold inactivation leads to dimer (Shirahashi et al., 1978) or monomer formation (Hatch, 1979). The reaction catalysed by maize and sugar cane PPDK proceeds via a two-step bi bi uni uni mechanism in which the β-phosphate of ATP phosphorylates a catalytic site histidine and is then transferred to pyruvate to form PEP; the γ-phosphate of ATP appears in PPi (Andrews and Hatch, 1969; Jenkins and Hatch, 1985).

At pH 8.1 K_m values for substrates of the forward reaction of Z. mays PPDK are; pyruvate 82 μM, ATP 32 μM, Pi 380 μM, and for the reverse reaction at pH 7.4, PEP 160 μM, AMP 0.85 μM and PPi 46 μM (Jenkins and Hatch, 1985). Similar values were obtained with the sugar cane leaf enzyme (Andrews and Hatch, 1969). The enzyme also requires free Mg^{2+} in addition to Mg^{2+} to bind with ATP (Hatch and Slack, 1968) and is stimulated several-fold by NH_4^+ or K^+ in both directions (Jenkins and Hatch, 1985). The pH optimum for PEP formation is about 8.3 and in the reverse direction is about 6.8. At the respective pH optima the reverse reaction rate is about twice that of the forward reaction (Jenkins and Hatch, 1985).

5. Regulation

In the forward direction the products PEP, AMP and PPi are inhibitors with K_i values of 150, 130 and 320 μM, respectively (Jenkins and Hatch, 1985). For the reverse reaction pyruvate and Pi are inhibitors and this inhibition is apparently competitive with PEP and PPi, respectively (Andrews and Hatch, 1969). PPDK activity in leaves is regulated by the light and can vary as much as 20-fold within a few minutes when plants are transferred from the dark to the light (Hatch and Slack, 1968; Hatch, 1981). The detailed mechanism of the light/dark-mediated regulation of PPDK has been described previously (Burnell and Hatch, 1985a, 1985b) and is discussed in detail in Section II.K.

C. NADP-Malate Dehydrogenase (EC 1.1.1.82)

$$OAA + NADPH + H^+ \rightleftharpoons Malate + NADP^+$$

NADP-dependent malate dehydrogenase (NADP-MDH) reduces OAA, the primary product of CO_2 assimilation in C_4 photosynthesis, to malate within the mesophyll cell chloroplast. This step is on the major path of carbon flux in NADP-ME-type C_4 photosynthesis but is at least quantitatively less important in NAD-ME-type and PCK-type C_4 plants (see Fig. 3.1 and Table 3.1).

1. Extraction for assay

To obtain the maximum potential activity in a leaf extract (see Section II) the enzyme must be fully reduced by incubating the extract under N_2 with 10–100 mM DTT at 25°C for an hour. In crude extracts endogenous thioredoxin will catalyse this reduction and 10 mM DTT is sufficient, but with the partially purified (and thioredoxin-depleted) enzyme the higher DTT concentrations may be necessary. Because of the low E_0' of the enzyme regulatory thiols (−0.33 V: Rebeille and Hatch, 1986a) 2-mercaptoethanol is not an effective activator of the enzyme. To measure the actual prevailing activity (activation status) of the enzyme in leaves the enzyme should be extracted in the presence of 5 mM DTT (to prevent inactivation of the enzyme by oxidised thioredoxin) and at 0°C to slow the DTT- and reduced thioredoxin-dependent reductive activation (Kagawa and Hatch, 1977) and then assayed immediately.

2. Assay

The assays described are modified from those originally developed by Johnson and Hatch (1970). NADP-MDH activity is measured spectrophotometrically in the direction of OAA reduction by following the oxidation of NADPH at 340 nm in 1 ml reaction mixtures containing 25 mM Tricine-KOH, 70 mM KCl, 1 mM EDTA, 1 mM DTT, 1 mM OAA and 0.2 mM NADPH at pH 8.3. For the reverse reaction NADP reduction is followed spectrophotometrically in reactions containing 25 mM Tricine-KOH, 1 mM EDTA, 1 mM DTT, 50 mM malate, 1 mM NADP at pH 8.5. The initial activity in this direction may be maintained constant by including 20 mM glutamate and 1 unit glutamate-oxaloacetate aminotransferase to continuously remove OAA by conversion to aspartate.

Maximum NADP-MDH activity in leaf extracts ranges between 10 and 18 µmol min^{-1} (mg Chl)$^{-1}$ for NADP-ME-type C_4 species and 0.8 to 4 µmol min^{-1} (mg Chl)$^{-1}$ for NAD-ME and PCK-type species (see Table 3.1).

3. Purification

This method is based on the procedures described by Ashton and Hatch (1983a) and Jenkins et al. (1986). All steps in the purification are conducted at 4°C. Maize leaf tissue is homogenised in 3 vols (w/v) of 25 mM Hepes-KOH, pH 7.5, 10 mM $MgSO_4$, 1 mM EDTA, 50 mM 2-mercaptoethanol and 20 g Polyclar-AT per 100 g of leaves. No thiol reducing agents are included in subsequent buffers so the enzyme reverts to the more stable oxidised form and portions of fractions must be reduced with DTT (see above) to detect enzyme activity. The homogenate is clarified by filtration through cheesecloth and Miracloth and then centrifugation at $10\,000 \times g$ for 15 min. Polyethylene glycol (M_r 6000–8000) is added to the supernatant to give a 25% (weight for original volume) solution and the precipitated protein removed by centrifugation. The resulting supernatant is adjusted to pH 5.0 with acetic acid (previously partially neutralised to pH 4). The precipitate containing NADP-MDH is collected and dissolved in 50 mM Tris-HCl, 10 mM KCl, 1 mM EDTA and 2 mM $MgCl_2$, pH 7.5 (referred to below as Tris-KCl buffer) and applied to a column of Reactive Red 4-Sephacryl S-300 (generated as described by Ashton and Polya, 1978). Reactive Red 4 (available from Sigma Chemical Co.) was chosen as the ligand because it was the best inhibitor among the dye-dextrans listed by Ashton and Polya (1978). The enzyme can be eluted with either an ionic strength gradient (0–0.4 M KCl) or an NADP gradient (0–200 µM) in the Tris-KCl buffer. The protein in the active fractions was concentrated by adding $(NH_4)_2SO_4$ to give a 75% saturated solution and applied to a Sephacryl S-300 gel filtration column equilibrated with the Tris-KCl buffer. Usually a single protein peak, coincident with enzyme activity, emerges from the column. Specific activities of up to 500 µmol min^{-1} (mg protein)$^{-1}$ have been obtained. The enzyme can be stored frozen or in 50% (w/v) glycerol at $-20°C$. Other procedures for the purification of maize NADP-MDH have been reported (Jacquot et al. 1981; Kagawa and Bruno, 1988). The procedure of Kagawa and Bruno (1988) includes protease inhibitors to minimise proteolytic damage to the enzyme and yields enzyme with a specific activity between 600 and 1000 units per mg protein. In our procedure the earlier addition of polyethylene glycol and the speed with which the whole procedure can be completed may minimise proteolysis.

4. Properties

Purified *Zea mays* NADP-MDH has subunits of about 42 kDa (Jenkins et al., 1986; Kagawa and Bruno, 1988) and the native enzyme apparently occurs as either a tetramer (Ashton and Hatch, 1983a) or a dimer (Kagawa and Bruno, 1988; Ferte et al., 1986). The enzyme prepared by the procedure described above contains six cysteine residues per subunit, four of which form two disulphide bonds during oxidative inactivation of the enzyme (Jenkins et al., 1986). Kagawa and Bruno (1988) reported nine cysteine residues per subunit and concluded that six of these can be oxidised to form three disulphide bonds but that reduction of only one of these is critical for activation. Decottignies et al. (1988) have now shown that Cys-10 and Cys-15 are reduced during the light activation of NADP-MDH. The oxidised enzyme from maize is essentially inactive under all conditions tested.

Zea mays NADP-MDH has an alkaline pH optimum. When assayed at pH 8.5 estimates of K_m values fall in the range 18–56 µM for OAA, 24–32 mM for malate, 24–45 µM for NADPH and 45–70 µM for NADP (Ashton and Hatch 1983a; Kagawa and Bruno, 1988). The reduction of OAA is strongly inhibited by NADP (K_i of 50 µM). The reduced, active enzyme is strongly inhibited by heavy metals and other reagents reacting with thiols such as mercurials and alkylating agents (Ashton and Hatch, 1983a).

5. Regulation

Zea mays NADP-MDH is reversibly inactivated and reactivated by oxidation and reduction of enzyme cysteine residues (also see Sections III.C.1 and III.C.4 above) and this thioredoxin-mediated process is believed to account for the rapid dark/light-mediated changes in enzyme activity seen *in vivo* (see Edwards *et al.*, 1985). The interconversion of reduced and oxidised enzymes is strongly influenced by the NADPH : NADP ratio, a high NADPH : NADP ratio favouring a more active enzyme. Because high NADPH : NADP ratios enable more thioredoxin to be reduced and minimise the product inhibition of the enzyme by NADP, high rates of OAA reduction only occur at high NADPH : NADP ratios (Ashton and Hatch, 1983a, 1983b; Rebeille and Hatch 1986a, 1986b).

D. NADP-Malic Enzyme (EC 1.1.1.40)

$$\text{Malate} + \text{NADP}^+ \rightleftharpoons \text{CO}_2 + \text{Pyruvate} + \text{NADPH} + \text{H}^+$$

NADP-malic enzyme catalyses the decarboxylation of malate in NADP-ME-type C_4 species (see Section I). The enzyme is located in the chloroplasts of bundle sheath cells (Slack *et al.*, 1969).

1. Extraction for assay

General procedures for extractions of leaves are discussed in Section II. NADP-malic enzyme is usually extracted from C_4 leaves in buffer containing low concentrations of EDTA and 5 mM DTT though thiol compounds do not greatly enhance activity (Johnson and Hatch, 1970; Hatch and Mau, 1977a). The enzyme is stable in desalted leaf extracts for several hours and activity often increases during incubation of extracts at 25°C (Hatch and Mau, 1977a).

2. Assay

NADP-malic enzyme may be assayed in the forward (decarboxylating) direction by following NADP$^+$ reduction spectrophotometrically at 340 nm. The assay mixture (1 ml) contains enzyme, 25 mM Tricine-KOH, pH 8.3, 5 mM malate, 0.5 mM NADP$^+$, 0.1 mM EDTA, 2 mM MgCl$_2$ (Hatch and Mau, 1977a). Reactions can be initiated with enzyme, malate, or Mg^{2+}. In crude extracts, interfering low background rates due to NADP-malate dehydrogenase, which may sometimes be encountered, can be resolved by starting the reaction with Mg^{2+}. The enzyme can be assayed in the reverse direction by measuring NADPH oxidation at 340 nm (Jenkins *et al.*, 1987). Typical decarboxylation activities in extracts from NADP-ME-type C_4 leaves range between 10 and 16 µmol min^{-1} (mg Chl)$^{-1}$.

3. Purification

Similar procedures for purifying NADP-malic enzyme from maize leaves to a high specific activity have been reported recently by Häusler *et al.* (1987) and Thorniley and Dalziel (1988); the former method is described here. Maize leaves are frozen in liquid nitrogen and ground to powder, then homogenised at 4°C in 100 mM Hepes-NaOH, pH 7, containing 2.5 mM MgCl$_2$, 5 mM 2-mercaptoethanol, 10 mM diethyldithiocarbamate and 1% (w/v) Polyclar AT. Subsequent procedures are carried out at 4°C. After filtration and centrifugation the supernatant is fractionated to obtain the protein which precipitates between 45 and 65% saturation of $(NH_4)_2SO_4$. The protein is collected by centrifugation then resuspended in a small volume of medium containing 20 mM Hepes-NaOH, pH 7.5, 5 mM $MgCl_2$, 2 mM DTT (referred to below as buffer medium) and applied to a Sephacryl S-300 column (2.6 × 68 cm, elution rate 80 ml h^{-1}) previously washed with the same buffer. Active fractions are pooled and NADP-malic enzyme concentrated by precipitation with $(NH_4)_2SO_4$ (65% saturation). The protein, recovered by centrifugation, is dissolved in the buffer medium and applied to a 2′,5′-ADP-Sepharose 4B affinity column. After washing, the enzyme is eluted with 6 mM NADP$^+$ in the buffer medium, concentrated by precipitation with $(NH_4)_2SO_4$, and stored in solution containing 50% glycerol at $-20°C$. Under these conditions the enzyme remains stable for at least several months. This procedure resulted in a 37-fold purification and 13% recovery of enzyme with a specific activity of about 60 units per mg protein (Häusler et al., 1987).

An alternative purification procedure involving double ion-exchange chromatography followed by hydroxy-apatite chromatography and affinity chromatography on a Blue Sepharose CL-6B column gave a preparation (specific activity of about 70 units per mg protein) which was homogeneous as assessed by SDS-polyacrylamide gel electrophoresis and analytical ultracentrifugation (Asami et al., 1979a). Purification procedures used in earlier studies (Johnson and Hatch, 1970; Coombs et al., 1973b, Persanov et al., 1976; Pupillo and Bossi, 1979), which were similar but did not include an affinity chromatography step, generally resulted in enzyme with a lower specific activity.

4. Properties

Native NADP-malic enzyme from maize leaf has a sedimentation coefficient ($s_{20,w}$) of 9.7 S, a molecular mass of 230–280 kDa, and consists of four identical 63 kDa subunits (Asami et al., 1979a; Häusler et al., 1987; Pupillo and Bossi, 1979; Thorniley and Dalziel, 1988).

The kinetic mechanism of the enzyme from maize (Pupillo and Bossi, 1979) and *Pennisetum purpureum* (Coombs et al., 1973b) is sequential but other mechanistic details are lacking for the enzyme from C_4 leaves. Divalent metal ions, Mg^{2+} or Mn^{2+}, are essential for activity with K_m values of 55 µM and 4 µM, respectively, for the maize enzyme at pH 8.5 (Johnson and Hatch, 1970). Carbon dioxide, not HCO_3^-, is the reaction product (Häusler et al., 1987; Jenkins et al., 1987). Like NADP-malic enzyme from other sources, the C_4 leaf enzyme also catalyses decarboxylation of OAA (Johnson and Hatch, 1970).

Maize leaf NADP-malic enzyme shows a pH optimum at pH 8.5 with saturating malate, but this is lowered to about 7.5 at subsaturating malate concentrations (Johnson and Hatch, 1970). The enzyme also shows low rates of NAD$^+$-dependent activity which, at pH 7 is about 50% of the maximum NADP$^+$-dependent rate, but declines to less than 1% at pH 8.3 (Hatch and Mau, 1977a). This NAD$^+$-dependent activity is strongly inhibited by very low concentrations (about 1 µM) of NADP$^+$ or NADPH (Hatch and Mau, 1977a).

For the forward direction, in the presence of Mg^{2+} at pH 8, the K_m values are 150 µM for malate and 25 µM for NADP$^+$ for the maize enzyme. The K_m for malate increases at higher pH (400 µM at 8.5). At a pH of 8 and below, substrate inhibition is observed with malate above about 1 mM (Johnson and Hatch, 1970; Asami et al., 1979b). At pH 8, with Mg^{2+}, the forward direction rate is four-fold greater than the rate in the reverse direction (Johnson and Hatch, 1970). K_m values of 340 μM for malate and 4 μM for $NADP^+$ were reported for the enzyme from *Pennisetum purpureum*, at pH 8 in the presence of Mn^{2+} (Coombs et al., 1973b).

5. Regulation

The complex responses of the enzyme to varying pH and malate concentration, as outlined above, may have a significant regulatory role in bundle sheath chloroplasts, but this is yet to be defined. Changes in the $NADPH:NADP^+$ ratio and the divalent metal ion concentration have also been implicated in regulation (Asami et al., 1979b). The reaction product CO_2 inhibits the enzyme at concentrations likely to occur in bundle sheath cells (around 0.5 mM) and this may provide negative feedback of physiological relevance (Jenkins et al., 1987). However, the most important factor integrating decarboxylation via NADP malic enzyme with CO_2 assimilation via the PCR cycle is probably the tight coupling of this reaction with 3-PGA reduction in chloroplasts (see Section I and Fig. 3.1).

E. NAD-Malic Enzyme (EC 1.1.1.39)

$$\text{Malate} + NAD^+ \rightleftharpoons \text{Pyruvate} + CO_2 + NADH + H^+$$

NAD-malic enzyme catalyses the decarboxylation of malate in the mitochondria of bundle sheath cells of NAD-ME-type and also PCK-type C_4 species (see Fig. 3.1. and Table 3.1.).

1. Extraction for assay

The general procedure for extraction of enzymes from leaves is described in Section II. Extracted NAD-malic enzyme activity is rapidly lost unless the extraction and storage medium includes Mn^{2+} (2 mM) and DTT (5 mM) (Hatch et al., 1974). It is recommended that PVP-40 (0.25% w/v) is also included (Hatch and Kagawa, 1974). Triton X-100 or digitonin (0.5% w/v) have been added before centrifugation to ensure complete release of the enzyme from mitochondria (Hatch et al., 1982). In Sephadex G-25-treated leaf extracts (see Section II) the enzyme is more stable stored at room temperature and in some cases significant increases in activity are observed under these conditions (Hatch and Kagawa, 1974). Activity is lost if solutions of enzyme are frozen (Hatch et al., 1974; Burnell, 1987).

2. Assay

NAD-malic enzyme is most conveniently assayed in the direction of malate decarboxylation by measuring NADH formation as a change in absorbance at 340 nm (Hatch et al., 1974, 1982). Reactions of 1 or 3 ml contain the following components at the final concentrations indicated: 25 mM Hepes-KOH, pH 7.2, 5 mM malate, 2 mM NAD, 4 mM $MnCl_2$, 0.2 mM EDTA, 5 mM DTT, 0.1 mM CoA (allosteric activator) and enzyme. With leaf extracts and other enzyme samples containing significant NAD malate dehydrogenase activity the inclusion in the reaction of 25 μM NADH and 1 unit of malate dehydrogenase eliminates transient interference as the reaction catalysed by malate dehydrogenase equilibrates (Hatch et al., 1982). The most common procedure is to add all the reaction components to a cuvette except enzyme, $MnCl_2$ and CoA or other activators of the reaction. After checking for any absorbance change at 340 nm with the addition of enzyme, the reaction is commenced by adding $MnCl_2$. Additional activity due to the addition of the activator CoA (or other activators, see Section II.E.4 below) is then determined. Although thermodynamically feasible, it is difficult to detect activity in the carboxylating direction with NAD-malic enzymes from both C_4 leaves and other sources (Jenkins et al., 1987). Activities in leaf extracts for the decarboxylating direction range between 6 and 20 μmol min^{-1} (mg Chl)$^{-1}$ for NAD-ME-type C_4 species and 1 and 4 μmol min^{-1} (mg Chl)$^{-1}$ for PCK-type species (see Table 3.1. and Hatch et al., 1982, 1988).

3. Purification

Difficulty was encountered in purifying NAD-malic enzyme from the NAD-ME-type species *Amaranthus edulis* and *Panicum miliaceum* because of enzyme instability (Hatch et al., 1974). The enzyme from the NAD-ME-type species *Atriplex spongiosa* has been partially purified, and interfering enzymes removed, by a procedure involving extraction from isolated bundle sheath cells (see below) followed by fractionation by $(NH_4)_2SO_4$ precipitation and then chromatography on columns of Sephadex G-200 and DEAE-cellulose (Hatch et al., 1974). This partially purified enzyme is stable for several months when stored at 0°C as a suspension in 70% saturated $(NH_4)_2SO_4$ solution.

NAD-malic enzyme from the PCK-type C_4 species *Urochloa panicoides* has been purified to near homogeneity by the following procedure (Burnell, 1987). The enzyme is extracted from bundle sheath cell strands prepared from deribbed leaves (Burnell and Hatch, 1988). Wash the combined bundle sheath cell preparations (from 5 × 15 g lots of leaves) with cold 25 mM Hepes-KOH, pH 7.5, containing 1 mM $MnCl_2$ and 50 mM 2-mercaptoethanol (medium used throughout and kept under N_2 where possible). Extract the bundle sheath cell strands by vigorously grinding in a large cold mortar with a limited amount of the above medium so that a thick viscous slurry is maintained. Add additional buffer medium to give about 150 ml and then filter on Miracloth. Centrifuge the filtrate at 30 000 × g for 10 min, 0°C, and then fractionate the supernatant by adding solid $(NH_4)_2SO_4$ to precipitate the protein that sediments between 40% and 60% saturation of $(NH_4)_2SO_4$ (at 0°C). Collect the protein by centrifugation, dissolve in a minimum volume of buffer medium, and then apply to a Sephadex G-200 column (40 × 3 cm) previously equilibrated with the buffer medium at room temperature. Elute the column at a flow rate of 0.5 ml min^{-1} and pool the fractions with highest activity (activity emerges just after the excluded protein). Load the pooled fractions on to a 10 ml column of Matrex Gel Orange (Amicon) and then wash the column with 50 ml of the buffer medium. Elute the NAD malic enzyme activity by applying a solution of the buffer medium containing 0.1 M KCl. The specific activity of this preparation was about 10 μmol min^{-1} (mg protein)$^{-1}$. For long-term storage the enzyme can be kept at 0°C as a suspension in 65% saturated $(NH_4)_2SO_4$ solution.

4. Properties

Plant NAD-malic enzyme is a polymeric enzyme apparently existing in various aggregation states ranging between 115 and 490 kDa (Artus and Edwards, 1985). The subunit molecular mass is about 60 kDa and the purified enzyme from some sources apparently comprises two subunits of molecular mass about 58 and 62 kDa (Artus and Edwards, 1985; Burnell, 1987).

Complex and varied responses are seen with NAD-malic enzymes from different sources as substrate or activator levels, pH and divalent metal ion concentrations are varied (Artus and Edwards, 1985). C_4 enzymes give either highly sigmoidal responses with increasing malate or activator concentrations (enzymes from NAD-ME-type dicotyledonous plants: see Hatch *et al.*, 1974; Chapman and Hatch, 1977) or simple hyperbolic responses (enzymes from NAD-ME-type and PCK-type monocotyledonous plants: see Hatch *et al.*, 1974; Burnell, 1987). Correlating with these differences is a K-type allosteric activation by CoA and other activators with the former type of enzymes (activation by decreasing the K_m for malate), or V-type allosteric activation with the latter type (activation by increasing V_{max} rather than affinity for malate). The theory behind these differing allosteric responses is explained by Monod *et al.* (1965).

Information on substrate and cofactor specificity, responses to various activators, and K_m or $K_{0.5}$ values for NAD-malic enzyme from various C_4 plants is provided in three papers (Hatch *et al.*, 1974; Chapman and Hatch, 1977; Burnell, 1987). Some of the main features of this data are briefly summarised. Generally, enzymes show an absolute and specific requirement for Mn^{2+}. Some enzymes have significant activity with NADP. Increases in activity due to addition of activators range between 4- and 20-fold. Largest activation responses are seen with CoA, acetyl-CoA and fructose 1,6-bisphosphate, with optimal effects for each seen in the range from 40–100 μM. Various other activators are less effective. With different enzymes, K_m or $K_{0.5}$ values are in the range of 0.8–1.7 mM for malate, 0.25–0.55 mM for NAD, 0.2–0.8 mM for Mn^{2+} and 5–20 μM for the activator CoA.

5. Regulation

Those enzymes showing sigmoidal responses to malate (co-operative binding) and K-type responses to activators are strongly inhibited by the product CO_2 (or HCO_3^-) competitively with respect to malate (Hatch *et al.*, 1974; Chapman and Hatch, 1977). The other enzymes showing hyperbolic responses to malate were little affected by CO_2 (Chapman and Hatch, 1977; Burnell, 1987). K- or V-type allosteric activation of the enzyme, particularly by the PCR-cycle intermediate fructose 1,6-bisphosphate (see above), could be critical in regulation.

F. Phosphoenolpyruvate Carboxykinase (EC 4.1.1.49)

$$OAA + ATP \rightleftharpoons PEP + ADP + CO_2$$

PEP carboxykinase is the major decarboxylating enzyme in the PCK-type subgroup of C_4 plants and catalyses the ATP-dependent decarboxylation of OAA in the cytosol of the bundle sheath cells (see Section I).

1. Extraction for assay

Since this enzyme is located in the bundle sheath cells special care must be taken to thoroughly extract the leaf material to ensure complete recovery of activity (see Section II). $MnCl_2$ (2 mM) and DTT (5 mM) are usually included in the extraction buffer medium.

2. Assay

The PEP carboxykinase activity in Sephadex G-25-treated crude leaf extracts (see Section II) is measured spectrophotometrically by following the ATP-dependent decrease in OAA concentration at 280 nm (Hatch, 1973a; Hatch and Mau, 1977b). Quantitation of this assay is based on determining the molar extinction coefficient of OAA under the exact conditions of the assay. This is best done prior to each set of assays by measuring the absorbance at 280 nm of a known amount of freshly prepared OAA solution added to a reaction mixture lacking enzyme. Reaction mixtures (1 ml) contain 50 mM Hepes-KOH, pH 7.6, 0.1 M KCl, 0.5 mM OAA, 1 mM $MnCl_2$, 1 unit of pyruvate kinase and crude extract. After recording the blank rate due to the spontaneous decarboxylation of OAA, reactions are started by adding ATP (0.2 mM final concentration) and measuring the change in absorbance at 280 nm. The crude enzyme extract sample added to the reaction mixture must be limited so that the total absorbance at 280 nm remains below about 1.8 absorbance units. Using this assay, PEP carboxykinase activities of between 5 and 15 μmol min^{-1} $(mg\,Chl)^{-1}$ have been reported for whole leaf extracts of PCK-type plants (Hatch et al., 1975; Chapman and Hatch, 1983).

With partially purified enzyme essentially free of malate dehydrogenase and PEP carboxylase, OAA decarboxylation is measured by coupling PEP production to NADH oxidation via pyruvate kinase and lactate dehydrogenase. Reaction mixtures (1 ml) contain 50 mM Hepes-KOH, pH 7.6, 0.1 M KCl, 0.5 mM OAA, 0.5 mM $MnCl_2$, 0.25 mM ATP, 0.2 mM NADH, 2 units of pyruvate kinase and 2 units of lactate dehydrogenase. PEP carboxykinase activity is measured as the ATP-dependent decrease in absorbance at 340 nm. With this procedure malate dehydrogenase and PEP carboxylase in crude leaf extracts interfere with the assay of PEP carboxykinase. As with the above assay, activity must be corrected for the spontaneous decarboxylation of OAA in the presence on Mn^{2+}.

The enzyme can also be assayed by a radiotracer exchange reaction (Dittrich et al., 1973) and in the carboxylating direction (Jenkins et al., 1987). At pH 7.6 the forward reaction is about 2.6 times faster than the reverse reaction (Jenkins et al., 1987).

3. Purification

The following procedure has been used to purify PEP carboxykinase from *Panicum maximum*, *Chloris gayana* and *Urochloa panicoides* (Burnell, 1986). All operations are conducted at 0°C. About 50 g of deribbed finely sliced leaf material is homogenised in a Waring blender with 200 ml of buffer containing 50 mM Hepes-KOH, pH 7.5, 2 mM $MgCl_2$, 2 mM $MnCl_2$, 5 mM DTT and 1% (w/v) PVP-40. The residual leaf material obtained by filtering on Miracloth, is re-extracted in 30 ml of buffer by vigorously grinding in a mortar. The filtered extract is pooled with the Waring blender extract.

Following centrifugation, the supernatant is fractionated with $(NH_4)_2SO_4$ and the protein precipitating between 45 and 55% saturation is dissolved in a minimal volume of 25 mM Hepes-KOH, pH 7.5, containing 2.5 mM $MnCl_2$, 2 mM $MgCl_2$, and 5 mM DTT (referred to below as buffer medium) and desalted by Sephadex G-25 gel filtration. The desalted protein is applied to a column of DEAE-Sepharose CL-6B and, after washing with 100 ml of the buffer medium, the enzyme is eluted with a KCl gradient (50–400 mM in buffer medium). The peak fractions are collected, concentrated by $(NH_4)_2SO_4$ precipitation (65% saturation), and applied in a small volume to a column of Sephacryl S-300. Peak fractions are pooled and applied to a second column of DEAE-Sepharose CL-6B and the protein eluted with a linear KCl gradient from 0.1–0.2 M. Following the second anion exchange column the enzyme is homogeneous as judged by SDS-polyacrylamide gel electrophoresis. The specific activity ranges between 30 and 50 units per mg protein, depending on the species. The purification procedure results in greater than 120-fold purification with a recovery of about 50%. The enzyme from *Chloris gayana* has been purified to a similar specific activity by another procedure (Hatch and Mau, 1977b). The purified enzyme is stable for at least three months frozen at $-20°C$ provided the protein concentration is above 1 mg ml^{-1}; BSA can be added to stabilise the protein in dilute solutions. The protein may also be stored at 2°C as a suspension in 70% saturated $(NH_4)_2SO_4$ solution.

4. Properties

The molecular weight of the native PEP carboxykinase, determined by column chromatography, is about 380 kDa for all species so far examined (Burnell, 1986). The subunit molecular weights are about 64 kDa indicating that the enzyme is hexameric (Burnell, 1986).

The kinetic constants for PEP carboxykinase isolated from different PCK-type plants are similar with K_m values of between 12 and 25 µM for OAA and 16 and 25 µM for ATP (Burnell, 1986; Hatch and Mau, 1977b; Ray and Black, 1976). These studies have also provided the following information. The enzyme has an absolute requirement for Mn^{2+} which cannot be replaced by any other divalent metal cations. There are separate binding sites for OAA, MnATP^{2-} and free Mn^{2+}; binding of free Mn^{2+} increases the affinity of the enzyme for MnATP^{2-} but does not affect the V_{max} (Burnell, 1986). PEP carboxykinase does not show high specificity for its nucleotide substrate but ATP is the preferred substrate. The pH optimum is broad (between 7.4 and 8.2) and Cl$^-$ stimulates activity.

5. Regulation

PEP carboxykinase from all three PCK-type C_4 plants studied is inhibited by 3-phosphoglycerate (3-PGA), fructose 6-P, fructose 1,6-P$_2$ and dihydroxyacetone phosphate (DHAP) at concentrations between 1 and 5 mM (Hatch and Mau, 1977b; Burnell, 1986). The enzyme is also inhibited by the product CO_2 (Jenkins *et al.*, 1987).

G. Aspartate Aminotransferase (EC 2.6.1.1)

L-aspartate + 2-oxoglutarate $\rightleftharpoons$ OAA + L-glutamate

Aspartate aminotransferase catalyses the interconversion of aspartate and OAA in both the mesophyll and bundle sheath cells of PCK-type and NAD-ME type C_4 plants (see Section I and Fig. 3.1). This enzyme is also discussed by Blackwell et al. (Chapter 8) and Joy and Ireland (Chapter 17) in this volume.

1. Extraction for assay

A general procedure for the extraction of aminotransferases from leaves (Hatch and Mau, 1973) involves homogenisation in a mortar (see Section II) in a medium containing 50 mM Hepes-NaOH, pH 8, 10 mM DTT, 20 µg ml^{-1} pyridoxal phosphate and 1 mM EDTA, followed by filtration and centrifugation. Procedures for isolating aspartate aminotransferase originating from mesophyll and bundle sheath cells have been described (Hatch and Mau, 1973).

2. Assay

Aspartate aminotransferase is assayed in the forward direction in a reaction mixture containing 50 mM Hepes-NaOH (pH 8), 2.5 mM L-aspartate, 2.5 mM 2-oxoglutarate, 10 µg ml^{-1} pyridoxal phosphate, 2 mM EDTA, 0.2 mM NADH and 4 units ml^{-1} of the coupling enzyme malate dehydrogenase (Hatch and Mau, 1973). OAA production is measured spectrophotometrically as NADH oxidation at 340 nm. The enzyme can be assayed in the reverse direction in a similar reaction mixture but with the substrates oxaloacetate (1 mM) and glutamate (2.5 mM) added. Oxaloacetate disappearance is monitored as a decrease in absorbance at 260 nm using an extinction coefficient of 1.2×10^3 l mol^{-1} cm^{-1}. The latter assay is not suitable for crude leaf extracts due to their high absorbance at 260 nm. Assayed in the direction of OAA formation, activities in leaf extracts of NAD-ME-type and PCK-type C_4 species range from 25–60 µmol min^{-1} (mg Chl)$^{-1}$.

3. Purification

The following procedure has been described for the separation and partial purification of the major mesophyll and bundle sheath cell aspartate aminotransferases from *Atriplex spongiosa* (Hatch, 1973b). Sliced and deribbed leaf tissue is extracted in 5 vols (w/v) of 50 mM Tris-HCl, pH 8, containing 5 mM DTT, 2 mM EDTA and 20 µg ml^{-1} pyridoxal phosphate by homogenisation in a Sorvall Omnimixer. This mixture is filtered on Miracloth and residual tissue is further extracted in a large cold mortar in the same buffer medium. The extracts are pooled, centrifuged at 15 000 × g for 10 min and the supernatant fractionated with $(NH_4)_2SO_4$ (pH 7, 4°C) to give protein precipitating between 40 and 75% saturation. The precipitated protein is resuspended in the extraction buffer and desalted on a Sephadex G-25 column. This extract is then applied to a 40 ml DEAE-cellulose column equilibrated at 0°C with 10 mM Tris-HCl (pH 7.5) containing 1 mM EDTA. The protein is eluted with a linear NaCl gradient (0–0.4 M in the same buffer medium). It is advisable to include 20 µg ml^{-1} of pyridoxal phosphate in these solutions, particularly if alanine aminotransferase isoenzymes are also to be examined (see Section II.H and Hatch, 1973b). As active fractions emerge, they are supplemented with BSA to give a concentration of 2 mg ml^{-1}. For storage the protein should be precipitated by the addition of solid $(NH_4)_2SO_4$ to give 70% saturation and then kept at 0°C. As required the protein is redissolved in 50 mM Hepes-NaOH buffer containing 1 mM EDTA and 20 µg ml^{-1} pyridoxal phosphate.

Using this procedure, aspartate aminotransferase separates into two major peaks. Gel electrophoresis of these fractions shows a single major isoenzyme band in each, corresponding to the mesophyll and bundle sheath isoenzymes seen in leaf extracts and bundle sheath and mesophyll cell preparations (Hatch, 1973b). A similar procedure, using a combination of Sephadex G-200 and DEAE-Sephadex A-50 column chromatography has been reported to result in a 40-fold purification of aspartate aminotransferase from *Panicum miliaceum* and *P. antidotale* (Balkow and Wildner, 1982).

Aspartate aminotransferase is stable for several hours in solution at 0°C but may lose 40–70% of its activity over a two-month period when stored as a protein suspension in $(NH_4)_2SO_4$ (Hatch, 1973b). Some of the lost activity can be restored by incubation with pyridoxal phosphate (20 µg ml^{-1} for 90 min; Hatch, 1973b).

4. Properties

Plant aminotransferases require pyridoxal phosphate but, in contrast to the animal enzymes, it is difficult to dissociate this cofactor. Kinetic evidence suggests that plant aminotransferases operate through a ping-pong bi bi reaction mechanism similar to that demonstrated for the animal enzyme (see Givan, 1980).

The kinetic constants for the mesophyll and bundle sheath isoenzymes from *Atriplex spongiosa* have been determined by Hatch (1973b). For the major mesophyll and bundle sheath cell isoenzymes, respectively, the K_m values are 0.85 mM and 0.5 mM for aspartate, 0.08 mM and 0.12 mM for 2-oxoglutarate, 0.1 mM and 0.045 mM for oxaloacetate and 1.6 mM and 0.75 mM for glutamate. Both enzymes are highly substrate-specific (Hatch, 1973b). Substantially higher K_m values have been reported for the enzyme from *Panicum* sp. (up to 5 mM for aspartate and 0.4 mM for 2-oxoglutarate: Balkow and Wildner, 1982). However, it should be noted that with enzymes of this mechanistic type, the K_m of one substrate is highly dependent upon the concentration of the second substrate in the assay (see Segel, 1975) and the concentration of the fixed substrates in the latter study were about 10-fold higher than those used by Hatch (1973b).

5. Regulation

There is no evidence that aspartate aminotransferase is highly regulated *in vivo*. Only malate has a significant effect on activity, causing a 35% inhibition of the forward reaction at a concentration of 5 mM (Hatch, 1973b). Significantly, the major isoenzymes specific to mesophyll and bundle sheath cells are rapidly synthesised during light-induced greening of leaves (Hatch and Mau, 1973).

H. Alanine Aminotransferase (EC 2.6.1.2)

$$\text{L-alanine} + \text{2-oxoglutarate} \rightleftharpoons \text{pyruvate} + \text{L-glutamate}$$

Alanine aminotransferase catalyses the interconversion of pyruvate and alanine in the cytosol of both mesophyll and bundle sheath cells of PCK-type and NAD-ME-type C_4 species (see Section I and Fig. 3.1).

1. Extraction for assay

Leaf extracts are prepared as described for aspartate aminotransferase (Section III.G).

2. Assay

Alanine aminotransferase is assayed in the direction of pyruvate formation by following NADH oxidation spectrophotometrically in the presence of the coupling enzyme lactate dehydrogenase. Reaction mixtures contain 50 mM Hepes-NaOH, pH 7.5, 10 mM alanine, 2.5 mM 2-oxoglutarate, 10 µg ml^{-1} pyridoxal phosphate, 2 mM EDTA, 0.2 mM NADH and 4 units ml^{-1} lactate dehydrogenase (Hatch and Mau, 1973). The reaction can be assayed in the reverse direction by coupling 2-oxoglutarate formation to NADH oxidation by adding aspartate aminotransferase and malate dehydrogenase (Hatch, 1973b). For this assay, the reaction contains 25 mM Hepes-NaOH (pH 7.5), 2 mM EDTA, 20 µg ml^{-1} pyridoxal phosphate, 0.2 mM NADH, 1 mM aspartate, 2.5 mM glutamate, 2.5 mM pyruvate, 3 units ml^{-1} aspartate aminotransferase and 5 units ml^{-1} malate dehydrogenase. Assayed in the direction of pyruvate formation activities in leaf extracts from NAD-ME-type and PCK-type C_4 species range between 25 and 60 µmol min^{-1} (mg Chl)$^{-1}$ (see Table 3.1).

3. Purification

The major alanine aminotransferase isoenzymes associated with mesophyll and bundle sheath cells of *Atriplex spongiosa* are separated and partially purified by the same general procedure described in Section III.G for aspartate aminotransferase (Hatch, 1973b). Three peaks of alanine aminotransferase elute from the DEAE-cellulose column with the linear NaCl gradient. The first peak that emerges is a minor one and is believed not to be involved in photosynthesis. The second and third peaks originate specifically from the mesophyll and bundle sheath cells, respectively (Hatch, 1973b). The mesophyll enzyme loses activity during purification but most of this can be restored by incubation with pyridoxal phosphate. The partially purified enzyme can be stored as a suspension in 70% saturated $(NH_4)_2SO_4$ as described for aspartate aminotransferase (Section III.G).

4. Properties

The K_m values for substrate of the major mesophyll and bundle sheath cell isoenzymes from *A. spongiosa*, respectively, are: 3 and 3.1 mM for alanine; 0.09 and 0.03 mM for 2-oxoglutarate; 0.04 and 0.02 mM for pyruvate; 1.15 and 0.8 mM for glutamate (Hatch, 1973b). Notably, the K_m for alanine of the major non-photosynthetic isoenzyme is substantially lower than that for the photosynthetic enzymes (0.25 mM) while the K_m for pyruvate is higher (0.45 mM). As with aspartate aminotransferase (Section III.G), the K_m values determined for this enzyme are dependent on the concentration of the second substrate. The relative rates for the forward/reverse reactions for the *A. spongiosa* isoenzymes are between 2.2 and 2.8 and both have a broad pH optimum between 6.5 and 8.5.

I. Adenylate Kinase (EC 2.7.4.3)

$$ATP + AMP \rightleftharpoons 2ADP$$

Adenylate kinase in leaves of C_4 plants serves to recycle AMP produced in the pyruvate, Pi dikinase reaction. The high activity in C_4 leaves is largely confined to the mesophyll chloroplasts.

1. Extraction for assay

Leaf tissue is extracted as described in Section II with 25 mM Hepes-KOH, pH 7.2, containing 5 mM DTT. For purification at least, Kleczkowski and Randall (1986) recommend the inclusion of the protease inhibitors benzamidine (1 mM), aminocaproic acid (1 mM) and leupeptin (5 µM) in all extraction buffers.

2. Assay

Adenylate kinase in leaf extracts can be assayed in the direction of ADP formation by following NADH oxidation spectrophotometrically at 340 nm in the presence of the coupling enzymes pyruvate kinase and lactate dehydrogenase (see Hatch, 1982). Reactions of 1 ml contain enzyme, 25 mM Hepes-KOH, pH 7.8, 2 mM ATP, 0.5 mM AMP, 2.5 mM $MgCl_2$, 0.5 mM PEP, 0.25 mM NADH, 30 mM KCl and 3 units ml^{-1} each of the coupling enzymes pyruvate kinase and lactate dehydrogenase. A useful procedure is to check initially for direct conversion of ATP to ADP (due to ATPase or non-specific phosphatase) with ATP alone. The adenylate kinase assay is then initiated by the addition of AMP. In the direction of ATP and AMP formation, adenylate kinase is assayed by linking ATP production to NADP reduction with the coupling enzymes hexokinase and glucose 6-phosphate dehydrogenase. NADP reduction is measured spectrophotometrically at 340 nm in a medium containing 25 mM Hepes-KOH, pH 8, 1.25 mM ADP, 2 mM $MgCl_2$ and a coupling system consisting of 4 units ml^{-1} hexokinase, 1 unit ml^{-1} glucose 6-phosphate dehydrogenase, 5 mM glucose and 0.25 mM NADP (Hatch, 1982).

Activity in extracts of C_4 leaves is in the range of 40 to 80 µmol min^{-1} (mg Chl)$^{-1}$ (Hatch, 1982).

3. Purification

Adenylate kinase has been purified to homogeneity from *Zea mays* leaves (Kleczkowski and Randall, 1986). Leaves are homogenised in a Waring blender in 40 mM Tricine-KOH, pH 7.8, 100 mM 2-mercaptoethanol, 2 mM $MgCl_2$, 1 mM EDTA, 1 mM benzamidine, 1 mM aminocaproic acid, and 5 µM leupeptin. After filtering and centrifugation, the supernatant is treated with $(NH_4)_2SO_4$ to give protein precipitating between 35 and 70% saturation. The precipitated protein is recovered by centrifugation and resuspended in buffer A: 20 mM Tricine-KOH (pH 7.8), 2 mM $MgCl_2$ and 28 mM 2-mercaptoethanol. The enzyme is de-salted and concentrated using an Amicon diaflow

PM 10 filtration apparatus and then applied to a DEAE-cellulose column and eluted with a linear gradient of 0–0.5 M NaCl in buffer A. Adenylate kinase activity emerges at about 0.17 M NaCl. The most active fractions are then applied to a hydroxyapatite column and eluted with a 0–0.5 M sodium phosphate gradient in buffer B: 20 mM Tricine (pH 7.8) and 28 mM 2-mercaptoethanol. Adenylate kinase elutes at approximately 0.1 M sodium phosphate. After desalting and concentrating as before (using buffer C: 20 mM 3-(N-morpholino)propane sulphonic acid (Mops), pH 7, 2 mM $MgCl_2$ and 28 mM 2-mercaptoethanol), the protein is applied to a Sephadex G-75 SF column. Adenylate kinase elutes after the major peak of excluded protein. The active fractions are pooled and chromatographed on a green A dye–ligand column (Amicon) and eluted with a step gradient of 1 mM ATP in buffer A. All procedures are performed at 0–4°C and buffers are degassed with N_2 before use. This method gives a 60-fold purification after the gel filtration step and a final enrichment of up to 300-fold following the dye–ligand column. The resulting protein is apparently homogeneous (Kleczkowski and Randall, 1986). Dilute solutions of partially purified adenylate kinase may be stabilised by the addition of BSA (0.5 mg ml^{-1}). The enzyme is stable for several months when stored as a protein suspension in 65% $(NH_4)_2SO_4$ solution at 0°C (Hatch, 1982). A simple procedure giving about 100-fold purification of maize leaf adenylate kinase has also been described (Hatch, 1982).

4. Properties

The C_4 leaf adenylate kinase is monomeric with a molecular weight of about 30 kDa (Kleczkowski and Randall, 1986). The enzyme reaction mechanism, by analogy with enzyme from other sources, is random bi bi (see Noda, 1973). Adenylate kinase from maize leaves requires a divalent cation for catalysis (Hatch, 1982; Kleczkowski and Randall, 1986). Mg-ATP is the substrate for the reaction producing ADP, but interpretations vary about the Mg^{2+} requirement for AMP binding (see Hatch, 1982 and Kleczkowski and Randall, 1986). This inconsistency may be due to an inhibitory effect of free ATP on the enzyme (Kleczkowski and Randall, 1986). For the reverse reaction the maize leaf enzyme apparently binds one molecule of free ADP and one molecule of Mg-ADP (Kleczkowski and Randall, 1986).

With an ADP:Mg^{2+} ratio of about 1 the K_m for ADP is about 80 μM. The K_m for ATP is about 30 μM when Mg^{2+} is present in a several-fold excess over adenylates and the K_m for AMP is 140 μM with 2.5 mM Mg^{2+} and 2 mM ATP, falling to 50 μM with 10 mM Mg^{2+} (Hatch, 1982). Notably, AMP is a potent competitive inhibitor with respect to ADP (Hatch, 1982).

The V_{max} for adenylate kinase in the direction of ADP production is usually similar to the reverse direction V_{max} consistent with its usual role in equilibrating adenylate pools (Noda, 1973). For the C_4 enzyme, however, the V_{max} for the reaction producing ADP is about four times faster than the reverse reaction; this may be related to the special role of the enzyme in C_4 photosynthesis (Hatch, 1982). The maize leaf adenylate kinase has a broad pH optimum for both the forward and reverse reactions in the region of 7.8 to 8.5. The shape of the pH curve varies as the ratio of ATP to Mg^{2+} is varied (Hatch, 1982).

J. Pyrophosphatase (EC 3.6.1.1)

$$PPi + H_2O \longrightarrow 2 Pi$$

Pyrophosphatase operates in the mesophyll chloroplasts of C_4 plants to recycle pyrophosphate produced in the pyruvate,Pi dikinase reaction.

1. Extraction for assay

For extraction from leaves see Section II. Pyrophosphatase in crude leaf extracts and in a semi-purified form can be stored for several days at 0°C or frozen without an appreciable loss of activity (Bennett et al., 1973).

2. Assay

Pyrophosphatase is generally assayed by measuring Pi production. One satisfactory reaction mixture used contained 2 mM PPi, 8 mM $MgCl_2$, 50 mM Tris-HCl, pH 8.9, and enzyme (Hatch et al., 1969). After an appropriate period of incubation at 30°C, the reaction is stopped by the addition of trichloroacetic acid to give a final concentration of 2% (w/v). Centrifuged samples are then used to determine Pi production (Lowry, 1957; or other suitable procedures). Extracts of C_4 leaves contain pyrophosphatase activity in the range of 10–40 µmol of PPi hydrolysed min^{-1} $(mg Chl)^{-1}$

3. Purification

C_4 leaf pyrophosphatase has been partially purified by treatment of sugar cane or maize leaf extracts on a DEAE-cellulose column followed by gel filtration (Bucke, 1970; Bennett et al., 1973). Two isoenzymes have been identified by polyacrylamide gel electrophoresis in extracts from maize leaves, but they were not separated on DEAE-cellulose (Bennett et al., 1973). Rip and Rauser (1971) reported substantial purification of pyrophosphatase following $(NH_4)_2SO_4$ fractionation, calcium phosphate gel chromatography, and chromatography on DEAE cellulose.

4. Properties

The molecular weight of the native forms of the two pyrophosphatase isoenzymes from Zea mays leaves are about 32 kDa and 38 kDa (Bennett et al., 1973). The PPi-Mg^{2+} complex appears to be the substrate for pyrophosphatase from C_4 leaves, as is the case with the animal enzyme. The interaction between Mg^{2+} and the substrate may be responsible for the varying values reported for the K_m PPi. Bucke (1970) reported optimal activity for the sugarcane enzyme with a ratio of Mg^{2+} : PPi of 4 and no further stimulation with higher Mg^{2+}. He determined a K_m for PPi of 0.75 mM, measured at a constant Mg^{2+} : PPi ratio of 4. In contrast, Simmons and Butler (1969) report a much lower K_m for the maize enzyme (5.6 µM) measured in the presence of 10 mM Mg^{2+}. At alkaline pH (pH 8–9) Mg^{2+} is by far the most effective cation (Simmons and Butler, 1969; Bucke, 1970; Bennett et al., 1973) but Bennett et al. (1973) report substantial activities with Zn^{2+} at pH values below 7.

5. Regulation

Bucke (1970) reported a strong inhibition by the product Pi (100% inhibition at 7 mM Pi) but we have not been able to duplicate this result in our laboratory. Since Pi levels in the chloroplast stroma may be as high as 10 mM (Lilley et al., 1977), such inhibition has potential significance *in vivo*.

K. Pyruvate,Pi Dikinase Regulatory Protein

Inactivation of PPDK:

$$\text{PPDK}\genfrac{}{}{0pt}{}{\text{thr}}{\text{his-P}} + \text{ADP} \longrightarrow \text{PPDK}\genfrac{}{}{0pt}{}{\text{thr-P}}{\text{his-P}} + \text{AMP}$$
$$\text{(active)} \qquad\qquad\qquad \text{(inactive)}$$

Activation of PPDK:

$$\text{PPDK}\genfrac{}{}{0pt}{}{\text{thr-P}}{\text{his}} + \text{Pi} \longrightarrow \text{PPDK}\genfrac{}{}{0pt}{}{\text{thr}}{\text{his}} + \text{PPi}$$
$$\text{(inactive)} \qquad\qquad\qquad \text{(active)}$$

Pyruvate,Pi dikinase regulatory protein (PDRP) is a bifunctional enzyme catalysing the interconversion between an active form of pyruvate,Pi dikinase (PPDK), phosphorylated on a catalytic-site histidine, and an inactive form phosphorylated on a regulatory-site threonine residue (Burnell and Hatch, 1985b, 1986). These reactions account for the light-mediated activation and dark inactivation of PPDK in the mesophyll chloroplasts of C_4 plants (Ashton et al., 1984; Burnell and Hatch, 1985a,b).

1. Extraction for assay

The general procedure for the extraction of enzymes from leaves is described in Section II. However, due to its instability PDRP must be extracted in the presence of 5 mM DTT and 5 mM $MgCl_2$, and all operations must be conducted between 0 and 4°C. PDRP activity is stabilised by Pi, ATP and ADP (Hatch and Burnell, 1983) and by Blue dextran and its dye moiety Cibacron Blue F3GA (Burnell and Hatch, 1983; Nakamoto and Sugiyama, 1982).

2. Assay

The PDRP-mediated inactivation of PPDK is assayed in reaction mixtures (0.24 ml, 25°C) containing 50 mM Tris-HCl, pH 8.3, 5 mM $MgCl_2$, 0.1 mM Pi, 0.1 mM EDTA, 2 mM DTT, active PPDK (approximately 0.2 units), and PDRP (Burnell and Hatch, 1985a). Inactivation is initiated by adding 10 μl of a mixture of 25 mM ADP and 5 mM ATP. Samples (25 μl) are withdrawn at various time intervals and the PPDK activity determined as described in Section II.B except that assay mixtures contained 10 μM

Cibacron Blue F3GA to inhibit activation of PPDK during the assay (see Burnell and Hatch, 1986).

Activation of PPDK is measured in reaction mixtures containing (0.24 ml, 25°C) 50 mM Tris-HCl, pH 7.5, 5 mM $MgCl_2$, 0.1 mM EDTA, 2 mM DTT, dark-inactivated PPDK (about 0.5 potential units) and PDRP (Burnell and Hatch, 1984a). Activation is initiated by the addition of 10 µl 25 mM Pi. Samples (25 µl) are taken at varying time intervals and the PPDK activity determined as described in Section II.B but with the addition of Cibacron Blue (see above).

3. Purification

The following purification procedure is based on the method described by Burnell and Hatch (1985a). All procedures are conducted at between 0 and 4°C. Fifty grams of deribbed finely sliced maize leaves are homogenised in a Sorvall Omnimixer with 250 ml of buffer containing 100 mM Tris-HCl, 10 mM $MgCl_2$, 1 mM EDTA, 5 mM DTT, 2 mM Pi, pH 8.3. The homogenate is filtered through Miracloth, centrifuged, and the protein precipitating between 30 and 40% saturated $(NH_4)_2SO_4$ is taken up in a minimal volume of medium A (50 mM Tris-HCl, 10 mM $MgCl_2$, 1 mM EDTA, 2 mM Pi, 2 mM DTT, pH 8.3) and desalted through a column of Sephadex G-25 equilibrated with buffer medium A. The excluded protein is applied to a 25 ml column of Agarose-Blue dextran, the column washed with 100 ml of medium A, 100 ml of 0.2 M KCl in medium A, and then the PDRP is eluted with 0.5 M KCl in medium A. The peak fractions of PDRP activity are pooled and concentrated by $(NH_4)_2SO_4$ precipitation and applied to a 180 ml column of Sephadex G-200. Active fractions are pooled and the protein stored as a precipitate in 65% saturated $(NH_4)_2SO_4$ at 4°C. As required, the PDRP is recovered by centrifugation and dissolved in medium A minus EDTA and Pi and desalted on a column of Sephadex G-25. This procedure purifies PDRP 35-fold with a yield of about 12% and a final specific activity of 0.59 µmol PPDK activated min^{-1} (mg protein)$^{-1}$ (Burnell and Hatch, 1985a).

4. Properties

Purified native *Z. mays* PDRP has a molecular weight of 180 kDa at pH 8.3 and 90 kDa at pH 7.5 with a monomeric molecular weight of 45 kDa (Burnell and Hatch, 1983). The substrate for ADP-mediated inactivation is the form of the enzyme phosphorylated on a catalytic site histidine residue. This form of active PPDK is inactivated by phosphorylation of a threonine residue close to the histidine residue (Ashton and Hatch, 1983c; Burnell and Hatch, 1984a,b). Pyruvate inhibits ADP-dependent inactivation by removing the catalytic-site phosphate on the histidine residue. Activation of inactive PPDK is several times more rapid if the catalytic-site histidine residue on the PPDK is not phosphorylated (Burnell, 1984).

For the inactivation reaction catalysed by *Z. mays* PDRP the K_m for ADP is 50 µM and for active PPDK is 1.2 µM; for the activation reaction the K_m for the substrates Pi and inactive PPDK are 650 µM and 0.7 µM, respectively (Burnell and Hatch, 1985a). Roeske and Chollet (1987) obtained a value of 13 µM for ADP. The inactivation and activation reactions probably require Mg^{2+} (Burnell and Hatch, 1985a).

5. Regulation

ADP and AMP inhibit Pi-dependent activation of PPDK; ADP inhibits activation competitively with respect to inactive PPDK with a K_i of 84 μM and AMP inhibits competitively with respect to Pi with a K_i of 0.4 mM (Burnell and Hatch, 1985a). PPi competitively inhibits activation with respect to the inactive PPDK with a K_i of 160 μM and the active PPDK competitively inhibits activation with respect to the inactive PPDK, but only in its catalytically phosphorylated form. Inactivation of PPDK is competitively inhibited by inactive PPDK with respect to active PPDK. The extent of this inhibition is dependent upon the catalytic phosphorylation status of inactive PPDK, with inhibition being greater with the catalytically unphosphorylated form of the enzyme (Burnell and Hatch, 1985a). Pyruvate inhibits ADP-dependent inactivation of PPDK by removing the catalytic phosphate on the histidine residue of PPDK (Burnell and Hatch, 1984a).

III. ENZYMES OF THE PHOTOSYNTHETIC CARBON REDUCTION CYCLE IN C_4 PLANTS

In C_4 plants the reactions of the photosynthetic carbon reduction (PCR) cycle occur in the bundle sheath cell chloroplasts where CO_2 is concentrated by the C_4 acid cycle of the C_4 pathway (see Section I). Most of the enzymes of the PCR cycle are located exclusively in the bundle sheath cells but the enzymes responsible for reducing 3-PGA to triose phosphates [3-phosphoglycerate kinase, NADP glyceraldehyde-3-phosphate dehydrogenase (GAPDH) together with triose-P isomerase] occur in both cell types (Slack et al., 1969; Hatch and Kagawa, 1973; Ku and Edwards, 1975; Hatch and Osmond, 1976; Hatch, 1987).

A number of enzymes of the PCR cycle from *Zea mays* have been characterised. These enzymes, which are briefly described below, have properties generally similar to the corresponding enzymes found in C_3 plants. The PCR cycle enzymes from C_3 plants, and procedures for their assay, are described by Keys and Parry (Chapter 1) and Leegood (Chapter 2) in this volume.

A. Ribulose 1,5-Bisphosphate Carboxylase/Oxygenase

Ribulose 1,5-bisphosphate carboxylase/oxygenase (Rubisco) is restricted to the bundle sheath cells of C_4 plants as determined by activity measurements (see Hatch and Osmond, 1976) and immunochemical procedures (Hattersley et al., 1977; Ku et al., 1979). Both the activity and content of Rubisco in C_4 leaves averages less than 1/3 of the levels present in leaves of C_3 plants (Ku et al., 1979). The enzyme from maize leaves has been purified to homogeneity by Reger et al. (1983) and the sequence of both the large subunit gene (McIntosh et al., 1980) and the small subunit gene (W. C. Taylor, pers. commun.) have been determined.

Rubisco from C_4 plants has about twice the specific activity of Rubisco from C_3 plants (Seeman et al., 1984). On the other hand, the Rubisco from C_4 plants has a lower affinity for CO_2, the K_m (CO_2) values being about two-fold higher than for C_3 plants (28–63 μM for C_4 plants compared to 13–26 μM for C_3 plants: Yeoh et al., 1980).

However, the relative specificity of the enzyme from C_4 plants for CO_2 as opposed to O_2 is similar to those of C_3 plants (see Andrews and Lorimer, 1987); apparently the oxygenase specific activity and K_m (O_2) of C_4 Rubisco must also be increased. Recent comparisons of sequences for the large subunit gene from C_3 and C_4 species from the same genera suggest that several amino acid changes may be necessary to generate these changes in kinetic characteristics (G. Hudson et al., pers. commun.). The differing content and kinetic properties of Rubisco from C_4 species no doubt relates to the fact that the C_4 enzyme operates in an environment of higher CO_2 and higher ratios of CO_2 to O_2 (see above and Section I).

B. 3-Phosphoglycerate Kinase and NADP-Glyceraldehyde 3-Phosphate Dehydrogenase

As indicated above, these activities are distributed about equally between mesophyll and bundle sheath cells in C_4 plants. NADP-GAPDH from C_4 plants, like the enzyme from C_3 plants, is light activated (Steiger et al., 1971). The gene for maize chloroplast NADP-GAPDH has been sequenced (Quigley et al., 1988) and the deduced amino acid sequence shows strong similarity to the deduced amino acid sequence of cDNAs encoding tobacco NADP-GAPDH (Shih et al., 1986). Both sequences contain cysteine residues unique to chloroplast GAPDH isoenzymes which may be involved in the light–dark redox regulation of the chloroplast enzymes.

C. Fructose 1,6-Bisphosphate Aldolase

The chloroplast fructose 1,6-bisphosphate aldolase from maize has been purified to homogeneity (Krüger and Schnarrenberger, 1983). This enzyme cross-reacts with antibodies raised against the spinach chloroplast enzyme but not against antibodies to the spinach cytoplasmic aldolase. The maize and spinach chloroplast enzymes have similar specific activities and both are tetramers with subunits of M_r 35 kDa (Krüger and Schnarrenberger, 1983). The kinetic properties of chloroplast aldolases from maize and the C_3 plants wheat, spinach and pea are similar (Schnarrenberger and Krüger, 1986).

D. Fructose 1,6-Bisphosphatase and Sedoheptulose 1,7-Bisphosphatase

Chloroplast fructose 1,6-bisphosphatase (FBPase) has been shown to be localised in the bundle sheath cells by activity measurements (Slack et al., 1969; Hatch and Kagawa, 1973; Furbank et al., 1985) and immunochemical means (Broglie et al., 1984). The mRNA for sedoheptulose 1,7-bisphosphatase (SBPase) and therefore presumably SBPase itself, occurs only in bundle sheath cells (Sheen and Bogorad, 1987).

Both FBPase and SBPase have been purified to homogeneity from maize leaves (Nishizawa and Buchanan, 1981). The maize FBPase has similar properties to the FBPase purified from the C_3 plant spinach (Nishizawa et al., 1982) while the maize SBPase has similar properties to the best characterised C_3 leaf SBPase (Woodrow et al., 1984). Both maize FBPase and SBPase possess similar regulatory properties to the C_3 enzymes, i.e. light–dark mediated redox regulation and complex interactions between pH, substrate and divalent metal ions (Nishizawa and Buchanan, 1981; Wolosiuk et al, 1982; Woodrow et al., 1984).

E. Ribulose 5-Phosphate Isomerase and Ribulose 5-Phosphate Kinase

A specific bundle sheath cell location has been demonstrated for both these enzymes by activity measurements (see above) and for ribulose 5-phosphate kinase by immunochemical techniques (Broglie et al., 1984). Ribulose 5-phosphate kinase from maize leaves has been partially purified (Ashton, 1984a,b) and shows similar kinetic properties to those of the C_3 enzyme from wheat (Surek et al., 1985). The C_4 enzyme also shows similar redox-mediated regulatory properties that account for the light–dark regulation of the enzyme in leaves.

REFERENCES

Andreo, C. S., Gonzalez, D. H. and Iglesias, A. A. (1987). *FEBS Lett.* **213**, 1–8.
Andrews, T. J. and Hatch, M. D. (1969). *Biochem. J.* **114**, 117–125.
Andrews, T. J. and Lorimer, G. H. (1987). *In* "The Biochemistry of Plants" (M. D. Hatch and N. K. Boardman, eds), Vol. 10, pp. 131–218. Academic Press, New York.
Andrews, T. J., Johnson, H. S., Slack, C. R. and Hatch, M. D. (1971) *Phytochemistry* **10**, 2005–2013.
Arnon, D. I. (1949). *Plant Physiol.* **24**, 1–15.
Artus, N. N. and Edwards, G. E. (1985). *FEBS Lett.* **182**, 225–233.
Asami, S., Inoue, K., Matsumoto, K., Murachi, A. and Akazawa, T. (1979a). *Arch. Biochem. Biophys.* **194**, 503–510.
Asami, S., Inoue, K., and Akazawa, T. (1979b). *Arch. Biochem. Biophys.* **196**, 581–587.
Ashton, A. R. (1984a). *Biochem. J.* **217**, 79–84.
Ashton, A. R. (1984b). *In* "Thioredoxins: Structure and Function" (P. Gadal, ed.), pp. 245–250. CNRS, Paris.
Ashton, A. R. and Hatch, M. D. (1983a). *Arch Biochem. Biophys.* **227**, 406–415.
Ashton, A. R. and Hatch, M. D. (1983b). *Arch. Biochem. Biophys.* **227**, 416–424.
Ashton, A. R. and Hatch, M. D. (1983c). *Biochem. Biophys. Res. Commun.* **115**, 53–60.
Ashton, A. R. and Polya, G. M. (1978). *Biochem. J.* **175**, 501–506.
Ashton, A. R., Burnell, J. N. and Hatch, M. D. (1984). *Arch. Biochem. Biophys.* **230**, 492–503.
Balkow, C. and Wildner, G. F. (1982). *Planta* **154**, 477–484.
Bauwe, H. (1986). *Planta* **169**, 356–360.
Bennett, V. L., Ristrophe, D. L., Hamming, J. J. and Butler, B. C. (1973). *Biochim. Biophys. Acta* **191**, 232–241.
Bjorkman, O. and Gauhl, E. (1969). *Planta* **88**, 197–203.
Broglie, R., Coruzzi, G., Keith, B. and Chua, N.-H. (1984). *Plant Mol. Biol.* **3**, 431–444.
Bucke, C. (1970). *Phytochemistry* **9**, 1303–1309.
Budde, R. J. A. and Chollet, R. (1986). *Plant Physiol.* **82**, 1107–1114.
Burnell, J. N. (1984). *Biochem. Biophys. Res. Commun.* **120**, 559–565.
Burnell, J. N. (1986). *Austral. J. Plant Physiol.* **13**, 577–587.
Burnell, J. N. (1987). *Austral. J. Plant Physiol.* **14**, 517–525.
Burnell, J. N. and Hatch, M. D. (1983). *Biochem. Biophys. Res. Commun.* **118**, 65–72.
Burnell, J. N. and Hatch, M. D. (1984a). *Arch. Biochem. Biophys.* **231**, 175–182.
Burnell, J. N. and Hatch, M. D. (1984b). *Biochem. Biophys. Res. Commun.* **118**, 65–72.
Burnell, J. N. and Hatch, M. D. (1985a). *Arch. Biochem. Biophys.* **237**, 490–503.
Burnell, J. N. and Hatch, M. D. (1985b). *Trends Biochem. Sci.* **10**, 288–291.
Burnell, J. N. and Hatch, M. D. (1986). *Arch. Biochem. Biophys.* **245**, 297–304.
Burnell, J. N. and Hatch, M. D. (1988). *Arch. Biochem. Biophys.* **260**, 177–186.
Chapman, K. S. R. and Hatch, M. D. (1977). *Arch. Biochem. Biophys.* **184**, 298–306.
Chapman, K. S. R. and Hatch, M. D. (1983). *Plant Sci. Lett.* **29**, 145–154.
Coombs, J., Baldry, C. W. and Bucke, C. (1973a). *Planta* **110**, 95–107.

Coombs, J., Baldry, C. W. and Bucke, C. (1973b). *Planta* **110**, 109–120.
Decottignies, P., Schmitter, J. M., Migniac-Maslow, M., LeMaréchal, P., Jacquot, J. P. and Gadal, P. (1988). *J. Biol. Chem.* **263**, 11780–11785.
Dittrich, P., Campbell, W. H. and Black, C. C. (1973) *Plant Physiol.* **52**, 357–361.
Doncaster, H. D., and Leegood, R. C. (1987). *Plant Physiol.* **84**, 82–87.
Edwards, G. E. and Walker, D. A. (1983). "C_3, C_4: Mechanism and Cellular and Environmental Regulation of Photosynthesis." Blackwell, Oxford.
Edwards, G. E., Kanai, R. and Black, C. C. (1971). *Biochem. Biophys. Res. Commun.* **45**, 278–285.
Edwards, G. E., Ku, M. S. B. and Hatch, M. D. (1982). *Plant Cell Physiol.* **23**, 1185–1195.
Edwards, G. E., Nakamoto, H., Burnell, J. N. and Hatch, M. D. (1985). *Ann. Rev. Plant Physiol.* **36**, 255–286.
Ferte, N., Jacquot, J.-P. and Meunier, J.-C. (1986). *Eur. J. Biochem.* **154**, 587–595.
Furbank, R. T., Stitt, M. and Foyer C. H. (1985). *Planta* **164**, 172–178.
Givan, C. V. (1980). *In* "The Biochemistry of Plants" (P. K. Stumpf and E. E. Conn, eds), Vol. 5, pp. 329–355. Academic Press, New York.
Gutierrez, M., Gracen, V. E. and Edwards, G. E. (1974). *Planta* **119**, 279–300.
Hatch, M. D. (1973a). *Anal. Biochem.* **52**, 280–285.
Hatch, M. D. (1973b). *Arch. Biochem. Biophys.* **156**, 207–214.
Hatch, M. D. (1978). *Curr. Topics Cellular Reg.* **14**, 1–27.
Hatch, M. D. (1979). *Austral. J. Plant Physiol.* **6**, 607–619.
Hatch, M. D. (1981). *In* "Photosynthesis: Regulation of Carbon Metabolism" (G. Akoyunoglou, ed.), Vol. IV, pp. 227–236. Balaban International Scientific Services, Philadelphia.
Hatch, M. D. (1982). *Austral. J. Plant Physiol.* **9**, 287–296.
Hatch, M. D. (1987). *Biochem. Biophys. Acta* **895**, 81–106.
Hatch, M. D. and Burnell, J. N. (1983). *Austral. J. Plant Physiol.* **10**, 179–186.
Hatch, M. D. and Heldt, H. W. (1985). *Anal. Biochem.* **145**, 393–397.
Hatch, M. D. and Kagawa, T. (1973). *Arch. Biochem. Biophys.* **159**, 842–853.
Hatch, M. D. and Kagawa, T. (1974). *Austral. J. Plant Physiol.* **1**, 357–369.
Hatch, M. D. and Mau, S. L. (1973). *Arch. Biochem. Biophys.* **156**, 195–206.
Hatch, M. D. and Mau, S. L. (1977a). *Arch. Biochem. Biophys.* **179**, 361–369.
Hatch, M. D. and Mau, S. L. (1977b). *Austral. J. Plant Physiol.* **4**, 207–216.
Hatch, M. D. and Oliver, I. R. (1978). *Austral. J. Plant Physiol.* **5**, 571–580.
Hatch, M. D. and Osmond, C. B. (1976). *In* "Encyclopedia Plant Physiology," New Series (C. R. Stocking and U. Heber, eds), Vol. 3, pp. 144–184. Springer, Berlin.
Hatch, M. D. and Slack, C. R. (1966). *Biochem. J.* **101**, 103–111.
Hatch, M. D. and Slack, C. R. (1968). *Biochem. J.* **106**, 141–146.
Hatch, M. D. and Slack, C. R. (1969). *Biochem. J.* **112**, 549–558.
Hatch, M. D. and Slack, C. R. (1970). *Ann. Rev. Plant Physiol.* **21**, 141–162.
Hatch, M. D., Slack, C. R. and Johnson, H. S. (1967). *Biochem. J.* **102**, 417–422.
Hatch, M. D., Slack, C. R. and Bull, T. A. (1969). *Phytochemistry* **8**, 697–706.
Hatch, M. D., Mau, S. and Kagawa, T. (1974). *Arch. Biochem. Biophys.* **165**, 188–200.
Hatch, M. D., Kagawa, T. and Craig, S. (1975). *Austral. J. Plant Physiol.* **2**, 111–128.
Hatch, M. D., Tsuzuki, M. and Edwards, G. E. (1982). *Plant Physiol.* **69**, 483–491.
Hatch, M. D., Agostino, A. and Burnell, J. N. (1988). *Arch. Biochem.* **261**, 357–367.
Hattersley, P. W., Watson, L. and Osmond, C. B. (1977). *Austral. J. Plant Physiol.* **4**, 523–539.
Häusler, R. E., Holtum, J. A. M. and Latzko, E. (1987). *Eur. J. Biochem.* **163**, 619–626.
Huber, S. C. and Edwards, G. E. (1975). *Can. J. Bot.* **53**, 1925–1933.
Huber, S. C. and Sugiyama, T. (1986). *Plant Physiol.* **81**, 674–677.
Huber, S. C., Sugiyama, T. and Akazawa, T. (1986). *Plant Physiol.* **82**, 550–554.
Izui, K., Ishijima, S., Yamaguchi, Y., Katagiri, F., Murata, T., Shigesada, K., Sugiyama, T. and Katsuki, H. (1986). *Nucleic Acids Res.* **14**, 1615–1628.
Jacquot, J.-P., Buchanan, B. B., Martin, F. and Vidal, J. (1981). *Plant Physiol.* **68**, 300–304.
Jenkins, C. L. D. and Hatch, M. D. (1985). *Arch. Biochem. Biophys.* **239**, 53–62.
Jenkins, C. L. D., Anderson, L. E. and Hatch, M. D. (1986). *Plant Sci.* **45**, 1–7.
Jenkins, C. L. D., Burnell, J. N. and Hatch, M. D. (1987). *Plant Physiol.* **85**, 942–957.
Jiao, J.-A. and Chollet, R. (1988). *Arch. Biochem. Biophys.* **261**, 409–417.

Johnson, H. S. and Hatch, M. D. (1970). *Biochem. J.* **119**, 273–280.
Kagawa, T. and Bruno, P. L. (1988). *Arch. Biochem. Biophys.* **260**, 674–695.
Kagawa, T. and Hatch, M. D. (1977). *Arch. Biochem. Biophys.* **184**, 290–297.
Kleczkowski, L. A. and Randall, D. D. (1986). *Plant Physiol.* **81**, 1110–1114.
Kortschak, H. P., Hart, C. E. and Burr, G. O. (1965). *Plant Physiol.* **40**, 209–213.
Krüger, I. and Schnarrenberger, C. (1983). *Eur. J. Biochem.* **136**, 101–106.
Ku, M. S. B. and Edwards, G. E. (1975). *Z. Pflanzenphysiol.* **77**, 16–32.
Ku, M. S. B., Schmitt, M. R. and Edwards, G. E. (1979). *J. Exp. Bot.* **30**, 89–98.
Lilley, R. M., Chon, C. J., Mosbach, A. and Heldt, H. W. (1977). *Biochem. Biophys. Acta.* **460**, 259–272.
Lowe, J. and Slack, C. R. (1971). *Biochim. Biophys. Acta* **235**, 207–209.
Lowry, O. H. (1957). In "Methods in Enzymology" (S. P. Colowick and N. O. Kaplan, eds), Vol. 4, p. 373. Academic Press, New York.
McIntosh, L., Poulsen, C. and Bogorad, L. (1980). *Nature* **288**, 556–560.
Meyer, C. R., Rustin, P. and Wedding, R. T. (1988). *Plant Physiol.* **86**, 325–328.
Monod, J., Wyman, J. and Changeux, J. (1965). *J. Mol. Biol.* **12**, 88–118.
Nakamoto, H. and Sugiyama, T. (1982). *Plant Physiol.* **69**, 749–753.
Nimmo, G. A., McNaughton, G. A. L., Fewson, C. A., Wilkins, M. B. and Nimmo, H. G. (1987). *FEBS Lett.* **213**, 18–22.
Nishizawa, A. N. and Buchanan, B. B. (1981). *J. Biol. Chem.* **256**, 6119–6126.
Nishizawa, A. N., Yee, B. C. and Buchanan, B. B. (1982). In "Methods in Chloroplast Molecular Biology" (M. Edelman, R. B. Hallick and N.-H. Chua, eds), pp. 707–713. Elsevier, Amsterdam.
Noda, L. (1973). In "The Enzymes" (P. D. Boyer, H. Lardy and K. Myrbach, eds), Vol. 8, pp. 279–305. Academic Press, New York.
O'Leary, M. (1982). *Ann. Rev. Plant. Physiol.* **33**, 297–315.
Persanov, V. M., Voronova, E. A., Oparina, L. A. and Karpilov, Y. S. (1976). *Biochemistry* (*Eng. Trans.*) **41**, 758–761.
Pupillo, P. and Bossi, P. (1979). *Planta* **144**, 283–289.
Quigley, F., Martin, W. F. and Cerff, R. (1988). *Proc. Natl. Acad. Sci. USA* **85**, 2672–2676.
Ray, T. B. and Black, C. C. (1976). *Plant Physiol.* **58**, 603–607.
Rebeille, F. and Hatch, M. D. (1986a). *Arch. Biochem. Biophys.* **249**, 164–170.
Rebeille, F. and Hatch, M. D. (1986b). *Arch. Biochem. Biophys.* **249**, 171–179.
Reger, B. J., Ku, M. S. B., Potter, J. W. and Evans, J. J. (1983). *Phytochemistry* **22**, 1127–1132.
Rip, J. W. and Rauser, W. E. (1971). *Phytochemistry* **10**, 2615–2619.
Roeske, C. A. and Chollet, R. (1987). *J. Biol. Chem.* **262**, 12575–12582.
Schnarrenberger, C. and Krüger, I. (1986). *Plant Physiol.* **80**, 301–304.
Seemann, J. R., Badger, M. R. and Berry, J. A. (1984). *Plant Physiol.* **74**, 791–794.
Segel, I. H. (1975). "Enzyme Kinetics: Behaviour and Analysis of Rapid Equilibrium and Steady-State Enzyme Systems". Wiley, New York.
Sheen, J.-Y. and Bogorad, L. (1987). *Plant Mol. Biol.* **8**, 227–238.
Shih, M.-C., Lazar, G. and Goodman, H. M. (1986). *Cell* **47**, 73–80.
Shirahashi, K., Hayakawa, S. and Sugiyama, T. (1978). *Plant Physiol.* **62**, 826–830.
Simmons, S. and Butler, L. G. (1969). *Biochim. Biophys. Acta* **172**, 150–157.
Slack, C. R. and Hatch, M. D. (1967). *Biochem. J.* **103**, 660–665.
Slack, C. R., Hatch, M. D. and Goodchild, D. J. (1969). *Biochem. J.* **114**, 489–498.
Steiger, E., Ziegler, I. and Ziegler, H. (1971). *Planta* **96**, 109–118.
Stiborová, M. and Leblová, S. (1986). *FEBS Lett.* **205**, 32–34.
Sugiyama, T. (1973). *Biochemistry* **12**, 2862–2868.
Surek, B., Heilbronn, A., Austen, A. and Latzko, E. (1985). *Planta* **165**, 507–512.
Thorniley, M. S. and Dalziel, K. (1988). *Biochem. J.* **254**, 229–233.
Uedan, K. and Sugiyama, T. (1976). *Plant Physiol.* **57**, 906–910.
Vidal, J. and Gadal, P. (1983). *Physiol. Plant.* **57**, 124–128.
Wagner, R., Gonzalez, D. H., Podesta, F. E. and Andreo, C. S. (1987). *Eur. J. Biochem.* **164**, 661–666.

Wagner, R., Podesta, F. E., Gonzalez, D. H. and Andreo, C. S. (1988). *Eur. J. Biochem.* **173**, 561–568.
Walker, G. H., Ku, M. S. B. and Edwards, G. E. (1986a). *Arch. Biochem. Biophys.* **248**, 489–501.
Walker, G. H., Ku, M. S. B. and Edwards, G. E. (1986b). *Plant Physiol.* **80**, 848–855.
Wintermans, J. F. and DeMots, A. (1965). *Biochim. Biophys. Acta.* **109**, 448–453.
Wolosiuk, R. A., Hertig, C. M., Nishizawa, A. N. and Buchanan, B. B. (1982). *FEBS Lett.* **140**, 31–35.
Woodrow, I. E., Murphy, D. J. and Latzko, E. (1984). *J. Biol. Chem.* **259**, 3791–3795.
Yeoh, H. H., Badger, M. R. and Watson, L. (1980). *Plant Physiol.* **66**, 1110–1112.

ADDENDUM

The following relevant information has been reported in the literature since this chapter was submitted.

Phosphoenolpyruvate carboxylase: The *in vitro* phosphorylation of a serine residue on *Zea mays* PEP carboxylase by a soluble protein kinase simulates the dark–light induced changes in catalytic activity and malate sensitivity of the enzyme [Jiao, J.-A. and Chollet, R. (1989). *Arch. Biochem. Biophys.* **269**, 526–535]. Oligomerisation of the enzyme is not directly involved as part of the mechanism of dark–light regulation [McNaughton, A. L., Fewson, C. A., Wilkins, M. B. and Nimmo, H. G. (1989). *Biochem. J.* **261**, 349–355].

Pyruvate,Pi dikinase: The likely contributions of ATP, ADP, PEP and pyruvate to the dark–light regulation of the enzyme were compared [Roeske, C. A. and Chollet, R. (1989). *Plant Physiol.* **90**, 330–337].

NAD-Malic enzyme: The enzyme was purified from two NAD-malic enzyme-type C_4 species and properties determined [Murata, T., Ohougi, R., Matsuoka, M. and Nakamoto, H. (1989). *Plant Physiol.* **89**, 316–324].

Phosphoenolpyruvate carboxykinase: The kinetics and reaction mechanism of the enzyme from the PCK-type C_4 species *Panicum maximum* were determined [Urbina, J. A. and Avilan, L. (1989). *Phytochemistry* **28**, 1349–1353].

NADP malic enzyme: The enzyme from sugar-cane leaves was purified to homogeneity and the properties determined [Iglesias, A. and Andreo, C. S., (1989). *Plant Cell Physiol.* **30**, 399–405].

Aspartate aminotransferase: The enzyme from mesophyll cells of the PCK-type C_4 species *Panicum maximum* was purified and compared with the aspartate aminotransferase from bundle sheath cells [Numazawa, T., Yamada, S., Hase, T. and Sugiyama, T. (1989). *Arch. Biochem. Biophys.* **270**, 313–319].

c-DNA and gene nucleotide sequences: The c-DNA or gene nucleotide sequences have now been determined for the following enzymes: phosphoenolpyruvate carboxylase (gene) [Matsuoka, M. and Minami, E. (1989). *Eur. J. Biochem.* **181**, 593–598; Hudspath, R. L. and Grula, J. W. (1989). *Plant Mol. Biol.* **12**, 579–589]; pyruvate,Pi dikinase (cDNA) [Matsuoka, M., Ozeki, Y., Yamamoto, N., Hirano, H., Kano-Murakami, Y. and Tanaka, Y. (1988). *J. Biol. Chem.* **263**, 11080–11083]; NADP malate dehydrogenase (cDNA) [Metzler, M. C., Rothermel, B. A. and Nelson, T. (1989). *Plant Mol. Biol.* **12**, 713–722]; NADP malic enzyme (cDNA) [Rothermel, B. A. and Nelson, T. (1989). *J. Biol. Chem.* (in press)].

4 Enzymes of Sucrose Metabolism

LES COPELAND

Department of Agricultural Chemistry, University of Sydney, NSW 2006, Australia

I.	Introduction	73
II.	Sucrose metabolism in plants	74
	A. Sucrose biosynthesis	74
	B. Sucrose breakdown	75
	C. Synthesis of higher oligosaccharides and fructans	76
III.	Enzymes of sucrose metabolism	76
	A. Sucrose phosphate synthase	76
	B. Sucrose phosphatase	79
	C. Sucrose synthase	80
	D. Invertase	82
	References	83

I. INTRODUCTION

Sucrose is the most abundant oligosaccharide in plants. It is a major product of photosynthesis and the main carbohydrate translocated to non-photosynthetic tissues. Sucrose may accumulate in vacuoles in significant quantities for storage, and it is formed when reserves are mobilized in germinating seeds. This chapter deals with the enzymes involved in the synthesis and breakdown of sucrose in plant tissues. Sucrose metabolism will be discussed briefly, together with methods that have been used for the purification and assay of the enzymes concerned, and their properties.

It is pertinent to mention some chemical properties of sucrose that are relevant to biochemical studies. Sucrose is a disaccharide in which an α-D-glucopyranose is linked to a β-D-fructofuranose by an α(1–2) glycosidic bond. As the anomeric carbons of both monosaccharide units are involved in the linkage, sucrose is a non-reducing sugar and relatively inert in comparison to other sugars. Sucrose is stable in mildly alkaline solutions, but the glycosidic bond is very sensitive to dilute acids. Complete hydrolysis to an equimolar mixture of α-D-glucose and β-D-fructose occurs within 5 min in 0.005 N HCl at 98°C (Silberman, 1961). The free energy of hydrolysis is -6.6 kcal mol^{-1}. Sucrose is very soluble, with 1 g dissolving in 0.5 ml of water at 20°C and 0.2 ml of boiling water.

II. SUCROSE METABOLISM IN PLANTS

The metabolism of sucrose in plant tissues has been discussed extensively (Akazawa and Okamoto, 1980; Avigad, 1982; Hawker, 1985; Stitt and Steup, 1985; Sicher, 1986; Stitt *et al.*, 1987) and will be considered only briefly here.

A. Sucrose Biosynthesis

Essentially all of the information on the biosynthesis of sucrose has come from studies of photosynthetic tissues. The biosynthetic pathway in leaves is located in the cytoplasm and involves hydrolysis of fructose 1,6-bisphosphate (Fru-1,6-P_2) by fructose 1,6-bisphosphatase (EC 3.1.3.11) as the first irreversible step. The fructose 6-phosphate (Fru-6-P) formed is converted to glucose 6-phosphate (Glu-6-P), glucose 1-phosphate (Glu-1-P) and UDP-glucose, in a series of readily reversible reactions. The transfer of a glucosyl unit from UDP-glucose to Fru-6-P, in the reaction catalysed by sucrose phosphate synthase (EC 2.4.1.14), results in the formation of sucrose 6^F-phosphate (Reaction 1). Hydrolysis of sucrose 6^F-phosphate by sucrose phosphatase (EC 3.1.3.24) yields sucrose (Reaction 2).

$$\text{D-Fru-6-P} + \text{UDP-glucose} \rightleftharpoons \text{Sucrose 6}^F\text{-phosphate} + \text{UDP} \qquad (1)$$

$$\text{Sucrose 6}^F\text{-phosphate} \rightleftharpoons \text{Sucrose} + \text{Pi} \qquad (2)$$

The reaction catalysed by sucrose phosphate synthase is readily reversible (Barber, 1985), but as the sucrose phosphatase step is essentially irreversible, the flux through the pathway favours the formation of sucrose.

The biosynthesis of sucrose in leaves is regulated by a complex mechanism which ensures that the rate of sucrose formation is geared to the availability of photosynthate in the cytoplasm, but prevents triose phosphates being withdrawn too rapidly from the chloroplasts and thereby depleting the photosynthetic carbon reduction cycle of intermediates. Studies of fluctuations in the concentrations of various metabolites in the cytoplasm, in response to variations in the rate of sucrose synthesis, have indicated that fructose 1,6-bisphosphatase and sucrose phosphate synthase catalyse key regulator steps in the pathway of sucrose biosynthesis, and that the activities of these enzymes are controlled in a coordinated fashion (Sicher *et al.*, 1986; Gerhardt *et al.*, 1987; Usuda *et al.*, 1987). Fructose 1,6-bisphosphatase is strongly inhibited by the regulatory metabolite, Fru-2,6-P_2, with the degree of inhibition being influenced by the concentration of the substrate and various effectors including Fru-6-P, dihydroxyacetone-P, AMP and

Pi. The concentration of Fru-2,6-P_2 in the cytoplasm is determined by the activities of two enzymes, phosphofructo-2-kinase (EC 2.7.1.105) and fructose 2,6-bisphosphatase (EC 3.1.3.46), which are themselves subject to complex controls (Stitt, 1987). Sucrose phosphate synthase is activated by glucose 6-phosphate and inhibited by Pi and, in addition, may be regulated by covalent modification of the protein. The regulatory properties of sucrose phosphate synthase are described more fully in Section III.A.3.

The rate of sucrose synthesis increases at the beginning of the photoperiod. The translocation of triose phosphates out of the chloroplast into the cytoplasm leads to a decrease in the concentration of Fru-2,6-P_2, which in turn relieves the inhibition of the cytoplasmic fructose 1,6-bisphosphatase. As a consequence, the concentration of hexose phosphates in the cytoplasm rises, causing sucrose phosphate synthase activity to increase, through greater availability of both the substrate (Fru-6-P) and activator (Glu-6-P). As the capacity for sucrose utilisation or export is reached, the rate of sucrose synthesis declines and an increasing amount of photosynthate is retained in the chloroplast for conversion to starch (Gerhardt et al., 1987). The activity of sucrose phosphate synthase in leaves is known to decrease, and the concentration of Fru-2,6-P_2 to increase, during an extended photoperiod (Stitt et al., 1987), but the nature of the regulatory signals which causes these changes has not been identified.

B. Sucrose Breakdown

The first step in the breakdown of sucrose in plant tissues is cleavage of the glycosidic bond by either invertase (EC 3.2.1.26) (Reaction 3) or sucrose synthase (EC 2.4.1.13) (Reaction 4).

$$\text{Sucrose} \rightleftharpoons \text{D-glucose} + \text{D-fructose} \qquad (3)$$

$$\text{Sucrose} + \text{UDP} \rightleftharpoons \text{UDP-glucose} + \text{D-fructose} \qquad (4)$$

Plant tissues contain two types of invertases. One form, an acid invertase, has optimum activity near pH 5 and is present in vacuoles, the free space outside cells, and may also be associated with the cell wall. Acid invertase activity is high in tissues that are undergoing rapid growth and development. Plant tissues also contain an alkaline or neutral invertase, which is maximally active at about pH 7 to 7.5, and is located in the cytoplasm. The reaction catalysed by invertases is irreversible. Sucrose synthase is a cytoplasmic enzyme that catalyses a readily reversible reaction. However, as discussed elsewhere (Avigad, 1982; Hawker, 1985), this enzyme is generally considered to act only in the breakdown of sucrose *in vivo*. The glucose and fructose produced in these reactions are metabolised further following phosphorylation to form the corresponding hexose 6-phosphate. Plant tissues contain a number of hexose-phosphorylating enzymes which have specificity towards either glucose or fructose (Copeland and Turner, 1987). UDP-glucose, formed when sucrose is cleaved by sucrose synthase, may act as a precursor for other nucleoside diphosphate sugars and certain polysaccharides (Feingold and Avigad, 1980), or it may be metabolised in the glycolytic or pentose phosphate pathways after conversion to glucose 1-phosphate in the reaction catalysed by UDP-glucose pyrophosphorylase (EC 2.7.7.9).

In contrast to the biosynthesis of sucrose, there does not appear to be a high degree of fine control acting on the enzymes of sucrose cleavage. Their activity may be regulated mainly by the availability of substrate.

C. Synthesis of Higher Oligosaccharides and Fructans

Sucrose is a precursor of oligosaccharides such as raffinose and stachyose, which occur widely in plants. The biosynthesis of raffinose involves the transfer of a galactosyl unit from galactinol to the glucosyl moiety of sucrose. The reaction is catalysed by galactinol : sucrose galactosyl transferase (EC 2.4.1.82), and results in the formation of an α(1–6) link. Addition of further galactosyl units gives rise to stachyose and higher homologues. Sucrose also acts as the fructosyl donor for the synthesis of the trisaccharides 1-kestose (1^F-β-fructosylsucrose), 6-kestose (6^F-β-fructosylsucrose) and neokestose (6^G-β-fructosylsucrose). The addition of fructosyl units to these trisaccharides leads to the formation of fructans, which occur as major reserve polysaccharides in the vegetative tissues of a number of agriculturally important grasses. For information on enzymes involved in the synthesis of these metabolites, the reader is referred elsewhere (Kandler and Hopf, 1982; Pontis and Del Campillo, 1985; Pollock, 1986; Nelson and Spollen 1987).

III. ENZYMES OF SUCROSE METABOLISM

Although enzymes involved in the synthesis and breakdown of sucrose have been detected in a very wide range of plant tissues (e.g. see Avigad, 1982), detailed information on their physicochemical properties is relatively limited. This section includes procedures that have been used for the purification and assay of enzymes of sucrose metabolism in plants, as well as a brief discussion of their properties. It is by no means comprehensive and is intended to serve as a guide, rather than a source of detailed information on methods. As with all enzyme studies, it is advisable to optimise experimental conditions with respect to pH, concentrations of substrates, cofactors, etc., for the tissue under investigation. Discussion of the principles of methods used to measure enzyme activity, including the use of stopped and continuous (i.e. coupled) assays, is beyond the scope of this chapter. Information on the relevant techniques is given in most reference books on enzymology (e.g. Bergermeyer, 1978).

A. Sucrose Phosphate Synthase

1. Purification

Sucrose phosphate synthase has been partially purified from spinach leaves using chromatography on AH-Sepharose 4B (Harbron *et al.*, 1981), chromatography on an amino–hexyl–agarose column followed by gel filtration through Ultragel AcA 34 (Doehlert and Huber, 1983, 1985) or by fast protein liquid chromatography methods (Walker and Huber, 1989). The resulting preparations were free of interfering enzymes, including phosphoglucose isomerase (EC 5.3.1.9), phosphoglucomutase (5.4.2.2), invertase, sucrose synthase and non-specific phosphatases. Other partially purified preparations of sucrose phosphate synthase have been obtained from wheat germ (Mendicino, 1960; Salerno and Pontis, 1978b) and germinating rice seeds (Nomura and Akazawa, 1974). The lability of sucrose phosphate synthase, particularly in crude plant homogenates, has presented difficulties for the purification of the enzyme. Glycerol, ethylene glycol and high concentrations of reductants, such as dithiothreitol, dithioerythritol or 2-mercaptoethanol, have been included in extraction media to increase the stability of the enzyme (Salerno and Pontis, 1978b; Amir and Preiss, 1982; Doehlert and Huber, 1983).

2. Assay methods

Sucrose phosphate synthase activity can be assayed by monitoring the production of UDP. Reaction mixtures typically contain Hepes-NaOH buffer (pH 7–7.5), and concentrations of Fru-6-P, UDP-glucose and $MgCl_2$ between 5 and 10 mM. Reactions are stopped after an appropriate incubation time by heating in a boiling water bath for 1–2 min and the decrease in absorbance at 340 nm, due to the oxidation of NADH, measured after adding P-enolpyruvate, pyruvate kinase (EC 2.7.1.40) and lactate dehydrogenase (EC 1.1.1.27) (Preiss and Greenberg, 1969; Nomura and Akazawa, 1974; Stitt *et al.*, 1988). The production of UDP has been coupled to the oxidation of NADH in a continuous assay (Harbron *et al.*, 1980). However, the continuous assay is not suitable for use with crude enzyme extracts or preparations that contain appreciable amounts of non-specific phosphatase activity. UDP has also been determined using 2,4-dinitrophenylhydrazine, to trap pyruvate formed in the reaction catalysed by pyruvate kinase (Cabib and Leloir, 1958; Salerno and Pontis, 1978a; Huber, 1981). An assay method for sucrose phosphate synthase based on the measurement of UDP using HPLC has been described (Khayat, 1987).

Sucrose phosphate synthase activity has been assayed by measuring the Fru-6-P-dependent formation of sucrose 6^F-phosphate (+ sucrose) from UDP-glucose (Cardini *et al.*, 1955; Murata, 1972b; Huber *et al.*, 1987). Reactions are stopped, and unreacted Fru-6-P (as well as fructose) destroyed, by adding NaOH to a final concentration of c. 0.5 N and heating in a boiling water bath for 10 min. The sucrose 6^F-phosphate (+ sucrose) is determined colorimetrically by the resorcinol (Roe, 1934) or thiobarbituric acid methods (Percheron, 1962). It should be noted that the basis of these colorimetric methods is a reaction between the chromogenic reagents and fructose, which is formed from sucrose 6^F-phosphate (+ sucrose). Hence it is important that any residual substrate is destroyed after the reactions have been terminated. Sucrose phosphate synthase activity measured in this way is likely to be underestimated if there is significant invertase activity in the enzyme preparation, and overestimated in the presence of sucrose synthase and non-specific phosphatases. Colorimetric determination of sucrose 6^F-phosphate was found to give lower values for sucrose phosphate synthase activity than assays in which the release of UDP was monitored (Stitt *et al.*, 1988). Sucrose phosphate synthase activity has also been assayed by measuring the incorporation of radioactive label into sucrose from [^{14}C]Fru-6-P (Hawker, 1967) or UDP-[^{14}C]glucose (Salerno *et al.*, 1979; Amir and Preiss, 1982). The products of the reaction are treated with alkaline phosphatase and the [^{14}C]sucrose separated from the substrates by paper chromatography or ion-exchange methods.

3. Properties

Sucrose phosphate synthase has optimum activity between pH 6.5 and 7.5. The enzyme is specific for both UDP-glucose and Fru-6-P and, when free of phosphoglucose isomerase activity, displays hyperbolic kinetics with respect to both substrates (Harbron *et al.*, 1981; Doehlert and Huber, 1983). Sigmoidal kinetics have been reported in some studies when Fru-6-P was the varied substrate. However, these effects are likely to be due to the presence of phosphoglucose isomerase, which would produce increasing concentrations of the activator as the concentration of the substrate was increased. Most estimates of the K_m values range between 0.7 and 5.5 mM for Fru-6-P, and 1.3 and 7.5 mM for UDP-glucose. The partially purified enzyme from spinach leaves does not have a requirement for divalent cations, such as Mg^{2+} and Mn^{2+}, and is not inhibited by EDTA at concentrations up to 20 mM (Harbron *et al.*, 1981; Amir and Preiss, 1982; Doehlert and Huber, 1983). UDP is a strong competitive inhibitor with respect to UDP-glucose, with K_i values ranging between 0.7 and 3.6 mM (Murata, 1972b; Nomura and Akazawa, 1974; Harbron *et al.*, 1981). The spinach leaf enzyme is inhibited by Fru-1,6-P_2 and sucrose 6^F-phosphate (Harbron *et al.*, 1981; Amir and Preiss, 1982), but not by sucrose, as has been reported for the enzyme from wheat germ (Salerno and Pontis, 1978a).

The activator, Glu-6-P, increases the maximum velocity of sucrose phosphate synthase and its affinity for Fru-6-P. In the absence of Glu-6-P, the spinach leaf enzyme has a K_m of 3.2 mM for Fru-6-P, and this is lowered to 0.7 mM in the presence of 5 mM Glu-6-P (Doehlert and Huber, 1983). The concentration of Glu-6-P required to activate the enzyme, and the degree of the activation observed, decrease with increasing concentrations of Fru-6-P. The activation of sucrose phosphate synthase by Glu-6-P is countered by Pi. The spinach leaf enzyme is activated 50% by 0.85 mM Glu-6-P in the absence of Pi, whereas with 20 mM Pi, 10 mM Glu-6-P is required to produce a similar degree of activation (Doehlert and Huber, 1983). Pi has been reported to act as a competitive inhibitor with respect to both substrates of sucrose phosphate synthase from spinach leaves (Harbron *et al.*, 1981; Amir and Preiss, 1982). However, more recent evidence suggests that Glu-6-P and Pi compete for a regulatory site on the enzyme, and that this could account for most of the inhibitory effect of Pi (Doehlert and Huber, 1985). The regulatory site where Glu-6-P and Pi interact appears to contain a sulphydryl group, which is inactivated in the absence of reductant, causing a loss of regulatory properties, but not of catalytic activity.

Sucrose phosphate synthase has been shown to undergo diurnal fluctuations in numerous plants, in that considerably more enzyme activity can be extracted from illuminated leaves than from those kept in the dark. The enzyme may occur in leaves as a mixture of two forms which are interconverted by covalent modification of the protein. Differences in the properties of the respective forms of sucrose phosphate synthase isolated from illuminated and darkened leaves are retained after gel filtration or partial purification. The enzyme obtained from illuminated spinach, barley and maize leaves has higher affinity for the substrates and lower sensitivity to inhibition by Pi than the enzyme from darkened leaves (Sicher and Kremer, 1984, 1985; Stitt *et al.*, 1988). In contrast, the main difference between the two forms of enzyme in soybean leaves may be the maximum velocity (Kerr and Huber, 1987). The interconversion of the two forms of sucrose phosphate synthase, and presumably the diurnal fluctuations in its activity in leaves, appears to be regulated by changes in the levels of certain metabolites in the cytoplasm, rather than by light directly (Huber *et al.*, 1987; Stitt *et al.*, 1988).

Clearly, sucrose phosphate synthase has complex regulatory properties. In addition to fine control through the interaction of Glu-6-P and Pi, the enzyme appears to be regulated by protein modification reactions. However, much more needs to be known of the properties of the enzyme, and in particular the nature of the covalent modifications it undergoes, to clarify the mechanism of control. There is little information on the structure of sucrose phosphate synthase. Several estimates have been made of the molecular weight, and these range between 280 and 480 kDa.

B. Sucrose Phosphatase

1. Purification

Sucrose phosphatase activity has been detected in a wide range of plant tissues (Hawker and Smith, 1984), but very few studies of the enzyme have been carried out. The enzyme has been purified to apparent homogeneity from pea shoots by a combination of $(NH_4)_2SO_4$ fractionation, DEAE-cellulose chromatography and fast protein liquid chromatography on a Mono Q anion exchange column (Whitaker, 1984). The purified enzyme was most stable when stored at $-20°C$ in buffer which contained 50% glycerol and 1 mM dithiothreitol. Partially purified preparations of sucrose phosphatase have also been obtained from sugar cane stems and carrot roots (Hawker and Hatch, 1966).

2. Assay methods

Sucrose phosphatase activity has been assayed by measuring the formation of [^{14}C] sucrose from [^{14}C]sucrose 6^F-phosphate, in reaction mixtures which contain Tris-maleate buffer (pH 6.8), 5–10 mM $MgCl_2$ and the substrate (Hawker and Hatch, 1966, 1975). Reactions are terminated by heating in a boiling water bath for 5–10 min and the [^{14}C]sucrose separated from the substrate by paper chromatography. An alternative method, in which the release of Pi from sucrose 6^F-phosphate is monitored, is described by Whitaker (1984). Hydrolysis of sucrose 6^F-phosphate by non-specific phosphatases is minimised by using low concentrations of substrate and enzyme. Control reaction mixtures which contain EDTA should be included, to correct for non-specific phosphatases. EDTA inhibits sucrose phosphatase but has little effect on non-specific phosphatases.

Although enzymic (Hawker and Hatch, 1975) and chemical (Buchanan et al., 1972) methods have been described for the synthesis of sucrose 6^F-phosphate, the difficulty in obtaining the substrate remains a major obstacle to studies of the enzyme.

3. Properties

Sucrose phosphatase has optimum activity between pH 6.4 and 6.7. The purified enzyme from pea shoots is specific for sucrose 6^F-phosphate, and has no activity with other phosphate esters and very little with p-nitrophenyl phosphate. Pea shoot sucrose phosphatase has a K_m of 0.25 mM for sucrose 6^F-phosphate (Whitaker 1984) and a K_m value of approximate 0.15 mM has been determined for the enzymes of sugar cane and carrot roots (Hawker and Hatch, 1966). Divalent cations such as Mg^{2+} or Mn^{2+} are essential for activity, and EDTA, Pi, pyrophosphate and fluoride are inhibitory (Hawker and Hatch, 1966; Hawker and Smith, 1984). Sucrose, at a concentration of 0.1 M, caused 30–60% inhibition of sucrose phosphatase activity in crude extracts of a number of plants (Hawker and Smith, 1984), but had little effect on the purified enzyme from pea shoots (Whitaker, 1984).

The native molecular weight of sucrose phosphatase of pea shoots is close to 120 kDa. The enzyme appears to be a dimer made up of subunits with molecular weight of 55 kDa (Whitaker, 1984).

C. Sucrose Synthase

1. Purification

Sucrose synthase has been purified to apparent homogeneity from several sources including mung beans (Delmer, 1972), rice grains (Nomura and Akazawa, 1973), wheat germ (Salerno and Pontis, 1978b) and maize endosperm. Purification procedures used for the enzyme from maize include $(NH_4)_2SO_4$ fractionation, gel filtration and affinity chromatography on a UDP–hexanol–amino–agarose column (Su and Preiss, 1978). Other methods have involved affinity elution of the enzyme from a carboxymethyl cellulose column (Echt and Chourey, 1985) and ion-exchange chromatography on a Mono Q column (Doehlert, 1987). Highly purified preparations of sucrose synthase have also been obtained from bamboo shoots (Su et al., 1977), wheat leaves (Larsen et al., 1985) and the plant cytosol of soybean nodules (Morell and Copeland, 1985).

2. Assay methods

Numerous procedures have been used to assay sucrose synthase activity in the sucrose cleavage and synthesis directions. Only the more commonly used methods will be included here. In the cleavage direction, activity has been assayed by monitoring the formation of UDP-glucose, in reaction mixtures which typically contain Hepes-NaOH buffer (pH 6–7), 0.1–0.2 M sucrose and 1–2 mM UDP. Reactions are terminated by heating in a boiling water bath for 2 min and the increase in absorbance at 340 nm, due to the reduction of NAD^+, measured after the addition of UDP-glucose dehydrogenase (Su et al., 1977; Gross and Pharr, 1982). Two moles of NADH are formed per mole of UDP-glucose oxidised. The formation of UDP-glucose has been coupled to the reduction of NAD^+ in the presence of excess UDP-glucose dehydrogenase (EC 1.1.1.22) in a continuous assay (Avigad, 1964; Morell and Copeland, 1985). However, as UDP-glucose dehydrogenase is optimally active between pH 8.5 and 9, this method is not suitable below about pH 7.5. Experience in the author's laboratory indicates that it is advisable to add the coupling enzyme directly to each reaction mixture, rather than to a stock solution containing a number of the reagents. An analytical method for UDP-glucose using HPLC has been described (ap Rees et al., 1984).

Sucrose synthase activity in the cleavage direction can also be assayed by following the UDP-dependent release of fructose from sucrose using either the Nelson–Somogyi method for determining reducing sugar (Nelson, 1944), or by measuring the increase in absorbance at 340 nm due to the reduction of $NADP^+$, after the addition of hexokinase (EC 2.7.1.1), phosphoglucose isomerase and Glu-6-P dehydrogenase (EC 1.1.1.49) (Nomura and Akazawa, 1973; Morell and Copeland, 1985). Control reactions which do not contain UDP should be included to correct for invertase activity. The formation of UDP-[^{14}C]glucose from [^{14}C]sucrose has been used to assay cleavage activity of sucrose synthase (Delmer, 1972; Su and Preiss, 1978; Salminen and Streeter, 1987). After reactions are stopped, the UDP-[^{14}C]glucose is separated from unreacted [^{14}C]sucrose by adsorption onto DEAE-cellulose paper.

Activity in the direction of sucrose synthesis is usually assayed at pH 7–8 in reaction mixtures which contain Hepes-NaOH buffer, 10–15 mM fructose, 2–5 mM UDP-glucose and 5 mM $MgCl_2$. The sucrose produced can be measured colorimetrically as described for sucrose phosphate synthase. Other assay methods include monitoring the formation of UDP by measuring the decrease in absorbance at 340 nm due to the oxidation of NADH after the addition of P-enolpyruvate, pyruvate kinase and lactate dehydrogenase (Morell and Copeland, 1985). This method has been used as a continuous assay with partially purified enzyme preparations (Avigad, 1964; Morell and Copeland, 1985; Doehlert, 1987). The formation of [^{14}C]sucrose from [^{14}C]fructose or UDP-[^{14}C]glucose has been used to assay the activity of sucrose synthase in the synthesis direction. The [^{14}C]sucrose formed may be separated from unreacted substrate by paper or ion-exchange chromatography (Avigad, 1964; Salerno et al., 1979; Gross and Pharr, 1982).

3. Properties

Sucrose synthase has optimum activity in the cleavage direction between pH 6 and 7, and between pH 7.5 and 9 in the direction of sucrose synthesis. Hyperbolic kinetics have mostly been observed for the enzyme, in both the cleavage and synthesis directions, although sigmoidal kinetics have been reported in some studies (Murata, 1971; 1972a; Su and Preiss, 1978). Depending on the plant source, the K_m values are usually in the ranges 15–60 mM and 0.1–2 mM for sucrose and UDP, respectively. ADP can act as the glucosyl acceptor in the cleavage reaction, but both the affinity and maximum velocity are lower than with UDP. CDP and GDP are even less effective as substrates than ADP. In the synthesis direction, UDP-glucose is the preferred glucosyl donor. The K_m values are generally in the ranges 0.1–2 mM and 2–8 mM, for UDP-glucose and fructose, respectively. Divalent metal ions such as Mg^{2+} and Mn^{2+} stimulate activity in the synthesis direction but have little or no effect on the sucrose cleavage reaction (Tsai, 1974; Su and Preiss, 1978; Morell and Copeland, 1985).

Inhibition of sucrose cleavage by UDP-glucose and fructose occurs only at relatively high concentrations and is unlikely to be physiologically important. UDP is a strong inhibitor of sucrose synthesis (Gross and Pharr, 1982; Morell and Copeland, 1985). Sucrose synthase is very sensitive to heavy metals and cyanide. Activity in both the cleavage and synthesis directions is inhibited 50% by approximately 1 μM $HgCl_2$ (Su et al., 1977; Morell and Copeland, 1985). Tris buffer, at a concentration of 20 mM, inhibited the enzyme from soybean nodules by 25% in the cleavage direction, but had little effect on the synthesis reaction (Morell and Copeland 1985).

Sucrose synthase appears to be one of the major soluble proteins in plant tissues. It is present in relatively large amounts, compared to other proteins, in extracts of mung beans (Delmer, 1972), maize kernels (Su and Preiss, 1978), bamboo shoots (Su et al., 1977) and soybean nodules (Morell and Copeland, 1985). Multiple forms of sucrose synthase have been reported in rice grains (Nomura and Akazawa, 1973) and cucumber fruits (Gross and Pharr, 1982) and maize endosperm has been shown to contain two forms of the enzyme which are encoded in separate genes (Echt and Chourey, 1985). Sucrose synthase mRNA has been isolated from maize endosperm and translated using an *in vitro* translation system (Wöstemeyer *et al.*, 1981). Estimates of the molecular weight of sucrose synthase range between 280 and 1000 kDa, with most values being near 400 kDa. The enzyme appears to be a tetramer, with subunits of approximately 90 kDa.

D. Invertase

1. Purification

Acid invertase has been highly purified from wheat coleoptiles by a combination of $(NH_4)_2SO_4$ fractionation, DEAE-cellulose chromatography, affinity chromatography with Concanavalin A-Sepharose and gel filtration (Krishnan *et al.*, 1985). Highly purified preparations of acid invertase have also been obtained from grape berries (Arnold, 1965), tomato fruits (Nakagawa *et al.*, 1971), radish seedlings (Faye *et al.*, 1982), *Lilium* pollen (Singh and Knox, 1984) and carnation petals (Woodson and Wang, 1987).

There have been relatively few studies of the alkaline invertase from plant sources. The enzyme has been partially purified from the plant cytosol of soybean nodules (Morell and Copeland, 1984) and from sugar beet roots (Masuda *et al.*, 1987) using $(NH_4)_2SO_4$ fractionation, gel filtration and DEAE-cellulose chromatography.

2. Assay methods

Acid invertase activity is conveniently assayed by measuring the formation of reducing sugar from sucrose using the Nelson–Somogyi method. Reaction mixtures usually contain acetate or citrate buffer at pH 5 and 0.1 M sucrose. Reactions are stopped with the addition of the alkaline copper reagent, or by neutralising and heating in a boiling water bath for 2 min. Alternatively, the production of glucose can be monitored using the glucose oxidase method of Trinder (1969) as modified by Blakeney and Matheson (1984). Alkaline invertase is assayed at pH 7–7.5 by a similar procedure, except that phosphate or another suitable buffer is used. Tris(hydroxymethyl)aminomethane (Tris) buffer should be avoided in assays of alkaline invertase (see Section III.D.3).

3. Properties

The plant invertases are β-fructosidases, in contrast to the α-glucosidases that hydrolyse sucrose in the digestive tract of animals. Both acid and alkaline invertases display typical hyperbolic kinetics. The K_m values for sucrose fall mostly in the ranges 2.5–13 mM for acid invertase, and 10–25 mM for alkaline invertase. Raffinose and stachyose are hydrolysed by acid invertase. However, the affinity of the enzyme for these oligosaccharides is much lower than for sucrose and the activity is generally less than 50% of that with sucrose (Arnold, 1965; Krishnan *et al.*, 1985). Raffinose and stachyose are poor substrates of alkaline invertase, with the activity being less than 10% of that with sucrose (Matsushita and Uritani, 1974; Morell and Copeland, 1984).

Tris buffer is a strong inhibitor of alkaline invertase (Hatch *et al.*, 1963; Kato and Kubato, 1978; Morell and Copeland, 1984). At pH 7.5, the enzyme from the plant cytosol of soybean nodules is inhibited 50% by 0.7 mM Tris. Inhibition of acid invertases by Tris at pH 5 is much less marked. Acid and alkaline invertases are very sensitive to inhibition by heavy metals, but glucose and fructose inhibit only at relatively high concentrations. A protein which inactivates acid invertase has been reported in crude homogenates of potato tubers (Pressey, 1967) and maize endosperm (Jaynes and Nelson, 1971).

Estimates of the molecular weight vary between 50 and 400 kDa for acid invertase and 66 and 260 kDa for alkaline invertase. The large ranges for these estimates may be attributed to differences in the carbohydrate content of enzyme preparations, which may be as high as 25% (Arnold, 1965; Del Rosario and Santisopari, 1977). Multiple forms of invertase have been reported in a number of plants, but these may also be due to variations in the carbohydrate associated with the enzyme (Faye et al., 1982). Little is known of the nature of the carbohydrate associated with invertases of higher plants. One report suggests a polysaccharide structure with a high mannose content (Faye et al., 1986). The carbohydrate moiety of yeast invertase is known to be rich in mannosyl groups, some of which are phosphorylated (Smith and Ballou, 1974).

REFERENCES

Akazawa, T. and Okamoto, K. (1980). In "The Biochemistry of Plants" (J. Preiss, ed.), Vol. 3, pp. 199–220. Academic Press, New York.
Amir, J. and Preiss, J. (1982). *Plant Physiol.* **69**, 1027–1030.
ap Rees, T., Leja, M., Macdonald, F. D. and Green, J. H. (1984). *Phytochemistry* **23**, 2463–2468.
Arnold, W. N. (1965). *Biochim. Biophys. Acta* **110**, 134–147.
Avigad, G. (1964). *J. Biol. Chem.* **239**, 3613–3618.
Avigad, G. (1982). In "Encyclopedia of Plant Physiology", New Series (F. A. Loewus and W. Tanner, eds), Vol. 13A, pp. 217–347. Springer, Berlin.
Barber, G. A. (1985). *Plant Physiol.* **79**, 1127–1128.
Bergmeyer, H. U. (1978). "Principles of Enzymatic Analysis", Verlag Chemie, Weinheim.
Blakeney, A. B. and Matheson, N. K. (1984). *Stärke* **36**, 265–269.
Buchanan, J. G., Cummerson, D. A. and Turner, D. M. (1972). *Carbohydrate Res.* **21**, 283–292.
Cabib, E. and Leloir, L. F. (1958). *J. Biol. Chem.* **231**, 259–275.
Cardini, C. E., Leloir, L. F. and Chiriboga, J. (1955). *J. Biol. Chem.* **214**, 149–155.
Copeland, L. and Turner, J. F. (1987). In "The Biochemistry of Plants" (D. D. Davies, ed.), Vol. 11, pp. 107–128. Academic Press, San Diego.
Delmer, D. P. (1972). *J. Biol. Chem.* **247**, 3822–3828.
Del Rosario, E. J. and Santisopasri, V. (1977). *Phytochemistry* **16**, 443–445.
Doehlert, D. C. (1987). *Plant Sci.* **52**, 153–157.
Doehlert, D. C. and Huber, S. C. (1983). *Plant Physiol.* **73**, 989–994.
Doehlert, D. C. and Huber, S. C. (1985). *Biochim. Biophys. Acta* **830**, 267–273.
Echt, C. S. and Chourey, P. S. (1985). *Plant Physiol.* **79**, 530–536.
Faye, L., Berjonneau, C. and Ghorbel, A. (1982). *Arch. Biochem. Biophys.* **213**, 45–49.
Faye, L., Mouatassim, B. and Ghorbel, A. (1986). *Plant Physiol.* **80**, 27–33.
Feingold, D. S. and Avigad, G. (1980). In "The Biochemistry of Plants" (J. Preiss, ed.), Vol. 3, pp. 101–170. Academic Press, New York.
Gerhardt, R., Stitt, M. and Heldt, H. W. (1987). *Plant Physiol.* **83**, 399–407.
Gross, K. C. and Pharr, D. M. (1982). *Phytochemistry* **21**, 1241–1244.
Harbron, S., Woodrow, I. E., Kelly, G. J., Robinson, S. P., Latzko, E. and Walker, D. A. (1980). *Anal. Biochem.* **107**, 56–59.
Harbron, S., Foyer, C. and Walker, D. (1981). *Arch. Biochem. Biophys.* **212**, 237–246.
Hatch, M. D., Sacher, J. A. and Glasziou, K. T. (1963). *Plant Physiol.* **38**, 338–343.
Hawker, J. S. (1967). *Biochem. J.* **105**, 943–946.

Hawker, J. S. (1985). *In* "Biochemistry of Storage Carbohydrates in Green Plants" (P. M. Dey and R. A. Dixon, eds), pp. 1–51. Academic Press, London.
Hawker, J. S. and Hatch, M. D. (1966). *Biochem. J.* **99**, 102–107.
Hawker, J. S. and Hatch, M. D. (1975). *Meth. Enzymol.* **42**, 341–347.
Hawker, J. S. and Smith, G. M. (1984). *Phytochemistry* **23**, 245–249.
Huber, S. C. (1981). *Z. Pflanzenphysiol.* **102**, 443–450.
Huber, S. C., Ohsugi, R., Usuda, H. and Kalt-Torres, W. (1987). *Plant Physiol. Biochem.* **25**, 515–523.
Jaynes, T. A. and Nelson, O. E. (1971). *Plant Physiol.* **47**, 629–634.
Kandler, O. and Hopf, H. (1982). *In* "Encyclopedia of Plant Physiology" New Series (F. Loewus and W. Tanner, eds) Vol. 13A, pp. 348–383. Springer, Berlin.
Kato, T. and Kubato, S. (1978). *Physiol. Plant.* **42**, 67–72.
Kerr, P. S. and Huber, S. C. (1987). *Planta* **170**, 197–204.
Khayat, E. (1987). *Carbohydrate Res.* **162**, 329–332.
Krishnan, H. B., Blanchette, J. T. and Okita, T. W. (1985). *Plant Physiol.* **78**, 241–245.
Larsen, A. E., Salerno, G. L. and Pontis, H. G. (1985). *Physiol. Plant.* **67**, 37–42.
Masuda, H., Takahashi, T. and Sugawara, S., (1987). *Agric. Biol. Chem.* **51**, 2309–2314.
Matsushita, K. and Uritani, I. (1974). *Plant Physiol.* **54**, 60–66.
Mendicino, J. (1960). *J. Biol. Chem.* **235**, 3347–3352.
Morell, M. K. and Copeland, L. (1984). *Plant Physiol.* **74**, 1030–1034.
Morell, M. K. and Copeland, L. (1985). *Plant Physiol.* **78**, 149–154.
Murata, T. (1971). *Agric. Biol. Chem.* **35**, 1441–1448.
Murata, T. (1972a). *Agric. Biol. Chem.* **36**, 1815–1818.
Murata, T. (1972b). *Agric. Biol. Chem.* **36**, 1877–1884.
Nakagawa, H., Kawasaki, Y., Ogura, N. and Takehana, H. (1971). *Agric. Biol. Chem.* **36**, 18–26.
Nelson, N. (1944). *J. Biol. Chem.* **153**, 375–380.
Nelson, C. J. and Spollen, W. G. (1986). *Physiol. Plant.* **71**, 512–516.
Nomura, T. and Akazawa, T. (1973). *Arch. Biochem. Biophys.* **156**, 644–652.
Nomura, T. and Akazawa, T. (1974). *Plant Cell Physiol.* **15**, 477–483.
Percheron, F. (1962). *Compte Rend. Acad. Sci.* **255**, 2521–2522.
Pollock, C. J. (1986). *New Phytol.* **104**, 1–24.
Pontis, H. G. and Del Campillo, E. (1985). *In* "Biochemistry of Storage Carbohydrates in Green Plants", (P. M. Dey and R. A. Dixon, eds), pp. 205–227. Academic Press, London.
Preiss, J. and Greenberg, E. (1969). *Biochem. Biophys. Res. Commun.* **36**, 289–295.
Pressey, R. (1967). *Plant Physiol.* **42**, 1780–1786.
Roe, J. H. (1934). *J. Biol. Chem.* **107**, 15–22.
Salerno, G. L. and Pontis, H. G. (1978a). *FEBS Lett.* **86**, 263–267.
Salerno, G. L. and Pontis, H. G. (1978b). *Planta* **142**, 41–48.
Salerno, G. L., Gammundi, S. S. and Pontis, H. G. (1979). *Anal. Biochem.* **93**, 196–199.
Salminen, S. O. and Streeter, J. G. (1987). *Plant Physiol.* **83**, 535–540.
Sicher, R. C. (1986). *Physiol. Plant.* **67**, 118–121.
Sicher, R. C. and Kremer, D. F. (1984). *Plant Physiol.* **76**, 910–912.
Sicher, R. C. and Kremer, D. F. (1985). *Plant Physiol.* **79**, 695–698.
Sicher, R. C., Kremer, D. F. and Harris, W. G. (1986). *Plant Physiol.* **82**, 15–18.
Silberman, H. C. (1961). *J. Organ. Chem.* **26**, 1967–1969.
Singh, M. B. and Knox, R. B. (1984). *Plant Physiol.* **74**, 510–515.
Smith, W. L. and Ballou, C. E. (1974). *Biochemistry* **13**, 355–361.
Stitt, M. (1987). *Plant Physiol.* **84**, 201–204.
Stitt, M. and Steup, H. (1985). *In* "Encyclopedia of Plant Physiology" New Series (R. Douce and D. A. Day, eds), Vol. 18, pp. 347–390. Springer, Berlin.
Stitt, M., Gerhardt, R., Wilke, I. and Heldt, H. W. (1987). *Physiol. Plant.* **69**, 377–386.
Stitt, M., Wilke, I., Feil, R. and Heldt, H. W. (1988). *Planta* **174**, 217–230.
Su, J.-C. and Preiss, J. (1978). *Plant Physiol.* **61**, 389–393.
Su, J.-C., Wu, J.-L. and Yang, C.-L. (1977). *Plant Physiol.* **60**, 17–21.
Trinder, P. (1969). *J. Clin. Path.* **22**, 158–161.

Tsai, C.-Y. (1974). *Phytochemistry* **13**, 885–891.
Usuda, H., Kalt-Torres, W., Kerr, P. S. and Huber, S. C. (1987). *Plant Physiol.* **83**, 289–293.
Walker, J. L. and Huber, S. C. (1989). *Plant Physiol.* **89**, 518–524.
Whitaker, D. P. (1984). *Phytochemistry* **23**, 2429–2430.
Woodson, W. R. and Wang, H. (1987). *Physiol. Plant.* **71**, 224–228.
Wöstemeyer, J., Behrens, U., Merckelbach, A., Muller, M. and Starlinger, P. (1981). *Eur. J. Biochem.* **114**, 39–44.

5 Fructose 2,6-Bisphosphate

MARK STITT

Lehrstuhl für Pflanzenphysiologie, Universität Bayreuth, 8580 Bayreuth, FRG

I.	Introduction	87
II.	Extraction of fructose 2,6-bisphosphate	88
III.	Assay of fructose 2,6-bisphosphate	89
IV.	Extraction and assay of fructose 6-phosphate, 2-kinase	89
V.	Assay of fructose 2,6-bisphosphatase	90
VI.	Partial purification of potato tuber pyrophosphate : fructose 6-phosphate phosphotransferase	90
VII.	Calibration of fructose 2,6-bisphosphate standards	91
VIII.	Purification of fructose 6-phosphate, 2-kinase and fructose 2,6-bisphosphatase	91
	Acknowledgement	91
	References	91

I. INTRODUCTION

Fructose 2,6-bisphosphate (Fru-2,6-P_2) was discovered in liver as a low molecular weight activator of phosphofructokinase (PFK). Subsequent work has established its involvement in the regulation of glycolysis in a wide range of eukaryotic tissues (Van Schaftingen, 1987). Fru-2,6-P_2 is synthesised and degraded by specific enzymes, called Fru-6-P,2-kinase and Fru-2,6-bisphosphatase.

$$\text{Fru-6-P} + \text{ATP} \longrightarrow \text{Fru-2,6-}P_2 + \text{ADP}$$

$$\text{Fru-2,6-}P_2 \longrightarrow \text{Fru-6-P} + \text{Pi}$$

In liver, these activities are present on a single bifunctional protein (Van Schaftingen, 1987), but they are on separate proteins in other tissues including yeast (François et al., 1988).

Fructose 2,6-bisphosphatase is important for the control of photosynthetic sucrose synthesis in plants (Stitt et al., 1987). It also contributes to plant respiratory metabolism by activating PFP, which catalyses a reversible phosphorylation of Fru-6-P using PPi as a phosphoryl donor, but the significance of PFP is still controversial (Stitt, 1987). Fru-6-P,2-kinase and Fru-2,6-bisphosphatase are on separate proteins in plants (MacDonald et al., 1987a,b, 1989) and are regulated by metabolites. Plant Fru-6-P,2-kinase is activated by Fru-6-P and Pi, and inhibited by 3 glycerate-3-phosphate and dihydroxyacetone-phosphate (Cseke and Buchanan, 1983; Stitt et al., 1984; Larondelle et al., 1986) and PPi (MacDonald et al., 1987a). Fru-2,6-bisphosphatase is inhibited by Pi, Fru-6-P and Mg^{2+} (Stitt et al., 1984; MacDonald et al., in press).

This chapter describes the extraction and assay of $Fru-2,6-P_2$ and of the enzymes which are responsible for its synthesis and degradation in plants. Particular emphasis will be placed on the need to validate these measurements by carrying out suitable control experiments and by using internal standards to calibrate the measurements. Many published measurements are unreliable because these precautions have been omitted.

II. EXTRACTION OF FRUCTOSE 2,6-BISPHOSPHATE

Fructose 2,6-bisphosphate is extremely acid-labile, being hydrolysed in a few minutes at pH 3. This poses problems during the extraction, because most methods for deproteinisation use acidic conditions. The following approaches are recommended. In all cases, the plant material should be frozen in liquid N_2 or freeze-clamped if it is bulky, and then powdered in liquid N_2 prior to extraction.

(a) Extraction in weak alkali (0.2 g plant material plus 1 ml 0.1 N NaOH) at room temperature, or at 95°C for 5 min (Ball and ap Rees, 1989).
(b) Extraction in a $CHCl_3/CH_3OH$ mixture (Stitt and Heldt, 1985). Powdered material (0.1–0.2 g) is mixed at about −80°C with 4.2 ml extraction medium (1.2 ml $CHCl_3$, 2.4 ml CH_3OH, 0.6 ml of a buffer containing 20 mM N-(2-hydroxyethyl)piperazine-N' (Hepes), 20 mM ethyleneglycol-bis(β-aminoethyl ether)N,N,N',N' (EGTA), 50 mM NaF) and then held at 4°C for 30 min before adding 4 ml H_2O, shaking, and centrifuging (5 min, 5000 × g). The supernatant is dried down at 35°C in a rotary evaporator, resuspended in 1 ml H_2O, and frozen in liquid N_2 until assay. Denatured protein collects at the interface, and chlorophyll partitions into the lower phase.

Extraction with $CHCl_3/CH_3OH$ has the advantage that the pH can be adjusted to ensure complete extraction of other metabolites. EGTA is included to chelate Ca^{2+}, which may otherwise precipitate out with metabolites, and NaF is included to inhibit phosphatases. It may be necessary to alter the pH or composition of the extraction medium with some tissues to prevent loss of metabolites. In some unusual cases, $Fru-2,6-P_2$ partitions into the chloroform (lower) phase.

It is essential to check the recovery of added $Fru-2,6-P_2$. To do this, small amounts of

Fru-2,6-P_2 (equivalent to the amount found in the tissue) are added to the plant material in the extraction medium. Comparison with normal extracts (which do not receive any added Fru-2,6-P_2) reveals how much Fru-2,6-P_2 has been recovered. We obtain recoveries of 80–100% with spinach, maize, pea and barley leaves, with potato and carrot tubers, *Arum* spadix and pea roots. However, losses of 50% and 80% are found with almond and stinging nettle leaves. Over 90% is lost from unripe banana (Ball and ap Rees, 1989).

III. ASSAY OF FRUCTOSE 2,6-BISPHOSPHATE

Fructose 2,6-bisphosphate is assayed in a bioassay involving activation of potato tuber PFP (Van Schaftingen *et al.*, 1982b). The response of Fru-2,6-P_2 is modified by many other compounds (Van Schaftingen *et al.*, 1982b; Kombrink and Kruger, 1985). It is essential that an internal standard curve is carried out for each sample. With plant material which contains high levels of phenols or tannins, the extract can be pre-treated with activated charcoal or 0.5% (w/w) defatted bovine serum albumin can be included in the bioassay solutions. Extracts containing high concentrations of malate should be treated on anion-exchange columns (Fahrendorf *et al.*, 1987).

The recommended assay procedure is as follows. Pipette 20 µl 0.25 N HCl into each of five cuvettes, and add extract (20 µl) to four of the cuvettes; mix, and allow to stand 10 min at room temperature. A fifth aliquot of extract is held unacidified at 4°C. Meanwhile, prepare the bioassay cocktail by pipetting together 2750 µl 100 mM Tris-HCl buffer (pH 8.1) containing 10 mM $MgCl_2$, 50 µl 120 mM Fru-6-P, 50 µl 15 mM NADH, 50 µl coupling enzymes (equivalent to 1 U aldolase, 10 U triose phosphate isomerase and 17 U glycerol-3-phosphate dehydrogenase), 50 µl 30 mM $Na_2P_2O_7$, and 50–300 µl of PFP. Each cuvette then receives 570 µl of this bioassay cocktail. The four acid-treated samples receive 20 µl of 0.05 N NaOH containing 0, 0.5, 1.0 and 2.0 pmol of Fru-2,6-P_2, respectively. The fifth cuvette receives 20 µl of 0.05 N NaOH and the 20 µl aliquot of untreated sample, which has been held on ice. The reaction is followed as a decrease of E_{340}, using a spectrophotometer with an automatic cuvette changer. The Fru-2,6-P_2 in the extract is estimated by comparison with the internal standard curve. Since commercial preparations of Fru-6-P can contain Fru-2,6-P_2, it is advisable to pre-treat the stock Fru-6-P solution at pH 3 for 10 min.

IV. EXTRACTION AND ASSAY OF FRUCTOSE 6-PHOSPHATE,2-KINASE

Fructose 6-phosphate,2-kinase is assayed in a two-step procedure in which (a) Fru-6-P and ATP are converted to Fru-2,6-P_2 by Fru-6-P,2-kinase, and (b) the amount of Fru-2,6-P_2 is measured by using the PFP bioassay for Fru-2,6-P_2. To avoid artifacts due to Fru-2,6-P_2 already present in plant material, and due to PFP and PFK activity in the extract, the following procedure is recommended (Stitt *et al.*, 1985).

Plant material (0.3–0.6 g) is homogenised at 4°C in 1.5 ml Tricine-KOH buffer (pH 7.6), 4 mM $MgCl_2$, 2 mM EDTA, 2 mM benzamidine-HCl, 0.5 mM phenyl methyl sulphonyl fluoride and 10% glycerol, centrifuged (2.5 min, 18 000 × g) and desalted by passage through a small Sephadex G-25 column (Stitt et al., 1986). Extract (20 µl) is added to (final volume 100 µl) 100 mM Tris-HCl (pH 7.4), 5mM $MgCl_2$, 1 mM EDTA, 1 mM EGTA, 2 mM ATP, 5 mM Fru-6-P and 5 mM Pi. Aliquots of 20 µl are removed and mixed with 30 µl 0.25 M KOH at various times after starting the reaction. Normally, two samples are taken immediately after adding extract, and one of these receives 10 µl of a 0.1 µM Fru-2,6-P_2 solution to provide an internal standard. Two further samples are taken at increasing time intervals, selected to have produced between 0.25 and 2 pmol of Fru-2,6-P_2 per 20 µl aliquot. The alkalised samples should be stored on ice for at least 5 min before continuing the assay, and can be stored for several days at −40°C without loss of Fru-2,6-P_2. The Fru-2,6-P_2 is then assayed by preparing a PFP assay cocktail (see above), distributing 570 µl onto each cuvette, mixing, and following the decrease of extinction at 340 nm in a spectrophotometer with an automatic sample changer. The production of Fru-2,6-P_2 is estimated by reference to the internal standard.

V. ASSAY OF FRUCTOSE 2,6-BISPHOSPHATASE

Three methods are available to assay the degradation of Fru-2,6-P_2: (1) disappearance of Fru-2,6-P_2 monitored by the PFP bioassay; (2) release of [^{32}P] from [2^{32}P]Fru-2,6-P_2; and (3) formation of Fru-6-P measured in a coupled enzyme assay. Disappearance of Fru-2,6-P_2 can only be applied at low Fru-2,6-P_2 concentrations (0.1 µM), and can be complicated by non-specific loss of phosphate from the carbon-6-position. Assay by formation of Fru-6-P is only applicable at Fru-2,6-P_2 concentrations above 1 µM and requires a dual wavelength spectrophotometer.

(1) Disappearance of Fru-2,6-P_2 (Stitt et al., 1986). Aliquots (24 µl) of extract are added to a final volume of 120 µl containing 50 mM Tris-HCl (pH 7.4), 2 mM $MgCl_2$, 1 mM EDTA, 1 mM EGTA. An aliquot (20 µl) is removed and alkalised, before starting the reaction by adding 20 µl of 0.6 mM Fru-2,6-P_2 and mixing (final concentration = 0.1 µM). Duplicate aliquots of 20 µl are immediately alkalised, and further aliquots are alkalised at later times, selected to remove between 10 and 40% of the Fru-2,6-P_2. Samples are alkalised by adding 30 µl of 0.25 N KOH. The alkalised samples are then treated as in the Fru-6-P,2-kinase assay.
(2) The preparation of [2^{32}P]Fru-2,6-P_2 is described in El-Maghrabi et al. (1982) and Van Schaftingen et al. (1982a), and its application to the assay of plant enzymes is described in Larondelle et al. (1986).
(3) Fru-2,6-P_2 hydrolysis to Fru-6-P can be assayed in 100 mM Tris-HCl (pH 7.3), 0.1 U hexose phosphate isomerase, 0.2 U glucose 6-phosphate dehydrogenase, 0.5 mM NADP, 10 µM Fru-2,6-P_2 and 1 mM dithiothreitol. The change in absorbance is followed at 340 nm (Stitt et al., 1984; MacDonald et al., 1987a,b).

VI. PARTIAL PURIFICATION OF POTATO TUBER PYROPHOSPHATE: FRUCTOSE 6-PHOSPHATE PHOSPHOTRANSFERASE

The following procedure is recommended (Van Schaftingen et al., 1982b). Peeled potato tubers (330 g) are cut into pieces and homogenised in 700 μl 20 mM Hepes (pH 8.2), 20 mM Na-acetate, and 2 mM dithiothreitol, and then filtered through 4 layers of muslin and brought to 2 mM Na-pyrophosphate, 2 mM $MgCl_2$, pH 8.1. The filtrate is heated at 57°C for 5 min, cooled rapidly in a salt/water mixture and adjusted to pH 7.1, six grams PEG 2000 are added per 100 ml, stirred 20 min at 4°C, centrifuged (10 min, 4000 × g). The supernatant receives another 8 g PEG 2000 per 100 ml, is stirred for 20 min at 4°C, and recentrifuged. The pellet is taken up in 40 ml 20 mM Tris (pH 8.2), 20 mM KCl and 2 mM dithiothreitol and applied to a column (30 mm × 100 mm) of DE 52 (Whatman) pre-equilibrated with the same buffer. The column is washed with 100 ml of the buffer, before applying a gradient of 20–400 mM KCl (vol 300 ml). The enzyme elutes at about 120 mM KCl. To assay PFP during the purification, use 5 mM Fru-6-P, 20 mM Glu-6-P, 1 μM Fru-2,6-P_2, 100 mM Tris-HCl (pH 8.1), 5 mM $MgCl_2$, 0.15 mM NADH, 1 U aldolase, 10 U triose phosphate isomerase and 17 U glycerol-3-phosphate dehydrogenase, and start the reaction by adding 2.5 mM Na-pyrophosphate. The fractions containing PFP are pooled, brought to 50% glycerol, and stored at $-20°C$.

VII. CALIBRATION OF FRUCTOSE 2,6-BISPHOSPHATE STANDARDS

Stock solutions of Fru-2,6-P_2 are stored in 0.1 N NaOH at -20 or $-40°C$. Standard solutions of 1 or 10 mM can be calibrated by measuring the increase of Fru-6-P following 10 min treatment at pH 2. Fru-6-P is assayed by coupled enzymic assay in 100 mM Tris-HCl (pH 8.1), 5 mM $MgCl_2$, 1 mM NADP, 0.2 U ml^{-1} glucose 6-phosphate dehydrogenase. The reaction is started by adding 0.1 U phosphoglucose isomerase, and is followed at 340 nm on a spectrophotometer.

VIII. PURIFICATION OF FRUCTOSE 6-PHOSPHATE,2-KINASE AND FRUCTOSE 2,6-BISPHOSPHATASE

Partial purification of Fru-6-P,2-kinase has been described by Cseke and Buchanan (1983) and Larondelle et al. (1986). A more extensive purification allowing separation of Fru-6-P,2-kinase from Fru-2,6-bisphosphatase has been described by MacDonald et al. (1987a,b) and the latter enzyme has recently been purified to homogeneity (MacDonald et al., 1989). The reader is referred to these articles for further details. Considerable caution is needed to prevent partial degradation by proteases, which can alter the properties of these enzymes.

ACKNOWLEDGEMENT

The preparation of this manuscript was supported by the Deutsche Forschungsgemeinschaft (SFB 137). I am grateful to Fr. Feil for typing the manuscript.

REFERENCES

Ball, K. and ap Rees, T. (1989). *Eur. J. Biochem.* **177**, 637–641.

Cseke, C. and Buchanan, B. B. (1983). *FEBS Lett.* **155**, 139–142.
El-Maghrabi, M. T., Claus, T. H., Pilkis, J., Fox, E. and Pilkis, S. J. (1982). *J. Biol. Chem.* **257**, 7603–7607.
Fahrendorf, T., Holtum, J. A. M., Mukherjee, U. and Latzko, E. (1987). *Plant Physiol.* **84**, 182–187.
François, J., Van Schaftingen, E. and Hers, H.-G. (1988). *Eur. J. Biochem.* **171**, 599–602.
Kombrink, E. and Kruger, N. J. (1985). *Z. Pflanzenphysiol.* **114**, 443–453.
Larondelle, Y., Mertens, E., Van Schaftingen, E. and Hers, H.-G. (1986). *Eur. J. Biochem.* **161**, 351–357.
MacDonald, F. D., Cseke, C., Chou, Q. and Buchanan, B. B. (1987a). *Proc. Natl. Acad. Sci. USA* **84**, 2742–2746.
MacDonald, F. D., Chou, Q. and Buchanan, B. B. (1987b). *Plant Physiol.* **85**, 13–16.
MacDonald, F. D., Chou, Q., Buchanan, B. B. and Stitt, M. (1989). *J. Biol. Chem.* **264**, 5540–5544.
Stitt, M. (1987). *Plant Physiol.* **84**, 201–204.
Stitt, M. and Heldt, H. W. (1985). *Planta* **164**, 179–188.
Stitt, M., Cseke, C. and Buchanan, B. B. (1984). *Eur. J. Biochem.* **143**, 43–89.
Stitt, M., Cseke, C. and Buchanan, B. B. (1985). *Physiol. Veg.* **23**, 819–827.
Stitt, M., Mieskes, G., Söling, H.-D., Große, H. and Heldt, H. W. (1986). *Z. Naturforsch.* **41c**, 291–296.
Stitt, M., Huber, S. C. and Kerr, P. (1987). *In* "The Biochemistry of Plants; Photosynthesis (M. D. Hatch and N. K. Boardman, eds), Vol. 10, pp. 328–410. Academic Press, New York.
Van Schaftingen, E. (1987). *Adv. Enzymol.* **59**, 315–396.
Van Schaftingen, E., Davies, D. D. and Hers, H.-G. (1982a). *Eur. J. Biochem.* **124**, 143–149.
Van Schaftingen, E., Lederer, B., Batrons, R. and Hers, H.-G. (1982b). *Eur. J. Biochem.* **129**, 191–195.

6 Enzymes of Starch Synthesis

ALISON M. SMITH

Department of Applied Genetics, AFRC-IPSR, John Innes Institute, Colney Lane, Norwich, NR4 7UH, UK

I.	Introduction	93
II.	General checks and controls	94
	A. Measurement of flux through the pathway	94
	B. Extraction procedures and mixing experiments	94
	C. Optimisation of assay conditions	95
III.	ADP glucose pyrophosphorylase	95
	A. Extraction	95
	B. Assay	95
	C. Purification	97
IV.	Starch synthase	98
	A. Extraction	98
	B. Assay	98
	C. Purification	99
V.	Starch branching enzyme	99
	A. Extraction	99
	B. Assay	100
	C. Purification	101
	References	101

I. INTRODUCTION

In this chapter I shall discuss methods of measurement and purification of enzymes of the committed pathway of starch synthesis. I shall regard this pathway as the conversion of glucose 1-phosphate to amylopectin (ap Rees et al., 1984; Preiss 1988) as follows:

$$\text{Glucose 1-phosphate} \xrightarrow{(1)} \text{ADP glucose} \xrightarrow{(2)} \text{Amylose} \xrightarrow{(3)} \text{Amylopectin}$$

(1) ADP glucose pyrophosphorylase (EC 2.7.7.27)
(2) starch synthase (ADP glucose: 1,4-α-D-glucan-4-glycosyl transferase, EC 2.4.1.21)
(3) starch branching enzyme (1,4-α-D-glucan: 1,4-α-D-glucan-6-glycosyl transferase, EC 2.4.1.18)

I shall concern myself particularly with the measurement of maximum catalytic activity. Comparisons between enzyme activities, be they within the same pathway, in different organs or in the same organ under different conditions, and comparisons of enzyme activities with fluxes, are not valid unless the measured activity is the maximum catalytic activity (ap Rees, 1974). Knowledge of the general properties of these enzymes has recently been comprehensively reviewed (Preiss, 1988).

Measurement of maximum catalytic activity for enzymes of the pathway of starch synthesis is difficult because of various sorts of interference by other components of plant extracts. I shall describe these problems and suggest how they may be overcome. Extraction and assay methods are given as suggested starting points for the checks and optimisation procedures described below.

I shall discuss the purification of these enzymes only where generally useful steps have emerged. The problems of purification differ enormously from one plant organ to another, and it would be misleading to attempt to synthesise general purification strategies.

II. GENERAL CHECKS AND CONTROLS

A. Measurement of Flux Through the Pathway

Comparison of the measured activity of an enzyme with the flux through the pathway *in vivo* can reveal whether there are problems with the extraction or assay of the enzyme. A measured activity that is lower than the *in vivo* flux is unlikely to be an accurate reflection of maximum catalytic activity. A minimum estimate of the flux through the pathway of starch synthesis is provided by the rate of starch accumulation in the organ of interest. The rate of turnover of starch during its synthesis in leaves and in pea roots is extremely low (Kruger et al., 1983; Hargreaves and ap Rees, 1988). For these organs at least, starch accumulation is an accurate reflection of the total flux through the pathway of starch synthesis.

Because of the nature of the reaction it catalyses, it is not possible to make an estimate of the activity of starch branching enzyme that can be related to a flux *in vivo*. Measurement of the flux into starch *in vivo* therefore provides a useful check only for estimates of activity of ADP glucose pyrophosphorylase and starch synthase.

B. Extraction Procedures and Mixing Experiments

In order to measure the maximum catalytic activity of an enzyme, all of the activity must be extracted from the tissue. If a significant proportion of cells remain unbroken during extraction, activity will be underestimated. Tissue debris should be examined after extraction to check whether complete cell breakage has been achieved.

Inhibition or degradation of an enzyme during extraction will lead to underestimation of its maximum catalytic activity. The occurrence of such losses can be detected by mixing experiments in which the enzyme from an organ in which these problems are known to be minimal is exposed to the same hazards during extraction as the enzyme from an organ for which the magnitude of these problems is not known. Two, replicate samples of each of the two organs are prepared. One sample of each organ is extracted separately. The remaining two samples are mixed, then extracted together. Activity in the mixed extract is expressed as a percentage of that predicted from the two separate extracts. A low percentage recovery of activity in the mixed extract indicates that the 'unknown' organ contains substances that inhibit or degrade the enzyme during extraction. Extraction conditions must be altered to counter the problem, and the mixing experiment repeated to check whether the alteration has been successful. Claims that a measured activity reflects the maximum catalytic activity of an enzyme must be accompanied by proof of this sort that losses of activity are unlikely to have occurred during extraction.

C. Optimisation of Assay Conditions

In order to measure the maximum catalytic activity of an enzyme, conditions in the assay must be optimised to give the maximum rate. The rate in the assay must be dependent upon the presence of all of the appropriate substrates and cofactors, proportional to extract concentration over the range in which measurements are made, and linear with respect to time over the measurement period. The concentrations of all components and the pH of the assay must be optimised to give the maximum rate. All of these checks should be made for each organ and each treatment for which measurements are presented.

III. ADP GLUCOSE PYROPHOSPHORYLASE

A. Extraction

The activity of this enzyme in crude extracts of plant organs is frequently labile. Inclusion of ethanediol, glycerol, bovine serum albumin, phosphate and inhibitors of serine proteases (chymostatin, phenylmethylsulphonyl fluoride) in the extraction medium may help to stabilise activity (Smith *et al.*, 1989; Plaxton and Preiss, 1987). Metabolites, particularly hexose phosphates, present in crude extracts may interfere with assays, hence it is desirable to desalt extracts prior to assay.

Extraction medium. 50 mM Na phosphate (pH 7.0), 5 mM dithiothreitol, 1 mM EDTA, 1 mM phenylmethylsulphonyl fluoride, 10% (v/v) ethanediol.

B. Assay

The reaction catalysed by ADP glucose pyrophosphorylase is close to equilibrium *in vivo*, and assays exploiting both forward and reverse reactions have been published.

$$\text{ATP} + \text{glucose 1-phosphate} \longrightarrow \text{ADP glucose} + \text{pyrophosphate}$$

It is necessary to check that any accumulation of products in the assay does not lead to non-linear reaction rates, and hence to underestimation of activity.

The metabolite 3-phosphoglycerate, an activator of the purified enzyme, is frequently added to assays. This should not be done if maximum catalytic activities are to be estimated. First, the maximum catalytic activity should be a measure of the maximum capacity of the organ to catalyse that reaction, without regard to the many modulations of activity that may occur *in vivo*. Second, addition of 3-phosphoglycerate to assays of ADP glucose pyrophosphorylase on crude, desalted extracts may generate artifacts. For example, its addition actually reduces the apparent activity in the NADH-linked assay (see below) on desalted extracts of pea leaves (A. M. Smith and S. Rawsthorne, unpubl. res.).

1. Forward direction

The assay measures production of ADP [^{14}C]glucose from [^{14}C]glucose 1-phosphate. At the end of the incubation the reaction is stopped by boiling, remaining [^{14}C]glucose 1-phosphate is converted to [^{14}C]glucose by addition of phosphatase, and anion-exchange paper or resin is used to separate [^{14}C]glucose from ADP [^{14}C]glucose (Ghosh and Preiss, 1966; Dickinson and Preiss, 1969). Inorganic pyrophosphatase is frequently added to remove pyrophosphate as it is produced.

A major potential problem with the assay on crude extracts is that glucose 1-phosphate is likely to be equilibrated rapidly with other hexose phosphates. In assays on desalted extracts of pea embryos, up to 75% of added [^{14}C]glucose 1-phosphate was converted to other hexose phosphates within 5 min. The total decline in hexose phosphates in this period was greater than accounted for by the appearance of ^{14}C in the ADP glucose fraction (Smith *et al.*, 1989). Considerable work would be required to establish whether this assay is really measuring the total activity of ADP glucose pyrophosphorylase.

Protocol. The assay contains 100 mM *N*-(2-hydroxyethyl)piperazine-N'-(2-ethane sulphonic acid) (Hepes) (pH 7.5), 10 mM $MgCl_2$, 1.5 mM ATP, 0.2 units inorganic pyrophosphatase, 2 mM [U-^{14}C]glucose 1-phosphate at 20 GBq mol^{-1}, and extract in a final volume of 200 µl. Two blanks contain no ATP, and boiled extract, respectively. After 10 min incubation at 25°C, the mixture is heated to 100°C for 1 min then cooled, 3 units *Escherichia coli* alkaline phosphatase are added, and the mixture is incubated at 37°C for 2 h. The mixture is then applied to a 200 µl column of Dowex 1 anion-exchange resin (Cl$^-$-form), and washed through with 2 ml H_2O. The column is eluted with 1 ml 2 N HCl, and radioactivity in the eluate is determined.

2. Reverse direction

(a) The assay measures the production of [^{32}P]ATP from [^{32}P]pyrophosphate. The incubation is stopped by acidification and [^{32}P]ATP is separated from remaining [^{32}P]pyrophosphate by absorption onto activated charcoal (Shen and Preiss, 1964; Dickinson and Preiss, 1969).

Protocol. The assay contains 100 mM Hepes (pH 7.8), 2 mM ADP glucose, 10 mM $MgCl_2$, 10 mM NaF, 2 mM [^{32}P]pyrophosphate at 50 GBq mol^{-1} and extract in a final volume of 250 µl. Two blanks contain no ADP glucose, and boiled extract, respectively. After 10 min incubation at 25°C, 3 ml 5% (w/v) aqueous trichloroacetic acid containing 10 µmol unlabelled pyrophosphate is added followed by 0.1 ml of 150 mg ml^{-1} activated charcoal. The charcoal is washed twice with 3 ml trichloroacetic acid solution and once with 3 ml H_2O and radioactivity in it is determined (Shen and Preiss, 1964).

(b) In the continuous assay, the product glucose 1-phosphate is converted via phosphoglucomutase and glucose 6-phosphate dehydrogenase to 6-phosphogluconate, with concomitant NAD(P) reduction (Sowokinos, 1976; Turner, 1969). If NADP is used in the assay, NADPH production may not be stoichiometrically related to glucose 1-phosphate production because endogenous 6-phosphogluconate dehydrogenase may convert some or all of the 6-phosphogluconate to ribulose 5-phosphate, with concomitant NADP reduction. The best solution is probably to use NAD, and NAD-linked glucose 6-phosphate dehydrogenase from *Leuconostoc mesenteroides*. Plants do not possess an NAD-linked 6-phosphogluconate dehydrogenase.

Protocol. The assay contains 100 mM Hepes (pH 7.8), 5 mM $MgCl_2$, 2 mM ADP glucose, 0.5 mM NAD, 2 units phosphoglucomutase, 5 units NAD-linked glucose 6-phosphate dehydrogenase, 2 mM sodium pyrophosphate and extract. NAD reduction is monitored spectrophotometrically at 340 nm and 25°C.

C. Purification

The stability of the enzyme during purification varies from one plant organ to another. Limited proteolytic degradation during purification may result not only in loss of activity, but also in loss of allosteric properties (Plaxton and Preiss, 1987). Use of serine protease inhibitors like phenylmethylsulphonyl fluoride and chymostatin during purification is strongly recommended. Sucrose (20%, v/v), glycerol (10%, v/v), ethanediol (20%, v/v), and phosphate (30 mM) stabilise activity during purification from some plant organs (Copeland and Preiss, 1981; Plaxton and Preiss, 1987; C. M. Hylton and A. M. Smith, unpubl. data). Removal of pigmented material at an early stage of purification with polyvinylpolypyrrolidone improves the subsequent stability and yield of the enzyme from spinach leaves (Copeland and Preiss, 1981).

The enzyme from spinach leaves and potato tubers is relatively heat stable, and this provides a useful early stage of purification. Incubation of crude extracts of spinach leaves at 60°C, and potato tubers at 70°C, for 5 min, resulted in a 2.2-fold purification in both cases (Copeland and Preiss, 1981; Sowokinos and Preiss, 1982). Hydrophobic chromatography on ethyl and butyl agarose (Plaxton and Preiss, 1987) and aminopropyl sepharose (Copeland and Preiss, 1981), and affinity chromatography on 8-(6-aminohexyl)-amino-ATP agarose (Copeland and Preiss, 1981) have been used in purification as well as more conventional differential solubility, anion-exchange and hydroxyapatite steps. The enzyme appears to be a tetramer of about 200–250 kDa native molecular weight, composed of four identical monomers (Hannah and Nelson, 1975; Copeland and Preiss, 1981; Sowokinos and Preiss, 1982; Plaxton and Preiss, 1987).

IV. STARCH SYNTHASE

A. Extraction

Starch synthase exists in plant organs both as a soluble enzyme in the stroma of the plastid and as an enzyme tightly bound to starch granules (Frydman and Cardini, 1967; Tanaka et al., 1967; MacDonald and Preiss, 1985). The starch-granule-bound form contributes 40–60% of the total activity in maize endosperm (MacDonald and Preiss, 1983) and 60–80% of the total activity in developing pea cotyledons (Smith et al., 1989). The precise roles of these two forms in starch synthesis are not known. However, both are likely to contribute significantly to flux through the pathway. For the few organs for which reliable measurements are available, the total activity of starch synthase is only just sufficient to account for the rate of starch synthesis. For example, in developing *Arum* spadix total activity is 1.3 to 3 times higher (ap Rees et al., 1984), in soybean cell suspension cultures 3 times higher (ap Rees et al., 1984) and in developing pea embryos 2 to 4 times higher (Smith et al., 1989) than the minimum estimated rate of starch synthesis.

Measurement of total starch synthase activity must take into account both soluble and starch-granule-bound forms. This is done by assaying crude, unfractionated homogenates of plant organs. Measurements of soluble starch synthase can be made on extracts from which starch grains have been removed by centrifugation. Estimates of starch-granule-bound activity can be made by subtraction of soluble activity from that in the crude, unfractionated homogenates.

It is not clear whether measurements of starch synthase activity made on unfractionated homogenates are true reflections of the maximum catalytic activity. A proportion of the starch-granule-bound activity in several plant organs cannot be assayed when the starch grains are intact. Breakage of starch grains increases assayable activity. For example, activity assayable after breakage of purified starch grains from maize endosperm was three times greater than that assayable in intact grains (MacDonald and Preiss, 1983). It is not known whether the activity rendered assayable by breakage of starch grains contributes to starch synthesis *in vivo*.

Extraction medium: 100 mM Hepes (pH 7.5), 5 mM dithiothreitol, 1 mM EDTA.

B. Assay

Both soluble and starch-granule-bound starch synthases exist in multiple forms that differ in kinetic properties. In general, one class of forms shows high activity in the absence of added glucan primer when high concentrations of citrate are present in the assay, while the other class shows little or no activity in the absence of primer (Boyer and Preiss, 1979; Pollock and Preiss, 1980; MacDonald and Preiss, 1985). To ensure that both forms are assayed optimally, assays for total starch synthase usually contain citrate and rate-saturating concentrations of glucan primer.

The assay measures the incorporation of ^{14}C from ADP [^{14}C]glucose into methanol-insoluble polymer. At the end of the incubation, the reaction is stopped by boiling and the polymer is precipitated with methanol/KCl. The precipitate is washed to remove remaining ADP [^{14}C]glucose.

It may be advisable to check that the insoluble polymer formed in this assay is starch, particularly if the sugar-nucleotide specificity of the reaction is being investigated. Cell-free extracts of plants can incorporate glucose from UDP glucose into methanol-insoluble polymers that do not release glucose upon digestion with α-amylase and α-amyloglucosidase, and are therefore unlikely to be starch (e.g. Tyson and ap Rees, 1988).

Protocol. The assay contains 100 mM Bicine, 25 mM K acetate, 0.5 M Na citrate (pH 8.0), 8 mg ml^{-1} amylopectin, 2 mM ADP [U-^{14}C]glucose at 10 GBq mol^{-1} and unfractionated homogenate. Blanks contain boiled extract. After incubation at 25°C for 15 min the mixture is heated to 100°C for 1 min and 3 ml 1% (w/v) KCl in 75% (v/v) aqueous methanol is added. After 5 min the precipitate is collected by centrifugation, and the supernatant is discarded. The precipitate is dissolved in 0.3 ml water, reprecipitated with methanol/KCl and again collected by centrifugation. The precipitate is again dissolved in water, and radioactivity in it is determined.

C. Purification

1. Starch-granule-bound starch synthase

This enzyme is purified from starch grains which have been washed repeatedly to remove soluble proteins. It is often the major starch-granule-bound protein. Solubilisation of the enzyme in an active form is achieved by mechanical breakage of the dry, acetone-washed starch grains followed by digestion with α-amylase and α-amyloglucosidase. Forms of starch synthase can be separated from other starch granule proteins, and the added α-amylase and α-amyloglucosidase, by anion-exchange chromatography. This procedure has been described in detail for the enzyme from maize endosperm (MacDonald and Preiss, 1983, 1985).

2. Soluble starch synthase

Different forms of soluble starch synthase may be separated from each other by anion-exchange chromatography. Subsequent purification steps used for the separated forms have included hydrophobic chromatography (e.g. aminobutyl and aminopropyl Sepharose: Pollock and Preiss, 1980) and affinity chromatography on ADP-hexanolamine Sepharose (Hawker *et al.*, 1974; Pollock and Preiss, 1980). ADP-hexanolamine Sepharose has proved particularly useful in freeing starch synthase preparations of activity of starch branching enzyme (e.g. Pollock and Preiss, 1980). This is essential if the kinetic properties are to be studied.

V. STARCH BRANCHING ENZYME

A. Extraction

Estimates of starch branching enzyme activity are made on extracts from which starch has been removed by centrifugation, both because the enzyme is generally regarded as soluble rather than starch-granule-bound, and because starch interferes with the assay. However, activity of starch branching enzyme has been detected in protein solubilised from washed starch grains from maize endosperm (MacDonald and Preiss, 1985) and from developing pea embryos (A. M. Smith, unpubl. data), and it is possible that a significant fraction of this enzyme may be bound to starch grains. Estimates of activity made on crude extracts of plant organs may therefore not be true reflections of the total activity in the organ.

Extraction medium. As for starch synthase.

B. Assay

Starch branching enzyme catalyses the transfer of a short length of α-1,4-linked glucose residues from the non-reducing end of a glucan chain to form an α-1,6-linked branch on the side of the same or another chain. Little is known of the mechanism of action *in vivo*, and there is no direct means of assaying this activity *in vitro*. Two indirect assays have been used. In the first, extract is incubated with amylose and activity is measured as the decrease in iodine-binding of the glucan as it becomes branched (Hawker *et al.*, 1974; Borovsky *et al.*, 1975). In the second, activity is measured as the stimulation by extract of glucan synthesis from glucose 1-phosphate by phosphorylase a (Hawker *et al.*, 1974). The stimulation is caused by branching of the glucan, which provides more non-reducing chain ends on which phosphorylase can act. Activity is monitored as the incorporation of ^{14}C from glucose 1-phosphate into methanol-insoluble polymer.

Starch branching enzyme exists in several different forms, separable by ion-exchange chromatography, in many plant organs (e.g. maize endosperm: Boyer and Preiss, 1978; spinach leaves: Hawker *et al.*, 1974; sorghum seeds: Boyer, 1985; developing pea embryos: Smith, 1988). Different isoforms frequently display different relative activities in the iodine-binding and phosphorylase-stimulation assays (e.g. Smith, 1988). These assays do not in any case reproduce the conditions under which the enzyme acts *in vivo*. It seems likely that the measured activity of total starch branching enzyme in a crude extract may not accurately reflect the total capacity of the organ to catalyse starch branching *in vivo*. For this reason, comparisons of starch branching enzyme activity between, for example, different plant organs or the same organ in different species, should be treated with caution.

Both assays are liable to interference by components of crude extracts, particularly starch and amylases. Very small activities of either α or β-amylase in the iodine-binding assay will lead to overestimation of starch branching enzyme activity. This assay is not generally suitable for crude extracts of plants.

The phosphorylase-stimulation assay is less sensitive to interference by amylases. However, in extracts in which amylase activities are high this assay may also be unusable, rendering estimation of starch branching enzyme activity impossible. Starch interferes with the phosphorylase-stimulation assay because it provides a substrate for the phosphorylase. The following checks for interference by amylases and starch are recommended. First, boiled extract should be added to the complete assay. Stimulation of the rate by boiled extract would indicate that the extract contains soluble glucan that is acting as a substrate for phosphorylase (Smith, 1988). Second, ^{14}C-labelled glucan formed in the assay should be added to complete assays that contain unlabelled instead of radioactive glucose 1-phosphate. Release of [^{14}C]glucose from the polymer would indicate that amylases in the extract are leading to underestimation of branching enzyme activity (Edwards *et al.*, 1988).

The rate of polymer formation in the phosphorylase-stimulation assay is not linear with respect to time. There is a lag at the start of the assay because starch branching enzyme will not act on glucans of less than about 40 glucose residues. It is therefore necessary to measure the amount of polymer synthesised at several points in time during the assay to establish the rate of synthesis during the linear phase (see Hawker et al., 1974). If activity in the assay is too high, the rate of polymer formation may reach a peak and then decline rapidly, so that there is no linear phase in the assay.

Protocol. The assay contains either 200 mM 2-(N-morpholino)ethanesulphonic acid (Mes) (pH 6.6) or 100 mM citrate (pH 7.0), 50 mM [U-^{14}C]glucose 1-phosphate at 300 MBq mol^{-1}, 0.3 units rabbit muscle phosphorylase a and extract in a final volume of 0.5 ml. The two blanks contain no phosphorylase, and boiled extract, respectively. At 30 min intervals during the incubation at 25°C, aliquots of 0.1 ml are removed and heated at 100°C for 1 min. Glycogen (0.1 ml, 10 mg ml^{-1} in water) is added as a carrier, then the polymer is precipitated and washed precisely as described above for starch synthase. Activity is expressed as the stimulation by extract of glucose incorporation into methanol-insoluble polymer during the linear phase of the assay, as μmol glucose incorporated min^{-1}.

C. Purification

In addition to conventional methods of purification (e.g. differential solubility, anion-exchange and hydroxyapatite chromatography and gel filtration; Borovsky et al. 1975; Smith, 1988), hydrophobic chromatography has been used to separate and purify different forms of the enzyme (e.g. aminobutyl and aminopentyl Sepharose; Boyer and Preiss, 1978). Chromatography on ADP-hexanolamine Sepharose has been used to separate forms of starch branching enzyme from starch synthases (Hawker et al., 1974). The enzyme appears to be a monomer. Estimates of molecular weight range from 40 kDa in rice endosperm (Smyth, 1988), to 80–90 kDa in potato tuber (Borovsky et al., 1975), maize endosperm (Boyer and Preiss, 1978; Singh and Preiss, 1985) and spinach leaf (Hawker et al., 1974) and 100–114 kDa in developing pea embryo (Smith, 1988).

REFERENCES

ap Rees, T. (1974) *In*: "MTP International Review of Science: Plant Biochemistry" (D. H. Northcote, ed.), Vol. 11, pp. 89–127. Butterworth, London and University Park Press, Baltimore.
ap Rees, T., Leja, M., MacDonald, F. D. and Green, J. H. (1984). *Phytochemistry* **23**, 2463–2468.
Borovsky, D., Smith, E. E. and Whelan, W. J. (1975). *Eur. J. Biochem.* **59**, 615–625.
Boyer, C. D. (1985). *Phytochemistry* **24**, 15–18.
Boyer, C. D. and Preiss, J. (1978). *Carbohydr. Res.* **61**, 321–334.
Boyer, C. D. and Preiss, J. (1979). *Plant Physiol.* **64**, 1039–1042.
Copeland, L. and Preiss, J. (1981). *Plant Physiol.* **68**, 996–1001.
Dickinson, D. B. and Preiss, J. (1969). *Arch Biochem. Biophys.* **130**, 119–128.
Edwards, J., Green, J. H. and ap Rees, T. (1988). *Phytochemistry* **27**, 1615–1620.
Frydman, R. B. and Cardini, C. E. (1967). *J. Biol. Chem.* **242**, 312–317.
Ghosh, H. P. and Preiss, J. (1966). *J. Biol. Chem.* **241**, 4491–4504.
Hannah, L. C. and Nelson, O. E. (1975). *Plant Physiol.* **55**, 297–302.
Hargreaves, J. A. and ap Rees, T. (1988). *Phytochem.* **27**, 1627–1629.

Hawker, J. S., Ozbun, J. L., Ozaki, H., Greenberg, E. and Preiss, J. (1974). *Arch. Biochem. Biophys.* **160**, 530–551.
Kruger, N. J., Bulpin, P. V. and ap Rees, T. (1983). *Planta* **157**, 271–273.
MacDonald, F. D. and Preiss, J. (1983). *Plant Physiol.* **73**, 175–178.
MacDonald, F. D. and Preiss, J. (1985). *Plant Physiol.* **78**, 849–852.
Plaxton, W. C. and Preiss, J. (1987). *Plant Physiol.* **83**, 105–112.
Pollock, C. and Preiss, J. (1980). *Arch. Biochem. Biophys.* **204**, 578–588.
Preiss, J. (1988). *In* "The Biochemistry of Plants" (P. K. Stumpf and E. E. Conn, eds), Vol. 14, pp. 181–254. Academic Press, New York.
Shen, L. and Preiss, J. (1964). *Biochem. Biophys. Res. Commun.* **17**, 424–429.
Singh, B. K. and Preiss, J. (1985). *Plant Physiol.* **79**, 34–40.
Smith, A. M. (1988). *Planta* **175**, 270–279.
Smith, A. M., Bettey, M. and Bedford, I. D. (1989). *Plant Physiol.* **89**, 1279–1284
Smyth, D. A. (1988). *Plant Sci.* **57**, 1–3.
Sowokinos, J. R. (1976). *Plant Physiol.* **57**, 63–68.
Sowokinos, J. R. and Preiss, J. (1982). *Plant Physiol.* **69**, 1459–1466.
Tanaka, Y., Minagawa, S. and Akazawa, T. (1967). *Stärke* **7**, 206–212.
Turner, J. F. (1969). *Austral. J. Biol. Sci.* **22**, 1145–1151.
Tyson, H. and ap Rees, T. (1988). *Planta* **175**, 33–38.

7 Starch Degrading Enzymes

MARTIN STEUP

*Botanisches Institut der Westf., Wilhelms-Universität Münster,
Schloßgarten 3, D-4400 Münster, FRG*

I.	Introduction	103
II.	Enzyme activity assays	105
	A. Hydrolases	105
	B. Phosphorylases	107
III.	Multiplicity of starch degrading enzymes	110
	A. Hydrolases	110
	B. Glucan phosphorylases	112
	C. D-Enzyme	116
IV.	Purification of compartment-specific glucan phosphorylase forms	117
	A. Purification of Type I glucan phosphorylase	118
	B. Purification of Type II glucan phosphorylase	120
V.	Immunological techniques	122
	A. Immunotitration	122
	B. Removal of cross-reacting idiotypes	122
	C. *In situ* localisation of glucan phosphorylase forms	125
	Acknowledgements	127
	References	127

I. INTRODUCTION

The term 'starch degrading enzymes' usually designates a relatively wide range of enzyme activities which decrease the degree of polymerisation of any oligoglucan or polyglucan predominantly containing α-1,4-linked glucose units. This designation does not, however, exclude that the net reaction catalysed by one of these enzymes may result in a polymerisation of carbohydrates rather than in degradation, nor does it imply that all enzyme activities considered here exert their physiological function in a catabolic pathway.

Depolymerisation as catalysed by these enzyme activities is achieved by the cleavage of an α-glycosidic intersugar bond and a transfer of the resulting glycosyl residue to an acceptor. Depending on properties of the carbohydrate(s) acting as a glycosyl donor, of the target linkage(s), of the residue(s) transferred, and of the glycosyl acceptor(s) preferred, several types of starch degrading enzymes can be distinguished (Table 7.1).

TABLE 7.1. Starch metabolising enzyme activities from higher plants.

Enzyme	EC number	Main reaction catalysed
α-Glucosidase	3.2.1.20	Hydrolytic cleavage of maltose and related oligosaccharides to glucose
α-Amylase	3.2.1.1	Hydrolytic cleavage of internal α-1,4 bonds
β-Amylase	3.2.1.2	Liberation of maltose from the non-reducing end of an α-D-glucan
Debranching enzyme	3.2.1.10	Hydrolysis of α-1,6 bonds in starch or starch-like α-D-glucans
α-1,4-Glucan phosphorylase	2.4.1.1	Formation of glucose 1-phosphate and a residual glucan from an amylose-like α-D-glucan and orthophosphate
Maltose phosphorylase	2.4.1.8	Conversion of maltose and orthophosphate to glucose 1-phosphate and glucose
D-Enzyme	2.4.1.25	Maltosyl transfer from maltotriose to another α-1,4-D-oligoglucan and release of glucose

A common feature of hydrolases is the use of water as glycosyl acceptor. Hydrolases can be further subclassified depending upon the intramolecular position and type (α-1,4 or α-1,6) of the target linkage(s) and upon the glycosyl residue(s) transferred.

Glucan phosphorylases use orthophosphate (or arsenate) as glycosyl acceptor; in addition, both the target bond and the glycosyl residue transferred are precisely defined. By transferring a glucosyl residue from the free non-reducing end of an amylose-like chain to orthophosphate, glucose 1-phosphate is formed and the degree of polymerisation of the polysaccharide is lowered. Removal of glucosyl residues from the non-reducing end of the polyglucan chain continues until the glucan is converted to maltotetraose. In branched glucans, such as soluble starch, amylopectin or glycogen, glucosyl removal ceases in the vicinity of branch points. Maltose phosphorylase is broadly similar to glucan phosphorylases, but its action is restricted to maltose as glucosyl donor.

In a third group of starch degrading enzymes, carbohydrates function both as glycosyl donor and acceptor. Due to an intermolecular rearrangement of glucose units, depolymerisation can take place simultaneously with an increase of the degree of polymerisation in another product of the same reaction. As an example, this type of 'disproportionation' occurs in the action of the D-enzyme (disproportionating enzyme).

In this chapter simple activity assays are described for several starch degrading enzymes. Procedures for separation and characterisation of isozymes are given. Immunological techniques for *in situ* localisation of distinct enzyme forms are described.

Finally, some techniques for oligosaccharide characterisation are included which allow a more detailed analysis of carbohydrates formed by an enzymatic reaction.

II. ENZYME ACTIVITY ASSAYS

Most of the enzymes listed in Table 7.1 occur as multiple forms which catalyse the same reaction but differ in some physical properties and, therefore, can be resolved by electrophoretic or chromatographic techniques. If a sample contains multiple enzyme forms which differ significantly in kinetic properties (such as pH dependency of activity or glucan specificity), it is difficult (if not impossible) to measure the total catalytic capacity of the respective enzyme with accuracy. In this case it is meaningful to combine an estimation of enzyme activity with an evaluation of enzyme pattern. Starch-like carbohydrates can be easily stained. Therefore, for most of the starch degrading enzymes, simple activity staining procedures have been developed. They permit the detection of starch degrading enzyme forms once they have been resolved by non-denaturing electrophoresis or isoelectric focusing.

A. Hydrolases

A variety of assays are available which have been discussed previously (Steup, 1988). In the procedure described here reducing groups are monitored which have been formed by hydrolytic cleavage of intersugar bonds. In principle, this assay allows the determination of not only the total hydrolytic activity, but also that of distinct hydrolases such as exo-, endo-amylases and debranching enzyme.

1. Total hydrolytic activity

(a) *Extract.* Plant material (approximately 5 g fresh weight) is homogenised in 30 ml ice-cold grinding medium which consists of 50 mM 2-(N-morpholino)ethane-sulphonic acid (Mes), brought to pH 6.0 with NaOH, 5 mM $CaCl_2$, and 5% (v/v) glycerol. The homogenate is filtered through several layers of Miracloth or nylon net and the filtrate is cleared by centrifugation (15 min at $40\,000 \times g$). The supernatant is either applied directly to the reaction mixture or is passed through a Sephadex G-25 gel (PD 10 column of Pharmacia-LKB). The Sephadex gel has been previously equilibrated with grinding medium. Gel filtration removes endogenous oligosaccharides, such as sucrose, which interfere with the measurement of starch hydrolysing enzyme activity after enzymic hydrolysis.

(b) *Incubation mixture.* The incubation mixture contains 3 ml 2% (w/v) soluble starch and 3 ml extract (or of the eluate of the Sephadex G-25 column). If necessary, samples are diluted with grinding medium before addition to the incubation mixture. A blank is prepared by mixing 2% (w/v) soluble starch with an equal volume of grinding medium. If the hydrolysis of endogenous carbohydrates is to be estimated, the extract is diluted with an equal volume of H_2O (instead of soluble starch). Mixtures are incubated at 30°C. At intervals (5–60 min, depending on the concentration of the activity), 1 ml of the incubation mixtures is added to an equal volume of alkaline colour reagent, mixed thoroughly, and heated for 5 min in a boiling water bath. Samples are then cooled to room temperature and stored for at least 30 min. Absorbance at 546 nm is measured against a reference (1 ml blank plus 1 ml alkaline colour reagent, treated as above).

(c) *Alkaline colour reagent.* The alkaline colour reagent is prepared by dissolving 1 g 3,5-dinitrosalicylic acid in a mixture of 40 ml 1 N NaOH and approximately 30 ml H_2O at an elevated temperature. Solid potassium sodium tartrate (3 g) is added and dissolved. After cooling to room temperature, the mixture is brought to a final volume of 100 ml. Stored in darkness at room temperature, the solution is stable for several months.

(d) *Calibration.* For calibration, varying amounts of maltose are reacted with the alkaline colour reagent. Samples (1 ml each) are prepared which contain 0–1.5 µmol maltose, 100 µl 2% (w/v) soluble starch and buffer (as in the incubation mixtures). Each of the samples is mixed with 1 ml alkaline colour reagent and processed as described above.

(e) *Limitations.* Despite its simplicity, the assay has some limitations. Reducing agents, such as dithioerythritol, and some buffers (e.g. Imidazole) affect significantly the slope of the calibration curve (Fig. 7.1). Consequently, the composition of the calibration mixtures should be as similar as possible to that of the incubation mixtures. In addition, the validity of the measurements can be ascertained by adding known amounts of maltose to the assay mixtures.

2. *Endoamylase activity*

Endoamylase activity is determined as described above (see Section II.A.1); however, soluble starch is replaced by amylopectin or glycogen which has been subjected to exhaustive hydrolysis by β-amylase (e.g. from *Ipomoea batatas*). For β-amylase pretreatment, branched polyglucans are dissolved in 20 mM sodium acetate buffer (pH 5.0), mixed with β-amylase, and are then placed in a dialysis tube. By dialysing the glucan-β-amylase mixture against a large volume of the acetate buffer, maltose is removed which may act as an inhibitor of β-amylase. After cessation of the reaction (which can be monitored by the determination of the content of reducing groups), β-amylase is inactivated by heating (5 min in a boiling water bath) and is removed by centrifugation (5 min at 5000 × g). β-Limited glucans are precipitated with methanol (final concentration 50% [v/v]) and are collected by centrifugation (as above).

3. *Debranching enzyme activity*

Higher plant debranching enzymes catalyse the immediate hydrolytic cleavage of the α-1,6-linkage, which results in the liberation of the entire A-chain (direct debranching; for details see Steup, 1988). In kinetic properties higher plant debranching enzymes are often similar to the microbial pullulanase and, therefore, they are usually assayed using pullulan as substrate. Pullulan is a linear polyglucan which is essentially composed of 1,6-linked α-maltotriose residues. Because this polyglucan is relatively resistant to other α-glucan degrading enzyme activities, a selective determination of the debranching

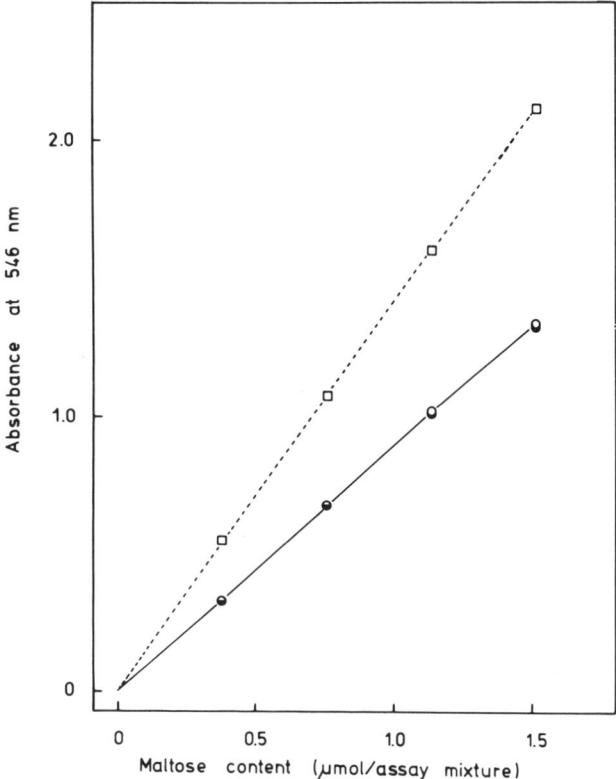

FIG. 7.1. Determination of reducing groups of maltose by the dinitrosalicylic acid technique as affected by buffer substances. Samples (1 ml each) were prepared which contained 2 mg soluble starch, 0–1.5 μmol maltose, and 25 μmol Imidazole (adjusted to pH 6.6 with HCl; □ - - - □) or 25 μmol Mes (adjusted to pH 6.6 with NaOH) plus 2.5 μmol $CaCl_2$ (● — ●). As a control, no buffer was applied (○ — ○). Samples were mixed with 1 ml alkaline colour reagent and processed as described (see text). A maltose-free sample of each series was used as blank.

enzyme is often possible even in plant crude extracts. The enzyme is assayed by monitoring the formation of reducing groups as described above (see Section II.A.1). However, soluble starch is replaced by pullulan.

If the samples contain a high α-glucosidase activity, maltotriose is converted to maltose and/or glucose and, therefore, debranching enzyme activity may be overestimated. This interference can be detected when the sample is incubated with maltotriose (final concentration 5–10 mM) instead of pullulan, and the formation of reducing groups is monitored. Alternatively, the degradation products can be analysed by thin layer or high pressure chromatographic separation of oligoglucans (see below).

B. Phosphorylases

In principle, both glucan and maltose phosphorylase activities are measured by the same assay procedure; however, both assays differ in the carbohydrate applied as substrate. Depending upon the glucose 1-phosphate : orthophosphate ratio, the net reaction as catalysed by phosphorylases is either glucan polymerisation (and orthophosphate release) or depolymerisation (and glucose 1-phosphate formation). Both glucan synthesis and depolymerisation are used to measure the enzyme activities.

1. Glucan phosphorylase activity

In a non-continuous assay, glucan synthesis is usually monitored as liberation of orthophosphate, using one of the numerous Pi determination techniques, or as ^{14}C transfer from labelled glucose 1-phosphate to glucans. However, the glucan synthesising and the phosphorolytic reaction of the phosphorylase can be measured by continuous photometric assays which possess both a high sensitivity and a far better time resolution. The continuous assay for the glucan polymerising reaction (Table 7.2) is based on an enzymatic colorimetric orthophosphate determination which has been recently described by Fossati (1985).

TABLE 7.2. Continuous assay for the glucan polymerising reaction.

$$G_n + \text{Glu-1-P} \rightleftharpoons G_{n+1} + \text{Pi} \quad (1)$$

$$\text{Inosine} + \text{Pi} \rightleftharpoons \text{Hypoxanthine} + \text{ribose 1-phosphate} \quad (2)$$

$$\text{Hypoxanthine} + 2\,\text{MTT} \xrightleftharpoons[\text{1-methoxy-PMS}]{\text{PMS or}} \text{Uric acid} + 2\,\text{formazan} \quad (3)$$

Enzymes: (1) glucan phosphorylase (EC 2.4.1.1); (2) purine-nucleoside phosphorylase (EC 2.4.2.1); (3) xanthine oxidase (EC 1.1.3.22).

In Table 7.2, Glu-1-P represents glucose 1-phosphate; G_n and G_{n+1} represent an α-linked glucan containing n and $(n+1)$ α-1,4-linked glucose units, respectively. The reduction of 3-(4′,5′-dimethyl-2-thiazolyl)-2,4-diphenyl-2H-tetrazolium bromide (MTT) to the coloured formazan is followed photometrically at 578 nm. Electron transfer is mediated by 1-methoxy phenazine methosulphate or, alternatively, by phenazine methosulphate (PMS). Because non-specific phosphatases interfere with this assay, its application is often restricted to highly purified phosphorylase preparations.

In the continuous assay of the phosphorolytic reaction, glucose 1-phosphate is converted to glucose 6-phosphate by phosphoglucomutase. NAD(P) reduction by glucose 6-phosphate dehydrogenase is followed photometrically at 340 or 365 nm. Because interference by other enzyme activities is low this assay is preferred in most cases. However, the V_{max} of the phosphorolytic reaction is lower than that of glucan synthesis.

(a) *Extract.* Plant material (4–8 g fresh weight) is homogenised in 30 ml grinding medium. For homogenisation, one of the following grinding media may be used: (a) 0.1 M Imidazole-HCl (pH 7.0); (b) 60 mM Tris, adjusted to pH 7.3 with orthophosphoric acid; (c) 50 mM sodium phosphate, 2 mM EDTA (pH 7.7) plus 10% (v/v) glycerol. Using the latter buffer, extracts can be frozen in liquid nitrogen and stored for a week at −70°C without noticeable changes in enzyme activity or pattern. To minimise proteolytic degradation, phenyl methyl sulphonyl fluoride (final concentration 0.4 mM) may be added to the grinding medium immediately before use (a stock solution is prepared in isopropanol). The homogenate is further processed as described above (see Section II.A.1).

(b) *Activity assay.* An assay mixture is prepared which, in a final volume of 950 μl, contains 25 μmol Imidazole (pH 7.0), 5 μmol $MgCl_2$, 1 μmol EDTA (pH 7.0), 5 μmol Na_2MoO_4, 0.6 μmol NADP, 2.5 nmol glucose 1,6-bisphosphate, 33 nkat phosphoglucomutase (from rabbit muscle; EC 5.4.2.2), 33 nkat glucose 6-phosphate dehydrogenase (from yeast; EC 1.1.1.49), 25 μmol sodium phosphate (pH 7.0), and 20–100 μl extract or enzyme preparation. The mixture is incubated at 30°C for 5 min and the phosphorolytic reaction is then initiated by adding 50 μl 2% (w/v) soluble starch. When glucose 6-phosphate dehydrogenase from *Leuconostoc mesenteroides* is used instead of that from yeast, NADP can be replaced by NAD, thus preventing a possible interference of 6-phosphogluconate dehydrogenase (see below). Sodium molybdate acts as a phosphatase inhibitor and therefore can be omitted if purified phosphorylases are to be assayed.

(c) *Limitations.* Endogenous soluble glucans can cause a significant rate of NAD(P) reduction before soluble starch is added to the assay mixture. In many cases, this reaction ceases within a few minutes of pre-incubation. Alternatively, phophorylase activity may be precipitated between 30 and 60% saturation of $(NH_4)_2SO_4$ to separate it from endogenous glucans. If present at a high concentration, endogenous 6-phosphogluconate dehydrogenase activity affects the stoichiometry of NADPH and glucose 1-phosphate formation and thereby results in an overestimation of phosphorylase activity. Under the respective assay conditions, activity of 6-phosphogluconate dehydrogenase can be easily measured by adding exogenous 6-phosphogluconate at a saturating concentration. Because 6-phosphogluconate dehydrogenase does not use NAD, an interference can be avoided as mentioned above.

2. Maltose phosphorylase activity

The phosphorolytic reaction catalysed by maltose phosphorylase is monitored continuously by the coupled assay described above (see Section II.B.1); however, soluble starch is substituted by maltose (final concentration 10 mM).

3. D-Enzyme Activity

In the reaction catalysed by D-enzyme the number of reducing groups of carbohydrates remains constant. However, the rate of maltotriose-dependent glucose formation can be followed photometrically via hexokinase/glucose 6-phosphate dehydrogenase.

(a) *Extract.* Plant material (4–8 g fresh weight) is homogenised in 30 ml chilled grinding medium which contains 50 mM Hepes-NaOH (pH 8.0), 2 mM EDTA-NaOH (pH 8.0), and 1 mM dithioerythritol. The homogenate is filtered through several layers of Miracloth or nylon net. Following centrifugation (15 min at $40\,000 \times g$; 4°C), an aliquot of the supernatant is added to the assay mixture.

(b) *Activity assay.* An assay mixture is prepared which, in a volume of 950 μl, contains 50 µmol Hepes (pH 7.5), 3.3 µmol $MgCl_2$, 3.2 µmol ATP, 0.45 µmol NAD(P), 12 nkat glucose 6-phosphate dehydrogenase (EC 1.1.1.49; from yeast or *Leuconostoc mesenteroides*), 25 nkat hexokinase (EC 2.7.1.1; from yeast), and 20–50 µl extract. The assay mixture is incubated for 5–10 min at 30°C. Reaction is initiated by adding maltotriose to give a final concentration of 5 mM. Absorbance is monitored at 340 or 365 nm.

(c) *Limitations.* If any of the assay components contain glucose NAD(P) reduction occurs. Glucose contamination of maltotriose preparations can be tested by thin layer chromatography or HPLC (see below). Any hydrolytic reaction resulting in the liberation of glucose (e.g. by α-glucosidase) interferes with the assay. It is, therefore, necessary to perform control experiments in which maltotriose is substituted by maltose.

III. MULTIPLICITY OF STARCH DEGRADING ENZYMES

Enzyme multiplicity can be studied by electrophoretic techniques if two assumptions are fulfilled: the catalytic activity of the enzyme forms being under investigation must be largely retained during electrophoretic migration, and a staining procedure must be available by which the catalytic activity can be localised within the gel. Most of the starch degrading enzymes are sufficiently stable under a relatively wide range of conditions and, therefore, can be resolved by several electrophoretic systems. Activity detection is, in most cases, based on the well-known iodine staining of starch or starch-like carbohydrates. α-1,4-Linked glucans form helices which bind iodine selectively. These iodine–glucan complexes have a blue or violet colour. Thus, any reaction which changes length and/or local concentration of glucan helices can be localised by a simple iodine staining of the gel. Sites of degradation of iodine-stained glucans are detected as unstained areas on a dark background ('negative staining'), whereas regions of glucan synthesis appear as dark zones on an unstained or less stained background. For enzymes acting on oligoglucans (such as D-enzyme) other staining techniques are available (see below).

Once the enzyme pattern is established, non-denaturing electrophoresis can provide information on some characteristics (such as apparent molecular size or glucan specificity) of distinct enzyme forms (Steup and Melkonian, 1981; Steup *et al.*, 1986). These investigations can be performed using crude extract samples.

A. Hydrolases

Glucan hydrolysing enzyme forms can be resolved by the discontinuous electrophoresis system described below (see Section III.B). Alternatively, isoelectric focusing may be applied. The latter technique is preferred if enzyme forms having a relatively high isoelectric point (IEP $\geqslant 6.5$) are to be resolved.

The accessibility of glucans to the enzyme forms can be achieved by several ways: by inclusion of the polysaccharides into the separation gel; by diffusion into the gel after electrophoresis; or by an electrophoretic transfer of the separated proteins into a glucan-containing polyacrylamide gel (Steup and Gerbling, 1983; Kakefuda and Duke,

1984). Incorporation of glucans into the separation gel ensures a highly sensitive activity staining. The major disadvantage is, however, a possible (or likely) effect on the electrophoretic separation. In a polyglucan-containing gel the electrophoretic mobility of a protein depends not only on its net charge, size, and shape, but also on the strength of binding to the immobilised carbohydrate. If a protein binds strongly to the glucan it is retarded or it may even be immobile. Therefore, proteins which have a high affinity towards the immobilised glucans but differ in other parameters are difficult to resolve. When various types of polyglucans (such as amylose, amylopectin or starch) are included in separation gels the retarding effect on a given enzyme form can be entirely different. Under these conditions different enzyme patterns can be obtained for the same sample and distinct enzyme forms are difficult to identify.

A further complication arises from enzymic glucan hydrolysis catalysed by the migrating enzyme forms. As a result, the local glucan concentration in the separation gel decreases. The extent of this decrease is related to the amount of hydrolytic activity applied per lane. In affinity electrophoresis, the retardation exerted by a given immobilised substrate is a function of its effective concentration unless saturation is achieved. Therefore, the amount of hydrolytic activity being analysed can affect the enzyme pattern obtained (MacGregor, 1977).

Because of these complications, electrophoresis in a glucan-free separation gel is preferred. If a moderate sensitivity of activity detection is acceptable Procedure I is used. A considerably higher sensitivity is achieved by Procedure II.

1. Detection Procedure I

Following electrophoresis, the separation gel is equilibrated for 1 h at 4°C in a suitable buffer (e.g. 100 mM citrate-NaOH, pH 6.0 or 50 mM Mes-NaOH pH 6.5 or 50 mM sodium acetate pH 5.0; one change after approximately 30 min). The buffer volume should be at least one order of magnitude larger than that of the gel. The separation gel is then transferred to a solution of 0.4% (w/v) soluble starch (or another polysaccharide) containing the same buffer as used for equilibration. $CaCl_2$ (final concentration 5 mM) may be added but is often not required. Following incubation at 37°C for an appropriate period of time the gel is carefully rinsed with water to remove any glucan attached to the gel surface. The gel is then stained for approximately 10 min in an iodine solution containing 10 mM I_2 and 14 mM KI in water. Conveniently, this solution is prepared by a 10-fold dilution of an iodine stock solution. The latter is obtained by dissolving KI (140 mM) and I_2 (100 mM) in water at room temperature (stirring overnight). The stock solution is then passed through a filter paper and stored in a darkened bottle at room temperature. It is stable for months. Following iodine staining, the polyacrylamide gel is transferred to a mixture containing 30% (v/v) methanol, 5% (v/v) acetic acid and water to reduce background staining. When stored in darkness, staining is maintained for weeks. If fading occurs, gels can be restained by the iodine solution. For documentation, gels are either scanned at 578 nm or photographed.

2. Detection Procedure II

Before electrophoresis is performed, an additional polyacrylamide gel is perpared which contains 0.4% (w/v) soluble starch or 0.1% (w/v) amylose but is otherwise identical to the separation gel. Following electrophoresis, both the separation gel and the glucan-containing gel are incubated for 30 min at 4°C in transfer buffer (288 mM glycine and 37.4 mM Tris, pH 8.6). After equilibration, the separation gel is placed side by side on the glucan-containing gel. Both polyacrylamide slabs are then sandwiched between wetted filter paper and Scotch-Brite pads and are placed into the transfer device (separation gel is oriented towards the cathode, the glucan-containing gel towards the anode). Chilled transfer buffer (as above) is used. Electrotransfer is performed under continuous cooling (4°C) at 50 V (370–470 mA; 7 cm electrode distance). The optimal transfer period (usually approximately 60 min) has to be carefully determined in order to recover the entire enzyme pattern. Following transfer, the separation gel is removed. It may be used for restaining. The glucan-containing slab is equilibrated for 60 min (4°C) in a suitable buffer (see Section III.A.1: Detection Procedure I). It is then incubated at 37°C in the same fresh medium. Iodine staining is performed as described above.

As an alternative to the transfer described here, transfer may be performed at a considerably lower voltage and at room temperature (for details see Kakefuda and Duke, 1984). In this case, glucan hydrolysis and electrophoretic transfer occur simultaneously.

B. Glucan Phosphorylases

In plants the glucan phosphorylase pattern is usually less complex than that of amylases. In most cases it is composed of 2–4 enzyme forms. Phosphorylase forms are mostly resolved by anionic polyacrylamide gel eletrophoresis. Isoelectric focusing often results in a loss of enzyme activity. Both continuous and discontinuous electrophoresis systems can be used provided highly alkaline conditions (pH $\geqslant 9.0$) are avoided. The latter tend to inactivate the enzyme activity rapidly and irreversibly. It should be taken into consideration that in anionic systems the separation gel becomes more alkaline (often by approximately one pH unit) when the electric field is applied.

Higher plant phosphorylases have been classified into two types which differ in kinetic properties (especially glucan specificity), monomer size, and intracellular location (for details see Steup, 1988). One group of enzymes (Type I) has a high affinity towards branched polyglucans (such as glycogen), whereas the other one (Type II) reacts very poorly with this glucose polymer. Both types can be easily identified by comparing enzyme patterns obtained by electrophoresis in either glycogen-free or glycogen-containing separation gels. In the latter case, binding of phosphorylase forms to the immobilised glucan is determined without interference by catalysis. The kinetics of glucan phosphorylases follow a rapid equilibrium random mechanism. Therefore, during electrophoresis glucan binding but no catalysis occurs except when orthophosphate or glucose 1-phosphate is present.

Type I and Type II phosphorylase forms differ in their intracellular location: Type I is situated outside, Type II inside the plastidic compartment (Fig. 7.2; Steup, 1988).

1. Continuous electrophoresis

The continuous electrophoresis system (Stegemann et al., 1973) described here has the advantage that a smearing of Type I enzyme forms, which occasionally occurs in other

7. STARCH DEGRADING ENZYMES

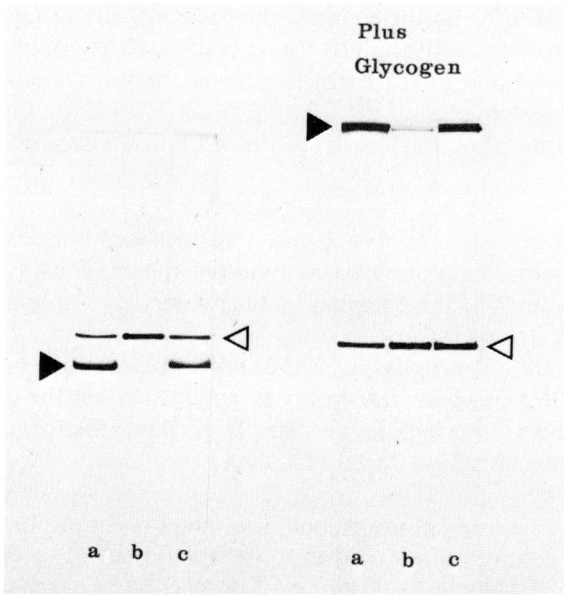

FIG. 7.2. Type I and Type II glucan phosphorylase from *Spinacia oleracea* L. Extracts of leaves (a), isolated chloroplasts (b) or a cell suspension culture (c) were resolved by non-denaturing polyacrylamide gel electrophoresis (continuous system; 6–15% (w/v) acrylamide). The separation gel contained no polysaccharide (left) or 0.1% (w/v) glycogen (right). Approximately 80 µg protein was applied to each lane. Electrophoresis was performed for 2400 volt hours. Migration direction is from top (cathode) to bottom (anode). For activity staining, gels were incubated (37°C) in a mixture containing 0.05% (w/v) soluble starch, 20 mM glucose 1-phosphate, and 100 mM citrate (pH 6.6). Incubation period was 150 min (left) or 90 min (right). Type I phosphorylase (▶) is detectable in leaf or cell suspension extract but not in that of isolated chloroplasts. Type I enzyme is strongly retarded by the immobilised polysaccharide, whereas the migration velocity of the Type II phosphorylase (◁) is essentially unaffected.

systems, is prevented. However, only small sample volumes ($\leqslant 20\,\mu l$ per lane) can be applied. Electrophoresis can be performed by using either a homogeneous or a gradient gel.

(a) *Polyacrylamide gel.* Two stock solutions are prepared. Monomer solution: 30 g acrylamide plus 0.8 g N,N'-methylene-bis-acrylamide (designated as bis-acrylamide) are dissolved in approximately 85 ml H_2O and passed through a filter paper. To the filtrate H_2O is added to give a final volume of 100 ml. When stored in a darkened bottle at 4°C, the monomer solution is stable for a month. Buffer concentrate: a solution of 0.6 M boric acid (H_3BO_3) is brought to pH 7.9 by addition of solid Tris. Stored at 4°C, the buffer concentrate is stable for several weeks. A homogeneous separation gel (7.5% (w/v) acrylamide) is prepared by mixing 3.75 ml buffer concentrate, 7.5 ml monomer solution, 17.25 ml H_2O, and 23 µl Temed. Following degassing at 4°C, polymerisation is initiated by adding 1.5 ml of 1.7% (w/v) freshly prepared ammonium peroxydisulphate. If a gradient gel is to be used, two polymerisation mixtures are prepared which contain 6% (w/v) and 15% (w/v) acrylamide, respectively. All other components are the same as used for the homogeneous gel, except that only half of the ammonium peroxydisulphate and one third of the Temed concentration is applied.

If electrophoresis is to be performed in a glycogen containing separation gel (affinity electrophoresis) a glycogen solution (2%; w/v) is dialysed overnight at room temperature against a large volume of H_2O. An aliquot of the retentate is then added to the polymerisation mixture to give a final concentration of 0.1% (w/v). Other polyglucans (soluble starch, amylopectin, amylose) are included into the separation gel by the same procedure.

(b) *Samples.* Plant material is homogenised as described above (see Section II.B.1). The grinding buffer consisting of 50 mM sodium phosphate, 2 mM EDTA (pH 7.7), and 10% (v/v) glycerol may be supplemented by phenyl methyl sulphonyl fluoride (PMSF) in a final concentration of 0.4 mM in order to minimise proteolytic degradation (see below). Crude extracts or partially purified enzyme preparations are passed through a nitrocellulose filter (0.2 μm pore size) prior to application on the gel. Purified enzyme preparations are diluted with grinding buffer. To facilitate sample application, bromophenol blue (final concentration 25 μg ml^{-1}) may be added.

(c) *Electrophoresis.* Electrophoresis buffer is prepared by diluting 1 vol of buffer concentrate with 7 volumes of ice-cold H_2O (no readjustment of the pH value). For a 'normal' size slab gel (140 mm × 90 mm × 1.5 mm) containing a homogeneous separation gel electrophoresis is performed for 1 h at 150 V (12 mA slab^{-1}) and then for 4 h at 350 V (20–10 mA slab^{-1}). A gradient gel of the same size is run for 1 h at 150 V and then for 4 h at 400 V. When a smaller electrophoresis device (e.g. Mini Protean II, BIO-RAD) is used electrophoresis is completed after approximately 500 volt hours. Electrophoresis is performed under permanent cooling (4°C). In the continuous system described here bromophenol blue reaches the end of the separation gel at less than half of the total volt hours.

(d) *Activity staining.* Following electrophoresis, the separation gel is equilibrated for 30–60 min with 0.1 M citrate-NaOH (pH 6.0; 4°C; one change of the buffer). The gel is then transferred to a mixture containing 0.1 M citrate-NaOH, 20 mM glucose 1-phosphate (pH 6.0), and 0.05% (w/v) soluble starch. When preparing this solution even a temporary acidification of glucose 1-phosphate should be avoided because otherwise hydrolysis of the phosphate ester may occur. A substrate deficiency control (minus glucose 1-phosphate) is also prepared. After between 0.5 and 3 h incubation at 37°C the gel is carefully washed with H_2O and stained with iodine (see Section III.A). Bands of phosphorylase activity appear as blue or violet zones on an essentially unstained background. They are strictly dependent on the presence of glucose 1-phosphate. In pherograms of chloroplast-containing plant materials iodine staining usually reveals one yellowish slowly migrating band which is due to a catalysis-independent staining of ribulose 1,5-bisphosphate carboxylase/oxygenase (Rubisco). The zymogram can be further confirmed by incubating another set of gel lanes in a mixture containing 50 mM sodium phosphate, 100 mM citrate-NaOH (pH 6.0), and 0.4% (w/v) soluble starch. A control lacking orthophosphate is also prepared. Upon iodine staining, glucan degrading enzyme activities are visible as unstained bands on a dark background. Those bands which are strictly orthophosphate-dependent coincide with the sites of glucose 1-phosphate-dependent glucan synthesis (for details see Steup and Latzko, 1979; Steup and Melkonian 1981).

2. Discontinuous electrophoresis

Plant phosphorylase forms from many plant species can be resolved by a discontinuous system which was originally developed for the separation of serum proteins (Jolley and Allen, 1965). This system separates under moderately alkaline conditions. Compared to the continuous system described above (see Section III.B.1) larger sample volumes (up to 200 μl per lane) can be applied without loss in resolution. However, in some cases Type I phosphorylase forms are poorly resolved if crude extract samples are used. This difficulty can be avoided by a fractionated ammonium sulphate precipitation (35–70% saturation) of the crude extract. Alternatively, the continuous system (see Section III.B.1) may be used.

(a) *Separation gel.* An homogeneous gel (8.5% (w/v) acrylamide) or a gradient gel (6–12% (w/v) acrylamide) is used. Stock solutions: the monomer solution is prepared as described above (see Section III.B.1). Buffer concentrate B: 0.88 M Tris-HCl (pH 7.2), containing 0.23% (v/v) N,N,N',N'-tetramethylethylenediamine (Temed). At 4°C, the concentrate is stable for at least one month. For preparation of the polymerisation mixture 2.5 ml H_2O, 3.75 ml buffer concentrate B, and 8.5 ml monomer solution are mixed and degassed. Then 15 ml of degassed, freshly prepared ammonium peroxydisulphate solution (1.4 mg ml^{-1}) is added. For preparing a gradient gel, two polymerisation mixtures are prepared which, in a final volume of 30 ml each, contain 6 and 12 ml of the monomer solution. All other constituents are the same as in the homogeneous gel.

(b) *Stacking gel.* The polymerisation solution is prepared by mixing 2 ml 0.5 M Tris, adjusted to pH 7.3 with orthophosphoric acid, 4 ml monomer solution (containing 10% (w/v) acrylamide plus 2.5% (w/v) bisacrylamide), 2 ml riboflavin (40 μg ml^{-1}), and 8 ml sucrose solution (400 mg ml^{-1}). Photopolymerisation (60–90 min) is terminated immediately before sample application.

(c) *Samples.* The ion content of the samples should be similar to that of the stacking gel. Therefore, 0.5 M Tris-phosphate (pH 7.3), diluted eight-fold and supplemented with PMSF (0.4 mM) and glycerol (final concentration 10%; v/v), is used as grinding or dilution buffer. Alternatively, samples are equilibrated with this buffer by gel filtration.

(d) *Electrophoresis.* A solution of 383 mM glycine, adjusted to pH 8.5 by the addition of solid Tris, is prepared (stock solution). Electrophoresis buffer is prepared by diluting one volume of the stock solution with nine volumes of ice-cold H_2O (no pH readjustment). Electrophoresis is performed at constant current (15 mA per slab gel; 'normal' size) and 260–470 V under permanent cooling. Electrophoresis is terminated before bromophenol blue reaches the end of the separation gel. If the anodic buffer chamber has a relatively small volume stirring or even an exchange of the anodic electrophoresis buffer may be required.

(e) *Activity staining.* See Section III.B.1.

(f) *Limitations.* A zymogram can both over- and underestimate the complexity of an enzyme heterogeneity. Complexity is underestimated if two (or more) enzyme forms are not resolved or if an enzyme form remains undetected during activity staining. The latter may be due to an instability of activity or, alternatively, to an interference with other comigrating proteins. As an example, a phosphorylase form is difficult to detect if it is located in the close vicinity of a highly active glucan hydrolase. Therefore, it is important to perform an electrophoretic separation under various experimental conditions (e.g. in homogeneous and gradient gels or in both continuous and discontinuous systems).

There are several possible reasons for an overestimation of complexity. Limited proteolysis of phosphorylases often gives rise to catalytically active fragments. Therefore, enzyme multiplicity can be artificially amplified (or even created) by the action of endogenous proteases. Oligomerisation of proteins may also cause additional bands of activity (for details see Steup, 1988).

C. D-Enzyme

The discontinuous electrophoresis system (see Section III.B.2) has been successfully used to determine the D-enzyme pattern in plant crude extracts. Detection of enzyme activity is based on a short reaction sequence which links the liberation of glucose (via hexokinase and glucose 6-phosphate dehydrogenase) to the formation of formazan.

1. Samples

Crude extract samples are prepared as described above (see Section III.B.2). However, dithioerythritol and EDTA (final concentrations 1 and 2 mM, respectively) are included in the grinding medium.

2. Separation gel and electrophoresis

A homogeneous separation gel containing 8.5% (w/v) acrylamide is prepared. For electrophoresis see Section III.B.2.

3. Activity staining

Following electrophoresis, the separation gel is incubated for 60 min in ice-cold 50 mM Hepes-NaOH, pH 7.5. It is then transferred to the staining solution containing 50 mM Hepes-NaOH (pH 7.5), 3.3 mM $MgCl_2$, 0.45 mM NADP, 3.2 mM ATP, 12 nkat glucose 6-phosphate dehydrogenase ml^{-1} (EC 1.1.1.49; from yeast), 25 nkat hexokinase ml^{-1} (EC 27.1.1.1; from yeast), 0.065 mM phenazine methosulphate and 0.48 mM 3(4,5-dimethyl-thiazolyl-2)-2,5-diphenyl tetrazolium bromide (MMT). The two auxiliary enzymes have been dialysed against 50 mM Hepes-NaOH (pH 7.5) prior to use. Phenazine methosulphate and MTT are dissolved in 10 mM Hepes-NaOH pH 7.5 immediately before adding to the staining mixture. As a control, polyacrylamide lanes are incubated in a staining mixture in which maltotriose is replaced by maltose (final concentration 10 mM). Following incubation for 2–3 h at 30°C gels are rinsed with H_2O and are photographed immediately.

4. Product analysis

Enzyme pattern as revealed by activity staining is confirmed by carbohydrate analysis. A gel slice containing the putative D-enzyme activity is incubated overnight at room temperature in 1 ml of the staining mixture from which phenazine methosulphate and MTT have been omitted. The gel slice is then discarded and the incubation mixture is passed through a coupled cation- and anion-exchange column (Dowex, analytical grade; AG 50W-X8 H^+-form, AGl-X8 acetate-form; each $1.8\,cm^2 \times 0.5\,cm$). The two columns are eluted with 25 ml H_2O. Following concentration by evaporation under reduced pressure (40°C) carbohydrates are analysed by HPLC using an HPX-42A column (Bio-Rad; 80°C; $0.6\,ml^{-1}\,min^{-1}$; approximately 40 bar). Sugars are monitored by refractive index determination (Fig. 7.3).

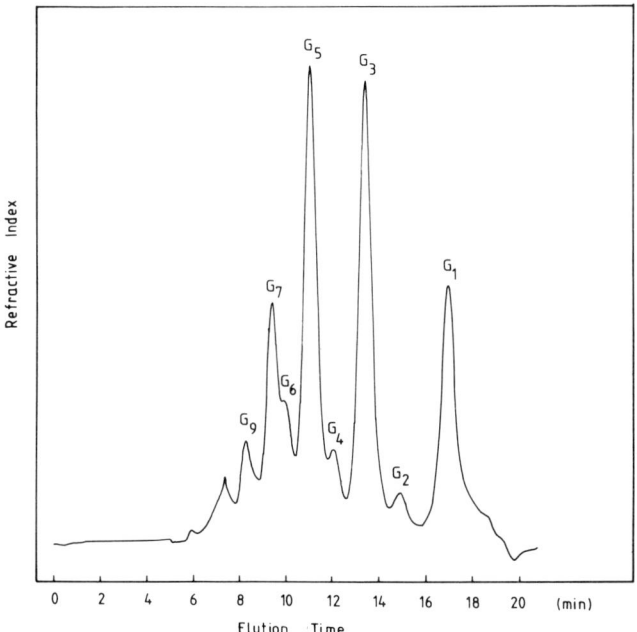

FIG. 7.3. Carbohydrate analysis by HPLC. A slice of a separation gel containing a putative D-enzyme activity was incubated in the presence of maltotriose (for details, see text). Following incubation, the mixture was deionised and the neutral fraction was chromatographed on a HPX-42A column (Bio-Rad). G_n: oligoglucans consisting of n glucose units.

Alternatively, oligoglucans can be separated by thin layer chromatography according to Kanaya *et al.* (1978). For a sensitive detection and quantification of carbohydrates on thin layer plates, see Bounias (1980).

IV. PURIFICATION OF COMPARTMENT-SPECIFIC GLUCAN PHOSPHORYLASE FORMS

The purification of compartment-specific glucan phosphorylase forms requires both the removal of non-phosphorylase proteins and the separation of the various phosphorylase forms. For several higher plants purification procedures have been developed which utilise two techniques, anion-exchange chromatography and affinity chromatography. Compared to Type II phosphorylases, the cytosolic enzyme form (Type I) binds less tightly to the ion-exchange gel. This enables the two enzyme types to be separated by ion-exchange chromatography, the cytosolic enzyme form being eluted first (Steup *et al.*, 1980a,b). In principle, both Type I and Type II phosphorylases bind to glucans covalently coupled to Sepharose (or another matrix). Thus, affinity chromatrography is applicable as a purification step. However, binding of the phosphorylase forms is significantly affected by the ligands which are coupled to the affinity gel during the deactivation procedure. Due to this peculiarity the binding specificity of the affinity gel can be easily modified.

The purification of Type I and Type II phosphorylase from spinach leaves is described below.

A. Purification of Type I Glucan Phosphorylase

The purification procedure takes advantage of the selective binding properties of a Sepharose-glucan gel which has been deactivated with glycine. By this deactivation procedure a Sepharose-glucan gel is obtained to which Type II phosphorylases from several plant species are unable to bind. Purification of the cytosolic phosphorylase form and separation of both phosphorylase types are thus achieved in a single step. The reason for the inability of the plastidic phosphorylase form(s) to bind is not yet clear. Deactivation with glycine results in a gel which, in addition to the bound α-glucan, carries negatively charged groups. It is possible that these charged groups prevent a binding of Type II phosphorylases.

1. Extract

Two hundred grams of spinach leaves are homogenised in 250 ml ice-cold 0.1 M Imidazole-HCl (pH 7.0) containing 0.4 mM phenylmethyl sulphonyl fluoride (PMSF). The homogenate is cleared by filtration through a nylon net followed by centrifugation (10 min at 25 000 × g). The supernatant (designated as crude extract) is subjected to a fractionated ammonium sulphate precipitation.

2. Ammonium sulphate precipitation

The crude extract is brought to 40% saturation with ammonium sulphate (saturated solution, adjusted to pH 7.0 with NH_4OH). Following centrifugation (10 min at 25 000 × g) the pellet is discarded. In the supernatant the ammonium sulphate concentration is raised to 70% saturation (as above). After centrifugation (10 min at 25 000 × g) the precipitate is dissolved in a buffer containing 10 mM Tris, 1 mM EDTA, 0.4 mM PMSF and brought to pH 7.0 with acetic acid. The solution is dialysed for 2 h against the same buffer. The retentate is then centrifuged for 10 min at 40 000 × g and the supernatant is used for affinity chromatography.

3. Affinity chromatography

The supernatant is degassed and applied to an affinity gel (Sepharose-dextrin-glycine;

for synthesis see below). The gel (5.3 cm^2 × 8 cm) has been equilibrated with 10 mM Tris-acetate (pH 7.0), 1 mM EDTA, 0.4 mM PMSF. The column is then washed with 100 ml buffer (as above), 100 ml 0.6M NaCl dissolved in buffer (as above), 100 ml buffer (as above) and, finally, with a linear gradient (60/60 ml) of low molecular weight glucans (for preparation see below). Flow rate throughout affinity chromatography is 8 cm h^{-1}. Type I phosphorylase is recovered in the eluate as a sharp peak of activity when the dextrin gradient is used as eluent. Fractions of the peak are pooled and applied to gel filtration.

4. Gel filtration

Glucans are removed from the enzyme preparation by gel filtration on Sephadex G-100 (5.3 cm^2 × 50 cm; flow rate 2.5 cm h^{-1}). The Sephadex gel has been equilibrated with 50 mM Imidazole-HCl (pH 7.0). Phosphorylase activity is recovered in the void volume (for details see Schächtele and Steup, 1986). The enzyme preparation is then concentrated by ultrafiltration (Amicon, PM$_{30}$ membrane). For storage, the concentrated enzyme preparation is mixed with an equal volume of 50 mM Imidazole-HCl (pH 7.0), containing 60% (v/v) glycerol. Aliquots of this mixture are frozen in liquid nitrogen and are stored at −70°C. The enzyme preparation is stable for months. Freezing in the absence of glycerol causes almost complete inactivation of the enzyme. Throughout the purification the temperature is kept at 2°C.

5. Synthesis of Sepharose-dextrin gel

Two hundred and fifty grams (fresh weight) of Sepharose (Pharmacia) is washed repeatedly with H$_2$O and is then suspended in 500 ml H$_2$O, previously brought to 10–12°C. Activation is performed in a fume hood and the instructions of the producer are carefully followed when handling BrCN. For activation of Sepharose, 25 g BrCN (dissolved in 25 ml dioxan) is slowly added to the gel suspension and the pH value is raised to 11–12 by adding 5 N NaOH. Throughout activation, the temperature is kept at 10–12°C and the pH value is maintained at 11–12 by dropwise addition of 5 N NaOH (approximately 50–70 ml over 20–30 min). After cessation of acidification the suspension is cooled to approximately 5°C by addition of ice. Immediately after cooling the gel is washed (on a Büchner funnel) with approximately 2 litres chilled H$_2$O, followed by 1 litre of chilled 10 mM potassium phosphate buffer (pH 8.0). The activated gel is then resuspended in a dextrin solution. The dextrin solution has been prepared by dissolving 25 g commercial dextrins (e.g. Sigma, product no. D-2006) in 400 ml 10 mM potassium phosphate buffer (pH 8.0) and passing the solution through a filter paper. Dextrins are coupled to the activated Sepharose gel overnight at room temperature under continuous agitation. Following coupling, 1.25 mol glycine, suspended in 250 ml 0.1 M NaHCO$_3$ and adjusted to pH 9.5 with 5 N NaOH, is added to the Sepharose-glucan mixture. After 2.5 h continuous agitation at room temperature, the gel is washed (on a Büchner funnel) with 2 litres H$_2$O, 2 litres 1 M NaCl, brought to pH 12.5 with NaOH, 2 litres H$_2$O, 2 litres 1 M NaCl, adjusted to pH 2.5 with acetic acid, and then again with 2 litres H$_2$O. Finally, the gel is equilibrated with a buffer containing 10 mM Tris-acetate, 1 mM EDTA (pH 7.0), and 0.1% (w/v) sodium azide.

When stored at 4°C, the affinity gel is stable for months (cf. Section IV.B). If samples containing a significant glucan hydrolysing activity are applied to the affinity gel, it is recommended that the gel is discarded after use.

6. Low molecular weight glucans

Commercial dextrin preparations usually contain some high molecular weight carbohydrates. When such dextrin preparations are applied as eluens for affinity chromatography, it is difficult to remove the high molecular weight glucans from the eluted Type I phosphorylase. Therefore, commercial dextrin preparations are freed from high molecular weight contaminants. To 15 g dextrin, dissolved in 100 ml H_2O, ethanol is added dropwise (under stirring) to give a final ethanol concentration of 75% (v/v). The mixture is stirred for a further 15 min at room temperature and is then centrifuged (10 min at 26 000 × g). The pellet, which contains glucans with a degree of polymerisation of more than 15, is discharged. The supernatant is evaporated to dryness and the residue is dissolved in 50 ml 10 mM Tris-acetate, 1 mM EDTA (pH 7.0). After readjustment of the pH value to 7.0 the solution is brought to a final volume of 60 ml.

B. Purification of Type II Glucan Phosphorylase

The entire purification procedure consists of four steps: (1) fractionated ammonium sulphate precipitation; (2) anion-exchange chromatography (by which Type I and Type II enzyme forms are separated); (3) affinity chromatography; and (4) gel filtration (which removes low molecular weight glucans (see Section IV.A).

In addition to the compounds stated all solutions listed below contain 0.4 mM PMSF and 15 μM thymol. Throughout the purification procedure the temperature is kept at 2°C.

1. Extract and fractionated ammonium sulphate precipitation

One kilogram of spinach leaves is homogenised in 500 ml 0.1 M Imidazole-HCl buffer (pH 7.0). The homogenate is filtered through a nylon net and then cleared by centrifugation (10 min at 25 000 × g). From the supernatant (designated as crude extract) protein is precipitated at 45–60% saturation with $(NH_4)_2SO_4$ (for details, see Section IV.A). The precipitate is dissolved in 20 mM citrate-NaOH (pH 6.5) and is then dialysed against the same buffer for at least 10 h. The retentate, cleared by centrifugation (10 min at 40 000 × g) is applied to the ion-exchange gel.

2. Anion-exchange chromatography

The supernatant is applied to a DEAE-Sephacel column (5.3 cm² × 15 cm, previously equilibrated with 20 mM citrate-NaOH pH 6.5). The column is washed with 100 mM citrate-NaOH (20 mM; pH 6.5) and then with a linear gradient (130/130 ml) of 20–60 mM citrate-NaOH (pH 6.5). The citrate gradient elutes Type I phosphorylase which is discarded. The Type II phosphorylase form is eluted from the column by a linear gradient (130/130 ml) of 0–1.2 M NaCl, dissolved in 60 mM citrate-NaOH (pH 6.5). Fractions containing more than 1 nkat phosphorylase activity (phosphorolytic assay, see Section II.B.1) are pooled and concentrated by ammonium sulphate precipitation

(70% saturation). The precipitate is dissolved in 10 mM Tris-acetate, 1 mM EDTA (pH 7.0; buffer I), centrifuged again (10 min at 40 000 × g) and the supernatant is degassed.

3. Affinity chromatography

The degassed supernatant is applied to a Sepharose-starch gel (for synthesis see below). The gel (2 cm² × 10 cm) is previously equilibrated with buffer I. Following sample application, the gel is washed with 100 ml of buffer I, then with a linear NaCl gradient (50/50 ml; 0–1 M NaCl dissolved in buffer I) and with 200 ml 1 M NaCl in buffer I. Type II phosphorylase is recovered from the affinity gel by applying a linear dextrin gradient (0–12%; see below), dissolved in 1 M NaCl and buffer I. Eluate fractions containing phosphorylase activity are pooled and used for gel filtration. Throughout affinity chromatography, flow rate is 25 cm h^{-1}.

4. Gel filtration

A Sephadex G-100 column (5.3 cm² × 55 cm) is prepared and equilibrated with 50 mM Hepes-NaOH, 2 mM EDTA (pH 8.0). The pooled fractions are applied to the Sephadex column. Enzyme activity is recovered in the void volume. It is finally concentrated by ultrafiltration (see Section IV.A) and frozen in liquid nitrogen. When stored at −70°C the enzyme is stable for months.

5. Synthesis of the Sepharose-starch gel

Activation of Sepharose is performed as described earlier (see Section IV.A). For glucan coupling, 15 g soluble starch is suspended in 400 ml 10 mM potassium phosphate buffer (pH 8.0) and heated under continuous stirring. The mixture is cooled to room temperature and centrifuged (10 min at 20 000 × g). The pellet is discarded. The activated, washed Sepharose is suspended in the starch solution (supernatant) and is agitated overnight at room temperature. For deactivation, a mixture of 1.25 mol butylamine, 50 ml dimethylformamide and 50 ml 0.1M NaHCO$_3$, adjusted to pH 9.5 with HCl, is added to the Sepharose-starch suspension and is agitated for a further 2.5 h at room temperature. The gel is then washed as described earlier (see Section IV.A) but an additional washing step is performed using 1 litre of a 1:1 mixture of dioxin and H$_2$O.

Sepharose-starch gel particles are intensively stained with iodine. For a quantitative test, the Sepharose-starch gel (or Sepharose-dextrin, see Section IV.A) is subjected to exhaustive hydrolysis by amyloglucosidase from *Aspergillus niger*. The liberated glucose is monitored by the hexokinase/glucose 6-phosphate dehydrogenase assay. For freshly prepared Sepharose-starch gels, 20–30 μmol glucose released from each gram fresh weight is found. Glucose release from Sepharose-dextrin gels is one order of magnitude lower. When the affinity gels are stored for one year at 4°C, approximately half of the amount of the bound glucose is lost.

6. Low molecular weight glucans

Ten grams of commercial dextrins are freed from high molecular weight compounds as described above (see Section IV.A.1). After evaporation to dryness, dextrins are dissolved in a mixture containing 1 M NaCl, 10 mM Tris-acetate (pH 7.0), 0.4 mM PMSF, 15 µM thymol. The pH value is readjusted to 7.0 and the solution is brought to a final volume of 30 ml.

V. IMMUNOLOGICAL TECHNIQUES

In this section immunological techniques are described which are useful for studies on compartment-specific enzyme forms. Standard methods (such as antibody production or IgG purification) are not considered.

A. Immunotitration

The titre of antisera and the cross-reaction with a heterologous enzyme form is conveniently determined by immunotitration. The procedure consists of four steps: (1) incubation of a purified enzyme form with a serial dilution of an antiserum or an IgG preparation; (2) addition of polyethylene glycol in order to precipitate otherwise soluble immune complexes; (3) centrifugation; (4) photometric assay of the enzyme activity recovered in the supernatant.

A serial dilution of serum or purified IgG is prepared using Hepes-buffered saline (150 mM NaCl, 50 mM Hepes-NaOH, 2 mM EDTA, 0.1% (w/v) NaN_3, 0.6% (w/v) bovine serum albumin, pH 8.0) as diluting medium. One and a half micrograms of purified Type I or Type II phosphorylase, dissolved in 100 µl 50 mM Hepes-NaOH, 2 mM EDTA (pH 8.0), is mixed with 100 µl of each serum of IgG dilution. As a control, 100 µl enzyme solution is mixed with 100 µl of serum- or IgG-free Hepes-buffered saline. After 60 min incubation at room temperature, 100 µl of 24% (w/v) polyethylene glycol (approximate molecular weight 8000 kDa; Sigma P-2139), dissolved in 50 mM Hepes-NaOH, 2 mM EDTA (pH 8.0) is added slowly and under continuous agitation to each incubation mixture. Samples are then kept for 16 h at 4°C. Following centrifugation (10 min at 18 000 × g) phosphorylase activity in the supernatant is determined. Between 80 and 100% of the enzyme activity applied to each incubation mixture is recovered in control and pre-immune samples.

B. Removal of Cross-reacting Idiotypes

Compartment-specific enzyme forms catalyse the same reaction. Therefore, a limited structural similarity is likely to be observed. Although Type I and Type II phosphorylase forms are remarkably dissimilar in peptide pattern and immunological properties (Steup and Schächtele, 1986; Conrads et al., 1986), polyclonal antibodies raised against one enzyme type may contain some idiotypes which cross-react with the other. Cross-reactivity can be estimated by immunotitration (see Section V.A).

In the case described here, antibodies directed against a Type II phosphorylase from Pisum sativum L. were raised in rabbits and the IgG fraction was isolated from the antisera. The IgG preparations obtained reacted strongly with the homologous antigen (i.e. Type II phosphorylase) and weakly with the heterologous (Type I) enzyme form. This cross-reaction was detectable by immunotitration (see Section V.A) but not by conventional 'Western Blotting', according to Towbin et al. (1979). Cross-reaction was, however, clearly observed when following SDS-electrophoresis the electrotransfer was performed under partially renaturing conditions (Dunn, 1986). This result implies that the monospecificity of a polyclonal antibody preparation is more vigorously tested by the latter transfer technique. Cross-reacting antibodies were removed from the IgG preparations by the following procedure: a sufficient amount of the heterologous antigen (Type I enzyme) is coupled to tresyl-activated Sepharose; following deactivation the gel is incubated with the IgG preparation and non-binding idiotypes are recovered from the gel suspension. Binding of antibodies to the Sepharose particles can be monitored directly by fluorescence microscopy.

1. Coupling of the heterologous antigen

Two grams of tresyl-activated Sepharose (Pharmacia No. 03873) are hydrated according to the instructions of the producer. To the hydrated gel 1.2 mg purified Type I phosphorylase from *Pisum sativum* is added. (The enzyme preparation has previously been equilibrated with a solution containing 0.1 M $NaHCO_3$ and 0.5 M NaCl.) The gel suspension is incubated with the enzyme for 16 h at 4°C under continuous agitation (250 rpm).

2. Deactivation

Coupling of the heterologous antigen is terminated by adding 30 ml 0.1 M Tris-HCl (pH 8.0). Deactivation is performed for 4 h under agitation (4°C; 250 rpm). The gel is then washed with 40 ml of 0.1 M sodium acetate buffer (pH 4.0) containing 0.5 M NaCl and then with 40 ml of 0.1 M Tris-HCl (pH 8.0) containing 0.5 M NaCl. Each of the two washing steps is repeated twice. Finally, the gel is equilibrated with a mixture containing 170 mM NaCl, 20 mM Tris-HCl (pH 7.5), and 0.1% (w/v) NaN_3.

3. Incubation with IgG

Four millilitres of purified IgG solution are added to the gel suspension (10 ml). After agitation for 90 min at room temperature (250 rpm) the gel suspension is applied to sintered glass. Non-binding antibodies are recovered in the filtrate. They are concentrated by addition of an equal volume of saturated $(NH)_4SO_4$ solution and then centrifugation. The pelleted antibodies are dissolved in 4 ml HPB (50 mM Hepes-NaOH, 2 mM EDTA pH 8.0, 150 mM NaCl, and 0.1% (w/v) NaN_3). Following dialysis against HBS antibodies are stored frozen. They do not exhibit a noticeable cross-reaction with Type I phosphorylase (Fig. 7.4(a)).

4. Fluorescence microscopy

Cross-reactivity can be monitored by fluorescence microscopic observation of the Sepharose particles. The technique is essentially the same as described below for *in situ* localisation (see Section V.C); however, antigen fixation is omitted and all washing or incubation steps are performed using Eppendorf vials. After each step gel particles are pelleted by centrifugation for 1–2 s in a minifuge (Beckman). Sepharose particles containing covalently bound Type I phosphorylase are equilibrated with phosphate buffered saline (PBS) buffer (137 mM NaCl, 1.5 mM KH_2PO_4, 8 mM Na_2PO_4, 3 mM KCl, adjusted to pH 7.4 with NaOH) and an IgG solution, diluted in PBS, is added. The gel suspension is incubated for 60 min at room temperature (occasionally agitated). The gel suspension is then washed five times (5 min for each washing step) with PBS, containing 0.5% (w/v) BSA and 0.1% (w/v) NaN_3. Following washing, gel particles are incubated for 60 min at room temperature in a solution of anti-rabbit-IgG–FITC conjugate (for details see Section V.C) and are then washed five times as before. Finally gel particles are transferred into a mixture composed of 1 ml PBS plus 0.5% (w/v) BSA and 2 ml glycerol and containing 0.1% (w/v) p-phenylenediamine. Fluorescence microscopy is performed as described below (see Section V.C).

When anti-Type I-phosphorylase-IgG are applied, gel particles exhibit a strong fluorescence (Fig. 7.4(b)). A somewhat lower fluorescence is observed when anti-Type II-IgG are used. After removal of cross-reacting idiotypes no fluorescent gel particles are detected. These results are confirmed by 'Western-Blot' analysis (Fig. 7.4(a)).

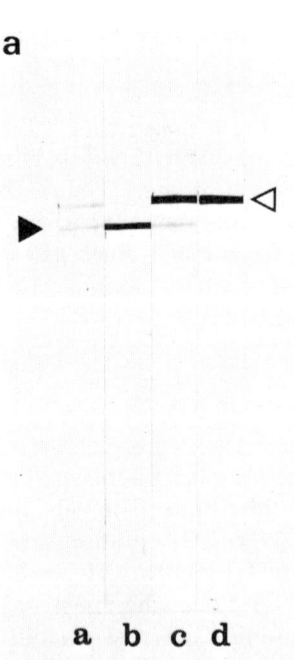
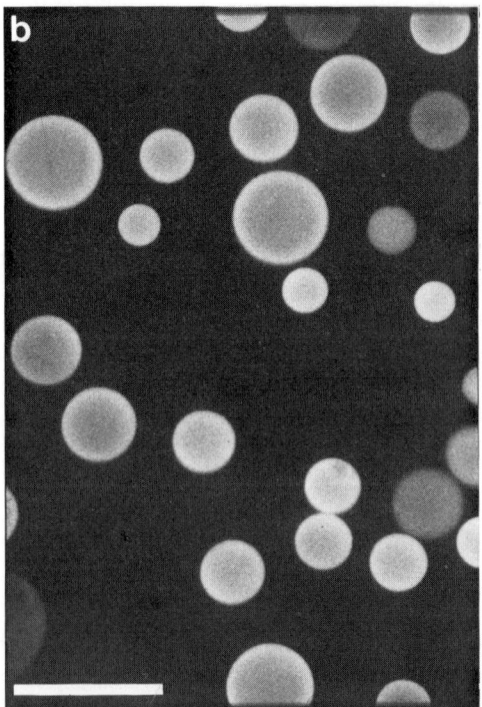

FIG. 7.4. (a) Removal of cross-reacting antibodies from an IgG preparation. Equal amounts of purified Type I (▶) and Type II (◁) glucan phosphorylase from *Pisum sativum* L. were denatured, mixed, and separated by denaturing polyacrylamide gel electrophoresis (7.5–15% (w/v) monomer concentration; 0.4 µg of each enzyme form per lane). Following electrophoresis, proteins were partially renatured and transferred to nitrocellulose according to Dunn (1986). Nitrocellulose was stained for protein (lane a) or was immunostained using peroxidase-coupled secondary antibodies (lanes b–d). For immunostaining a polyclonal anti-type I-IgG (lane b; no cross-reaction detectable) or an anti-type II-IgG (lanes c and d) was used. Lane c, IgG preparation containing cross-reacting idiotypes; lane d, IgG preparation after removal of cross-reacting idiotypes (see text). (b) FITC-labelled Sepharose particles which contain covalently bound Type I glucan phosphorylase from *Pisum sativum* L. Sepharose particles were immunolabelled by primary antibodies (anti-type I-phosphorylase-IgG) and a fluorochrome–IgG conjugate. Bar: 95 µm.

C. *In situ* Localisation of Glucan Phosphorylase Forms

Multiple forms of an enzyme can be visualised *in situ* by immunochemical techniques if they differ significantly in their immunological properties and monospecific antibodies are available (Schächtele and Steup, 1986; Conrads *et al.*, 1986). In the following, indirect immunofluorescence is described. This technique is based on light-microscopic detection of a fluorochrome which is covalently bound to a secondary antibody. *In situ* labelling of the antigen is achieved by the following steps: (a) fixation of the plant material (in order to immobilise the antigen to be localised); (b) sectioning of the fixed plant material; (c) incubation of the thin sections with the primary antibody (i.e. the antibody directed against the antigen to be localised); (d) labelling of the sections with the fluorochrome-conjugated secondary antibody. Non-bound primary antibodies or fluorochrome–IgG conjugates are removed from the specimen by washing. Washing and incubation of the thin sections are performed on a slide at room temperature.

1. *Tissue fixation*

Plant material is cut in small pieces (approximately 1 mm × 2 mm × 2 mm) using a fresh razor blade and is immediately transferred into 10 ml of a freshly prepared fixation medium. The latter contains 4–6% (w/v; see below) paraformaldehyde, dissolved in 50 mM sodium phosphate buffer. Paraformaldehyde is solubilised in the phosphate buffer by heating (up to 70°C; water bath is placed in a fume hood) and adding a few drops of 1 N NaOH. Following cooling to room temperature, the solution is passed through a filter paper and the pH value of the filtrate is readjusted. Optimal fixation conditions have to be determined for each plant material. Leaf samples are usually fixed for 2–3 h at room temperature in 4% (w/v) paraformaldehyde. They are slightly degassed at the onset of, and occasionally agitated during, the fixation period. Care is taken that cells are not destroyed during degassing. For other tissues (such as pea cotyledons or potato tubers) a fixaton in 6% (w/v) paraformaldehyde results in better tissue preservation. In this case, degassing is omitted but the fixation period is extended to 16 h (at room temperature). Following fixation samples are washed for several hours at room temperature in PBS (see Section V.B). During this period the washing medium is continuously replaced by fresh PBS. Finally, the fixed plant tissue is transferred to a solution containing 1 M sucrose in PBS and is incubated for 10–30 min at room temperature. This incubation tends to enhance the stability of the thin sections.

2. *Cryosectioning*

After treatment with sucrose the tissue samples are frozen and cryosectioned. Thin sections (16–20 μm thickness) are transferred into PBS containing 0.5% (w/v) BSA (PBS-BSA) and are washed with the same medium three times (10 min each).

3. *Immunolabelling of the thin sections*

Specimens are incubated for 1 h in a solution of primary antibodies (IgG fraction; diluted with PBS-BSA). After repeated washes (as above), thin sections are incubated for 1 h in darkness with fluorescence isothiocyanate (FITC) covalently bound to secondary antibodies. The fluorochrome conjugate is previously diluted with PBS-BSA

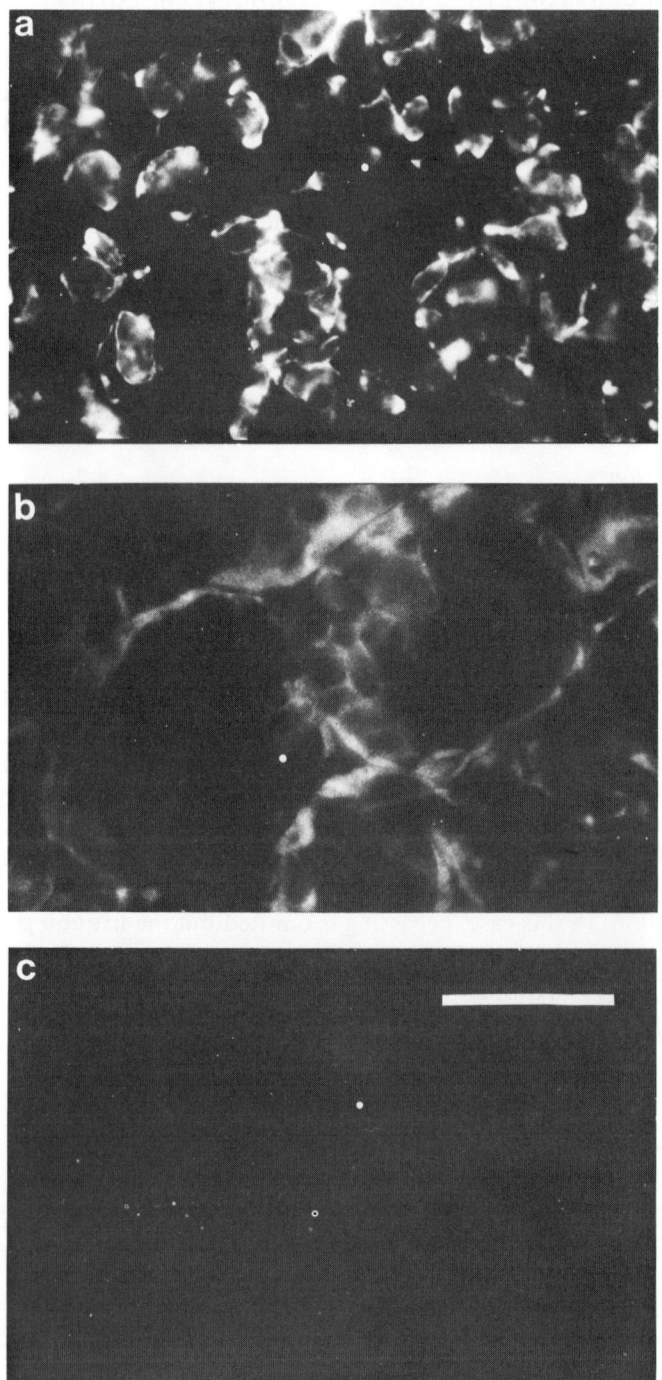

FIG. 7.5. *In situ* localisation of Type I and Type II glucan phosphorylases in cotyledons of *Pisum sativum* L. by indirect immunofluorescence (for details see text). (a) Plastidic enzyme form (Type II). (b) Cytosolic enzyme (Type I). (c) Pre-immune control. Bar: 40 μm.

which, in addition, may contain sodium azide in a final concentration of 0.1% (w/v). Suitable concentrations of primary antibodies and of the FITC conjugate have to be determined empirically. Specimens are then washed (as above) and are transferred to a mixture composed of 1 ml PBS and 2 ml glycerol and containing 0.1% (w/v) *p*-phenylenediamine (which retards bleaching of the fluorochrome). Specimens are covered with a cover glass, sealed with nail varnish and kept in darkness at room temperature. Microscopic evaluation and photographic documentation should be complete during the next 24 h.

4. Light microscopic evaluation and documentation

For fluorescence microscopy (epifluorescence) FITC-specific filter systems are used: excitation filter 450–490 nm; dichroic reflector 510 nm; barrier filter with transmittance between 520 and 570 nm. For micrographs a colour reversal film (e.g. Fujichrome 400) is used.

Any immunolabelling observed is relevant only if it is strictly dependent upon the primary antibody and is not detectable in the pre-immune control. Therefore, it is absolutely necessary to include a pre-immune control in every immunolabelling experiment. The pre-immune control is processed in exactly the same manner as the other specimens. In addition, it is often useful to confirm the specificity of the immunolabelling obtained by preparing specimens in which the primary antibody or fluorochrome conjugate has been omitted.

The localisation of Type I and Type II glucan phosphorylases in cotyledons of *Pisum sativum* is shown in Fig. 7.5.

ACKNOWLEDGEMENTS

Research in the author's laboratory has been made possible by grants from the Deutsche Forschungsgemeinschaft and from the Minister für Wissenschaft und Forschung des Landes Nordrhein-Westfalen.

REFERENCES

Bounias, M. (1980). *Anal. Biochem.* **106**, 291–295.
Conrads, J., van Berkel, J., Schächtele, C. and Steup, M. (1986). *Biochim. Biophys. Acta* **882**, 452–463.
Dunn, S. D. (1986). *Anal. Biochem.* **157**, 144–153.
Fossati, P. (1985). *Anal. Biochem.* **149**, 62–65.
Jolley, W. B. and Allen, H. W. (1965). *Nature (London)* **208**, 390–391.
Kakefuda, G. and Duke, S. H. (1984). *Plant Physiol.* **75**, 278–280.
Kanaya, K., Chiba, S. and Shimomura, T (1978). *Agric. Biol. Chem.* **42**, 1947–1948.
MacGregor, A. W. (1977). *Anal. Biochem.* **79**, 605–609.
Schächtele, C. and Steup, M. (1986). *Planta* **167**, 444–451.
Stegemann, H., Francksen, H. and Macko, V. (1973). *Z. Naturforschung* **28c**, 722–732.
Steup, M. (1981). *Biochim. Biophys. Acta* **659**, 123–131.
Steup, M. (1988). *In* "The Biochemistry of Plants" (J. Preiss ed.), Vol. 14, 255–296. Academic Press, New York.

Steup, M. and Gerbling, K.-P. (1983). *Anal. Biochem.* **134**, 96–100.
Steup, M. and Latzko, E. (1979). *Planta* **145**, 69–75.
Steup, M. and Melkonian, M. (1981). *Physiol. Plant* **51**, 343–348.
Steup, M. and Schächtele, C. (1986). *Planta* **168**, 222–231.
Steup, M., Schächtele, C. and Latzko, E. (1980a). *Z. Pflanzenphysiol.* **96**, 365–374.
Steup, M., Schächtele, C. and Latzko, E. (1980b). *Planta* **148**, 168–173.
Steup, M., Schächtele, C. and Melkonian, M. (1986). *Physiol. Plant.* **66**, 234–244.
Towbin, H., Staehelin, T. and Gordon, J. (1979). *Proc. Natl. Acad. Sci. USA* **76**, 4350–4354.

8 Enzymes of the Photorespiratory Carbon Pathway

RAY D. BLACKWELL[1], ALAN J. S. MURRAY[2] and PETER J. LEA[1]

[1]*Division of Biological Sciences, University of Lancaster, Lancaster, LA1 4YQ, UK and*
[2]*William Grant and Sons, The Distillery, Girvan, Ayrshire, KA26 9PT, UK*

I.	Introduction	130
II.	Phosphoglycolate phosphatase	132
	A. Reaction	132
	B. Assay	132
	C. Activity stain on native polyacrylamide gels	133
	D. Purification	133
	E. Regulation	134
III.	Glycolate oxidase and catalase	134
	A. Reaction	134
	B. Assay	134
	C. Purification	135
	D. Regulation	135
IV.	Aminotransferases	136
	A. Reaction	136
	B. Assay	136
	C. Purification	137
	D. Regulation	137
V.	Glycine–serine conversion	138
	A. Reaction	138
	B. Assay	138
	C. Purification	139

VI.	Hydroxypyruvate reductase	140
	A. Reaction	140
	B. Assay	140
	C. Activity stain on native polyacrylamide gels	140
	D. Purification	141
VII.	Glycerate kinase	141
	A. Reaction	141
	B. Assay	141
	C. Purification and regulation	141
	References	142

I. INTRODUCTION

The photorespiratory cycle of higher plants involves at least 11 different enzymes, all of which have been purified to varying extents. In this chapter we have decided to describe the methods available for assaying all of the enzymes in relatively crude extracts. Due to the limitations on space it has not been possible (unlike other chapters) to describe in detail full methods of purification. Wherever possible we have given the reader sufficient bibliographic detail to allow further study of each individual enzyme.

The photorespiratory cycle was initially detected as a series of physiological responses to prevailing light conditions. Probably the first clear indication of photorespiration was the demonstration of a reproducible burst of CO_2 evolution in the dark, from leaves of tobacco, immediately after a period of photosynthesis. Photorespiration has been shown, unlike dark respiration, to be inhibited by low levels of O_2 (1–2%) or high levels of CO_2 (Zelitch, 1971). Subsequently the process was shown to occur by a unique reaction mechanism totally distinct from dark respiration (Lorimer, 1981).

The carbon flow through photorespiration (Fig 8.1) involves three cellular organelles: the chloroplast, the peroxisome and the mitochondrion. Carbon enters the cycle as a result of the oxygenase reaction of ribulose 1,5-bisphosphate carboxylase/oxygenase (Rubisco). Oxygenation produces one molecule of the Calvin cycle substrate phosphoglycerate and one molecule of phosphoglycolate, which then enter the photorespiratory cycle. The metabolism of phosphoglycolate by this pathway permits 75% recovery of the carbon with 25% being liberated as CO_2. The function of photorespiration has, however, still not been fully identified; possible functions have been listed by Keys (1983).

The importance of certain aspects of the photorespiratory cycle remained unclear for a number of years with results varying depending on the investigative techniques employed. For example, only within the last 10 years has unequivocal evidence been presented that phosphoglycolate is the sole carbon source for photorespiration. Much of this evidence has come from the use of conditionally lethal mutants isolated using a screen devised by Somerville and Ogren (1979) working with the cruciferous plant *Arabidopsis*. This novel method allowed for the fact that when C_3 plants were grown at elevated levels of CO_2 the kinetics of Rubisco permitted little or no phosphoglycolate to be synthesised. Under these conditions any plant containing mutant non-functional enzymes of the photorespiratory cycle would grow normally. However, on exposure to air the photorespiratory cycle would not operate, with mutants showing symptoms of

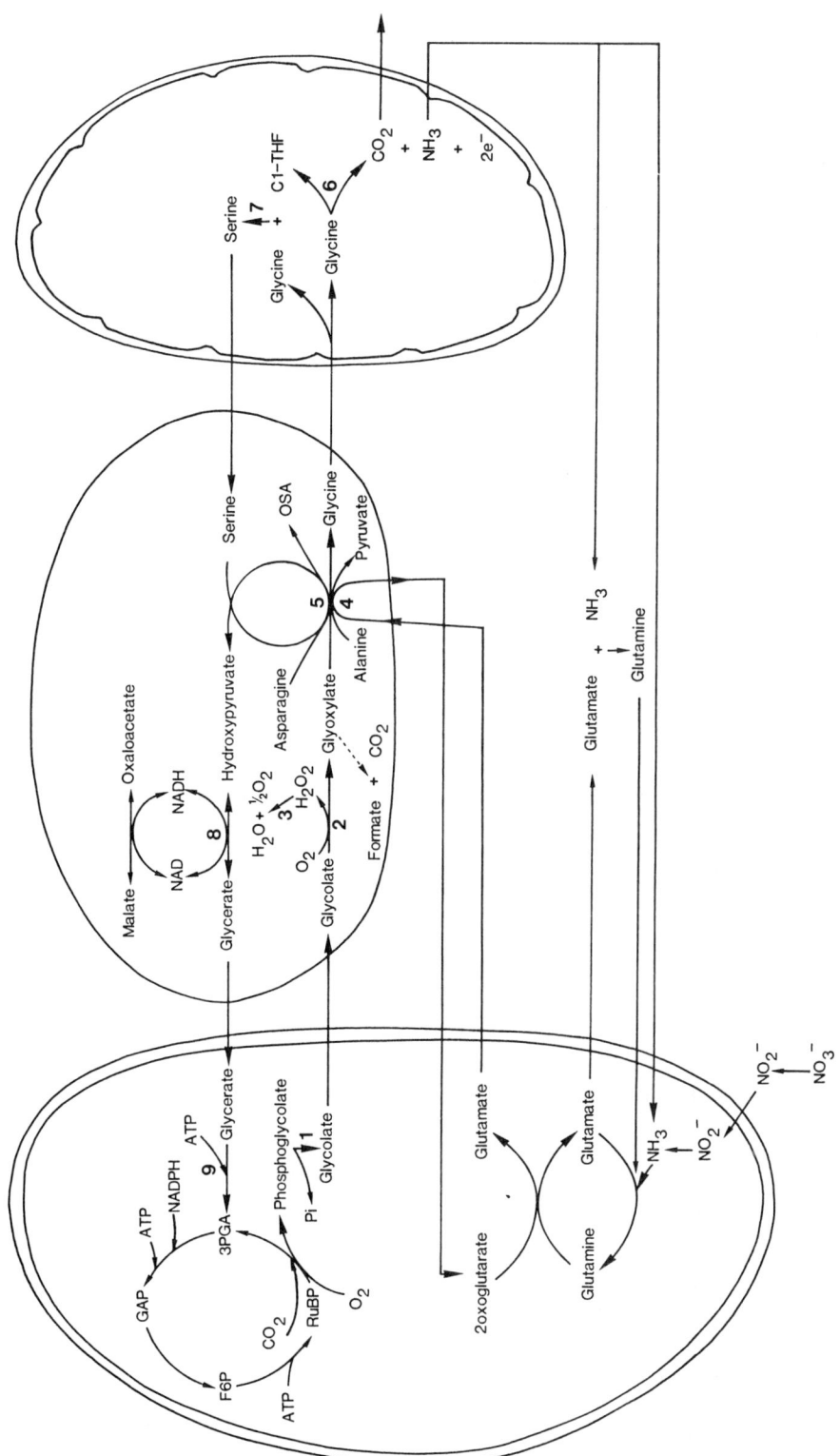

FIG. 8.1. Photorespiratory carbon and nitrogen cycles in C$_3$ plants. The numbers are for the name of the enzyme: 1, phosphoglycolate phosphatase; 2, glycolate oxidase; 3, catalase; 4, glutamate:glyoxylate aminotransferase; 5, serine:glyoxylate aminotransferase; 6, glycine decarboxylase; 7, serine transhydroxymethylase; 8, hydroxypyruvate reductase; 9, glycerate kinase.

stress, resulting in death on continued exposure to air. By regular examination of plants, mutants lacking enzymes of the photorespiratory cycle could be 'rescued' by placing them back in an atmosphere containing elevated levels of CO_2. Subsequent work with mutants of barley and pea has confirmed and extended this work into agronomically important crops. At present mutants are available in all but three steps of the pathway (for review, see Blackwell *et al.*, 1988a). Photorespiratory carbon metabolism is intimately connected with nitrogen metabolism through glutamine synthetase (GS) and glutamate synthase activities (Fig. 8.1). Due to the large volume of literature on these two enzymes alone, they are considered separately (see Lea *et al.*, Chapter 15, this volume). All of the remaining activities can be assayed in crude extracts of barley prepared in the following extraction buffer: 40 mM Tris pH 7.8 containing 2 mM $MgCl_2$, 1 mM ethyldiaminetetraacetic acid (EDTA), 2% (w/v) polyvinyl pyrrolidone (PVP), 1 mM phenylmethylsulphonyl fluoride (PMSF), 1 mM aminocaproic acid and 1.4 mM dithiothreitol (DTT).

II. PHOSPHOGLYCOLATE PHOSPHATASE

A. Reaction

In photosynthetic organisms, phosphoglycolate phosphatase (EC 3.1.3.18) specifically catalyses the hydrolysis of 2-phosphoglycolate, a product of the oxygenation reaction of Rubisco, to produce glycolate and inorganic phosphate. Synthesis of phosphoglycolate and its subsequent hydrolysis are the first two steps of the photorespiratory cycle and occur in the chloroplast. Phosphoglycolate phosphatase has the important role of removing phosphoglycolate, a potent inhibitor of the Calvin cycle enzyme triose phosphate isomerase (Wolfenden, 1969, 1970; Anderson, 1971; Boyle and Keys, 1982).

B. Assay

The assay most commonly used to detect phosphoglycolate phosphatase activity is the two-step method described by Anderson and Tolbert (1966):

(1) Production of glycolate and inorganic phosphate from phosphoglycolate.
(2) Measurement of inorganic phosphate using trichloroacetic acid (TCA)/acid molybdate (Ames, 1965).

1. Production of glycolate and inorganic phosphate

The 360 μl reaction mixture contains 10 mM $MgSO_4$, 40 mM cacodylic acid (pH 6.3) and 10 mM phosphoglycolate. The reaction is initiated by the addition of crude extract and terminated after 5 min incubation at 30°C by the addition of 120 μl 10% (w/v) TCA. The mixture is centrifuged to remove precipitated protein and the supernatant used for the determination of inorganic phosphate.

2. Determination of inorganic phosphate

Reagents required:

(1) 0.5% (w/v) TCA
(2) Acid molybdate reagent

Prepare 100 ml of 4% (w/v) ammonium molybdate in distilled water and separately 100 ml of 2.5 M sulphuric acid. Mix these two solutions and make up to 250 ml with distilled water. Before use, $FeSO_4$ is added to this solution to make final concentration of 5% (w/v).

The reaction mixture contains 0.5 ml of 0.5% TCA, 0.5 ml acid molybdate/$FeSO_4$ reagent and 0.5 ml supernatant from the phosphoglycolate phosphatase assay. This mixture is incubated for 15 min at 20°C and the absorbance of the blue colour is measured at 740 nm. Phosphate concentration and hence enzyme activity is then calculated by reference to a calibration graph prepared using phosphate standards made up in 2.5% (w/v) TCA.

C. Activity Stain on Native Polyacrylamide Gels

Phosphoglycolate phosphatase activity can also be detected following non-denaturing polyacrylamide gel electrophoresis using an activity stain devised by Christeller and Tolbert (1978), which visualises the region of enzyme activity by the formation of a white precipitate of lead acetate. The stain is made with two freshly prepared solutions:

(a) 200 mM cacodylic acid, pH 6.3 containing 12 mM lead acetate and 12 mM magnesium acetate;
(b) 25 mM phosphoglycolate in distilled water.

The gel is incubated at 30°C with a solution made up of equal volumes of (a) and (b), mixed immediately prior to addition.

Band enhancement can then be achieved through incubation of the gels with 5% (w/v) ammonium sulphide for 2 min and then washing in water (Christeller and Tolbert, 1978).

D. Purification

Phosphoglycolate phosphatase has been purified to near homogeneity from tobacco and spinach (Christeller and Tolbert, 1978; Husic and Tolbert, 1984) and partially purified from a range of other species including wheat, peas and *Phaseolus* (Yu *et al.*, 1964; Kerr and Gear, 1974; Vérin-Vergeau *et al.*, 1980). Using ion-exchange chromatography and chromatofocusing, Belanger and Ogren (1985) purified phosphoglycolate phosphatase from *Nicotiana rustica*. Using this preparation they raised polyclonal antibodies which have been shown to cross-react with *Nicotiana tabaccum*, *Glycine max* and *Arabidopsis thaliana* but not with a mutant of *A. thaliana* lacking phosphoglycolate phosphatase activity (Somerville and Ogren, 1979).

Two isoenzymes of phosphoglycolate phosphatase activity have been detected in

Phaseolus, using acetone fractionation and DEAE-cellulose chromatography (Vérin-Vergeau *et al.*, 1979), and barley (Hall *et al.*, 1987), both of which have a high specificity for phosphoglycolate. It has been shown that the native molecular weight of the enzyme is between 81.6 and 86 kDa and that SDS-PAGE shows a single band with a molecular weight between 19 and 22.4 kDa, suggesting that the enzyme is a tetramer (Christeller and Tolbert, 1978). By contrast, phosphoglycolate phosphatase from *Nicotiana* (Belanger and Ogren, 1985) has been shown to have a molecular weight of 58 kDa with two subunits of 32 kDa.

Enzyme activity during purification and storage may be maintained by the presence of isocitrate (Christeller and Tolbert, 1978).

E. Regulation

There appears to be no variation of the enzyme activity *in vivo* (Husic *et al.*, 1987) although enzyme from all sources requires a divalent cation for activity. Any regulation that occurs seems to be via the production of the substrate, phosphoglycolate, by the enzyme Rubisco (Husic *et al.*, 1987). Few inhibitors of phosphoglycolate phosphatase are known, although ribose-5-phosphate has been shown to bind to the allosteric site on the enzyme (Christeller and Tolbert, 1978), the significance of which has not been fully elucidated.

III. GLYCOLATE OXIDASE AND CATALASE

A. Reaction

During photorespiration, the peroxisomal enzyme glycolate oxidase (EC 1.1.3.15) catalyses the essentially irreversible oxidation of glycolate by O_2 to form glyoxylate. During this reaction hydrogen peroxide is produced which is subsequently degraded to water and oxygen by catalase (EC 1.11.1.6). Isolation of a barley plant lacking catalase activity confirmed that the major role for catalase in C_3 plant leaves is to detoxify hydrogen peroxide produced during photorespiration (Kendall *et al.*, 1983). Levels of glycolate oxidase in C_3 plants have been reported to be between 20–170 μmol glycolate oxidised h^{-1} (mg Chl)$^{-1}$ while levels have been shown to be at least four times lower in C_4 species (Zelitch, 1971). Rates of catalase activity in tobacco were as high as 600 μmol min^{-1} (mg protein)$^{-1}$ (Havir and McHale, 1987).

B. Assay

Glycolate oxidase activity is most simply measured by monitoring the rates of O_2 uptake using an oxygen electrode with the following assay conditions:

The reaction mixture of 800 μl is made up with 50 mM Tris (pH 8.3) containing 1 mM sodium azide and 0.1 mM FMN with crude leaf extract; the reaction is initiated by adding 100 μl of 10 mM glycolate in 50 mM Tris (pH 8.3). Sodium azide is essential to inhibit catalase activity. Methods of assaying the enzyme have also been developed based on the ability to quantify glycolate-dependent production of H_2O_2 by oxidation of dyes (Frigerio and Harbury, 1958; Zelitch, 1971). Glyoxylate production can also be quantified by monitoring the phenylhydrazone derivative (Tolbert, 1985). However, this reaction is not totally specific for glyoxylate, since acids such as L-lactic acid may also react (Tolbert et al., 1949; Kenten and Mann, 1952).

Catalase activity can be assayed spectrophotometrically by monitoring the change in absorbance at 240 nm based on a method originally devised by Lück (1962). Into a quartz cell are placed 3.0 ml of 10 mM Tris-HCl (pH 8.5) and 0.1 ml of 0.88% H_2O_2 in 100 mM Tris-HCl (pH 8.5). The reaction is started by adding 0.2 ml crude extract and the change in A_{240} monitored against a blank of Tris-HCl and extract. It is important to use fresh reagents for this assay and to keep them at room temperature. Inhibition of this activity at 4°C (or kept on ice) is known.

Catalase activity can be detected with gel electrophoresis using starch gels, after the method of Scandalios (1968). Following separation on 15% starch gels, staining is achieved by soaking for 1 min in 0.5% hydrogen peroxide solution. The gel is then washed with distilled water and immersed in 1% (w/v) potassium iodide (acidified with glacial acetic acid). Peroxide releases iodine which stains the starch gel dark blue, except where the catalase activity has destroyed the peroxide.

The above method is restricted to starch gels and Gregory and Fridovich (1974) have described other methods based on the use of diaminobenzidine. Catalase activity may also be detected in electron microscope sections of peroxisomes (Parker and Lea, 1983).

C. Purification

Glycolate oxidase has been purified from the leaves of a number of species including spinach (Zelitch, 1955; Lindqvist and Branden, 1985), pea (Kerr and Groves, 1975) and also from pumpkin and cucumber cotyledons (Behrends et al., 1982; Nishimura et al., 1983). Kerr and Groves (1975) described a six-step method which utilised the high isoelectric point of the enzyme. Using glycolate oxidase purified to homogeneity from spinach as a basis for the construction of synthetic oligonucleotides, Volokita and Somerville (1987) have identified a cDNA clone that encodes for glycolate oxidase.

Hall et al. (1985) also describe a rapid three-stage partial purification of glycolate oxidase from a number of species including wheat, spinach, pea and tobacco. These preparations are then used to compare the quaternary structure of the enzyme amongst these species.

Catalase has been purified to homogeneity from lentil (Scheiffer et al., 1976) and cucumber (Lamb et al., 1978). The native enzyme has a molecular weight of 225 kDa and is composed of four equal subunits of 54 kDa.

D. Regulation

Relatively high concentrations of glycolate oxidase are found in green leaves (Zelitch, 1971). Activity is inducible by both light (Tolbert and Burris, 1950; Huang et al., 1983) and glycolate (Huang et al., 1983). Such substrate activation is typical of several peroxisomal enzymes (Huang et al., 1983).

Multiple forms of catalase have been detected in spinach (Galston et al., 1951), mustard (Drumm and Schopfer, 1974) barley (Kendall et al., 1983) and tobacco (Havir and McHale, 1987). In maize three genetically distinct catalase proteins have been purified to homogeneity and characterised biochemically (Scandalios et al., 1984).

Mutants lacking specific maize catalase isoenzymes have been examined in detail (Tsaftaris et al., 1983).

A rapid screen for catalase, glycolate oxidase and phosphoglycolate phosphatase activities has been devised by Turner and Hall (1988). The assays depend on monitoring the production of a red dye from aminoantipyrene, hydrogen peroxide and phenol utilising peroxidase as a catalyst.

IV. AMINOTRANSFERASE REACTIONS

A. Reaction

Plants contain a multitude of aminotransferase reactions (Givan, 1980; Ashton et al., Chapter 3 and Ireland and Joy, Chapter 16, this volume) most of which, apart from serine : glyoxylate aminotransferase (SGAT) are thought to be reversible. The two activities involved in photorespiration, i.e. glutamate : glyoxylate aminotransferase (GGAT; EC 2.6.1.4) and SGAT (EC 2.6.1.45) are carried on different protein molecules, both of which are located in the peroxisomes. Most aminotransferase enzymes are not specific for one single amino/keto acid pair. The proteins responsible for GGAT also carry glutamate : pyruvate aminotransferase (GPAT) and alanine : glyoxylate aminotransferase activities (Rehfeld and Tolbert, 1972). Analysis of a barley plant lacking SGAT (Murray et al., 1987) confirmed that this enzyme is also responsible for asparagine : glyoxylate aminotransferase and serine : pyruvate aminotransferase activities as previously suggested (Rehfeld and Tolbert, 1972; Ireland and Joy, 1983).

B. Assay

Enzyme activities can be assayed providing an acceptable method of separating the amino acids, reactant and product, is available. This can be accomplished simply using TLC (Murray et al., 1987) to monitor the production of glycine during the assay of SGAT and GGAT. It is, however, difficult to quantify the results and for more precise work it is desirable to separate and quantify glycine using a more sensitive amino acid analysis technique, e.g. HPLC separation of o-phthaldialdehyde (OPA) derivatives (Fleury and Ashley, 1983). Whichever method of amino acid analysis is chosen, the following assay method can be employed.

In a final volume of 100 µl, 50 mM Hepes (pH 7.0) should contain 10 mM glyoxylate and either 100 mM glutamic acid for the assay of GGAT or 100 mM serine for the assay of SGAT. The reaction is started by adding 50 µl crude enzyme extract. After incubation at 30°C for 90 min the reaction is stopped by adding 450 µl of absolute alcohol and the mixture centrifuged to precipitate the protein.

A suitable semi-quantitative TLC method for separating the amino acid product glycine from the reaction is given below.

Apply 10 µl of the assay mix to a cellulose TLC plate and develop the plate in one direction using n-butanol–acetone–diethylamine–water (70:70:14:35; v/v). After allowing the acetone to evaporate in air, place the plate in a hot (120°C) oven for 60 min to reduce background interference from amines and then spray the plate evenly with a 5% solution of ninhydrin in acetone.

Rehfeld and Tolbert (1972) describe a spectrophotometric assay for SGAT. In principle, the production of hydroxypyruvate from serine by SGAT is coupled to the subsequent reduction of hydroxypyruvate by added hydroxypyruvate reductase (HPR). Since the latter reaction is coupled to NADH oxidation a change in A_{340} can be detected. However, commercial preparations of HPR also contain substantial levels of glyoxylate, an alternative substrate for the HPR reaction (see later). Since the K_m of HPR for glyoxylate (15 µM in spinach) is much higher than for hydroxypyruvate (120 µM), with careful attention to substrate concentration, low background rates of NADH oxidation can be achieved.

For this assay the reaction mix (1 ml) contains 70 mM Hepes (pH 7.0), 0.15 mM NADH, 1 mM glyoxylate, 0.05 units of hydroxypyruvate reductase and enzyme extract. The reaction is initiated with 20 mM serine and NADH oxidation monitored at 340 nm.

GGAT activity can be assayed in a similar way using one of the alternative activities (GPAT) following the method of Biekmann and Feierabend (1982). The assay mix (1 ml) contains 90 mM phosphate buffer (pH 7.5), 50 mM L-alanine, 0.2 mM NADH, 5 µl (0.5 mg ml^{-1} in 50% glycerol) of lactate dehydrogenase and enzyme extract. The reaction is initiated with 6.7 mM 2-oxoglutarate (pH 7.0) and the NADH oxidation followed at 340 nm.

GGAT activity can be determined specifically by the use of an assay utilising [1-^{14}C]glyoxylate. For the full assay procedure see Aminotransferases (Ireland and Joy, Chapter 16, this volume).

C. Purification

SGAT has been partially purified from leaves of oats (Brock et al., 1970), beans (Smith, 1973), peas (Ireland and Joy, 1983) and spinach peroxisomal fraction (Nakamura and Tolbert, 1983). SGAT has also been purified to homogeneity from cotyledons of cucumber using two-step ammonium sulphate precipitation, gel filtration and DEAE-cellulose chromatography (Hondred et al., 1985). SDS-PAGE revealed two bands and antibodies raised to these proteins showed no cross-reactivity with other peptides from the cotyledon homogenate. For a review of the kinetic properties and substrate specificity, see Husic et al. (1987).

Similarly GGAT has been partially purified from a number of sources including spinach peroxisomes (Nakamura and Tolbert, 1983), *Euglena* (Foley and Beale, 1982) and *Chlorella* (Shioi et al., 1984).

D. Regulation

It is suggested from the multiplicity of substrates for these aminotransferases that the photorespiratory nitrogen cycle may not, in fact, be a closed one as originally proposed by Keys et al. (1978), but may involve an input of nitrogen from alanine via GGAT. Evidence to support this view came originally from ^{15}N labelling studies (Betsche, 1983; Ta and Joy, 1986) and more recently from studies with plants lacking GS/glutamate synthase (Blackwell et al., 1988b). The rate of CO_2 fixation by leaves of these mutants declines dramatically with onset of photorespiration; however, supplying alanine can overcome this effect. It is still not clear to what extent alanine enters the cycle *in vivo*.

V. GLYCINE–SERINE CONVERSION

A. Reaction

Glycine, produced in plants during photorespiration from the transamination of glyoxylate, is metabolised by a complex series of reactions in the mitochondria (Walker and Oliver, 1986; Bourguignon et al., 1988). It has recently been proposed (Sarojini and Oliver, 1983; Bourguignon et al., 1988) that the mechanism of glycine oxidation in plants is similar to that elucidated for animal liver and microorganisms (Kikuchi, 1973).

Overall, the sequence involves two stages and utilises two molecules of glycine (Fig. 8.2). In the first reaction, glycine decarboxylase (the glycine cleavage system; EC 2.1.2.10) catalyses the oxidation of glycine to yield CO_2, ammonia and 5,10-methylene tetrahydrofolate. The mitochondrial isoenzyme of serine transhydroxymethylase (STHM; EC 2.1.2.1) then transfers the C_1 group of 5,10-methylene tetrahydrofolate to a second glycine molecule to yield a molecule of serine. Glycine decarboxylase also catalyses the exchange between the carboxyl carbon of glycine and bicarbonate (Clandinin and Cossins, 1975).

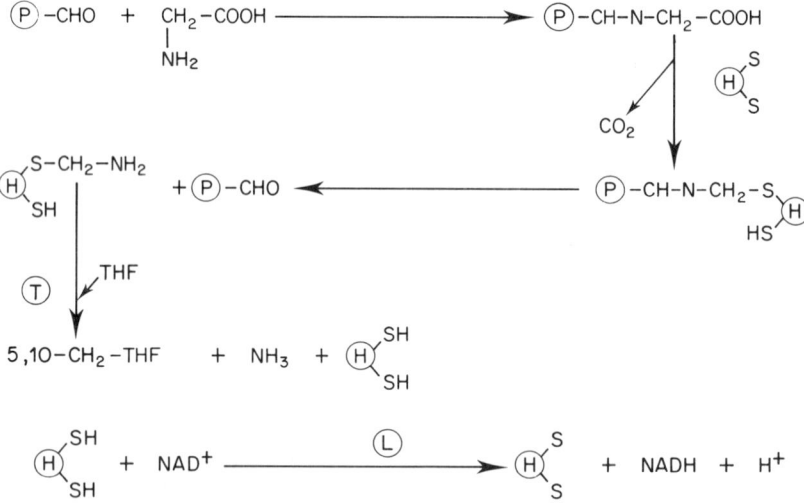

FIG. 8.2. Scheme for the reaction of reversible glycine cleavage. P, H, L and T in the circles represent the respective proteins (Kikuchi, 1973).

B. Assay

1. Glycine decarboxylase

The properties of glycine decarboxylase that allow exchange between the carboxyl carbon and bicarbonate, enable an estimate of the activity of the enzyme complex to be measured by the use of [^{14}C]bicarbonate using a method similar to that of Walker et al. (1982).

The reaction mixture made up in 20 mM MOPS/KOH (pH 7.0), contains 0.1 mM pyridoxal phosphate, 20 mM glycine, 2 mM DTT and crude enzyme extract, is initiated by the addition of 4 μmol NaH^{14}CO$_3$ (55 mCi mol^{-1}). This is incubated for 30 min at 30°C and terminated with 25 μl glacial acetic acid.

After termination, dissolved ^{14}CO$_2$ is driven off by drying the samples under an infrared lamp and acid stable [^{14}C]glycine estimated by scintillation counting.

Glycine decarboxylation can be measured in a manner similar to that of Somerville and Ogren (1982) and Walker and Oliver (1986) by trapping ^{14}CO$_2$ liberated in an assay system supplied with [^{14}C]glycine. Crude enzyme extract is incubated with 8 mM [^{14}C]glycine (3.5 μCi μmol^{-1}) in a suitable assay buffer (20 mM MOPS/KOH (pH 7.0), 30 mM pyridoxal phosphate, 0.5 mM tetrahydrofolate, 1 mM NAD and 2 mM DTT). The reaction mixture is placed in small cups suspended over 10% (v/v) triethanolamine in 20 ml scintillation vials, and the reaction initiated by the addition of glycine. The reaction is terminated after 30 min by the addition of 100 μl 6 M acetic acid and allowed to stand for 16 h to permit trapping of all ^{14}CO$_2$. The amount of trapped ^{14}C radioactivity can then be determined by scintillation counting.

Glycine oxidation can also be assayed by measuring the formation of NADH or serine that is dependent on the presence of both glycine and tetrahydrofolate by the method of Cossins (1987) and Bourguignon *et al.* (1988), although this method is more complex than those described above.

2. Serine transhydroxymethylase

The simplest estimation of serine transhydroxymethylase activity is based on the production of a radioactive C-1 unit in 5,10-methylene tetrahydrofolate from [3-^{14}C]serine using the reverse reaction of serine transhydroxymethylase (Taylor and Weissbach, 1965).

The complete assay system made up in 75 mM potassium phosphate buffer, pH 7.4 comprises 0.25 mM [3-^{14}C]serine (6.56 μCi μmol^{-1}), 0.25 mM pyridoxal phosphate, 2 mM DL-tetrahydrofolate, 10 mM mercaptoethanol and enzyme extract.

The reaction mixture (except serine) is incubated for 5 min at 37°C and the reaction initiated by the addition of [3-^{14}C]serine. After 15 min incubation at 37°C the reaction is terminated by the addition of 0.3 ml of 1 M sodium acetate (pH 4.5). This is followed by the addition of 0.2 ml 0.1 M formaldehyde and 0.3 ml of 0.4 M dimedon (in 50% ethanol) and boiled for 5 min to accelerate formation of the formaldehyde/dimedon derivative. After cooling, the dimedon compound is extracted by shaking with 5 ml toluene at 20°C and the phases separated by centrifugation. The upper phase is removed for quantification of radioactivity in 5,10-methylene tetrahydrofolate by scintillation counting.

C. Purification

The glycine decarboxylase complex has recently been isolated in an active form from spinach mitochondria (Neuburger *et al.*, 1986) and pea leaf mitochondria (Walker and Oliver, 1986; Bourguignon *et al.*, 1988).

Characterisation of the complex from pea has demonstrated that at least four proteins are involved in the oxidation of glycine. These have been designated P, H, T, and L proteins in plants, equivalent to P1, P2, P3 and P4 proteins in bacteria (Klein and Saggers, 1966). The glycine cleavage system in plants is confined to the mitochondria (Walker and Oliver, 1986) and studies of the mitochondrial matrix proteins from pea leaves have shown that these proteins are separable by SDS-PAGE (Douce, 1985).

Serine transhydroxymethylase has been partially purified from a wide range of sources including wheat and spinach leaves and maize seedlings (Woo, 1979) and purified to homogeneity from mung bean seedlings (Rao and Appaji Rao, 1982).

VI. HYDROXYPYRUVATE REDUCTASE

A. Reaction

Hydroxypyruvate produced through the transamination of serine in the peroxisome, is reduced to D-glycerate by a very active peroxisomal enzyme, hydroxypyruvate reductase (EC 1.1.1.29: Kohn and Warren, 1970; Tolbert et al., 1970). This enzyme activity is reversible and can also utilise glyoxylate as a substrate, although only at high unphysiological concentrations. The reductant with which hydroxypyruvate reductase is commonly associated is NADH, which reacts at least 10 times faster than NADPH in the reduction of hydroxypyruvate. Recently, however, a novel hydroxypyruvate reductase, preferring NADPH to NADH as a cofactor, has been isolated in spinach (Kleczkowski and Randall, 1988a). Further work by Kleczkowski and colleagues (Kleczkowski et al., 1988) and work with a mutant of barley lacking NADH-dependent hydroxypyruvate reductase activity (Murray et al., 1989) has demonstrated the presence, and started to examine the physiological significance, of this enzyme in other species. This NADPH-preferring enzyme has been shown to be localised in the cytoplasm (Givan et al., 1988; Kleczkowski et al., 1988).

B. Assay

The assay for NADH or NADPH-dependent hydroxypyruvate reductase activity is based on that described by Kohn and Warren (1970) modified by Stabenau (1974) and Kleczkowski and Randall (1988a). The assay follows the oxidation of either NADH or NADPH spectrophotometrically at 340 nm.

The 1 ml assay mixture contains 0.5 mM hydroxypyruvate, 25 mM phosphate buffer (pH 6.3), 0.2 mM NADH or NADPH and varying amounts of enzyme extract. Reactions are initiated by the addition of hydroxypyruvate and control assays contain the components except hydroxypyruvate to correct for non-specific oxidation of NADH or NADPH.

C. Activity Stain on Native Polyacrylamide Gels

Hydroxypyruvate reductase activity can be detected on non-denaturing polyacrylamide gels using the reverse reaction, glycerate oxidation, which results in the appearance of reduced Nitroblue tetrazolium in the gel as a purple precipitate (Titus et al., 1983).

Activity staining is carried out for 20–30 min in the dark after non-denaturing polyacrylamide gel electrophoresis using a solution comprising: 200 mM Tris (pH 8.9),

58 mM DL-glycerate, 0.33 mM Nitroblue tetrazolium, 2.2 mM NAD, 0.081 mM phenazine methosulphate.

The reaction is stopped by the addition of 7.5% (v/v) acetic acid, which also removes the background yellow colour from the gel.

D. Purification

The high activity and simple assay procedure of hydroxypyruvate reductase activity have made it a useful marker for peroxisomal fractions, and also led to detailed studies of the enzyme. NADH-hydroxypyruvate reductase has been purified from spinach (Zelitch, 1953), cucumber (Titus et al., 1983) and partially purified from Chlamydomonas (Husic and Tolbert, 1985) and NADPH-hydroxypyruvate reductase activity has been purified from spinach (Kleczkowski and Randall, 1988a).

The native molecular weight of NADH-hydroxypyruvate reductase has been shown to be 90–95 kDa (Kohn et al., 1970; Titus et al., 1983; Kleczkowski et al., 1987) with a subunit mass of between 41 and 43 kDa (Kleczkowski et al., 1986). NADPH-hydroxypyruvate reductase has a molecular weight of approximately 70 kDa comprising two subunits, each of molecular mass 38 kDa (Kleczkowski and Randall, 1988a).

VII. GLYCERATE KINASE

A. Reaction

Glycerate kinase (EC 2.7.1.31) is located in the chloroplast where it catalyses the phosphorylation of glycerate, returning carbon from the photorespiratory pathway into the sugar phosphate pool. Activities of glycerate kinase extracted from leaves are generally high (Schmidt and Edwards, 1983) and are known to exceed the maximal estimated rate of glycerate transport into chloroplasts (Robinson, 1982).

B. Assay

Glycerate kinase activity can be measured spectrophotometrically by linking 3-PGA and/or ADP formation to NADH oxidation in a coupled assay (Kleczkowski and Randall, 1988b).

The most suitable set of coupling enzymes (Fig. 8.3) use a reaction mixture (vol 1.1 ml) placed in a cuvette containing 0.2 mM NADH (pH 7.0), 5 mM ATP (pH 7.0), 10 mM $MgCl_2$ and 100 mM Tris (pH 7.8), 5 units of phosphoglycerate phosphokinase and 5 units of glyceraldehyde 3-phosphatase and varying amounts of crude enzyme extract. The reaction is initiated by adding 0.1 ml of 100 mM glycerate and the change in A_{340} noted.

C. Purification and Regulation

Glycerate kinase has been purified from peas and rye (Schmidt and Edwards, 1983), spinach (Kleczkowski and Randall, 1983; Chagatura, 1985; Kleczkowski et al., 1985) and maize (Kleczkowski and Randall, 1988b), and shown to be a monomer of around

40 kDa with Mg-ATP as the preferred phosphate donor. No specific control mechanisms are thought to exist. The availability of substrate within the chloroplast stroma probably determines the activity. Activity stimulation by thiols and light have been suggested (Kleczkowski and Randall, 1985). Physiological changes in pH, Mg^{2+} concentration or energy charge within the chloroplast have little effect on the activity of the enzyme.

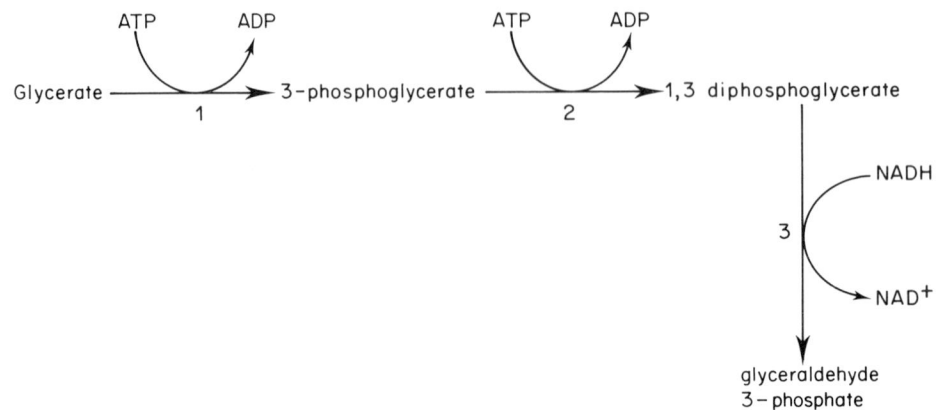

FIG. 8.3. Reaction mechanism for glycerate kinase assay. 1, glycerate kinase; 2, phosphoglycerate phosphokinase; 3, glyceraldehyde 3-phosphatase dehydrogenase system.

REFERENCES

Ames, B. N. (1965). *Meth. Enzymol.* **8**, 115–118.
Anderson, D. E. and Tolbert, N. E. (1966). *Meth. Enzymol.* **9**, 646–650.
Anderson, L. E. (1971). *Biochim. Biophys. Acta* **235**, 237–244.
Behrends, W., Rausch, U., Löffler, H.-G. and Kindl, H. (1982). *Planta* **156**, 566–571.
Belanger, F. C. and Ogren, W. L. (1985). *Plant Physiol.* **77**, 601S.
Betsche, T. (1983). *Plant Physiol.* **71**, 961–965.
Biekmann, S. and Feierabend, J. (1982). *Biochim. Biophys. Acta* **721**, 268–279.
Blackwell, R. D., Murray, A. J. S., Lea, P. J., Kendall, A. C., Hall, N. P., Turner, J. C. and Wallsgrove, R. M. (1988a). *Photosynth. Res.* **16**, 155–176.
Blackwell, R. D., Murray, A. J. S., Lea, P. J. and Joy. K. W. (1988b). *J. Exp. Bot.* **39**, 845–858.
Bourguignon, J., Neuburger, M. and Douce, R. (1988). *Biochem. J.* **255**, 169–178.
Boyle, F. A. and Keys, A. J. (1982). *Photosynth. Res.* **3**, 105–111.
Brock, B. L. W., Wilkinson, D. A. and King, J. (1970). *Can. J. Biochem.* **48**, 486–492.
Chagatura, R. (1985). *Physiol. Plant.* **63**, 19–24.
Christeller, J. T. and Tolbert, N. E. (1978). *J. Biol. Chem.* **253**, 1780–1785.
Clandinin, M. T. and Cossins, E. A. (1975). *Phytochemistry* **14**, 387–391.
Cossins, E. A. (1987). *In* "Biochemistry of Plants, Biochemistry of Metabolism" (D. D. Davies, ed.), Vol. 11, pp. 317–353. Academic Press, New York.
Douce, R. (1985). "Mitochondria in Higher Plants, Structure, Function and Biogenesis." Academic Press, New York.
Drumm, H. P. and Schopffer, P. (1974). *Planta* **120**, 13–30.
Fleury, M. O. and Ashley, D. V. (1983). *Anal. Biochem.* **133**, 330–335.
Foley, T. and Beale, S. I. (1982). *Plant Physiol.* **70**, 1495–1502.
Frigerio, N. A. and Harbury, H. A. (1958). *J. Biol. Chem.* **231**, 135–157.

8. ENZYMES OF THE PHOTORESPIRATORY CARBON PATHWAY

Galston, A. W., Bonnichsen, R. F. and Arnon, D. I. (1951). *Acta Chem. Scand.* **5**, 781–790.
Givan, C. V. (1980). *In* "Biochemistry of Plants: Amino Acids and their Derivatives" (B. J. Miflin, ed.), Vol. 5, pp. 329–357. Academic Press, New York.
Givan, C. V., Tsutakawa, S., Hodgson, J. M., David, N. and Randall, D. D. (1988). *J. Plant Physiol.* **132**, 593–599.
Gregory, E. M. and Fridovich, I. (1974). *Anal. Biochem.* **58**, 57–62.
Hall, N. P., Reggiani, R. and Lea, P. J. (1985). *Phytochemistry* **24**, 1645–1648.
Hall, N. P., Kendall, A. C., Lea, P. J., Turner, J. C. and Wallsgrove, R. M. (1987). *Photosynth. Res.* **11**, 89–96.
Havir, E. A. and McHale, N. A. (1987). *Plant Physiol,* **84**, 450–455.
Hondred, D., Hunter, J. McC., Keith, R., Titus, D. E. and Becker, W. M. (1985). *Plant Physiol.* **79**, 95–102.
Huang, A. H. C., Trelease, R. N. and Moore, T. S. (1983). "Plant Peroxisomes". Academic Press, New York.
Husic, D. W., Husic, H. D. and Tolbert, N. E. (1987). *CRC Critical Reviews in Plant Science* **5**, 45–100.
Husic, H. D. and Tolbert, N. E. (1984). *Arch. Biochem. Biophys.* **229**, 64–72.
Husic, H. D. and Tolbert, N. E. (1985). *Plant Physiol.* **77**, 25S.
Ireland, R. J. and Joy, K. W. (1983). *Arch. Biochem. Biophys.* **223**, 291–296.
Kendall, A. C., Keys, A. J., Turner, J. C., Lea, P. J. and Miflin, B. J. (1983). *Planta* **159**, 505–511.
Kenten, R. H. and Mann, P. J. G. (1952). *Biochem. J.* **52**, 130–134.
Kerr, M. W. and Gear, C. F. (1974). *Biochem. Soc. Trans.* **2**, 338–340.
Kerr, M. W. and Groves, D. (1975). *Phytochemistry* **14**, 359–362.
Keys, A. J. (1983). *Pest. Sci.* **19**, 313–316.
Keys, A. J., Bird, I. F., Cornelius, M. J., Lea, P. J., Wallsgrove, R. M. and Miflin, B. J. (1978). *Nature* **275**, 741–743.
Kikuchi, G. (1973). *Mol. Cell. Biochem.* **1**, 169–187.
Kleczkowski, L. A. and Randall, D. D. (1983). *FEBS Lett.* **158**, 313–316.
Kleczkowski, L. A. and Randall, D. D. (1985). *Plant Physiol.* **79**, 274–277.
Kleczkowski, L. A. and Randall, D. D. (1988a). *Biochem. J.* **250**, 145–152.
Kleczkowski, L. A. and Randall, D. D. (1988b). *Planta* **173**, 221–229.
Kleczkowski, L. A., Randall, D. D. and Zahler, W. L. (1985). *Arch. Biochem. Biophys.* **236**, 185–194.
Kleczkowski, L. A., Randall, D. D. and Blevins, D. G. (1986). *Biochem. J.* **239**, 653–659.
Kleczkowski, L. A., Randall, D. D. and Blevins, D. G. (1987). *In* "Progress in Photosynthesis" (J. Biggins, ed.), Vol. 3, pp. 565–568. Martinus Nijhoff, The Hague.
Kleczkowski, L. A., Givan, C. V., Hodgson, J. M. and Randall, D. D. (1988). *Plant Physiol.* **88**, 1182–1185.
Klein, S. M. and Saggers, R. D. (1966). *J. Biol. Chem.* **241**, 197–205.
Kohn, L. D. and Warren, W. A. (1970). *J. Biol. Chem.* **245**, 3831–3839.
Kohn, L. D., Warren, W. A. and Carroll, W. R. (1970). *J. Biol. Chem.* **245**, 3821–3830.
Lamb, J. E., Reizman, H., Becker, W. M. and Leaver, C. J. (1978). *Plant Physiol.* **62**, 754–760.
Lindqvist, Y. and Branden, C. I. (1985). *Proc. Natl. Acad. Sci. USA* **82**, 6855–6859.
Lorimer, G. H. (1981). *Ann. Rev. Plant Physiol.* **32**, 349–383.
Lück, H. (1962). *In* "Methods of Enzymatic Analysis" (H. U. Bergmeyer, ed.), pp. 885–894. Verlag Chemie, Weinheim.
Murray, A. J. S., Blackwell, R. D., Joy, K. W. and Lea, P. J. (1987). *Planta* **172**, 106–113.
Murray, A. J. S., Blackwell, R. D. and Lea, P. J. (1989). *Plant Physiol.* **91**, 395–400.
Nakamura, Y. and Tolbert, N. E. (1983). *J. Biol. Chem.* **258**, 7631–7638.
Neuburger, M., Bourguignon, J. and Douce, R. (1986). *FEBS Lett.* **207**, 18–22.
Nishimura, M., Akhmedov, Y. D., Strzalka, K. and Akazawa. T. (1983). *Arch. Biochem. Biophys.* **222**, 397–402.
Parker, M. L. and Lea, P. J. (1983). *Planta* **159**, 512–517.
Rao, D. N. and Appaji Rao, N. (1982). *Plant Physiol.* **69**, 11–18.
Rehfeld, D. W. and Tolbert, N. E. (1972). *J. Biol. Chem.* **247**, 4803–4811.
Robinson, S. P. (1982). *Plant Physiol.* **70**, 1032–1038.

Sarojini, G. and Oliver, D. J. (1983). *Plant Physiol.* **72**, 194–199.
Scandalios, J. G. (1968). *Ann. New York Acad. Sci.* **151**, 274–293.
Scandalios, J. G., Tsaftaris, A. S., Chandler, J. M. and Skadsen, R. W. (1984). *Dev. Genet.* **4**, 281–293.
Scheiffer, S., Teifel, W. and Kindl, H. (1976). *Hoppe-Seyler's Z. Physiol. Chem.* **357**, 163–175.
Schmidt, M. R. and Edwards, G. E. (1983). *Arch. Biochem. Biophys.* **224**, 332–341.
Shioi, Y., Nagamine, M. and Sasa, T. (1984). *Arch. Biochem. Biophys.* **234**, 117–124.
Smith, I. K. (1973). *Biochim. Biophys. Acta* **321**, 156–164.
Somerville, C. R. and Ogren, W. L. (1979). *Nature* **280**, 833–836.
Somerville, C. R. and Ogren, W. L. (1982). *Biochem. J.* **202**, 373–380.
Stabenau, H. (1974). *Plant Physiol.* **54**, 921–924.
Ta, T. C. and Joy, K. W. (1986). *Planta* **169**, 118–122.
Taylor, R. T. and Weissbach, H. (1965). *Anal. Biochem.* **13**, 80–84.
Titus, D. E., Hondred, D. and Becker, W. M. (1983). *Plant Physiol.* **72**, 402–408.
Tolbert, N. E. (1985). *In* "Nitrogen Fixation and CO_2 Metabolism" (P. W. Ludden and J. E. Burrs, eds), pp. 333–341. Elsevier, Amsterdam.
Tolbert, N. E. and Burris, R. H. (1950). *J. Biol. Chem.* **186**, 791–804.
Tolbert, N. E., Clagett, C. O. and Burris, R. H. (1949). *J. Biol. Chem.* **181**, 905–914.
Tolbert, N. E., Yamazaki, R. K. and Oeser, A. (1970). *J. Biol. Chem.* **245**, 5129–5136.
Tsaftaris, A. S., Bosabalidis, A. M. and Scandalios, J. G. (1983). *Proc. Natl. Acad. Sci. USA* **80**, 4455–4459.
Turner, J. C. and Hall, N. P. (1988). *J. Exp. Bot.* **39**, 345–351.
Vérin-Vergeau, C., Baldy, P. and Cavalié, G. (1979). *Phytochemistry* **18**, 1279–1282.
Vérin-Vergeau, C., Baldy, P., Puech, J. and Cavalié, G. (1980). *Phytochemistry* **19**, 763–767.
Volokita, M. and Somerville, C. R. (1987). *J. Biol. Chem.* **262**, 15825–15828.
Walker, G. H., Sarojini, G. and Oliver, D. J. (1982). *Biochem. Biophys. Res. Commun.* **107**, 856–861.
Walker, J. L. and Oliver, D. J. (1986). *J. Biol. Chem.* **261**, 2214–2221.
Wolfenden, R. (1969). *Nature* **223**, 704–705.
Wolfenden, R. (1970). *Biochemistry* **9**, 3404–3407.
Woo, K. C. (1979). *Plant Physiol.* **63**, 783–787.
Yu, Y. L., Tolbert, N. E. and Orth, G. M. (1964). *Plant Physiol.* **39**, 643–647.
Zelitch, I. (1953). *J. Biol. Chem.* **201**, 719–726.
Zelitch, I. (1955). *J. Biol. Chem.* **216**, 553–575.
Zelitch, I. (1971). "Photosynthesis, Photorespiration and Plant Productivity." Academic Press, London, New York.

9 Glycolysis

WILLIAM C. PLAXTON

Department of Biology, Queen's University, Kingston, Ontario, K7L 3N6, Canada

I.	Introduction	145
II.	ATP-dependent phosphofructokinase	147
	A. Assay	147
	B. Extraction and purification	147
	C. Properties	149
III.	PPi-dependent phosphofructokinase	150
	A. Assay	151
	B. Extraction and purification	151
	C. Properties	152
IV.	Aldolase	153
	A. Assay	153
	B. Extraction and purification	153
	C. Properties	154
V.	Triose phosphate isomerase	154
	A. Assay	155
	B. Extraction and purification	155
	C. Properties	156
VI.	Glyceraldehyde 3-phosphate dehydrogenase	156
	A. Assay	157
	B. Extraction and purification	157
	C. Properties	159
VII.	Phosphoglycerate kinase	159
	A. Assay	159
	B. Extraction and purification	160
	C. Properties	161
VIII.	Phosphoglycerate mutase	161
	A. Assay	161

	B. Extraction and purification	162
	C. Properties	163
IX.	Enolase	163
	A. Assay	163
	B. Extraction and purification	164
	C. Properties	165
X.	Pyruvate kinase	165
	A. Assay	165
	B. Extraction and purification	166
	C. Properties	168
XI.	Phospho*enol*pyruvate phosphatase	169
	A. Assay	170
	B. Extraction and purification	170
	C. Properties	171
	Acknowledgements	172
	References	172

I. INTRODUCTION

Glycolysis was the first major metabolic pathway to be elucidated, an achievement that was instrumental in the development of many experimental and conceptual aspects of modern biochemistry. Glycolysis fulfils two fundamental roles: it catabolises hexoses to generate ATP and pyruvate, and it provides building blocks for anabolism. Plant glycolysis is unique in that the sequential conversion of hexose to pyruvate can occur in two subcellular compartments; the cytosol and the plastid. Work initiated in the laboratory of D. T. Dennis has led to the view that plastid and cytosolic glycolytic reactions are catalysed by isoenzymes. The relative amounts of the respective isoenzymes depends upon the type of tissue, and its developmental stage. It is now apparent that plastidic and cytosolic glycolytic isoenzymes generally differ in immunological properties, often diverge in kinetic and physicochemical characteristics, and are probably encoded by separate nuclear genes (Copeland and Turner, 1987).

'Classical' glycolysis is generally thought to be the set of reactions which converts one mole of glucose into two moles of pyruvate. It has been argued, however, that the initial step of plant glycolysis should be considered to be the conversion of fructose 6-phosphate (Fru-6-P) to fructose 1,6-bisphosphate (Fru-1,6-P_2) (Dennis and Greyson, 1987). This chapter will therefore review the assay, purification and properties of the various plant plastidic and/or cytosolic glycolytic (iso)enzymes catalysing the sequence of reactions which converts Fru-6-P to pyruvate. It is highly recommended that readers who wish to employ the following purification protocols as a basis for their own purification schemes, exploit some of the recent advances in protein purification technologies. The current availability of fast flow ion-exchange and size exclusion resins, as well as fast protein liquid chromatography (FPLC) and high pressure liquid chromatography (HPLC) has now made it possible to purify to homogeneity, and thus more thoroughly characterise, several plant glycolytic enzymes which are notorious for their instability (i.e. ATP-dependent phosphofructokinase (PFK) and pyruvate kinase (PK)). A variety of excellent reviews on plant glycolysis have appeared over the past decade (Turner and Turner, 1980; Dennis and Miernyk, 1982; ap Rees, 1985; ap Rees *et al.*, 1985; Dennis *et al.*, 1985; Black *et al.*, 1987; Copeland and Turner, 1987; Dennis and Greyson, 1987; Sung *et al.*, 1988; Miernyk, 1989).

II. ATP-DEPENDENT PHOSPHOFRUCTOKINASE

(EC 2.7.1.11; ATP : D-fructose 6-phosphate 1-phosphotransferase)

$$\text{Fru-6-P} + \text{MgATP} \longrightarrow \text{Fru-1,6-P}_2 + \text{MgADP}$$

The conversion of Fru-6-P to Fru-1,6-P$_2$ is catalysed by PFK or PFP (PPi-dependent phosphofructokinase). These enzymes are therefore under tight regulatory control because they catalyse reactions that can drain metabolites from the hexose-P pool (Dennis and Greyson, 1987). PFK catalyses a reaction which is considered to be thermodynamically irreversible under physiological conditions. Plastidic and cytosolic PFK isoenzymes (abbreviated as PFK$_p$ and PFK$_c$, respectively) have recently been purified to homogeneity from a variety of non-photosynthetic plant tissues. Partial purifications of both isoenzymes from several photosynthetic tissues have also been performed.

A. Assay

PFK is assayed by coupling the formation of Fru-1,6-P$_2$ with the aldolase (ALD), triose phosphate isomerase (TPI), and glycerol 3-phosphate dehydrogenase reactions and monitoring NADH oxidation continuously at 340 nm. Standard assay conditions for developing castor bean endosperm PFK$_p$ or carrot root PFK$_c$ are 50 mM Tris-HCl (pH 8.0), 2 mM Fru-6-P, 5 mM MgCl$_2$, 0.1 mM NADH, 1 mM EDTA, 1 U ALD, 10 U TPI, 1 U glycerol 3-phosphate dehydrogenase, and enzyme extract in a final volume of 1 ml (Garland and Dennis, 1980a; Wong *et al.*, 1987).

Notes: (a) In crude preparations, rates of NADH oxidation in the absence of Fru-6-P or ATP should be subtracted from the total activity.

(b) Two molecules of NADH are oxidised for every molecule of Fru-1,6-P$_2$ produced.

B. Extraction and Purification

1. PFK$_c$

The following procedure for carrot storage roots (described fully by Wong *et al.*, 1987) yields about 5 µg of homogeneous PFK$_c$. Carrot roots contain very low amounts of PFK$_p$, but provide a fairly active preparation of PFK$_c$. All operations should be carried out at 0–5°C.

(a) *Crude extract.* Diced carrot roots (1 kg), are homogenised using a Waring blender with an equal volume of 50 mM Tris-HCl (pH 8.0), containing 10% (v/v) glycerol, 2 mM EDTA, 14 mM 2-mercaptoethanol, 20 mM sodium diethyldithiocarbamate, 5 mM MgCl$_2$, 50 mM KF, 0.1 mM ATP, 0.5 mM PMSF (phenylmethylsulphonyl fluoride), 2 mM α-amino-α-caproic acid, 2 mM benzamidine hydrochloride, and 1.5% (w/v) insoluble PVP (polyvinylpolypyrrolidone). After filtration through nylon mesh, the filtrate is clarified by centrifugation (30 min, 10 000 × g) and filtered through glass wool to remove lipid material.

(b) *Polyethylene glycol (PEG) fractionation.* PEG 8000 is added to 5% (w/v); after centrifugation as described above, the supernatant is adjusted to 15% (w/v) PEG. The precipitate is collected by centrifuging.

(c) *DEAE-cellulose chromatography.* The 5 to 15% (w/v) PEG pellets are dissolved with a minimal volume of 50 mM Tris-HCl (pH 8.0), containing 10% (v/v) glycerol, 1 mM EDTA, 14 mM 2-mercaptoethanol, 80 mM KCl, and 0.1 mM ATP (buffer A), centrifuged (20 min, 27 000 × g) to clarify, and absorbed onto a column (2.5 × 47 cm) of DEAE-cellulose (Whatman DE-52) pre-equilibrated with buffer A. PFK activity is eluted in a single peak with 700 ml of a linear gradient of 80 to 500 mM KCl in buffer A. Pooled peak fractions are concentrated by overnight dialysis against 20 mM Tris-HCl (pH 8.0), containing 50% (v/v) glycerol, 1 mM EDTA, 14 mM 2-mercaptoethanol, and 5 mM Fru-6-P (buffer B).

(d) *Sepharose 4B gel filtration.* The concentrated extract is diluted with an equal volume of buffer C (buffer C = buffer B, but with 10% (v/v) glycerol in place of 50% (v/v) glycerol) and applied to a Sepharose 4B column (3.5 × 60 cm) pre-equilibrated with buffer C. PFK activity is resolved into a major Pi-activated peak (representing PFK_c; ~95% of total PFK recovered) eluting just after the void volume, and a minor Pi-inhibited peak (representing PFK_p; ~95% of total PFK recovered) eluting in the middle of the elution profile. Pooled PFK_c peak fractions are concentrated by dialysis against buffer B as above.

(e) *Blue Sepharose affinity chromatography.* Concentrated Sepharose 4B pooled fractions containing PFK_c are desalted on a column (2.5 × 12 cm) of Sephadex G-25 pre-equilibrated with 20 mM Tris-HCl (pH 8.0), containing 10% (v/v) glycerol, 0.1 mM EDTA, 0.05 mM Fru-1,6-P_2, 2.5 mM DTT, and 25 mM KF (buffer D). Pooled fractions are combined and absorbed onto a column (1.5 × 14 cm) of Blue Sepharose CL pre-equilibrated with buffer D. The column is washed sequentially with buffer D and buffer D containing 0.15 mM ADP. PFK_c eluted with buffer D containing 10 mM $MgCl_2$ and 5 mM ATP. Pooled peak fractions are concentrated by overnight dialysis against buffer B, and stored at $-20°C$.

Notes: (a) Carrot root PFK_c is purified about 800-fold to a final specific activity of 80 U mg^{-1} and an overall recovery of 1.4%.

(b) The final preparation is relatively unstable; more than 50% of the original activity is lost within 5 days when stored at $-20°C$ in 50% (v/v) glycerol and 5 mM Fru-6-P.

2. PFK_p

The following procedure (described fully by Knowles et al., 1989) is a modification of that reported by Garland and Dennis (1980a) and allows the purification to apparent homogeneity of about 40 µg of PFK_p from the developing endosperm of the castor oil plant (*Ricinus communis*). All operations are performed at 0–4°C unless otherwise noted.

(a) *Crude extract.* Developing castor bean endosperm tissue (80 g; 32–40 days post-pollination) is homogenised with a Polytron in 1.5 vols of 50 mM Tes-NaOH (pH 7.5) containing 0.5 M sucrose. Cell debris is removed by centrifugation (5 min, 500 × g). Plastids are pelleted by centrifuging the supernatant (15 min, 10 000 × g).

(b) *Acetone fractionation.* An acetone powder of the plastid pellet is prepared by resuspending and washing the pellet in acetone at −20°C then drying in a stream of air. About 20 mg of acetone powder is obtained per gram of starting material. The acetone powder is resuspended (20 mg ml^{-1}) in 20 mM Imidazole-HCl (pH 7.2), containing 80 mM KCl (buffer A), using a Teflon homogeniser. The resuspension is centrifuged (20 min, 27 000 × g) and the supernatant used as the source of PFK_p.

(c) *DEAE-Sephacel chromatography.* The acetone powder supernatant is applied to a DEAE-Sephacel column (1.5 × 28 cm) pre-equilibrated with buffer A, and PFK_p eluted with 90 ml of a linear 80–500 mM KCl gradient in buffer A. Pooled peak fractions are concentrated to about 5 ml with an Amicon PM-30 ultrafilter and desalted on a column (1.5 × 16 cm) of Sephadex G-25 pre-equilibrated with 15 mM KPi (pH 7.0), containing 1 mM EDTA, 1 mM DTE, 5 mM $MgCl_2$, and 15% (v/v) glycerol (buffer B).

(d) *FPLC Mono-Q chromatography.* The desalted extract is absorbed onto a Mono Q HR 5/5 column pre-equilibrated with buffer B. PFK_p is eluted with 50 ml of a linear gradient of 15 mM KPi (pH 7.0) to 500 mM KPi (pH 7.5) in buffer B.

(e) *FPLC Phenyl Superose chromatography.* Pooled peak fractions are adjusted to 30% (saturation) $(NH_4)_2SO_4$ and absorbed onto a Phenyl Superose HR 5/5 column pre-equilibrated with 40 mM KPi (pH 7.0), containing 1 mM EDTA, 1 mM DTE, and 30% (saturation) $(NH_4)_2SO_4$ (buffer C). The column is then eluted in a stepwise fashion with decreasing concentrations of buffer C and simultaneously increasing concentrations of 20 mM KPi (pH 7.0), 1 mM EDTA, 1 mM DTE, and 15% (v/v) glycerol (buffer D). PFK_p is eluted following a step from 50 to 75% buffer D (50–25% buffer C). Pooled peak fractions are concentrated to about 250 μl using an Amicon YM-30 ultrafilter, frozen with liquid N_2, and stored at −80°C.

Notes: (a) The enzyme is stable for several months when stored frozen.
(b) The enzyme is purified about 1600-fold to a final specific activity of 48 U mg^{-1} and an overall recovery of 50%.

C. Properties

1. PFK_c

Homogeneous PFK_c from carrot storage roots is composed of a single type of 60 kDa subunit (Wong et al., 1987). These data contrast with the multiple polypeptides associated with apparently homogeneous cucumber seed or potato tuber PFK_p or PFK_c (Botha et al., 1988; Kruger and Hammond, 1988), but are similar to that reported for castor bean endosperm PFK_p (Knowles et al., 1989). The native homogeneous carrot root enzyme can apparently exist in three major forms that range in molecular weights from 240 to 5000 kDa. These different forms can reversibly interconvert by addition or removal of appropriate metabolites (i.e. ATP, citrate, or Fru-6-P). The ability to undergo metabolite mediated aggregation–disaggregation has also been reported for several other plant and non-plant PFKs.

The three major forms of carrot root PFK_c all show hyperbolic substrate saturation kinetics and, like other plant PFKs, are inhibited by PEP. The different forms of carrot PFK_c, however, differ markedly in their response to ATP, citrate, Pi, and pH (Wong et al., 1987). It has been speculated that metabolite-dependent subunit association–dissociation may regulate the in vivo activity of carrot root PFK_c, and the consequent rate of cytosolic glycolysis.

Developing castor bean endosperm and cucumber seed PFK_c have pH optima of about 7.0 to 7.2, exhibit hyperbolic saturation kinetics, and have an absolute dependence for a divalent metal cation (Mg^{2+} preferred). The respective K_m values of the castor bean and cucumber seed PFK_c for Fru-6-P are 0.025 and 0.7 mM; and for ATP, 10 and 5 µM, respectively (Garland and Dennis, 1980b; Botha et al., 1988).

2. PFK_p

SDS-PAGE of purified developing castor bean endosperm PFK_p reveals a single 57 kDa protein staining band; a native molecular weight of 220 kDa is determined by gel filtration (Knowles et al., 1989). These data suggest that castor bean PFK_p is a homotetrameric protein. Different aggregation states of the enzyme have not been observed.

PFK_p from leaf or developing endosperm of castor oil plant and from germinating cucumber seeds have a pH optimum from pH 7.6 to 8.0, are activated by Pi, and show an absolute requirement for a divalent metal cation (Mg^{2+} preferred) (Garland and Dennis, 1980a; Botha et al., 1988; Knowles et al., 1989). Pi activates PFK_p by increasing V_{max} and affinity for substrates. Phospho*enol*pyruvate (PEP) is a potent inhibitor of PFK_p of castor oil plant developing endosperm and leaf; I_{50} values for PEP range from 0.1 to 0.5 µM (Garland and Dennis 1980b; Knowles et al., 1989). PFK_p from leaf or developing endosperm of castor oil plant and from germinating cucumber seed always show hyperbolic ATP saturation kinetics, but exhibit hyperbolic Fru-6-P saturation kinetics at pH 8.0–8.2, and sigmoidal Fru-6-P saturation kinetics at pH 7.0–7.2. At pH 8.0–8.2 the respective K_m values of the castor oil plant endosperm and cucumber seed PFK_ps for Fru-6-P are about 0.1 and 1.7 mM, respectively; for ATP, 9 and 13 µM. Pea leaf PFK_p has been reported to be light inactivated, possibly by a thioredoxin-mediated dithiol–disulphide interconversion (Heuer et al., 1982). Light-promoted depletion and elevation of chloroplastic Pi and PEP levels, respectively, may also play a role in the light inactivation of PFK_p.

III. PPi-DEPENDENT PHOSPHOFRUCTOKINASE

(EC 2.7.1.90; inorganic pyrophosphate:fructose 6-phosphate 1-phosphotransferase)

$$\text{Fru-6-P} + \text{MgPPi} \longleftrightarrow \text{Fru-1,6-P}_2 + \text{Mg}^{2+} + \text{Pi}$$

In contrast to PFK, PFP is restricted to the cytoplasm and catalyses a freely reversible reaction. The enzyme has been described in a number of plants and bacteria. The observation of Sabularse and Anderson (1981) that the activity of the mung bean enzyme is greatly stimulated by the key regulatory metabolite Fru-2,6-P_2 indicated an important role for PFP in the control and integration of plant energy metabolism. The enzyme has since been purified and characterised from a variety of photosynthetic and non-photosynthetic plant sources.

A. Assay

The PFP assay is very similar to the PFK assay. Fru-1,6-P_2 production is coupled to the ALD, TPI, and glycerol 3-phosphate dehydrogenase reactions and assayed continuously at 25°C by monitoring NADH oxidation at 340 nm. Standard assay conditions for potato tuber PFP are 50 mM Tris-HCl (pH 8.0), 5 mM Fru-6-P, 1 μM Fru-2,6-P_2, 5 mM MgCl$_2$, 0.2 mM NADH, 1 mM EDTA, 0.4 U ALD, 6 U TPI, 1.2 U glycerol 3-phosphate dehydrogenase, and enzyme extract in a final volume of 1 ml (Yuan et al., 1988). The reaction is initiated by the addition of 2.5 mM NaPPi following a 5 min preincubation.

Notes: (a) In crude preparations, rates of NADH oxidation in the absence of Fru-6-P or PPi should be subtracted from the total activity.

(b) Two molecules of NADH are oxidized for every molecule of Fru-1,6-P_2 produced.

B. Extraction and Purification

The following procedure for potato tubers (described fully by Yuan et al., 1988) yields up to 1.5 mg of nearly homogeneous enzyme. All steps are carried out at 4°C.

1. Crude extract

Peeled potato tubers (2 kg) are chopped into small pieces and homogenised in a Waring blender with 2 vols of 20 mM Hepes-NaOH (pH 8.2), containing 20 mM K-acetate, 2 mM DTT, 1 mM PMSF, and 1 mM EDTA. The homogenate is filtered through cheesecloth. To the filtrate is added NaPPi (0.989 g l^{-1}) and 2 ml l^{-1} of 1 M MgCl$_2$, and the pH is adjusted to 8.2 by the addition of NaOH.

2. Heat treatment

The extract is heated to 59°C in a 75°C water bath, and maintained at this temperature for 5 min. The extract is cooled on ice to 4°C and the pH adjusted to 7.1 by the addition of HCl.

3. PEG fractionation

PFP is precipitated between 6 and 18% (w/v) PEG 8000 and the precipitated protein is collected by centrifugation (20 min, 10 000 × g). Pellets are dissolved in a minimal volume of 5 mM Hepes-NaOH (pH 6.6), containing 2 mM DTT, and 1 mM EDTA (buffer A), and centrifuged (20 min, 27 000 × g) to remove insoluble material.

4. Phosphocellulose chromatography

The extract is absorbed onto a column (2.5 × 17 cm) of Cellex-P pre-equilibrated with buffer A. PFP is eluted with buffer A containing 20 mM NaPPi. Pooled peak fractions are subjected to repeated passes through a centrifugal concentrator (Amicon Centricon YM-30) to both concentrate the solution and to equilibrate it with 20 mM Tris-HCl (pH 8.2), containing 20 mM KCl, 2 mM DTT, and 1 mM EDTA (buffer B).

5. DEAE-BioGel chromatography

The solution is applied to a column (2.5 × 17 cm) of DEAE-BioGel pre-equilibrated with buffer B. PFP is eluted with 400 ml of a linear gradient of 80–600 mM KCl in buffer B. Pooled peak fractions are concentrated to about 1.5 ml as described above.

6. FPLC Superose 6

The enzyme solution in 0.2 ml aliquots is eluted through a FPLC Superose 6 HR 10/30 column pre-equilibrated with 20 mM Tris-HCl (pH 8.2), containing 2 mM DTT and 1 mM EDTA. The first protein peak emerging from the column represents PFP that is about 95% homogeneous.

Notes: (a) The final preparation is adjusted to contain 25% (v/v) glycerol, then frozen with liquid N_2 and stored at $-80°C$. The enzyme is stable for several months when stored frozen.

(b) Potato tuber PFP is purified about 600-fold to a final specific activity of 41 U mg^{-1} and an overall yield of 14%.

C. Properties

All plant PFPs studied to date are composed of two types of immunologically distinct α and β subunits generally having M_rs in the range of 68 to 65, and 60 kDa, respectively. The native potato tuber PFP purified by Yuan et al. (1988) has the structure $α_4β_4$ (535 kDa). This contrasts with the $α_2β_2$ structure previously reported for a preparation of potato tuber PFP (Kruger and Dennis, 1987). These differences may simply reflect differences in the conditions of the gel filtration used by the respective research groups.

Plant PFPs are heat stable, show similar pH optima for both the forward and reverse reactions of pH 7.3 to 7.8, and usually exhibit hyperbolic substrate saturation kinetics. In the absence of Fru-2,6-P_2 the K_m value of plant PFP for Fru-6-P ranges from 2.5–9 mM; for PPi, 35–60 μM; for Mg^{2+}, 185 μM; in the reverse direction for Fru-1,6-P_2, 90–240 μM; for Pi, 140–310 μM (Kombrink et al., 1984; Botha et al., 1986). Fru-2,6-P_2 is a potent activator of all plant PFPs studied to date. K_a values for Fru-2,6-P_2 usually range from 5–50 nM; the major effect of Fru-2,6-P_2 is to increase the V_{max} from 10- to 20-fold, lower K_ms for substrates, and relieve inhibition of the forward reaction by the product Pi (Sabularse and Anderson, 1981; Kombrink et al., 1984; Botha et al., 1986). The recent finding that black mustard (*Brassica nigra*) suspension cell PFP appears to be inducible by Pi starvation suggests that one important role for the enzyme may be to

9. GLYCOLYSIS   153

bypass the ATP-dependent PFK reaction during extended periods of Pi starvation when intracellular ATP and Pi pools become greatly depleted (Duff et al., 1989a).

IV. ALDOLASE

(EC 4.1.2.13; fructose 1,6-bisphosphate D-glyceraldehyde 3-phosphate lyase)

$$\text{Fru-1,6-P}_2 \longleftrightarrow \text{DHAP} + \text{G3P}$$

Aldolase (ALD) catalyses the reversible aldol cleavage of Fru-1,6-P_2 yielding two triose phosphates, dihydroxyacetone phosphate (DHAP) and glyceraldehyde 3-phosphate (G3P). There are two classes of ALD. Class I ALDs are homotetramers which have an essential sulphydryl-group and Schiff base reaction mechanisms, whereas Class II ALDs are smaller monomeric proteins which require K^+ and divalent metal cations for activity. Animal and higher plant ALDs are Class I, while bacteria and fungi contain Class II ALDs. Eukaryotic algae have either Class I or Class II ALDs.

Plant cytosolic and/or plastidic ALD isoenzymes (abbreviated as ALD_c and ALD_p, respectively) have been purified to homogeneity from wheat, pea, spinach and maize leaves, rice bran, carrot storage roots and the green alga *Chara foetida*. In several instances it was not determined whether it was the cytosolic or plastid isoenzyme which was purified.

A. Assay

The ALD reaction is normally coupled to the TPI, and glycerol 3-phosphate dehydrogenase reactions and assayed by monitoring NADH oxidation at 340 nm. The assay mixture (1.0 ml) contains 50 mM Tris-HCl (pH 7.5), 2 mM Fru-1,6-P_2, 1 mM EDTA, 0.2 mM NADH, 1 U each of TPI and glycerol 3-phosphate dehydrogenase, and enzyme extract, which is added to start the reaction.

Notes: (a) Assays should be corrected for contaminating NADH oxidase activity by omitting Fru-1,6-P_2 from the reaction mixture.
(b) Two NADH molecules are oxidised for every Fru-1,6-P_2 which is cleaved.

B. Extraction and Purification

The following protocol (described fully by Kruger and Schnarrenberger, 1983) yields up to 21 and 83 mg of homogeneous ALD_c and ALD_p, respectively, from 1 kg of spinach leaves. All procedures are conducted at 4°C.

1. Crude extract

Washed and deribbed spinach leaves (1 kg) are homogenised using a Waring blender or Polytron with two volumes of 20 mM K_2HPi (pH 8.6), containing 10 mM 2-mercaptoethanol (buffer A). The homogenate is filtered through cheesecloth, the filtrate pH is readjusted to 8.6 with 5 M KOH, and the material is centrifuged (20 min, 20 000 × g).

2. DEAE-cellulose chromatography

The crude supernatant is diluted with 10 mM 2-mercaptoethanol (pH 8.6) to a final conductivity of 2.5 mS and proteins absorbed batchwise onto 1.5 l of wet packed DEAE-cellulose (Whatman DE-32) pre-equilibrated with buffer A. After suspending in grinding medium the DEAE-cellulose is packed into a column 8 cm in diameter. Salt gradient formation and subsequent fractionation of ALD isoenzymes is achieved by applying 4 l of a 0–400 mM KCl gradient in buffer A. The early and later eluting peaks of ALD activity correspond to ALD_c and ALD_p, respectively.

3. Phosphocellulose chromatography

Respective extracts are concentrated by precipitation with 70% (saturation) $(NH_2)_4SO_4$ and, after solubilisation, dialysed overnight against 10 mM Tris-HCl (pH 7.3), containing 1 mM EDTA, and 10 mM 2-mercaptoethanol (buffer B). Proteins are absorbed onto a column (4 × 10 cm) of P11 phosphocellulose (Whatman) pre-equilibrated with buffer B. ALD is eluted with 1 mM Fru-1,6-P_2 in buffer B. Pooled peak fractions are concentrated by dialysis against Aquacide II (Calbiochem).

Notes: ALD_c is purified about 65-fold to a final specific activity of about 7.2 U mg^{-1} and an overall yield of about 9%. ALD_p is purified about 71-fold to a final specific activity of about 7.8 U mg^{-1} and an overall yield of about 47%.

C. Properties

The physicochemical and catalytic properties of plant ALDs are similar to those of mammalian Class I ALDs. In all cases the plant enzyme is homotetrameric with subunit M_r values of 36 to 40 kDa. Leaf ALD_c is relatively heat stable, whereas ALD_p is heat labile. Optimal activity occurs at about pH 7.0. Both isoenzymes show hyperbolic substrate saturation kinetics. The K_m value of plant ALD_c for Fru-1,6-P_2 ranges from 5–20 μM, whereas that for ALD_p ranges from 20–68 μM (Anderson, 1982; Moorhead and Plaxton, unpubl. res.). Plant ALD_c shows a lower V_{max}, but an equivalent K_m for sedheptulose 1,7-bisphosphate. PEP has recently been demonstrated to inhibit homogeneous carrot root ALD_c non-competitively with respect to Fru-1,6-P_2, suggesting that the enzyme may have allosteric properties (Moorhead and Plaxton, unpubl. res.).

V. TRIOSE PHOSPHATE ISOMERASE

(EC 5.3.1.1; D-glyceraldehyde 3-phosphate ketol isomerase)

$$DHAP \longleftrightarrow G3P$$

Triose phosphate isomerase (TPI) catalyses the interconversion of G3P and DHAP. Plastid and/or cytosolic TPI isoenzymes (abbreviated as TPI_p and TPI_c, respectively) have been purifed to homogeneity from spinach, lettuce, celery, and rye leaves.

A. Assay

The TPI reaction is routinely assayed at 30°C by coupling the production of G3P through G3P dehydrogenase and monitoring NAD^+ reduction at 340 nm. The assay mixture (3.0 ml) for the lettuce or spinach leaf enzyme contains 20 mM Tris-acetate (pH 7.2), 2 mM sodium arsenate, 0.06 mM NAD^+, 3 U G3P dehydrogenase, and enzyme preparation. The reaction is initiated by the addition of 2.5 mM DHAP (Pichersky and Gottlieb, 1984).

Notes: (a) A minus DHAP blank should be conducted so as to account for any contaminating NAD^+ reductase activity.

(b) The enzyme can also be assayed in the direction of DHAP formation with G3P as substrate by coupling with glycerol 3-phosphate dehydrogenase and monitoring NADH oxidation at 340 nm (Pichersky and Gottlieb, 1984).

B. Extraction and Purification

The following procedure for lettuce (described fully by Pichersky and Gottlieb, 1984) yields up to 1.3 and 0.5 mg of homogeneous of TPI_c and TPI_p, respectively. All procedures should be carried out at 4°C.

1. Crude extract

Approximately 200 g of deribbed lettuce leaves are homogenised in a Waring blender in 4 vols of 50 mM Tris-HCl (pH 7.6), containing 1 mM EDTA, 14 mM 2-mercaptoethanol, and 8% (w/v) insoluble PVP. The homogenate is filtered through Miracloth (Calbiochem) and the filtrate centrifuged (20 min, 30 000 × g).

2. Ammonium sulphate fractionation

$(NH_4)_2SO_4$ is added to the crude supernatant to 55% (saturation); after centrifugation the supernatant is adjusted to 75% (saturation). The precipitate is collected by centrifuging as above.

3. Hydroxyapatite chromatography

The 55 to 75% (saturation) $(NH_4)_2SO_4$ pellets are resuspended with a minimal volume of 5 mM KPi (pH 6.0), containing 25 mM NaCl, and 14 mM 2-mercaptoethanol (buffer A), and dialysed overnight against the same buffer. The dialysate is centrifuged as above to clarify and loaded onto a column (2 × 24 cm) of hydroxyapatite pre-equilibrated in buffer A. TPI activity is eluted by applying 400 ml of a 5–200 mM KPi (pH 6.0) linear gradient in buffer A. Pooled peak fractions are concentrated in a dialysis sack against solid PEG 8000 to about 3 ml.

4. Sephacryl S-200 gel filtration

Sucrose is added to a concentration of 10% (w/v), and the sample is loaded at 50 ml h^{-1}

onto a column (2.5 × 100 cm) of Sephacryl S-200 pre-equilibrated with 10 mM EDTA-HCl (pH 7.2), containing 25 mM NaCl and 42 mM 2-mercaptoethanol.

5. DEAE-cellulose chromatography

Pooled peak S-200 fractions are dialysed against 15 mM Mes-KOH (pH 5.5), containing 14 mM 2-mercaptoethanol (buffer B) and absorbed onto a column (1 × 30 cm) of DEAE-cellulose (Whatman DE-53) pre-equilibrated with buffer B. Salt gradient formation and subsequent fractionation of TPI isoenzymes is achieved by applying 400 ml of a linear 0–250 mM KCl gradient in buffer B. The early and later eluting peaks of TPI activity correspond to the cytosolic and plastid isoenzymes, respectively. Pooled peak fractions are adjusted to pH 7.2 and dialysed against 1 litre of 25 mM KPi (pH 7.2), containing 30% (v/v) glycerol for 5 h.

Notes: (a) Final preparations can be stored without loss of activity for several days at 4°C.

(b) TPI_c is purified about 770-fold to a final specific activity of about 10 200 U mg^{-1} and an overall yield of about 63%. TPI_p is purified about 700-fold to a final specific activity of about 9200 U mg^{-1} and an overall yield of about 21%.

C. Properties

The physicochemical and catalytic properties of plant TPIs are generally very similar to the enzyme from other sources. Rye leaf TPI_c is much more heat labile than is the corresponding plastid isoenzyme (Kurzok and Feierabend, 1984). Rye, lettuce, spinach and celery leaf TPI_c and TPI_p are homodimeric proteins with each subunit having a molecular weight of 27 kDa (Kurzok and Feierabend, 1984; Pichersky and Gottlieb, 1984). Although TPI_c from a variety of species can be distinguished immunologically from TPI_p, both isoenzymes do share some common antigenic determinants.

Optimum pH values for rye leaf TPI_c and TPI_p are about 7.3 and 8.4, respectively (Kurzok and Feierabend, 1984). Both isoenzymes show hyperbolic substrate saturation kinetics. The respective K_m values of rye leaf TPI_c and TPI_p for DHAP are 0.6 and 0.68 mM; for G3P, 1.5 and 2.5 mM. Both rye leaf isoenzymes are inhibited competitively by 2-phosphoglycolate and sodium arsenate. The K_i values of rye leaf TPI_c and TPI_p for sodium arsenate are 9 and 26 mM, respectively; the K_i value for 2-phosphoglycolate is 0.2 mM for both isoenzymes (Kurzok and Feierabend, 1984).

VI. GLYCERALDEHYDE 3-PHOSPHATE DEHYDROGENASE

(EC 1.2.1.12 and EC 1.2.1.13; D-glyceraldehyde 3-phosphate: NAD(P)$^+$ oxidoreductase [phosphorylating])

$$G3P + Pi + NAD(P)^+ \longleftrightarrow 1,3\text{-Diphosphoglycerate} + NAD(P)H$$

Higher plants contain two distinct phosphorylating G3P dehydrogenases. One form (EC 1.2.1.12) is NAD-specific and is localised in the cytosol. The other form (EC

9. GLYCOLYSIS

1.2.1.13) is active with $NADP^+$ or NAD^+ as the coenzyme, is localised in plastids, and functions in photosynthetic tissue as part of the Calvin cycle. There is also an NADP-dependent non-phosphorylating G3P dehydrogenase (EC 1.2.1.9) which is localised in the cytoplasm and catalyses the irreversible conversion of G3P to 3-phosphoglycerate (3-PGA). The non-phosphorylating enzyme may be involved in shuttling reducing equivalents, protons and 3-PGA from the chloroplast to cytosol in illuminated leaf tissue (Iglesias et al., 1987). What follows is the assay and isolation of homogeneous NAD- and NADP-dependent phosphorylating G3P dehydrogenases from white mustard (*Sinapis alpa*) seedlings (Cerff, 1982).

A. Assay

G3P dehydrogenase can be assayed in either direction by monitoring the increase or decrease of NAD(P)H absorbance at 340 nm. However, the reaction in the glycolytic direction ($NAD(P)^+$ reduction) may be seriously interfered with by the presence of TPI and non-phosphorylating G3P dehydrogenase. Moreover, G3P is very unstable at the pH optimum of the enzymes. It is therefore recommended to perform the assay in the gluconeogenic direction by monitoring NAD(P)H oxidation at 25°C. The substrate 1,3-P_2-glycerate is commercially unavailable and must be synthesised *in situ* from 3-PGA and ATP in a coupled reaction with 3-PGA kinase (PGK) as the coupling enzyme. The assay mixture (1.0 ml) contains 100 mM Tris-HCl (pH 7.8), 4.5 mM 3-PGA, 0.32 mM NAD(P)H, 2 mM ATP, 8 mM $MgSO_4$, 2 mM dithioerythritol (DTE), 1 mM EDTA, 1.8 U PGK and enzyme preparation (Cerff, 1982). The assay is initiated by the addition of NAD(P)H.

Notes: (a) Blanks containing all the reactants except 3-PGA should be performed with impure extracts.

(b) In crude extracts of some plant tissues, the NADP-dependent enzyme is rapidly inactivated owing to oxidation of catalytically important dithiol groups. The activity can be restored by incubation of the crude extract in the assay medium (containing 2 mM DTE) at 25°C for 5 min prior to initiation of the reaction with NADPH.

B. Extraction and Purification

The following procedure for white mustard seedlings (described fully by Cerff, 1982) describes the isolation of 4–10 mg of homogeneous NAD-G3P dehydrogenase, NADP-G3P dehydrogenase isoenzyme I, and NADP-G3P dehydrogenase isoenzyme II. All procedures should be carried out at 4°C.

1. Crude extract

White mustard seedlings (500 g, 7 days old, grown under constant light) are homogenised in a Waring blender with 3.5 vols of 100 mM 3-(N-morpholino)propanesulphonic acid (Mops)-NaOH (pH 6.9), containing 5 mM DTE, 7 mM EDTA, 0.1 mM NAD^+, 50 µg ml^{-1} PMSF, and 2.5% (w/v) insoluble PVP. The extract is clarified by centrifugation (30 min, 10 000 × g), and the pH of the crude supernatant adjusted to 7.5 with 0.1 M NaOH.

2. Ammonium sulphate fractionation

$(NH_4)_2SO_4$ is added to the crude supernatant to 55% (saturation). After centrifugation the supernatant is adjusted to 70% (saturation) and the solution centrifuged as above. The 70% (saturation) supernatant is adjusted to 95% (saturation) $(NH_4)_2SO_4$ and centrifuged as above. The 55–70% and 70–95% $(NH_4)_2SO_4$ pellets are dissolved in 30 ml and 10 ml, respectively, of 100 mM Tris-HCl (pH 7.4), containing 2 mM DTE, 1 mM EDTA, 0.2 mM NAD^+, 25 µg ml^{-1} PMSF, and 0.02% (w/v) NaN_3 (buffer A). Solubilised pellets are desalted on a column (2.5 × 20 cm) of Sephadex G-25 pre-equilibrated with buffer A.

3. Acetone fractionation

To the desalted 55–70% $(NH_4)_2SO_4$ fraction is added an equal volume of 50% (v/v) cold ($-20°C$) acetone with stirring and the mixture maintained at $-12°C$. The precipitate is collected by centrifugation at $-20°C$, the pellet resuspended in 10 ml buffer A, and the resultant suspension clarified by centrifugation (20 min, 23 000 × g).

4. Sephadex G-200 chromatography

The acetone fraction and the 70–90% $(NH_4)_2SO_4$ fraction are eluted through separate columns (2.6 × 100 cm) of Sephadex G-200 pre-equilibrated with buffer A. A single peak of NAD-dependent G3P dehydrogenase is recovered following gel filtration of the 70–90% $(NH_4)_2SO_4$ fraction and is further purified by affinity chromatography (see below). Two peaks of NADP-G3P dehydrogenase activity are recovered from each column. The early eluting activity peaks represent 'aggregated' NADP-linked isoenzymes 1 (acetone fraction) and 2 (70–90% $(NH_4)_2SO_4$ fraction). These 'early eluting' fractions are concentrated to about 10 ml with an Amicon PM-30 ultrafilter and rechromatographed on Sephadex G-200 as described, with the exception that the elution and column pre-equilibration buffer now contains 0.05 mM $NADP^+$ in place of 0.2 mM NAD^+. As $NADP^+$ causes deaggregation of the $NADP^+$-linked enzymes, NADP-linked isoenzymes 1 and 2 elute much later when rechromatographed on *Sephadex G-200 in the presence of* $NADP^+$. Respective pooled peak fractions, containing homogeneous NADP-linked isoenzymes 1 or 2, are concentrated as described above.

5. Blue Sepharose CL-6B affinity chromatography of NAD-G3P dehydrogenase

The partially purified NAD-specific enzyme is concentrated as described above to about 5 ml and eluted through a column (2 × 16 cm) of Sephadex G-25 pre-equilibrated in 20 mM Tris-HCl (pH 7.4), containing 0.5 mM EDTA, and 0.05 mM DTE (buffer B). The preparation is then applied to a column (2 × 16 cm) of Blue Sepharose CL-6B pre-equilibrated with buffer B. The column is washed sequentially with 5–10 bed vols of buffer B, 50 ml of buffer B adjusted to pH 8.5, and the NAD-linked enzyme eluted with buffer B (pH 8.5) containing 5 mM NAD^+. The purified enzyme is adjusted to contain 100 mM NaPi (pH 7.5) and 2 mM DTE prior to concentration as described above.

Notes: (a) If stored in buffer A at 0°C and sealed under N_2 the NADP-linked enzymes will remain fully active for at least 2-3 weeks. All final preparations can be frozen in liquid N_2, lyophilised, and stored indefinitely at $-20°$.

(b) NADP-linked isoenzymes I and II are each purified about 60-fold, to a final specific activity of about 120 U mg^{-1}, and overall yields of 17 and 9%, respectively. NAD-linked G3P dehydrogenase is purified 34-fold to a final specific activity of 66 U mg^{-1} and an overall yield of 7%.

C. Properties

Chloroplastic NADP-G3P dehydrogenase isoenzyme I contains two separate subunits of non-identical M_r values arranged as A_2B_2 in the native enzyme (Cerff, 1982). Isoenzyme II is a homotetramer which contains a single subunit type identical in M_r with subunit A of isoenzyme I. The M_r values of subunits A/B are in the range of 38 to 42 kDa in white mustard, spinach, peas, and barley. The cytosolic NAD-specific enzyme from green seedlings of spinach, peas and mustard is a homotetramer composed of four 39 kDa subunits (Cerff, 1982).

Optimum pH values for the various G3P dehydrogenases range from 7.5 to 8.5. All G3P dehydrogenases show hyperbolic substrate saturation kinetics. The K_m values of the various plant NAD-linked G3P dehydrogenases for G3P are about 0.3 mM; for NAD$^+$, about 0.15 mM; and for Pi, 0.3-2.0 mM. The chloroplastic NADP-linked isoenzymes are believed to be light-activated by a thioredoxin mediated disulphide–dithiol interconversion and can utilise NAD(H) as the coenzyme. The NADPH:NADH activity ratio for the white mustard isoenzymes is about 2. The K_m values are 23 μM and 300 μM for NADPH and NADH, respectively (Cerff, 1982).

VII. PHOSPHOGLYCERATE KINASE

(EC 2.7.2.3; ATP:3-phospho-D-glycerate 1-phosphotransferase)

$$1,3\text{-PGA} + \text{MgADP} \longleftrightarrow 3\text{-PGA} + \text{MgATP}$$

Phosphoglycerate kinase (PGK) catalyses the first ATP-generating step of glycolysis. Cytosolic and plastid isoenzymes have been partially purified from developing castor bean endosperm (Miernyk and Dennis, 1982). Spinach leaf and silver beet leaf PGKs have been purified to homogeneity and extensively characterized (Cavell and Scopes, 1976; Kuntz and Krietsch, 1982).

A. Assay

PGK activity is routinely assayed in the reverse reaction by coupling with the G3P dehydrogenase reaction and monitoring NADH oxidation at 340 nm. The assay mixture for the silver beet leaf enzyme (1.0 ml) contains 30 mM triethanolamine-HCl (pH 7.5), 5 mM 3-PGA, 1 mM ATP, 5 mM MgSO$_4$, 50 mM KCl, 0.2 mM EDTA, 0.15 mM NADH, 0.2 mg ml^{-1} BSA, 5 U ml^{-1} G3P dehydrogenase, and enzyme extract which is used to start the reaction (Cavell and Scopes, 1976).

Notes: In crude preparations, rates of NADH oxidation in the absence of 3-PGA or ATP should be subtracted from the total activity.

B. Extraction and Purification

The following procedure for silver beet (*Beta vulgaris*) leaves (described fully by Cavell and Scopes, 1976) yields homogeneous chloroplastic PGK. All procedures should be carried out at 4°C.

1. Crude extract

Deribbed and diced silver beet leaves (1 kg) are homogenised with a Waring blender in 2 vols of 30 mM triethanolamine-HCl (pH 7.5), containing 10 mM EDTA, 10 mM 2-mercaptoethanol, and 50% (saturation) $(NH_4)_2SO_4$. The homogenate is filtered through cheesecloth.

2. Ammonium sulphate fractionation

The filtrate (about 2400 ml), which is assumed to be 35% (saturation) $(NH_4)_2SO_4$, is adjusted to contain 50% (saturation) $(NH_4)_2SO_4$, and centrifuged (45 min, 4500 × g). The 35–50% supernatant is adjusted to contain 80% (saturation) $(NH_4)_2SO_4$, centrifuged as above, and the pellets dissolved in 30 mM triethanolamine-HCl (pH 7.5) containing 1 mM EDTA, and 10 mM 2-mercaptoethanol to a volume of about 150 ml. The extract is desalted on a column (2.3 × 90 cm) of Sephadex G-25 pre-equilibrated with 20 mM KPi (pH 7.0), containing 1 mM EDTA and 10 mM 2-mercaptoethanol (buffer A).

3. DEAE-cellulose chromatography

The desalted extract is absorbed onto a column (1.1 × 15 cm) of DEAE-cellulose (Whatman DE-52) pre-equilibrated with buffer A. PGK is eluted following the application of 300 ml of a linear 20–80 mM KPi gradient. Pooled peak fractions are concentrated by salting out at 85% (saturation) $(NH_4)_2SO_4$, centrifuging (20 min, 24 000 × g) and dissolving resultant pellets with about 10 ml of buffer A.

4. Cibacron-blue Sephadex chromatography

The sample is absorbed onto a column (0.8 × 5 cm) of Cibacron-blue Sephadex pre-equilibrated with buffer A. The enzyme is eluted using the same buffer containing 500 mM $(NH_4)_2SO_4$. Pooled peak fractions are salted out with 5.5 g of $(NH_4)_2SO_4$ per 10 ml and collected by centrifugation as above. Resultant pellets are dissolved with about 10 ml of 5 mM KPi (pH 7.0), 100 mM $NaSO_4$ (buffer B).

5. Sephadex G-75 gel filtration

The sample is eluted at 50 ml h^{-1} through a column (1.1 × 60 cm) of Sephadex G-75 pre-equilibrated with buffer B. Pooled peak fractions are concentrated to about 3 ml using an Amicon YM-10 ultrafilter.

Notes: (a) The final preparation is stable for at least 1 year when stored at 4°C as a precipitate in 3.3 M $(NH_4)_2SO_4$.

(b) Silver beet chloroplastic PGK is purified about 142-fold to a final specific activity of about 1000 U mg^{-1} and an overall yield of 34%.

(c) The same protocol has been used to purify a spinach leaf PGK to apparent homogeneity (Cavall and Scopes, 1976).

C. Properties

In general, the kinetic and physicochemical properties of plant PGKs are very similar to those of the yeast or mammalian enzymes. All plant PGKs studied to date are monomers with M_rs of about 47 kDa.

The pH optimum for silver beet and spinach leaf PGK is about 7.5. Hyperbolic substrate saturation kinetics are observed. The K_m value of homogeneous spinach leaf PGK for MgADP is about 0.3 mM; for 1,3-diphosphoglycerate, about 0.2 μM (Kuntz and Krietsch, 1982). The respective K_m values of the purified silver beet and spinach leaf enzymes for 3-PGA are 0.7 and 3 mM; for MgATP, 0.17 and 0.3 mM (Cavall and Scopes, 1976; Kuntz and Krietsch, 1982). Much additional research is required before our understanding of the biochemistry of plant phosphoglycerate kinase is complete. Of great interest will be the determination of the comparative molecular, immunological and kinetic properties of the purified cytosolic and plastidic PGK isoenzymes.

VIII. PHOSPHOGLYCERATE MUTASE

(EC 5.4.2.1; D-phosphoglycerate 2,3-phosphomutase)

$$3\text{-PGA} \longleftrightarrow 2\text{-PGA}$$

There are two classes of phosphoglycerate mutases (PGMs), those found in mammals and yeast which catalyse an intermolecular phosphoryl group transfer, utilising 2,3-bisphosphoglycerate as a cofactor, and those occurring in insects, plants and fungi which catalyse an intramolecular phosphoryl group transfer, but have a 2,3-bisphosphoglycerate-independent mechanism. Both types of reactions involve a phosphoryl–enzyme intermediate.

The enzyme has been purified to homogeneity from wheat and rice germ, but in neither case was the cellular localisation or the possible presence of isoenzymes investigated. What follows is the assay and isolation of highly purified cytosolic and plastid PGM isoenzymes (abbreviated as PGM$_c$ and PGM$_p$, respectively) from developing castor bean endosperm (Botha and Dennis, 1986).

A. Assay

The PGM reaction is routinely assayed in the direction of 2-PGA formation by coupling to the enolase, PK and LDH (lactate dehydrogenase) reactions and monitoring NADH oxidation at 340 nm. The assay mixture (1.0 ml) contains 100 mM N-tris(hydroxymethyl)methyl-2-aminoethanesulphonic acid (Tes)-NaOH (pH 7.2), 10 mM $MgCl_2$, 0.05 mM NADH, 2.7 mM ADP, 1 U enolase, 5 U PK, and 6 U LDH. The reaction is initiated by the addition of 20 µl of 150 mM 3-PGA.

Note: PGM can also be assayed by coupling with enolase and measuring the increase in A_{240} resulting from conversion of 2-PGA to PEP (Botha and Dennis, 1986).

B. Extraction and Purification

The following procedure for developing castor bean endosperm tissue (described fully by Botha and Dennis, 1986) yields highly purified preparations of PGM_c and PGM_p. All procedures should be carried out at 4°C.

1. Crude extract

Developing castor bean endosperm (25 g; 30–40 days post-pollination) is homogenised with a mortar and pestle in 1.5 vol of 50 mM KPi (pH 7.5), containing 400 mM sorbitol, 2 mM $MgCl_2$, and 2 mM DTT. The homogenate is filtered through cheesecloth and centrifuged (5 min, 500 × g). The supernatant is centrifuged at 10 000 × g for 20 min. The resulting supernatant and pellet are used for the purification of PGM_c and PGM_p, respectively. The pellet is resuspended in 20 ml of 10 mM KPi (pH 7.5), containing 400 mM sorbitol and 2 mM DTT and centrifuged as above. The supernatant is discarded and the pellet resuspended in the above buffer. This is layered on a 12 ml sucrose cushion containing 10 mM KPi (pH 7.5) and 15% (w/v) sucrose, and centrifuged (30 min, 27 000 × g). The resulting pellet, used for the purification of PGM_p, is resuspended in a minimum volume of 10 mM KPi (pH 7.0). Identical purification schemes are subsequently used for both PGM isoenzymes.

2. Phenyl Sepharose chromatography

The extract is adjusted to 45% (saturation) $(NH_4)_2SO_4$ and centrifuged as above. The supernatant is applied to a Phenyl Sepharose column (1.5 × 20 cm) pre-equilibrated with 10 mM KPi (pH 6.8), containing 27.7% (w/v) $(NH_4)_2SO_4$. PGM is eluted (20 ml h^{-1}) with a 90 ml linear gradient of 27.7–0% (w/v) $(NH_4)_2SO_4$ and 0–20% (v/v) ethylene glycol in 10 mM KPi (pH 6.8). Pooled peak fractions are dialysed against 3 × 3 l of 10 mM KPi (pH 7.5).

3. DEAE-Sephacel chromatography

The extract is absorbed onto a column (1.5 × 12 cm) of DEAE-Sephacel pre-equilibrated with 10 mM KPi (pH 7.5). PGM is eluted (30 ml h^{-1}) with 100 ml of a linear gradient of 0–300 mM KCl in 20 mM KPi (pH 6.3). Pooled peak fractions are adjusted to pH 6.8 with 1 M KPi (pH 7.0), and dialysed against three changes (1 litre each) of 10 mM KPi (pH 6.8).

4. HA-Ultragel chromatography

The extract is then absorbed onto a column (1.5 × 12 cm) of HA-Ultragel pre-equilibrated with 10 mM KPi (pH 6.3). PGM is eluted (20 ml h^{-1}) with 90 ml of a linear 10–300 mM KPi (pH 6.3) gradient. Pooled peak fractions are adjusted to pH 7.2 and dialysed against 1 litre of 25 mM KPi (pH 7.2) containing 30% (v/v) glycerol for 5 h.

Notes: (a) Final preparations are stored at −20°C, and are stable for at least one month.
(b) PGM$_p$ is purified about 280-fold to a final specific activity of about 84 U mg^{-1} and an overall yield of about 20%. PGM$_c$ is purified about 741-fold to a final specific activity of about 620 U mg^{-1} and an overall yield of about 24%.

C. Properties

There is considerable consistency in the physical and catalytic properties of plant PGMs. In all cases the enzyme is monomeric with subunit M_r values of about 63 kDa. Castor bean endosperm PGM$_c$ is relatively heat stable, whereas PGM$_p$ is heat labile (Botha and Dennis, 1986). Optimum pH values for castor bean PGM$_p$ and PGM$_c$ are about 7.5 and 7.0, respectively, in the direction of 2-PGA formation. When assayed in the reverse direction, PGM$_p$ has a pH optimum of 7.2, whereas the pH optimum for PGM$_c$ is 7.0 (Botha and Dennis, 1986). Both isoenzymes show hyperbolic substrate saturation kinetics. The K_m values of castor endosperm PGM$_p$ and PGM$_c$ for 3-PGA are 430 and 330 µM, respectively; for 2-PGA, 112 and 60 µM, respectively.

IX. ENOLASE

(EC 4.2.1.11; 2-phospho-D-glycerate hydro-lyase)

$$2\text{-PGA} \xleftrightarrow{Mg^{2+}} PEP$$

Enolase is a metalloenzyme which catalyses the freely reversible conversion of 2-PGA to PEP. Although the enzyme has been purified to apparent homogeneity from potato tubers and spinach leaves, in neither case was it determined if it was the plastidic or cytosolic enolase which was purified. What follows is the assay and isolation of highly purified plastidic and cytosolic enolase isoenzymes from developing castor bean endosperm (Miernyk and Dennis, 1984).

A. Assay

The enolase reaction is routinely assayed at room temperature in the direction of PEP formation by coupling with the PK and LDH reaction and monitoring NADH oxidation at 340 nm. The reaction mixture (1 ml) contains 88 mM Tes-NaOH (pH 7.5), 10 mM MgCl$_2$, 0.05 mM NADH, 2.7 mM ADP, 5 U PK, 6 U LDH, and enzyme extract in a final volume of 1.0 ml (Miernyk and Dennis, 1984). The reaction is initiated by addition of 20 µl of 25 mM 2-PGA.

Note: When interference with the coupling enzymes is a possibility, enolase can be assayed directly by measuring the increase in A_{240}.

B. Extraction and Purification

The following procedure for developing castor seed endosperm tissue (described fully by Miernyk and Dennis, 1984) yields about 10 and 20 µg of plastid (~75% pure) and cytosolic (~90% pure) enolase, respectively. All steps are conducted at 4°C unless otherwise noted.

1. Crude extract

Developing castor bean endosperm (28 g; 32–40 days post-pollination) is homogenised in 2 vols of 100 mM KPi (pH 6.9), containing 5 mM $MgCl_2$, and centrifuged (20 min, 27 000 × g). The supernatant, after removal of the floating lipid layer, is used directly for ion-filtration chromatography.

2. Ion-filtration chromatography

The crude supernatant is loaded onto a column (2.6 × 65 cm) of DEAE-Sephadex A-25 pre-equilibrated with 10 mM Imidazole-HCl (pH 7.0). Salt gradient formation and subsequent fractionation of enolase isoenzymes is achieved by applying 500 ml of 500 mM KH_2Pi (pH 6.7) at 4 ml min^{-1}. The early and later eluting peaks of enolase activity correspond to the plastid and cytosolic isoenzymes, respectively.

3. Ammonium sulphate and ethanol fractionation

Combined peak fractions are made 10 mM with respect to $MgCl_2$, and 1 mM with 2-PGA. Each isoenzyme is precipitated between 40 and 60% (saturation) $(NH_4)_2SO_4$, then centrifuged as above. Respective pellets are resuspended in a minimal volume of 20 mM Imidazole-HCl (pH 6.7), containing 5 mM $MgCl_2$ and 1 mM 2-PGA, and are dialysed overnight against 400 vol of the same buffer. Absolute ethanol at −20°C is added to the dialysed enolase isoenzymes, and the protein fraction precipitating between 45 and 70% (v/v) ethanol is collected by centrifugation as described above.

4. Sephacryl S-200 gel filtration

The ethanol-fractionated enolase isoenzymes are dissolved with a minimum volume of 10 mM Imidazole-HCl (pH 6.7), containing 50 mM KCl, 10 mM $MgCl_2$, and 0.1 mM 2-PGA, and applied to a column (2.5 × 116 cm) of Sephacryl S-200.

5. Hydroxyapatite chromatography

Following gel filtration, plastidic enolase is absorbed onto a column (1.5 × 10.5 cm) of hydroxyapatite pre-equilibrated with 20 mM KPi (pH 6.9), containing 10 mM $MgCl_2$. Enolase is eluted with 200 ml of a linear 10–500 mM KPi gradient developed by pumping in an upward direction. Hydroxyapatite chromatography is conducted at room temperature to minimise precipitation of MgPi.

Following gel filtration, cytosolic enolase is chromatographed a second time on an ion-filtration column, but this time at pH 7.2. Pooled peak activity fractions are further purified by hydroxyapatite chromatography as described above.

Notes: (a) Final preparations are adjusted to 20% (v/v) with ethylene glycol, stored at $-20°C$, and are stable for at least one month.

(b) Plastidic enolase is purified about 188-fold to a final specific activity of about 200 U mg^{-1} and an overall yield of about 4%. Cytosolic enolase is purifed about 240-fold to a final specific activity of about 250 U mg^{-1} and an overall yield of about 11%.

C. Properties

Enolase from all sources thus far examined are dimeric proteins composed of identical subunits of approximately 55 kDa. Cytosolic enolase from developing castor bean endosperm is more heat stable than its plastidic counterpart (Miernyk and Dennis, 1984). The p*I* values for cytosolic and plastid enolase are about 5.2 and 4.8, respectively. Antibodies to enolase from baker's yeast cross-react with plastid enolase, but not with the cytosolic isoenzyme.

Optimum pH value for both castor bean isoenzymes is about 7.5 in the direction of PEP formation. In contrast, when assayed in the reverse direction, cytosolic enolase has a pH optimum of 7.8, whereas the pH optimum for the plastid isoenzyme is 6.0 (Miernyk and Dennis, 1984). Both isoenzymes have an absolute requirement for a divalent metal cation (Mg^{2+} preferred), are inhibited by F^- in the presence of Pi, and show hyperbolic substrate saturation kinetics. The K_m values of the plastid and cytosolic castor bean endosperm enolase for 2-PGA are 25 and 61 μM, respectively; for PEP, 2.6 and 7.2 mM, respectively; and for Mg^{2+}, 1.4 and 1.2 mM, respectively.

X. PYRUVATE KINASE

(EC 2.7.1.40; ATP: pyruvate O^2-phosphotransferase)

$$PEP + MgADP \xrightarrow{K^+} Pyruvate + MgATP$$

Pyruvate kinase (PK) is considered to be a key regulatory enzyme of the glycolytic pathway and has been highly purified and characterised from a wide variety of non-plant sources. The enzyme catalyses a thermodynamically irreversible reaction under physiological conditions. The complete purification of the enzyme from a plant source has been delayed owing to the general lability of the plant enzyme. Recent advances in protein purification technologies, however, have now made it possible to purify both cytosolic and plastid PK isoenzymes (abbreviated as PK_c and PK_p, respectively) to apparent homogeneity.

A. Assay

The PK reaction is normally coupled to the LDH reaction and assayed at 30°C by monitoring NADH oxidation at 340 nm. The assay mixture (1.0 ml) contains 50 mM Hepes-NaOH (pH 6.9), 2 mM PEP, 2 mM ADP, 50 mM KCl, 10 mM $MgCl_2$, 0.15 mM NADH, 0.2 mg ml^{-1} BSA, 2 mM DTE, 2.0 U LDH, and enzyme extract, which is added to start the reaction (Plaxton, 1988). Identical assay conditions can be used for PK_p, with the exception that the assay pH may need to be adjusted to pH 7.9 (Ireland *et al.*, 1980).

Notes: (a) Assays must be corrected for contaminating PEP phosphatase activity by omitting ADP from the reaction mixture.

(b) The enzyme can also be assayed by coupling the production of ATP with yeast hexokinase and *Leuconostoc* glucose 6-phosphate dehydrogenase in the presence of 5 mM glucose, and monitoring the reduction of NAD^+ at 340 nm (Ireland *et al.*, 1980). In this case, minus PEP controls must be performed so as to account for contaminating adenylate kinase activity.

B. Extraction and Purification

1. PK_c

The following procedure for germinating castor bean endosperm tissue (described fully by Plaxton, 1988) yields up to 1.1 mg of homogeneous enzyme. Germinating castor bean endosperm contains very low amounts of PK_p, but provides a fairly active preparation of PK_c. All operations should be carried out at 0–5°C.

(a) *Crude extract.* Quick-frozen 5-day old germinating castor bean endosperm dissected free of seed coat and cotyledons (500 g) is homogenised using a Waring blender or Polytron with 2 vols of 50 mM KPi (pH 7.6), containing 2 mM EDTA, 2 mM EGTA, 50 mM NaF, 2 mM DTE, 2.5 mM $MgCl_2$, 2.5% (w/v) insoluble PVP, 20% (v/v) glycerol, 0.1 mM PEP, 1 mM PMSF, $5\,\mu g\,ml^{-1}$ chymostatin, and $5\,\mu g\,ml^{-1}$ leupeptin. After centrifugation (20 min, $17\,000 \times g$) the crude supernatant is filtered through Miracloth (Calbiochem) to remove the fat layer.

(b) *Heat treatment.* The crude supernatant is divided equally between two 1 litre flasks, heated to 60°C in a 65°C water bath, and maintained at this temperature for 5 min. The extract is cooled on ice to 4°C and centrifuged as above to remove copious precipitated protein. Any contaminating activity of the heat labile PK_p is removed by this treatment.

(c) *PEG fractionation.* PEG 8000 is added to 3% (w/v); after centrifugation as above the supernatant is adjusted to 8.5% (w/v). The precipitate, containing PK_c, is collected by centrifuging as above and the pellets are stored overnight at 4°C.

(d) *Q-Sepharose chromatography.* The 3–8.5% (w/v) PEG pellets are resuspended with 110 ml of 10 mM KPi (pH 7.1), containing 1 mM EDTA, 5 mM $MgCl_2$, 2 mM DTE, and 20% (v/v) glycerol (buffer A), supplemented with $5\,\mu g\,ml^{-1}$ leupeptin and $5\,\mu g\,ml^{-1}$ chymostatin, and centrifuged (20 min, $27\,000 \times g$) to clarify. The clear supernatant is absorbed at $2.5\,ml\,min^{-1}$ onto a column of Q-Sepharose ($2.5 \times 26\,cm$) pre-equilibrated with buffer A and connected to the FPLC system. After washing with buffer A containing added 20 mM Pi, PK_c is eluted in a sharp peak following a step from 30 to 60 mM Pi. Pooled peak fractions are concentrated to about 25 ml with an Amicon PM-30 ultrafilter, and then adjusted to 17% (w/v) PEG. The precipitate, containing PK_c, is collected by centrifugation as above and pellets stored at 4°C overnight.

(e) *ADP-agarose chromatography.* The PEG pellets are resuspended in 12 ml of

25 mM Hepes-NaOH (pH 7.1), containing 1 mM EDTA, 5 mM $MgCl_2$, and 2 mM DTE (buffer B), supplemented with 5 μg ml^{-1} leupeptin and 5 μg ml^{-1} chymostatin, and centrifuged as above to clarify. The supernatant is absorbed at 0.6 ml min^{-1} onto a column (1.5 × 5.6 cm) of ADP-agarose pre-equilibrated with buffer B. After washing to remove unbound protein, PK_c is eluted in a sharp peak with buffer B containing 5 mM Pi, 2 mM ADP, and 10% (v/v) glycerol.

(f) *FPLC Mono-Q.* Pooled peak ADP-agarose fractions are absorbed at 0.5 ml min^{-1} onto a Mono-Q HR 5/5 column which had been pre-equilibrated with buffer A. PK_c is eluted with 20 ml of a 30–120 mM Pi gradient.

(g) *FPLC Phenyl Superose.* The pooled peak Mono-Q fractions are adjusted to 30% (saturation) with $(NH_4)_2SO_4$, and the solution adsorbed at 0.4 ml min^{-1} onto a Phenyl Superose HR 5/5 column pre-equilibrated with 25 mM KPi (pH 7.1), containing 0.5 mM EDTA, 1 mM $MgCl_2$, 1 mM DTE, and 30% (saturation) $(NH_4)_2SO_4$ (buffer C). The column is eluted at 0.25 ml min^{-1} in a stepwise fashion with decreasing concentrations of buffer C and simultaneously increasing concentrations of 15 mM KPi (pH 7.1), containing 1 mM EDTA, 2 mM $MgCl_2$, 2 mM DTE, and 40% (v/v) ethylene glycol (buffer D). The enzyme elutes in a sharp peak following the step from 70 to 100% buffer D (30 to 0% buffer C). Pooled peak fractions are concentrated to about 0.5 ml using an Amicon YM-30 ultrafilter, frozen with liquid N_2, and stored at $-80°C$.

Notes: (a) The enzyme is purified about 3100-fold to a final specific activity of 200 U mg^{-1} and an overall recovery of up to 30%.

(b) Although glycerol (or ethylene glycol), Pi, Mg^{2+}, and DTE afford some stabilisation of PK_c during its extraction and purification, even in their presence the enzyme is very unstable when protein concentration decreases below about 0.4 mg ml^{-1}. FPLC effectively prevents dilution of the enzyme during the latter stages of purification, when the total amount of protein is very low. Consequently, there is little reduction in yield during the final two FPLC-utilising purification steps.

2. PK_p

The following procedure is a modification of that described by Lin *et al.* (1989a) and allows the purification to apparent homogeneity of about 0.4 mg of PK_p from the green alga, *Selenastrum minutum* (V. L. Knowles, D. T. Dennis, and W. C. Plaxton, unpubl. res.). All operations are performed at 4°C.

(a) *Crude extract.* Quick-frozen cells (110 g) are rapidly thawed in 2 vols of 50 mM KPi (pH 7.5), containing 1 mM EDTA, 1 mM EGTA, 1 mM DTE, 30 mM NaF, 2 mM $MgCl_2$, 0.1 mM PEP, 10% (v/v) glycerol, 2% (w/v) soluble PVP, 1 mM PMSF, 5 μg ml^{-1} chymostatin, and 5 μg ml^{-1} leupeptin and passed through a French press at 18 000 psi ($\sim$124 MPa). The homogenate is clarified by centrifugation (20 min, 24 000 × g) prior to precipitation and chromatography.

(b) *Ammonium sulphate fractionation and treatment with Cell Debris Remover.* The enzyme is precipitated between 35 and 50% (saturation) $(NH_4)_2SO_4$ then centrifuged as above. The pellet is resuspended in 32 ml of 15 mM KPi (pH 7.5), containing 1 mM EDTA, 1 mM EGTA, 1 mM DTE, 30 mM NaF, 2 mM $MgCl_2$, and 10% (v/v) glycerol (buffer A), and dialysed for 3 h against 2×2 l of buffer A. Cell Debris Remover (2 g; Whatman) is added with stirring to the dialysed extract for 5 min and the solution is then percolated with suction through a layer (7×0.5 cm) of Cell Debris Remover pre-equilibrated with buffer A.

(c) *Q-Sepharose chromatography.* The extract is then diluted to about 110 ml with buffer A prior to being absorbed at 3 ml min^{-1} onto a column (2.6×38 cm) of Q-Sepharose pre-equilibrated with buffer A and connected to the FPLC system. PK activity is eluted in two peaks following the application of 1000 ml of a 15–400 mM KPi linear gradient. The first activity peak containing PK_p (Lin *et al.*, 1989a) is concentrated to about 20 ml with an Amicon YM-30 ultrafilter.

(d) *Butyl-agarose chromatography.* The pooled Q-Sepharose PK_p concentrated peak fractions are adjusted to 30% (saturation) $(NH_4)_2SO_4$, centrifuged as above, and absorbed at 0.7 ml min^{-1} onto a column (1.5×8 cm) of butyl-agarose pre-equilibrated with 40 mM KPi (pH 7.0), containing 1 mM EDTA, 1 mM EGTA, 1 mM DTE, 30 mM NaF, and 30% (saturation) $(NH_4)_2SO_4$ (buffer B), and connected to the FPLC system. The column is then eluted in a stepwise fashion with decreasing concentrations of buffer B and simultaneously increasing concentration of 20 mM KPi (pH 7.0), containing 1 mM EDTA, 1 mM EGTA, 1 mM DTE, 30 mM NaF, and 15% (v/v) glycerol (buffer C). PK_p elutes in a sharp peak following the step from 30–50% buffer C (70–50% buffer B). Pooled peak fractions are concentrated to about 1.0 ml using an Amicon YM-30 ultrafilter.

(e) *Superose 6 FPLC.* The extract is then loaded at 0.25 ml min^{-1} onto a column (1.6×50 cm) of Superose 6 ('prep' grade) pre-equilibrated with 50 mM Hepes-NaOH (pH 7.0), containing 1 mM EDTA, 1 mM EGTA, 1 mM DTE, 50 mM KCl, 2.5 mM $MgCl_2$, 30 mM NaF, and 15% (v/v) glycerol, and connected to the FPLC system. The pooled peak fractions are concentrated to about 0.5 ml as described above, frozen in liquid N_2, and stored at $-80°C$.

Notes: (a) The enzyme is stable for several months when stored frozen.
(b) The enzyme is purified about 700-fold to a final specific activity of 217 U mg^{-1} and an overall recovery of 10% (V. L. Knowles, D. T. Dennis, and W. C. Plaxton, unpubl. res.).

C. Properties

1. PK_c

Homogeneous PK_c from 5-day old germinating castor bean endosperm appears to be a 240 kDa heterotetramer, apparently composed of two types of subunits which migrate as 57 and 56 kDa proteins upon SDS-PAGE (Plaxton, 1988). Both proteins have been detected on a Western blot of a crude extract prepared under denaturing conditions from 5-day old germinating castor bean endosperm which was probed with rabbit anti-germinating castor endosperm PK_c IgG (Plaxton, 1989). This indicates that both subunits exist *in vivo*. CNBr peptide mapping and amino acid composition analyses demonstrated that the two germinating castor bean endosperm PK_c subunits are distinct proteins which share a high degree of structural homology (Plaxton, 1989). Although PK_c from developing castor seed endosperm and leaf were found to be antigenically very similar to PK_c from germinating endosperm, leaf and developing endosperm PK_c were both found to be 230–240 kDa homotetramers composed of a single type of 56 kDa subunit (Plaxton, 1989). PK_c from the green alga *S. minutum* is immunologically closely related to the germinating castor seed endosperm enzyme, but appears to be an apparent homodecamer composed of a single type of 57 kDa subunit (Lin *et al.*, 1989a).

All PK_cs examined to date are relatively heat stable, have a pH optimum of about pH 7.0, and show hyperbolic substrate saturation kinetics (Ireland *et al.*, 1980; Lin *et al.*, 1989b). The K_m for PEP is 0.03–0.09 mM; for ADP 0.03–0.05 mM; for K^+, 0.5–3 mM; and for Mg^{2+}, 0.45–0.85 mM (Ireland *et al.*, 1980; Lin *et al.*, 1989b). *S. minutum* PK_c was shown to be allosterically activated by DHAP, and inhibited by Pi and glutamate (Lin *et al.*, 1989b). This provides a rationale for the activation of algal PK_c which occurs during periods of enhanced NH_4^+ assimilation.

2. PK_p

SDS-PAGE of purified *S. minutum* PK_p revealed a single, 210 kDa, protein staining band; a native M_r of 235 kDa was determined by gel filtration (V. L. Knowles, D. T. Dennis and W. C. Plaxton, unpubl. res.). These data suggest that, in contrast to preliminary findings (Lin *et al.*, 1989a), *S. minutum* PK_p is a monomeric protein. Further studies are needed to determine if higher plant PK_ps show this novel subunit structure.

All PK_ps are heat labile, but may differ in their optimal pH (Ireland *et al.*, 1979, 1980; Lin *et al.*, 1989a,b). PK_p from castor bean leaf or developing endosperm has been reported to have a sharp pH optimum of about pH 8.0 (Ireland *et al.*, 1979, 1980). It was therefore suggested that the leaf enzyme may be light activated since stromal pH rises to 8 upon illumination. In contrast, *S. minutum* PK_p may be light inactivated since it shows a broad pH optimum of about 6.5, and was inhibited by metabolites whose stromal concentrations are probably elevated in the light (e.g. malate, phosphoglycolate, and Calvin cycle intermediates: Lin *et al.*, 1989b). Plant PK_ps generally have lower affinities for substrates than do their cytosolic counterparts: the K_m for PEP is 0.05–0.2 mM; for ADP, 0.2–0.3 mM; for K^+, 2–4 mM; and for Mg^{2+}, 0.6–1.6 mM (Ireland *et al.*, 1980; Lin *et al.*, 1989b).

XI. PHOSPHO*ENOL*PYRUVATE PHOSPHATASE

(EC not yet assigned; phospho*enol*pyruvate phosphohydrolase)

$$PEP + H_2O \longrightarrow Pyruvate + Pi$$

The presence of a plant phosphatase specific for PEP has been inferred for many years owing to the substantial 'PEP phosphatase' activity which often interferes with the determination of plant PK activity. However, until very recently the enzyme had never been purified to allow characterisation of its individual biochemical and kinetic properties. Thus, 'PEP phosphatase' activity was generally thought to be due to the presence of non-specific acid phosphatase(s) (Davies, 1979).

A. Assay

The PEP phosphatase assay is very similar to the PK assay. Pyruvate production is routinely coupled to the LDH reaction and assayed continuously at 25°C by monitoring NADH oxidation at 340 nm. Standard assay conditions for black mustard suspension cell PEP phosphatase are 50 mM sodium acetate (pH 5.6), containing 1 mM PEP, 4 mM $MgCl_2$, 0.2 mM NADH, and 3 U dialysed rabbit muscle LDH in a final volume of 1 ml (Duff et al., 1989b). The reaction is initiated by the addition of enzyme extract.

Notes: (a) Assays must be corrected for contaminating NADH oxidase activity by omitting PEP from the reaction mixture. NADH, which is unstable at pH 5.6, should be added just before addition of the enzyme extract.

(b) For substrates other than PEP, Pi released by the phosphatase reaction must be measured colorimetrically using a stop-timed assay (Duff et al., 1989b).

B. Extraction and Purification

The following procedure for heterotrophic black mustard leaf petiole suspension cells (described fully by Duff et al., 1989b) yields up to 300 µg of homogeneous enzyme. All steps are carried out at 4°C.

1. Crude extract

Quick frozen cells (1 kg) are ground to a powder under liquid N_2. The powder is thawed in 500 ml of 50 mM sodium acetate (pH 5.6), containing 1 mM EDTA, 0.5 mM DTE, 5 mM thiourea, 1 mM PMSF, 10 µg ml^{-1} leupeptin, and 10 µg ml^{-1} chymostatin, sonicated for 10 min at maximum power (20 s intervals using a Bronson Sonic Power Co. Model W350 Sonifer), and centrifuged (20 min, 17 000 × g).

2. Ammonium sulphate fractionation

The enzyme is precipitated between 35 and 80% $(NH_4)_2SO_4$ (saturation) and then centrifuged as above. The pellets are resuspended in 150 ml of 30 mM sodium acetate (pH 4.9), containing 0.25 mM EDTA, 0.5 mM DTE, and 0.5 mM PMSF (buffer A), dialysed overnight against 10 l of buffer A, and centrifuged (20 min, 27 000 × g) to remove insoluble material.

3. S-Sepharose chromatography

The clear supernatant is absorbed at 2 ml min^{-1} onto a column (2.5 × 20 cm) of S-Sepharose pre-equilibrated in buffer A minus PMSF, and connected to a FPLC system. PEP phosphatase is eluted with 450 ml of a 0–500 mM KCl gradient in buffer A minus PMSF. Peak activity fractions are pooled and concentrated in a dialysis sack against solid PEG 8000.

4. Chelating Sepharose 6B chromatography

The concentrated S-Sepharose peak fractions are diluted two-fold with 25 mM Tris-HCl (pH 7.1), containing 500 mM NaCl buffer C (buffer B), adjusted to pH 7.1 with solid Tris, and absorbed at 0.75 ml min^{-1} onto a column (1.5 × 10 cm) of Chelating Sepharose which had been charged with 18 mM $CuSO_4$, and pre-equilibrated with buffer B. The column is sequentially washed with buffer B, 25 mM sodium acetate (pH 4.0), containing 500 mM NaCl, 150 mM NH_4Cl, and 50 mM Imidazole (buffer C), and buffer C containing 10 mM EDTA. Two peaks of phosphatase activity are eluted. The first activity peak elutes with buffer C and is designated phosphatase A, while the second activity peak elutes with buffer C containing 10 mM EDTA and is designated phosphatase B. A blue coloured Cu^{2+}-EDTA complex co-elutes with the phosphatase B activity peak. To remove this complex, the pooled phosphatase B peak fractions are adjusted to 100% (saturation) $(NH_4)SO_4$. The solution is stirred for 20 min and centrifuged as above. The precipitate is collected by centrifugation as above, and the resulting pellets are resuspended in 20 ml of 5 mM KPi (pH 6.5), containing 100 mM KCl, 0.5 mM DTE, 1 μm $CaCl_2$, and 1 μm $MnCl_2$ (buffer D), and centrifuged as above to clarify.

5. Concanavalin A chromatography

The resuspended pellet is absorbed at 0.5 ml min^{-1} onto a column (0.5 × 5 cm) of Concanavalin A Sepharose pre-equilibrated with buffer D and connected to the FPLC system. PEP phosphatase is eluted in a single peak with 30 ml of a 0–500 mM mannopyranoside linear gradient in buffer D (gradient volume = 30 ml). Pooled peak activity fractions are concentrated to 1.5 ml using an Amicon YM-10 ultrafilter.

6. FPLC Superose 12 gel filtration

Concentrated Concanavalin A peak activity fractions are applied at 0.25 ml min^{-1} onto a column (1.6 × 50 cm) of Superose 12 ('prep' grade) pre-equilibrated with 30 mM Mes-HCl (pH 6.0), containing 1 mM DTE, 1 mM EDTA, 50 mM KCl, and 0.03% (w/v) NaN_3, and connected to the FPLC system. Pooled peak activity fractions are concentrated to about 0.5 ml as described above, frozen with liquid N_2, and stored at $-80°C$.

Notes: (a) The purified enzyme is stable for at least 6 months when stored frozen.
(b) The above scheme purifies PEP phosphatase about 1700-fold to a final specific activity of 380 U mg^{-1}, and an overall recovery of about 10%.

C. Properties

Black mustard PEP phosphatase, a 56 kDa monomer, is relatively heat stable and displays a broad pH optimum of about pH 5.6 (Duff *et al.*, 1989a). Binding of the enzyme to Concanavalin A suggests that PEP phosphatase is a glycoprotein. The final preparation exhibits a broad substrate selectivity, showing high activity towards PEP, *p*-nitrophenyl phosphate, ADP, ATP, and gluconate 6-phosphate, and moderate activity towards several other organic phosphates. PEP phosphatase possesses at least a five-fold and six-fold greater affinity and specificity constant, respectively, for PEP (K_m = 50 μM) than for any other non-artificial substrate. This enzyme thus joins the ranks of

3-PGA phosphatase and phytase as a class of plant acid phosphatases which exhibit preferential, but non-absolute, substrate selectivity. The enzyme is activated 1.7-fold by 4 mM Mg^{2+} ($K_a = 1.2$ mM), but shows potent inhibition by Pi ($K_i = 85$ μM). It has been postulated that PEP phosphatase functions to bypass the ADP-dependent PK reaction during extended periods of Pi starvation when intracellular ADP and Pi pools become greatly reduced (Duff *et al.*, 1989b). Consistent with the 'PK bypass' hypothesis is the finding that when Pi-fed black mustard suspension cells are transferred to Pi-free Murashige and Skoog media for 7 days total PEP phosphatase specific activity increases about 10-fold (Duff *et al.*, 1989a). The subcellular localisation of black mustard PEP phosphatase has not yet been established. Although its acid pH optimum is suggestive of a vacuolar localisation, significant PEP hydrolysing activity is observed at neutral pH. Examination of the distribution, subcellular localisation, molecular properties, and 'coarse' and 'fine' metabolic regulation of this 'new' plant glycolytic enzyme should prove to be a fruitful area for future research.

ACKNOWLEDGEMENTS

I am grateful to Professor D. T. Dennis for his helpful suggestions and criticism of the manuscript. Thanks to Mr Greg Moorhead and Mr Steve Duff for their comments on the manuscript.

REFERENCES

Anderson, L. E. (1982). *In* "Methods in Chloroplast Molecular Biology" (M. Edelman, R. B. Hallick and N.-H. Chua, eds), pp. 715–722. Elsevier Biomedical, Amsterdam.
ap Rees, T. (1985). *In* "Encyclopedia of Plant Physiology" (R. Douce and D. A. Day, eds), Vol. 18, pp. 391–414. Springer, Berlin.
ap Rees, T., Morrell, S., Edwards, J., Wilson, P. M. and Green, J. H. (1985). *In* "Regulation of Carbon Partitioning in Photosynthetic Tissue" (R. L. Heath and J. Preiss, eds), pp. 76–92. American Society of Plant Physiologists, Rockville, MD.
Black, C. C., Mustardy, L., Sung, S. S., Kormanik, P. P., Xu, D.-P. and Paz, N. (1987). *Physiol. Plant.* **69**, 387–394.
Botha, F. C. and Dennis, D. T. (1986). *Arch. Biochem. Biophys.* **245**, 96–103.
Botha, F. C., Small, J. G. C. and de Vries, C. (1986). *Plant Cell Physiol.* **27**, 1285–1295.
Botha, F. C., Cawood, M. C. and Small, J. G. C. (1988). *Plant Cell Physiol.* **29**, 415–421.
Cavell, S. and Scopes, R. K. (1976). *Eur. J. Biochem.* **63**, 483–490.
Cerff, R. (1982). *In* "Methods in Chloroplast Molecular Biology" (M. Edelman, R. B. Hallick and N.-H. Chua, eds), pp. 683–694. Elsevier Biomedical, Amsterdam.
Copeland, L. and Turner, J. F. (1987). *In* "The Biochemistry of Plants" (P. K. Stumpf and E. E. Conn, eds), Vol. 11, pp. 107–128. Academic Press, New York.
Davies, D. D. (1979). *Ann. Rev. Plant Physiol.* **30**, 131–158.
Dennis, D. T. and Greyson, M. F. (1987). *Physiol. Plant.* **69**, 395–404.
Dennis, D. T. and Miernyk, J. A. (1982). *Ann. Rev. Plant Physiol.* **33**, 27–50.
Dennis, D. T., Hekman, W. E., Thomson, A., Ireland, R. J., Botha, F. C. and Kruger, N. J. (1985). *In* "Regulation of Carbon Partitioning in Photosynthetic Tissue" (R. L. Heath and J. Preiss, eds), pp. 127–146. American Society of Plant Physiologists, Rockville, MD.
Duff, S. M. G., Moorhead, G. B. G., Lefebvre, D. D. and Plaxton, W. C. (1989a). *Plant Physiol.* **90**, 1275–1278.
Duff, S. M. G., Lefebvre, D. D. and Plaxton, W. C. (1989b). *Plant Physiol.* **90**, 734–741.

Garland, W. J. and Dennis, D. T. (1980a). *Arch. Biochem. Biophys.* **204**, 302–309.
Garland, W. J. and Dennis, D. T. (1980b). *Arch. Biochem. Biophys.* **204**, 310–317.
Heuer, B., Hansen, M. J. and Anderson, L. E. (1982). *Plant Physiol.* **69**, 1404–1406.
Iglesias, A. A., Serrano, A., Guerrero, M. G. and Losada, M. (1987). *Biochim. Biophys. Acta.* **925**, 1–10.
Ireland, R. J., De Luca, V. and Dennis, D. T. (1979). *Plant Physiol.* **63**, 903–907.
Ireland, R. J., De Luca, V. and Dennis, D. T. (1980). *Plant Physiol.* **65**, 1188–1193.
Knowles, V. L., Greyson, M. F. and Dennis, D. T. (1989). *Plant Physiol.* in press.
Kombrink, E., Kruger, N. J. and Beevers, H. (1984). *Plant Physiol.* **74**, 395–401.
Kruger, I. and Schnarrenberger, C. (1983). *Eur. J. Biochem.* **136**, 101–106.
Kruger, N. J. and Dennis, D. T. (1987). *Arch. Biochem. Biophys.* **256**, 273–279.
Kruger, N. J. and Hammond, J. B. W. (1988). *Plant Physiol.* **86**, 645–648.
Kuntz, G. W. K. and Krietsch, W. K. G. (1982). *In* "Methods in Enzymology" (W. A. Wood, ed.), Vol. 90, pp. 110–114. Academic Press, New York.
Kurzok, H.-G. and Feierabend, J. (1984). *Biochim. Biophys. Acta.* **788**, 214–221.
Lin, M., Turpin, D. H. and Plaxton, W. C. (1989a). *Arch. Biochem. Biophys.* **269**, 219–227.
Lin, M., Turpin, D. H. and Plaxton, W. C. (1989b). *Arch. Biochem. Biophys.* **269**, 228–238.
Miernyk, J. A. (1989). *In* "Advanced Plant Physiology: Integration and Control of Carbon and Nitrogen Metabolism" (D. T. Dennis and D. H. Turpin, eds), Longman, London (in press).
Miernyk. J. A. and Dennis, D. T. (1982). *Plant Physiol.* **69**, 825–828.
Miernyk. J. A. and Dennis, D. T. (1984). *Arch. Biochem. Biophys.* **233**, 643–651.
Pichersky, E. and Gottlieb, L. D. (1984). *Plant Physiol.* **74**, 340–347.
Plaxton, W. C. (1988). *Plant Physiol.* **86**, 1064–1069.
Plaxton, W. C. (1989). *Eur. J. Biochem.* **181**, 443–451.
Sabularse, D. C. and Anderson, R. L. (1981). *Biochem. Biophys. Res. Commun.* **103**, 848–855.
Sung, S. S., Xu, D.-P., Galloway, C. M. and Black, C. C. (1988). *Physiol. Plant.* **72**, 650–654.
Turner, J. F. and Turner, D. H. (1980). *In* "The Biochemistry of Plants" (D. D. Davies, ed.), Vol. 2, pp. 279–316. Academic Press, New York.
Wong, J. H., Yee, B. C. and Buchanan, B. B. (1987). *J. Biol. Chem.* **262**, 3185–3191.
Yuan, X.-H., Kwiatkowska, D. and Kemp, R. G. (1988). *Biochem. Biophys. Res. Commun.* **154**, 113–117.

10 The Mitochondrial Pyruvate Dehydrogenase Complex

DOUGLAS D. RANDALL and JAN A. MIERNYK

Department of Biochemistry, University of Missouri-Columbia, Columbia, MO 65211 USA

I.	Introduction	176
II.	Measurement of pyruvate dehydrogenase complex activity	177
	A. Continuous measurement of NADH formation	177
	B. Measurement of $^{14}CO_2$ release from [1-^{14}C]-pyruvate	178
	C. Other methods	178
	D. Pyruvate dehydrogenase, PDH, E_1	179
	E. Dihydrolipoyl transacetylase, LTA, E_2	179
	F. Dihydrolipoamide dehydrogenase, flavoprotein, E_3	179
	G. Pyruvate dehydrogenase kinase, PDH kinase	179
	H. Phospho-pyruvate dehydrogenase phosphatase, PDH-P phosphatase	182
III.	Methods of purification	182
	A. Plant mitochondria	182
	B. The pyruvate dehydrogenase complex	183
	C. Pyruvate dehydrogenase, PDH, E_1	184
	D. Dihydrolipoyl transacetylase, LTA, E_2	184
	E. Dihydrolipoamide dehydrogenase, LAD or E_3	184
	F. Pyruvate dehydrogenase kinase, PDH kinase	185
	G. Phospho-pyruvate dehydrogenase phosphatase, PDH-P phosphatase	186
IV.	Physicochemical and catalytic properties	186
	A. Structure of the complex	186
	B. Kinetic properties, effectors and regulation	187
	C. Properties of component enzymes	189
	D. Properties of PDH kinase	189
	E. PDH-P phosphatase	190
	F. Pyruvate oxidation and assay of PDC activity *in situ*	190

Acknowledgement 191
References 191

I. INTRODUCTION

The mitochondrial pyruvate dehydrogenase complex (PDC) is a very large multi-enzyme complex that catalyses the irreversible oxidative decarboxylation of pyruvate, esterification of the two carbon acetyl unit to coenzyme A, and ultimately the reduction of NAD^+ (Miernyk et al., 1985, 1987b). Figure 10.1 illustrates the overall reaction and relationship among the component enzymes, pyruvate dehydrogenase (EC 1.2.4.1), dihydrolipoyl acetyltransferase (EC 2.3.1.12) and lipoyl dehydrogenase (EC 1.6.4.3), and the required cofactors, Mg^{2+}, thiamine pyrophosphate (TPP), FAD, lipoic acid, NAD^+ and coenzyme A. The mitochondrial PDC (mPDC) also has two associated regulatory enzymes, pyruvate dehydrogenase (PDH) kinase (EC 2.7.1.99) and PDH-P phosphatase (EC 3.1.3.43). A component, which as yet has not been described for the plant PDCs, is the 'X-protein' that may have a role in anchoring other subunits to the transacetylase core of the complex (Yeaman, 1986), and in the catalytic function of the complex (Gopalakrishnan et al., 1989).

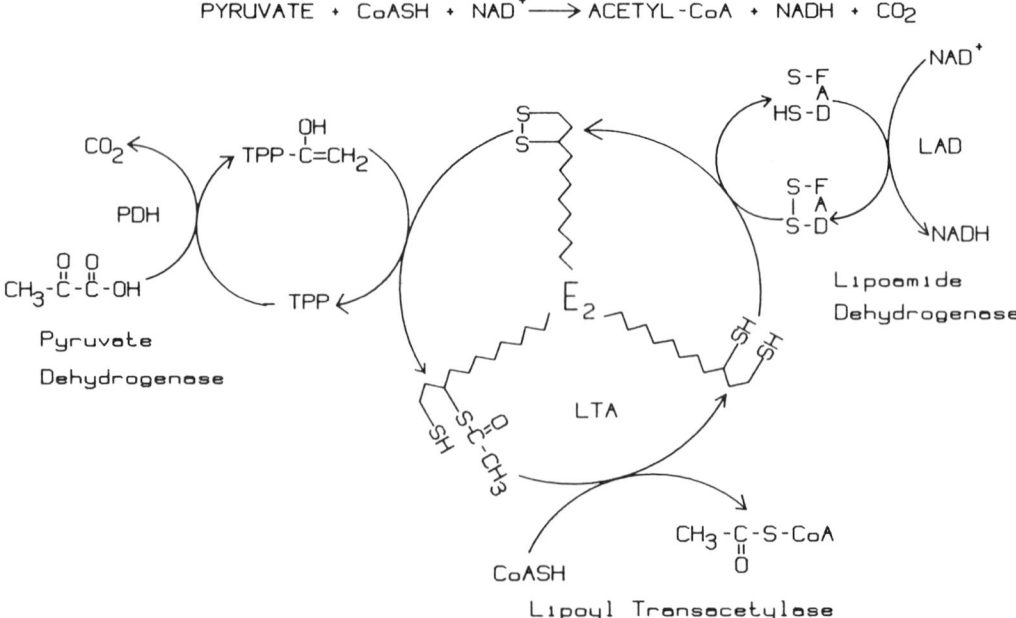

FIG. 10.1. Pyruvate dehydrogenase complex reaction sequence and relationship among components.

Carbon flows into the citric acid cycle through PDC. The irreversible nature of the PDC-catalysed reaction and the cross-roads position of pyruvate in metabolism point to the PDC as an important regulatory site. The regulation of PDC is multi-tiered, including product inhibition, metabolite effects and reversible phosphorylation (Miernyk et al., 1985, 1987b). The PDCs of plants are likely points for regulating carbon flow into the Krebs cycle during photosynthesis. Plants are unique in having two distinct, spatially separated types of PDCs, one mitochondrial and the other in the plastids. Each type of PDC has characteristic structural, catalytic and regulatory properties (Camp and Randall, 1985, Miernyk et al., 1985, 1987b). Therefore, when measuring PDC activity it is necessary to keep in mind that approximately 25% of the PDC activity of a leaf and up to 50% of the activity in a developing oilseed can be from the plastid PDC (pPDC) (Williams and Randall, 1979; Rapp et al., 1987). The assays and procedures described below will provide ways for dealing with the problem of distinguishing the two types of complex in tissue homogenates. However, in working with isolated mitochondria or plastids, contamination by the other organelle can be estimated through the use of marker enzymes.

II. MEASUREMENT OF PYRUVATE DEHYDROGENASE COMPLEX ACTIVITY

A. Continuous Measurement of NADH Formation

The preferred method of measuring PDC activity is to follow the formation of NADH (6.22×10^3 M^{-1} cm^{-1}) at 22–30°C with a recording spectrophotometer at 340 nm. An assay mixture should contain 50–85 mM N-tris(hydroxymethyl)methyl-2-aminoethanesulphonic acid (Tes) buffer, pH 7.6, 2.5 mM $MgCl_2$, 2 mM NAD^+, 0.12 mM lithium Coenzyme A in 2.6 mM cysteine-HCl, 0.2 mM TPP, 1.5 mM pyruvate, and enzyme in a total volume of 1.0 ml. The buffer, magnesium, TPP and NAD can be made up in a stock solution and stored frozen in aliquots usable for one day. The CoA/cysteine solution (cysteine protects the —SH of CoA from oxidation) should be prepared fresh daily and kept on ice until added to the cuvette. Pyruvate can be made up as a solution of the K^+ or Na^+ salt, and frozen in daily aliquots. Since pyruvate dimerises and polymerises it should be discarded at the end of each day. Assays should be initiated by the addition of pyruvate after establishing the background rate. With Triton X-100 lysed mitochondria, however, initiation with enzyme will yield slightly higher rates. To ensure that the activity being measured is PDC, it must be dependent upon NAD, CoA and pyruvate. In homogenates, it is also necessary to determine if NADH oxidase activity is present, e.g. add 0.2 mM NADH to the reaction mixture in the absence of CoA to test for oxidation of NADH. The spectrophotometric measurement of PDC activity by monitoring NADH production in crude homogenates is generally unsuccessful. The best results are obtained using this assay with isolated mitochondria or the partially purified complex. Mitochondria can be lysed by freeze/thaw cycles, sonication, or detergent treatment, e.g., Triton X-100 at 0.1%. If one is also studying the regulatory enzymes associated with the complex, however, 0.02% Triton X-100 will permeabilise the mitochondria without inhibiting PDH-P phosphatase activity.

Some PDCs we have studied are inhibited by certain buffers or buffer concentrations (Williams and Randall, 1979, Budde and Randall, 1987). Consequently, it is best to evaluate the system thoroughly when studying new tissues, and we recommend not using glycylglycine or Mops buffers.

B. Measurement of $^{14}CO_2$ Release from [1-^{14}C]Pyruvate

This assay is very effective when there is interference with determination of NADH formation, or when it is necessary to assay PDC activity in crude homogenates. Most components of the assay are the same as in method A. To determine the steady state activity of PDC a quick extraction technique is also necessary. A small sample of leaf material can be quickly harvested and placed in a Hughes press (a heavy steel cylinder with an 0.2 mm hole in the bottom and tight fitting piston) with a piece of cheesecloth in the bottom of the cylinder. With the piston in place, a drill press can be used to apply pressure and the filtered extract collected in a microfuge tube. Fifty μl of extract can then be injected into a serum stoppered glass scintillation vial, equipped with hanging basket containing a 2.5 cm² piece of Whatman 3 MM paper wetted with 70 μl of 5 M ethanolamine to trap CO_2 and the vial placed in a shaking water bath at 30°C. The vial should also contain the following assay components: 80 mM Tes-NaOH (pH 7.6), 0.5 mM $MgCl_2$, 0.2 mM TPP, 2 mM NAD, 0.12 mM LiCoA, 2 mM cysteine and 1 mM [1-^{14}C]pyruvate (1000 dpm nmol^{-1}) in a final volume of 1.0 ml. After two minutes, the reaction can be quenched by addition of 50 μl of 6 M HCl. After twenty minutes the vials can be opened and the paper used to trap CO_2 removed and placed in a scintillation vial for analysis by liquid scintillation spectrometry. For each analysis, control assays should be performed with boiled plant extract or by initiating the reaction with buffer instead of extract. Performing this assay at pH 7.6 and 0.5 mM $MgCl_2$ minimises the interference by pPDC which has optimal activity at pH 8 and 5 mM $MgCl_2$. The extraction procedure is rapid enough to permit the initiation of the assay within 10–15 s. Consequently, this is the assay of choice for determining the steady-state level of PDC activity (Budde and Randall, 1989).

C. Other Methods

Alternatives to NAD reduction or $^{14}CO_2$ release have been developed and should be considered when the above, more simple assays are not suitable (e.g. tissues which contain significant lactate dehydrogenase). Such methods include the coupling of the reaction to INT ([*p*-iodophenyl]-3-*p*-nitrophenyl-5-phenyl-tetrazolium chloride) reduction ($\varepsilon = 12.4 \times 10^3$ M^{-1} cm^{-1}) through the redox carrier phenazine methosulphate (Hinman and Blass, 1981) and the measurement of acetyl-CoA production by coupling through citrate synthase (EC 4.1.3.7) (Szutowicz *et al.*, 1981), or arylamine acetyltransferase (EC 2.3.1.5) (Hoffmann *et al.*, 1978). Acetyl-CoA formation can also be determined by the HPLC method of Ingebretson and Farstad (1980). An additional method, which exploits the ability of acetyl-CoA to acetylate dithiothreitol nonenzymatically, has been developed by Liedvogel (1985) and measures PDC by quantitating the radioactivity from [2-^{14}C]pyruvate which partitions into trichloromethane. Most of these methods are less convenient and more time consuming than Methods A or B, but under certain conditions are suitable alternatives. The INT method is more sensitive than direct NADH measurement, however, the reagents are light sensitive and the dye and redox carrier must be added immediately before putting the cuvette in the spectrophotometer. Additionally, it is necessary to reduce the concentration of thiol reagents in the assay buffer to about 0.3 mM.

D. Pyruvate Dehydrogenase (E_1)

This component enzyme can be measured by $^{14}CO_2$ release from [1-^{14}C]pyruvate as described above, or by monitoring $Fe(CN)_6^{3-}$ reduction at 430 nm ($\varepsilon = 1030 \text{ M}^{-1} \text{ cm}^{-1}$) using a recording spectrophotometer (Schwartz and Reed, 1970). The reaction mixture should contain 50 mM Tes-KOH, pH 7.6, 5 mM potassium pyruvate, 10 mM $MgCl_2$, 0.2 mM TPP and 1.8 mM $K_3Fe(CN)_6$ and enzyme. The reaction can be initiated with either enzyme or pyruvate. This assay is less sensitive than the NADH assay for the overall complex activity but is a suitable alternative to the radioisotope assay.

E. Dihydrolipoyl Transacetylase (E_2)

Reid *et al.* (1977) report measuring the dihydrolipoyl transacetlyase component of PDC by the transfer of the [^{14}C]acetyl group from [1-^{14}C]acetyl-CoA to dihydrolipoamide. The reaction can be performed in 1.5 ml capped microfuge tubes with a 0.5 ml reaction mixture containing 1.2 mM dihydrolipoamide, 1 mM acetyl-CoA with (*c.* 50 nCi ^{14}C), 25 mM Tes-KOH (pH 7.3) and enzyme. The reaction should be initiated with dihydrolipoamide and terminated after 15 s by addition of 1 ml benzene followed by vigorous shaking of the tube to extract the acetylated dihydrolipoamide. The two phases can be separated by centrifugation followed by the transfer of 500 µl of the upper benzene layer to a scintillation vial for quantitation. A minus-enzyme control is necessary to correct for unreacted [1-^{14}C]acetyl-CoA extracted in the benzene phase. An alternative assay (Schwartz and Reed, 1969) uses a 1 ml reaction mixture containing 50 mM Tes-KOH (pH 7.3), 10 mM acetyl-phosphate, 2 units phosphotransacetylase, 4 mM dihydrolipoamide, 0.13 mM CoA, and enzyme. Thioester bond formation is monitored at 232 nm ($\varepsilon = 4400 \text{ M}^{-1} \text{ cm}^{-1}$) at 25°C using a recording spectrophotometer.

F. Dihydrolipoamide Dehydrogenase (E_3)

Dihydrolipoamide dehydrogenase can be measured spectrophotometrically in either direction, although dihydrolipoamide-dependent NAD reduction is preferred. The assay components for NAD reduction are 50 mM Mes-KOH buffer (pH 7.0), 2 mM dihydrolipoamide, 2 mM NAD, 0.05 mM NADH, and enzyme. The NADH must be added to overcome a significant lag in the rate of reaction. The reaction should be initiated with dihydrolipoamide after establishing a background rate. DL-Dihydrolipoamide (DL-6,8-thioctic acid amide) can be prepared from DL-lipoamide by reduction with either dithionite or $NaBH_4$ (Reed *et al.*, 1958). For NADH oxidation, a buffer at pH 8.5 is recommended and the NADH concentration should be adjusted to 0.2 mM (J. A. Miernyk, unpublished). Reid *et al.* (1977) and Reed and Willms (1966) offer variations of this assay, but we find their pH values less than optimal.

G. Pyruvate Dehydrogenase Kinase (PDH Kinase)

Pyruvate dehydrogenase kinase is an intrinsic component of the PDC, and phosphorylation of PDH by the kinase inactivates the complex.

1. Method 1

The most facile method for measuring PDH kinase activity is to incubate the sample with ATP followed by assaying for PDC activity. Typical reaction mixtures should contain 10–100 mM Tes-KOH buffer, pH 7.5, 25–500 µM MgATP, and the preparation of PDC to be assayed. Aliquots can be removed at 30 s intervals and either assayed immediately for PDC activity, or stopped by the addition of a mixture of glucose (20 mM) and yeast hexokinase (10 units) to remove the ATP, then kept on ice and assayed when convenient. If lysed mitochondria are being used, the inactivated PDC-P will begin to reactivate if not kept on ice or if allowed to stand more than 30 min. Reactivation will not occur at Mg^{2+} levels less than 0.5 mM when partially purified complex is used. The initial amount of PDC in the assay should be varied so that inactivation is complete within 5–10 minutes (Fig. 10.2). The true substrate for PDH kinase is MgATP, and free magnesium should be minimised as it both inhibits kinase activity and activates PDH-P phosphatase (Miernyk and Randall, 1987b; Budde and Randall, 1988). It should be noted that, unlike mPDC, pPDC is not inactivated by phosphorylation (Miernyk et al., 1985). Thus, any contamination of mPDC with the plastid complex will underestimate kinase activity.

2. Method 2

The second method for assaying PDH kinase activity exploits our observation that, under the assay conditions described above, the $E_{1\alpha}$ subunit of PDH is the major phosphopeptide in the matrix protein fraction of plant mitochondria, when magnesium levels are maintained at ≤ 0.5 mM. However, control SDS-PAGE and autoradiography should be done to ensure that the 43 kDa band ($E_{1\alpha}$) is the primary protein labelled. Preparations of $[\gamma\text{-}^{32}P]$ATP can be obtained from commercial sources or made enzymatically by the method of Johnson and Walseth (1979). By incubating PDC with $[\gamma\text{-}^{32}P]$ATP, then stopping the reaction with an equal volume of ice-cold 20% (w/v) trichloroacetic acid, incubating on ice for at least 1 hour, centrifuging the insoluble material, and finally quantitating the radioactivity in the pellet, it is possible to obtain a relative measure of PDH-kinase activity. Radioactivity can be measured either by liquid scintillation or Cerenkov counting. In order to use this method effectively it is necessary to correlate the incorporation of ^{32}P with the degree of inactivation of the PDC (Randall and Rubin, 1977; Randall et al., 1977, 1981; Rubin and Randall, 1977a). A variation of this method involves stopping the reaction with an excess of glucose and hexokinase, followed by immunoprecipitation of the PDC. Immunoprecipitation can be accomplished by incubating the PDC with antibodies at 0°C, for at least 8 h. Addition of 20 mM KF following the glucose and hexokinase will ensure that PDH-P phosphatase does not reactivate the PDC. The IgG-PDC super complex is so large that it is easily pelleted by centrifugation (10 000 × g for 10 min) without the addition of protein-A or other aids. While this method takes longer and is somewhat more cumbersome than acid precipitation, it is much more specific. It does, however, require antibodies to the PDC. It is also possible to quantitate ^{32}P incorporation on autoradiograms of SDS gels of PDH kinase reaction mixtures (Budde and Randall, 1988; Ludlow et al., 1986).

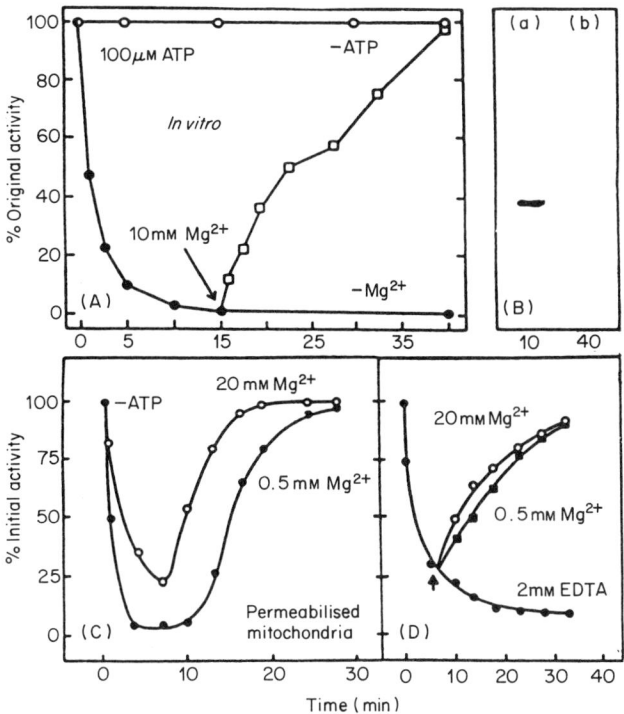

FIG. 10.2. ATP dependent inactivation–phosphorylation of PDC and the effect of magnesium on reactivation–dephosphorylation. (A) PDC was partially purified by lysing the mitochondria, removing the membranes, pelleting the PDC and resuspending in buffer. The addition of 100 μM ATP resulted in inactivation that was reversible by the addition of 10 mM MgCl$_2$. (B) Autoradiogram of SDS-PAGE of samples in (A) taken at 14 min (a) and 40 min (b) after addition of 100 μM [^{32}P]ATP at 0 min. Band in (a) is 43 kDa E_{1a} subunit of PDC. (C) Intact mitochondria were permeabilised by 0.01% (v/v) Triton X-100 prior to addition of 200 μM ATP in the presence of 20 or 0.5 mM MgCl$_2$. (D) Intact mitochondria were permeabilised with 0.01% (v/v) Triton X-100 prior to the addition of 200 μM ATP and at the 6 min point either 20 mM MgCl$_2$, 0.5 mM MgCl$_2$ or 2 mM EDTA was added.

3. Method 3

The third method for assaying PDH kinase activity is based upon that described by Braun *et al.* (1983). This assay utilises small peptides as phosphoryl acceptors, rather than the macromolecular protein substrates. The use of small peptides as protein kinase substrates has gained considerable popularity in recent years, due in no small part to the relative ease and flexibility of the assays (e.g. Kemp *et al.*, 1983). This assay depends upon phosphorylation of the substrate by [γ-^{32}P] ATP, hydrolysis of the unused nucleotide substrate to ^{32}Pi by boiling for 20 min in 1 N HCl, extraction of this ^{32}Pi with isobutanol–benzene in the presence of ammonium molybdate, then comparing the [^{32}P] ATP initially used as the phosphoryl donor to the ^{32}P from the hydrolysis and extraction procedure. Most phosphomonoesters (e.g. phospho-serine) are stable to mild acid hydrolysis, making this assay generally suitable.

Small peptides have been effective as substrates in assaying mammalian PDH-kinase (Davis et al., 1977; Mullinax et al., 1985). The best of these substrates, a tetradecapeptide, is not, however, a substrate for plant PDH kinase (Miernyk and Randall, 1989). This is presumably due to differences in the sequence of the phosphorylation sites of the complexes from the two different sources.

H. Phospho-Pyruvate Dehydrogenase Phosphatase (PDH-P Phosphatase)

PDH-P phosphatase is also an intrinsic component of the PDC and catalyses the divalent cation dependent reactivation of PDC-P (Miernyk and Randall, 1987b). The mPDC can be inactivated by incubation with 200 μM MgATP at room temperature and stored on ice. Removal of excess ATP can be accomplished by adding 20 mM glucose and 10 units of hexokinase following the inactivation of PDC. PDH-P phosphatase is assayed by incubating an aliquot of the ATP-inactivated PDC in 50–80 mM Tes-KOH (pH 7.5) in the presence of 10–20 mM $MgCl_2$ for various time periods and removing an aliquot to assay PDC as described above. Alternatively, [^{32}P]ATP can be used to inactivate PDC. Inactive, [^{32}P]-PDC is then incubated with Tes-KOH buffer, pH 7.5, and $MgCl_2$ at 10–20 mM. At various times aliquots are removed and injected into a microfuge tube containing an equal volume of ice-cold 20% trichloroacetic acid. After incubation on ice for 1 h, samples should be centrifuged and the pellets washed twice by resuspension in 10% trichloroacetic acid. The washed, acid insoluble material can be solubilised by heating (70°C) in 4 M urea, 1% SDS, and 2% 2-mercaptoethanol. Triplicate aliquots of the solubilised protein can then be quantitated by Cerenkov counting or in a liquid scintillation spectrometer. With pea mitochondria, we found that more than 90% of the ^{32}P was in the $E_{1\alpha}$ of the PDH (M_r 43 000), but it was necessary to do the labelling/inactivation of the PDC with minimal Mg^{2+} (<0.5 mM).

III. METHODS OF PURIFICATION

A. Plant Mitochondria

Several procedures have been reported over the last few years for the isolation and purification of plant mitochondria (Bergman et al., 1980; Gardestrom et al., 1978; Day et al., 1985; Jackson et al., 1979; Nash and Wiskich, 1983; Neuburger et al., 1982;). We have developed a procedure for isolating mitochondria from green pea seedlings that results in very pure, well coupled mitochondria that have excellent respiratory control, and are devoid of chlorophyllous material (Fang et al., 1987).

All isolation steps should be performed at 4°C. Light-grown, green pea seedlings (14 days old), in 100 g portions, are homogenised in 200 ml of chilled grinding medium (0.3 mannitol, 50 mM Tes, pH 7.2, 1 mM EDTA, 1 mM $MgCl_2$, 0.2% (w/v) defatted BSA, 0.5% (w/v) PVP-40, 4 mM cysteine, and 10 mM freshly prepared 2-mercaptoethanol) using a Braun homogeniser modified to use single-edged razor blades, with 3 × 10 s bursts (new razor blades are used each day). Homogenates are filtered through four layers of cheesecloth plus one layer of Miracloth and then centrifuged at 3300 × g for 5 min. The mitochondria in the supernatant are then pelleted by centrifuging at 18 000 × g for 20 min. Each pellet is gently resuspended and evenly dispersed using a loose-fitting Teflon pestle in a glass Potter–Elvehjem homogeniser, in 8 ml resuspending medium A (RM-A) (0.3 M mannitol, 20 mM Tes, (pH 7.2), 2 mM K-phosphate, 1 mM EDTA, 0.1% (w/v) defatted BSA, 2 mM $MgCl_2$ and 14 mM 2-mercaptoethanol). About 6–8 ml of resuspended mitochondria are layered on each 36 ml discontinuous gradient composed of 10 ml 47%, 12 ml 26% and 6 ml 21% (v/v) Percoll. All Percoll solutions contain 0.25 M sucrose, 0.1% defatted BSA and 10 mM Tes (pH 7.2). The gradients are centrifuged at $58\,500 \times g$ for 45 min in a Beckman SW 28 rotor. The Percoll layers above the mitochondrial band are aspirated away and the mitochondria located at the interface between the 26% and the 47% Percoll layers removed using a 35 ml syringe with a 13-gauge needle, then diluted with an equal volume of resuspending medium B (RM-B) containing 0.3 M mannitol, 20 mM Tes (pH 7.2), 1 mM EDTA, 0.1% defatted BSA, 2 mM $MgCl_2$, and 2 mM DTT. Approximately 16 ml portions of the diluted, enriched mitochondrial fraction are layered onto a second series of Percoll gradients composed of 10 ml 47% (v/v) and 10 ml 26% (v/v) Percoll prepared as described for the first gradient, and then centrifuged at $58\,500 \times g$ (max.) for 30 min. The purified mitochondria, which band at the 26% to 47% Percoll interface, are carefully removed and slowly diluted with 5 to 10 vols of RM-B. The diluted mitochondria are centrifuged at $18\,800 \times g$ for 20 min, the pellets gently resuspended in RM-B, and then centrifuged again at $7720 \times g$ (max.) for 5 min to remove any remaining Percoll. Finally, the mitochondrial pellets are resuspended in RM-B (made to 2 mM in DTT) to a concentration of 20–40 mg protein ml^{-1} and stored at 4°C. These mitochondria are 98% intact, functional, well coupled and stable to storage for up to 3 days. It is, however, necessary to add all cofactors (CoA, TPP, NAD^+) prior to pyruvate oxidation studies with intact mitochondria.

B. The Pyruvate Dehydrogenase Complex

Purification of plant PDC has been very limited and only achieved for the mitochondrial complex from broccoli and cauliflower by Rubin and Randall (1977b) and Randall *et al.* (1977). The very large size of PDC (M_r $5–6 \times 10^6$) is a factor in all purification procedures. Isolated mitochondria (5–10 mg protein ml^{-1}) are broken by freezing, thawing and homogenising in 2–5 volumes of 25 mM potassium phosphate buffer (pH 6.7) containing 5 mM dithiothreitol. Yields are often enhanced by including 0.2 M NaCl, 1 mM NAD and 0.1 mM TPP. The mitochondrial homogenate should be centrifuged at $40\,000 \times g$ for 30 min to remove membranes. Nucleic acids may be removed by dropwise addition of 0.01 volume of 0.2% (w/v) protamine sulphate, equilibration for 15 min and centrifugation at $40\,000 \times g$ for 20 min. The PDC can then be pelleted by centrifugation at $204\,000 \times g$ for 3 h, and the pellet resuspended in a minimal volume buffer A: 25 mM Tes-KOH (pH 7), containing 1 mM NAD, 0.1 mM TPP and 5 mM DTT. Polyethylene glycol precipitation is performed at room temperature, with the protein concentration at 4–5 mg ml^{-1}, by dropwise addition of 0.1 volume of 50% (w/v) PEG 6000. The solution should be equilibrated for 15 min then centrifuged at $18\,800 \times g$ for 15 min to pellet the PDC. The PEG pellet is redissolved in buffer A plus 1 mM EDTA and recentrifuged to remove the insoluble material. The complex can be further purified by rate-zonal centrifugation on linear gradients of 10–40% (v/v) glycerol in buffer A. The gradients are developed by centrifuging for 9 h using a Beckman SW28 rotor at $72\,000 \times g$. The order of the above steps, after the removal of the mitochondrial membranes, can be varied but these have been purification steps that are consistently successful. Reid et al. (1977), made acetone powders of mitochondria prior to partial purification by a similar procedure. The mitochondrial complex appears to be stable during purification, but the regulatory enzymes either dissociate or are inactivated. If there is dissociation of the mitochondrial complex, addition of exogenous E_3 has no effect, in contrast to the plant 2-oxoglutarate dehydrogenase complex (Poulsen and Wedding, 1970). Most studies on the mPDC have been performed using enzyme enriched by lysing the mitochondria, removing the membranes and pelleting the PDC by ultracentrifugation. The PDH-kinase and PDH-P phosphatase usually remain active and with the complex through removal of the membranes. The phosphatase activity will be lost by additional pelleting, and the kinase activity is lost during further purification procedures.

C. Pyruvate Dehydrogenase PDH (E_1)

There are no reports for the purification of this component for any plant PDC. Yields of pure plant PDC are so small that it is prohibitive to start from this point. Dissociation of the complex following the removal of the mitochondrial membranes could be a potential starting point. The PDH has been resolved and characterised from mammalian mPDC (Linn et al., 1972) and E. coli (Reed and Willms, 1966). Resolution of the mammalian complex is achieved by dissociating PDC with 10 mM DTT, 1 M NaCl, and 0.1 M glycine buffer (pH 9), at room temperature for 1 h, followed by chromatography on Sepharose 4B using 0.1 M glycine (pH 9), containing 1 M NaCl, 1 mM EDTA, 2 mM DTT and 1 mM $MgCl_2$. The gel permeation column yields two major peaks, the first containing E_2 and PDH-kinase, and the second peak containing E_1, E_3, PDH-kinase and PDH-P phosphatase. The PDH can be further purified by adding DTT to 10 mM and precipitating the enzyme with 20–35% saturated ammonium sulphate. The PDH is then dissolved in 20 mM phosphate buffer (pH 7.5), containing 1 mM DTT, 0.1 mM $MgCl_2$ and 0.01 mM EDTA. While this procedure has not been used to obtain plant PDH, a modification has yielded an enriched E_3 fraction (J. A. Miernyk and D. D. Randall, unpublished).

D. Dihydrolipoyl Transacetylase LTA (E_2)

This component of PDC has not been resolved from PDC or purified from any plant tissues. The procedure described in III.C for resolution of mammalian PDH from the complex also yields E_2 (Linn et al., 1972). The E_2 component can be further purified after Sepharose 4B chromatography by precipitation by p-mercuribenzenesulphonate (Linn et al., 1972) to remove the kinase. Whether this approach will work for the plant enzymes has not yet been tested.

E. Dihydrolipoamide Dehydrogenase LAD (E_3)

The E_3 component is the only enzyme of plant PDCs that has been extensively purified and characterised. Matthews and Reed (1963) reported the first purification of E_3 from a plant source, green spinach leaves. Using ammonium sulphate fractionation, calcium phosphate gel/cellulose column chromatography, and DEAE-cellulose anion-exchange chromatography, they were able to obtain a preparation which was homogeneous by the criterion of sedimentation in the analytical ultracentrifuge. The specific activity of the pure spinach enzyme was 136 µmol min^{-1} (mg protein)$^{-1}$, somewhat lower than values obtained with preparations from pig heart muscle and *E. coli*. Yanagawa and Egami (1976) purified E_3 from etiolated asparagus shoots, starting with isolated mitochondria. The mitochondria were disrupted either by several freeze/thaw cycles or by incubation with detergents, and the membranes removed by centrifugation. It was observed that E_3 was most stable in phosphate buffer at pH 7. Unless otherwise noted, 67 mM phosphate buffer was used throughout the purification. The clarified mitochondrial supernatant was concentrated, then subjected to two cycles of Sephadex G-200 gel permeation chromatography. The combined peak fractions from gel permeation were diluted to a buffer concentration of 10 mM, applied to a DEAE-cellulose column previously equilibrated with the same buffer and then, after washing, eluted with a linear gradient of phosphate buffer increasing to 200 mM. The E_3 activity eluted as a single peak at a buffer concentration of 120 mM. The final preparation, purified 500-fold to a specific activity of 200 µmol min^{-1} (mg protein)$^{-1}$, was homogeneous as judged by both native- and SDS-PAGE.

It is also possible to purify the E_3 component by first purifying the complex, and then dissociating the components by incubation in 100 mM glycine buffer (pH 9.5), containing 10 mM DTT and 1.0 M KCl. After dissociation the different components can be separated by gel permeation chromatography, and E_3 further purified by anion-exchange chromatography. While this strategy appears more simple and efficient than beginning with whole tissue homogenates, the difficulties in purifying intact PDC at high yields from plant tissues must be considered.

In bacteria the E_3 components of the PDC and glycine decarboxylase are distinct proteins (Sokatch and Burns, 1984). In plants, however, it appears that there may be a single species of E_3 present within the mitochondrial matrix, and that this species functions as a component of the pyruvate- and 2-oxoglutarate dehydrogenase complexes, and of glycine decarboxylase (Walker and Oliver, 1986). Bourguignon *et al.* (1988) recently reported the purification of E_3 to homogeneity. Pea mitochondria were isolated, and the high molecular weight matrix proteins retained on an XM-300 Diaflo membrane. A glycine decarboxylase 'subcomplex' was then separated by gel permeation chromatography using a 2.5 × 35 cm column of Sephacryl S-300. Finally, pure E_3 was obtained by anion-exchange FPLC at pH 7.5, eluting with a 10–500 mM gradient of potassium phosphate. While the final preparation showed a single band upon SDS-PAGE, the specific activity reported was only 13.9 µmol min^{-1} (mg protein)$^{-1}$.

F. Pyruvate Dehydrogenase Kinase (PDH-Kinase)

Pyruvate dehydrogenase kinase has not yet been purified from any plant source. The kinase has, however, been purified from bovine kidney (Stepp *et al.*, 1983). The purification procedures begin with the isolation of mitochondria and purification of the PDC. Highly purified complex is suspended in NaCl plus glycine such that the final concentrations are 1.0 and 0.1 M, respectively, and the pH is 9. After 30 min at 0°C, the solution is applied to a 2.5 × 50 cm column of Sepharose 6B previously equilibrated with 0.1 M glycine (pH 9), 1.0 M NaCl, 1 mM MgCl$_2$ and 0.1 mM EDTA. The first major peak of A_{280} absorbing material eluting from the column contains the E_2-PDH kinase subcomplex. Fractions containing the subcomplex are combined, and solid ammonium sulphate added to 0.12 g ml^{-1}. The suspension is stirred for 15 min, then centrifuged at $30\,000 \times g$ for 15 min. The pellet is dissolved in a small volume of 0.05 M K-phosphate buffer (pH 7.5), containing 1 mM $MgCl_2$ and 0.1 mM EDTA, then dialysed against two 500 ml changes of the same buffer. The dialysed subcomplex should be diluted to a protein concentration of 5 mg ml^{-1} with 1.0 M NaCl, 0.1 M glycine (pH 9), 1 mM $MgCl_2$, 0.1 mM EDTA, plus 2 mM DTT, and dialysed against this same buffer for 6 h, followed by addition of 0.01 vols of 0.25 M monosodium p-hydroxymercuriphenyl sulphonate, incubation on ice for 30 min, and centrifugation at $30\,000 \times g$ for 20 min. The PDH kinase activity is in the supernatant fraction which is then dialysed for 16 h against two 1 litre changes of 0.01 M imidazole-asparagine buffer (pH 7.3), containing 0.1 mM $MgCl_2$ and 0.01 mM EDTA. After dialysis the solution is centrifuged twice at $144\,000 \times g$ for 2.5 h, then concentrated by vacuum dialysis. The kinase preparation is then applied to a 0.9×1.5 cm column of Whatman DE-52 DEAE- cellulose previously equilibrated with the imidazole-asparagine buffer. After washing with equilibration buffer, the column is developed with 4 ml portions of buffer containing 0.1, 0.2 and 0.5 M NaCl, the kinase eluting with 0.2 M NaCl. Beginning with 350 mg of purified PDC, the final yield is c. 4 mg of homogeneous kinase with a recovery of 32%.

G. Phospho-Pyruvate Dehydrogenase-Phosphatase, PDH-P Phosphatase

The PDH-P phosphatase that catalyses the reactivation of the phosphorylated PDĊ also has not been purified from any plant tissue or PDC. The phosphatase has been purified from bovine kidney and heart mitochondria (Teague et al., 1982) by dissociation of the phosphatase from partially purified PDC by treatment with 2 mM EGTA and pelleting the complex by ultracentrifugation for 3 h at $105\,000 \times g$. The phosphatase is further purified by precipitation using 0.002% protamine sulphate, redissolving in 20 mM Mops (pH 7), 5 mM $MgCl_2$, 10% glycerol, 0.5 mM DTT plus 0.16% yeast sodium ribonucleate. The critical stage in the purification is an affinity chromatography step using either E_2-Sepharose 4B, or PDC-Sepharose 4B. The affinity matrix is equilibrated with the Mops, $MgCl_2$, glycerol and DDT buffer given above plus 2 mM $CaCl_2$ and the protamine sulphate fraction is made 5 mM in $CaCl_2$. The phosphatase is bound to the affinity matrix, the column washed with buffer, then eluted with 2 mM EGTA. The affinity purification is very specific since the phosphatase binds only to the E_2 core of the complex, and this binding is greatly enhanced by Ca^{2+}.

The plant PDH-P phosphatase is apparently unstable and very easily lost from the complex. Any attempts to purify the phosphatase beyond the initial ultracentrifugation of the complex have failed. Phosphatase activity is inhibited by Triton X-100 at levels greater than 0.05%. To date Ca^{2+} does not appear to enhance the phosphatase binding to the plant PDC and concentrations of Ca^{2+} greater than 5–10 μM inhibit the phosphatase activity (Miernyk and Randall, 1987b).

IV. PHYSICOCHEMICAL AND CATALYTIC PROPERTIES

A. Structure of the Complex

The structures of non-plant PDCs have been the most thoroughly characterised.

Organization of the E. coli PDC is based on cubic symmetry with M_r of c. 5.3×10^6. It consists of 60 polypeptides, 12 E_1 (homodimer, $^{SU}M_r$ 96 000), 24 E_2 (monomer, M_r 65 000) and 6 E_3 (homodimer, $^{SU}M_r$ 56 000) (Reed, 1974). The mammalian mitochondrial PDC is larger, M_r 7–9×10^6, probably based on an icosohedric structure and consists of 20 E_1 (α_2,β_2-heterodimer, $^{SU}M_r$s 41 and 36 000), 60 E_2 (monomer, M_r 52 000), and 5–10 E_3 (homodimer, $^{SU}M_r$ 55 000) (Reed et al., 1985) plus the regulatory enzymes, c. 5 PDH kinases (α,β heterodimer, $^{SU}M_r$s 48 and 45 000) (Stepp et al., 1983) and c. 5 PDH-P phosphatases (α,β heterodimer, M_r 97 and 50 000) per complex (Teague et al., 1982). In both prokaryotic and eukaryotic PDCs the E_2 has a lipoic acid covalently attached to each subunit and the E_3 component has a single FAD moiety bound per subunit. The TPP associated with E_1 is not tightly bound and can be easily dissociated.

By comparison, very little is known about the size, organisation and subunit structure of plant PDCs. Plant PDC has only been purified to homogeneity twice (Randall et al., 1977; Rubin and Randall, 1977b) and of the components, only E_3 has been purified and characterised (Matthews and Reed, 1963; Bourguignon et al., 1988; J. A. Miernyk and D. D. Randall, unpublished). Sedimentation analysis of a plant mitochondrial PDC yielded a 59.3 S value (Rubin and Randall, 1977b) which is larger than the bacterial PDC but smaller than the mammalian complex. Immunochemical analysis showed that the pea mitochondrial complex has five major subunits of 97.7, 67.4, 58.1, 43.0 and 37.3 kDa, which are qualitatively similar to those of mammalian PDC components (Camp and Randall, 1985). The 58.1 kDa subunit cross-reacts with antiporcine E_3 antibodies, and the 43.0 kDa subunit is $E_{1\alpha}$.

B. Kinetic Properties, Effectors and Regulation

The kinetic constants of several plant PDCs are summarised in Table 10.1. The kinetic mechanism is a multi-site ping pong sequence for all PDCs studied (Randall et al., 1977). The pea leaf mitochondrial complex has a K_m for MgTPP of 80 nM and an additional Mg^{2+} requirement with a K_m of approximately 350 µM (Miernyk and Randall, 1987a). Mn^{2+} and Ca^{2+} also support substantial activity. Most of the plant PDCs have optimal activity at pH 7–8; in vitro activity of the pea leaf mitochondrial complex activity is optimal at pH 7.6 with 50% maximal activity at pH 6.8 and 8.5. The purified PDC is fairly specific for pyruvate, with 2-oxobutyrate giving 10–20% of the rate with pyruvate. Other 2-oxo acids are not utilised, and any activity with 2-oxoglutarate or branched chain oxoacids such as 2-oxoisovalerate indicates contamination by other complexes. The activity of PDC in enriched mitochondria is approximately 10–90 nmol min^{-1}(mg protein)$^{-1}$ whereas purified mitochondria yield activities of 200–350 nmol min^{-1}(mg protein)$^{-1}$. The activity of PDC in leaf extracts, determined by the [^{14}C]pyruvate assay, is about 1–2 nmol min^{-1} (mg protein)$^{-1}$ from tissue in the light and 8–10 nmol min^{-1} (mg protein)$^{-1}$ from tissue in the dark. The activity of PDC in extracts of illuminated tissue reflects mainly the pPDC, while measurements from tissue in the dark would include both pPDC and mPDC activities. The leaf mPDC is at least partially (if not entirely) inactivated in the light (Budde and Randall, 1989). If one uses a pH of 7.5 and 0.5 mM Mg^{2+} the contribution of pPDC can be minimised, but not eliminated, since the pPDC has a higher pH optimum (8.0) and requires 5–10 mM Mg^{2+} (Camp and Randall, 1985).

TABLE 10.1. Michaelis constants for PDCs from various plant tissues and organelles.

Plant tissue	K_m(Pyr)	K_m(NAD)	K_m(CoA)	K_i(NADH)	K_i(Acetyl-CoA)
Pea leaf mPDC[a]	57	122	4	18	10
Etiolate pea[b] shoot mPDC	73	238	5	17	14
Broccoli floret[c] mPDC	250	110	6	13	19
Cauliflower floret[d] mPDC	207	125	7	34	13
Castor seed[e] endosperm mPDC	90	51	3	15	20
Etiolate maize[f,h] shoot mPDC	79	60	4	8	18
Etiolated soybean[f] shoot mPDC	68	94	5	—	—
Pea leaf pPDC[g]	120	36	10	9	16
Castor seed[b] endosperm pPDC	62	130	6	27	23
Etiolated maize[f,h] shoot pPDC	120	16	4	12	18
Etiolated soybean[f] shoot pPDC	22	12	4	—	—

[a] Miernyk and Randall (1987a).
[b] Thompson et al. (1977).
[c] Rubin and Randall (1977b).
[d] Randall et al. (1977).
[e] Rapp et al. (1987).
[f] Cho et al. (1988).
[g] Camp et al. (1988).
[h] Miernyk (unpublished).

Table 10.1 shows that the PDC is inhibited by the reaction products, acetyl-CoA and NADH, and is very sensitive to the NAD:NADH ratio. The K_i(NADH) is lower than the K_m(NAD). On the other hand, *in situ* experiments have shown that the acetyl-CoA: CoA ratio has very significant regulatory influence because of the limited CoA pool size (Budde et al., 1989). Other than NADH and acetyl-CoA, the only metabolites directly affecting the leaf mPDC are ADP (competitive versus CoA, K_i 570 μM) and glyoxylate (competitive versus pyruvate, K_i 51 μM). Krebs cycle intermediates, amino acids, Pi, PPi, phospho-amino acids, polyamines, AMP and carbamyl-P are without effect (Miernyk and Randall, 1987a). TPP may have some regulatory potential in that it rapidly dissociates from some but not all complexes *in vitro* (Rubin et al., 1978). Calmodulin antagonists inhibit PDC very effectively and should be used with caution where PDC may be functioning (Miernyk et al., 1987a).

The most spectacular regulatory feature of mPDCs is the reversible inactivation by phosphorylation (Fig. 10.2). The mPDC was the first plant enzyme identified as being regulated by this form of covalent modification (Randall and Rubin, 1977; Randall et al., 1981). An intrinsic PDH kinase transfers the gamma phosphate of ATP to serine residues of $E_{1\alpha}$ (M_r 43 000) resulting in inactivation. The complex can be isolated from some tissues in a partially phosphorylated state, particularly from quiescent tissues (Miernyk et al., 1985; Rao and Randall, 1980; Rapp et al., 1987) and illuminated green leaves (Budde and Randall, 1989). To obtain total PDC activity it may be necessary to incubate the enzyme with 10–20 mM $MgCl_2$ for 10–30 min. The steady state activity of PDC is the result of the balanced activities of PDH kinase and PDH-P phosphatase. *In situ* studies have shown that this steady state activity is affected by metabolites, ATP levels and the respiratory state of the mitochondria (Budde and Randall, 1987, 1988; Budde et al., 1988).

C. Properties of Component Enzymes

The E_3 component is the only plant enzyme which has been extensively characterised. It was originally suggested that spinach E_3 is a large monomeric protein (Matthews and Reed, 1963). It is now clear, however, that plant E_3s are very similar to the mammalian and microbial enzymes, homodimers with a subunit M_r of approximately 58 000. While there is no sequence data available for any plant E_3, there is significant homology in that antibodies against both the pig heart and *E. coli* enzymes cross-react with pea mitochondrial E_3 on Western blots (J. A. Miernyk and D. D. Randall, unpublished observation). The mammalian enzyme has additionally been used to reconstitute an active cauliflower 2-oxoglutarate dehydrogenase complex (Poulsen and Wedding, 1970). The native M_r of plant E_3s is in the range 92–95 000, the $S_{20,w}$ 5.9, and the partial specific volume 0.733 (Yanagawa and Egami, 1976). The amino acid composition of asparagus E_3 is similar to that of the pig heart and *E. coli* enzymes, and the isoelectric point is 6.8.

Like E_3s from other sources, the plant enzymes are flavoproteins containing one mole of non-covalently bound FAD per mole of subunit (Matthews and Reed, 1963; Yanagawa and Egami, 1976). In addition to dehydrogenase activity, all E_3s, including those from plants, exhibit diaphorase activity. While the natural substrate for E_3 is protein bound (E_2) lipoyl-lysine, the enzyme will reduce lipoic acid or lipoamide *in vitro*. Purified E_3 will not, however, reduce the structurally related asparagusic acid (Yanagawa and Egami, 1976). In all instances NAD is the cofactor used, NADP being completely inactive. The E_3 reaction is freely reversible with pH optima of 6.0 in the direction of NADH oxidation and 8.5 in the direction of NAD reduction. Michaelis constants are 0.2 mM for dihydrolipoamide, 0.4 mM for lipoamide, 0.3 mM for NAD and 0.03 mM for NADH. There is a pronounced lag in initial rates unless both NAD and NADH are present in the assay mixtures (Matthews and Reed, 1963; Yanagawa and Egami, 1976). Thiol-directed reagents such as mercurials, arsenite, and N-ethylmaleimide, are potent inhibitors of E_3 activity, suggesting the occurrence of an essential sulphydryl group.

D. Properties of PDH Kinase

Bovine kidney PDH kinase is an α,β heterodimer. The M_r of the α subunit is 48 000 and that of the β subunit is 45 000. It has been suggested that α is the catalytic subunit and β is a regulatory subunit (Stepp *et al.*, 1983). The K_m of PDH kinase for MgATP is 20 μM and the K_m for the E_2-PDH subcomplex is 0.6 μM (Hucho *et al.*, 1972). Tryptic peptides isolated from PDH and synthetic peptides based upon the sequence of the tryptic peptides are both substrates for PDH kinase (Davis *et al.*, 1977; Yeaman *et al.*, 1978; Mullinax *et al.*, 1985). The K_m value for the best peptide substrate, the asp-8-tetradecapeptide, is 0.43 mM. The *in vitro* pH optimum for PDH kinase is 7.0–7.2. The activity of PDH kinase is inhibited by pyruvate and ADP, and may be modulated by the NADH:NAD and acetyl-CoA:CoA ratios (Hucho *et al.*, 1972; Kerbey *et al.*, 1979; Pettit *et al.*, 1975). It has also been suggested that mammalian PDH kinase activity could be modulated by thiol–disulphide exchange (Pettit *et al.*, 1982).

There have been preliminary studies of the PDH kinase activity from plant tissues (Rubin and Randall, 1977b; Rao and Randall, 1980; Rapp *et al.*, 1987); however, the pea seedling kinase has been characterised to the greatest extent (Miernyk and Randall, 1987c). The K_m for MgATP is 2.5 μM, considerably lower than the value for the mammalian kinase, and free magnesium is inhibitory. The *in vitro* pH optimum for the pea kinase is 7.5. Several mitochondrial metabolites are potential negative modulators of PDH kinase activity, including ADP, pyruvate plus TPP, acetyl-CoA, NADH, citrate and sodium ions. The NADH and acetyl-CoA inhibition is in contrast to their stimulation of the mammalian PDH kinase. Significant stimulation of PDH-kinase is observed with mM K^+ and μM NH_4^+ (Schuller and Randall, 1989). Photoaffinity labelling of pea mitochondrial PDC with 8-azido-[α-^{32}P]ATP followed by immunoprecipitation, SDS-PAGE, and autoradiography resulted in the specific labelling of a single subunit with an M_r of 53 000 (Miernyk and Randall, 1988). We believe this to be the catalytic subunit of pea mitochondrial PDH-kinase.

E. PDH-P Phosphatase

Bovine kidney PDH-P phosphatase is an α,β heterodimer (M_r 150 000). The larger subunit of M_r 97 000 contains an FAD moiety and the small subunit, M_r 50 000, has the phosphatase activity (Teague *et al.*, 1982). The free bovine phosphatase binds 1 mole of Ca^{2+} per mole enzyme but when associated with E_2 a second Ca^{2+} is bound. The phosphatase has a K_m for PDH-P of 58 μM which decreases to 2.9 μM in the presence of Ca^{2+}. Addition of Calmodulin has no effect on phosphatase activity while 200 mM KCl inhibits about 95%.

Initial studies of the plant PDH-P phosphatase have been done with PDC-associated phosphatase following lysis of the mitochondria and removal of the membranes (Miernyk and Randall, 1987b). The PDH-P phosphatase requires a divalent cation for activity with Mg^{2+} being more effective than Mn^{2+} or Co^{2+} (K_ms 3.8, 1.7 and 1.4 mM, respectively). Ca^{2+} does not activate or stimulate the phosphatase as it does with the mammalian enzyme and low concentrations (10–50 μM) antagonise the Mg^{2+}-dependent activation. Fluoride (1–20 mM) and orthophosphate (5–20 mM) inhibit the phosphatase but vanadate and molybdate do not. Krebs cycle intermediates, amino acids, phosphoamino acids, adenylate and pyridine nucleotides, and acetyl-CoA are all without effect on the phosphatase *in vitro*. The activity of the PDH-P phosphatase is about 15% of the PDH kinase, consequently there must be some sort of down-regulation of the kinase or stimulation of the phosphatase to control steady state PDC activity. *In situ* experiments with the purified intact mitochondria show that 0.5 mM Mg^{2+} is sufficient to activate fully the phosphatase as shown in Fig. 10.2 (Budde and Randall, 1988). It is quite obvious that there are some similarities but also major differences between the plant and animal regulatory enzymes acting on PDC.

F. Pyruvate Oxidation and Assay of PDC Activity *In situ*

Pyruvate oxidation by intact mitochondria can be measured using an O_2 electrode system. The reaction should contain 0.3 M mannitol, 20 mM Tes-KOH (pH 7.5), 3 mM $MgCl_2$, 1 mM EDTA, 0.1% defatted BSA, 10 mM KCl, 5 mM K-phosphate, 1 mM NAD, 0.1 mM CoA, 0.1 mM TPP and 0.1–0.5 mM 'sparker' malate plus mitochondria (0.1–0.2 mg protein per ml). Addition of the cofactors is essential since Percoll purification strips most cofactors from the mitochondria. Pyruvate (1–2 mM) is used to initiate the oxidation and 0.15 mM ADP is added to attain state 3 (coupled) rates of oxidation (Budde *et al.*, 1989). To determine the phosphorylation level of the PDC under various oxidation states and conditions, aliquots can be removed from the electrode chamber and added to a spectrophotometric assay cuvette with reaction components plus 0.2% Triton X-100. For total PDC activity the reaction mixture and sample should be incubated for 10 min in the presence of 1–10 mM $MgCl_2$ to allow the PDH-P phosphatase to fully activate the complex.

For the phosphorylation (inactivation) of PDC *in situ*, the incubation buffer should contain 10 mM Tes-NaOH (pH 7.5), 0.5 mM $MgCl_2$, 0.2 mM EDTA, 0.1% defatted BSA, 0.3 M mannitol and 0.5 mM DTT. Phosphorylation is initiated by addition of 20–200 mM ATP, and PDC activity is determined by taking aliquots for the spectrophotometric assay previously described. Purified mitochondrial preparations contain a few broken mitochondria that expose very active F_1-ATPase activity. Consequently, oligomycin should be added to inhibit the ATPase or an ATP regenerating system can be included. The inclusion of 10–100 mM NaF will inhibit PDH-P phosphatase *in situ*. Dephosphorylation (reactiviation) *in situ* does not require high Mg^{2+} levels, 0.5 mM Mg^{2+} is sufficient (Fig. 10.2). However, since the PDH kinase rates are *c*. six-fold greater than the PDH-P phosphatase rates, reactivation will not occur unless the kinase is inhibited or the ATP is removed (Budde and Randall, 1988).

ACKNOWLEDGEMENT

This is report number 10755 of the Missouri Agriculture Experiment Station: National Science Foundation Grants (Metabolic Biology Program) and the Missouri Agriculture Experiment Station supported the authors' research on this topic.

REFERENCES

Bergman, A., Gardestrom, P. and Ericson, I. (1980). *Plant Physiol.* **66**, 442–445.
Bourguignon, J., Neuburger, M. and Douce, R. (1988). *Biochem. J.* **255**, 169–178.
Braun, S., Abdel Ghany, M. and Racker, E. (1983). *Anal. Biochem.* **135**, 369–389.
Budde, R. J. A. and Randall, D. D. (1987). *Arch. Biochem. Biophys.* **258**, 600–606.
Budde, R. J. A. and Randall, D. D. (1988). *Plant Physiol.* **88**, 1026–1030.
Budde, R. J. A. and Randall, D. D. (1989). *Proc. Natl. Acad. Sci. USA*, in press.
Budde, R. J. A., Fang, T. K. and Randall, D. D. (1988). *Plant Physiol.* **88**, 1031–1036.
Budde, R. J. A., Fang, T. K., Randall, D. D. and Miernyk, J. A. (1989). *Plant Physiol.*, submitted.
Camp, P. J. and Randall, D. D. (1985). *Plant Physiol.* **77**, 571–577.
Camp, P. J., Miernyk, J. A. and Randall, D. D. (1988). *Biochim. Biophys. Acta* **933**, 269–275.
Cho, H-Y., Widholm, J. M. and Slife, F. W. (1988). *Plant Physiol.* **87**, 334–340.
Davis, P. F., Pettit, F. H. and Reed, L. J. (1977). *Biochem. Biophys. Res. Commun.* **75**, 541–549.
Day, D. A., Neuburger, M. and Douce, R. (1985). *Austral. J. Plant Physiol.* **12**, 219–228.
Fang, T. K., David, N. R., Miernyk, J. A. and Randall, D. D. (1987). *Curr. Top. Plant Biochem. Physiol.* **6**, 175.
Gardestrom, P., Ericson, I. and Larsson, C. (1978). *Plant Sci. Lett.* **13**, 231–239.
Gopalakrishnan, S., Rahmatullah, M., Radke, G. A., Powers-Greenwood, S. and Roche, T. E. (1989). *Biochem. Biophys. Res. Commun.* **160**, 715–721.
Hinman, L. M. and Blass, J. P. (1981). *J. Biol. Chem.* **256**, 6583–6586.
Hoffmann, G., Weiss, L. and Wieland, O. H. (1978). *Anal. Biochem.* **84**, 441–448.
Hucho, F., Randall, D. D., Roche, T. E., Burgett, M. W., Pelley, J. W. and Reed, L. J. (1972). *Arch. Biochem. Biophys.* **151**, 328–340.

Ingebretsen, O. C. and Farstad, M. (1980). *J. Chromatogr.* **202**, 439–445.
Jackson, C., Dench, J. E., Hall, D. O. and Moore, A. L. (1979). *Plant Physiol.* **64**, 150–153.
Johnson, R. A. and Walseth, T. F. (1979). *Adv. Cyclic Nucleotide Res.* **10**, 135–167.
Kemp, B. E., Pearson, R. B. and House, C. (1983). *Proc. Natl. Acad. Sci. USA* **80**, 7471–7475.
Kerbey, A. L., Radcliffe, P. M., Randle, P. J. and Sugden, P. H. (1979). *Biochem. J.* **181**, 427–433.
Liedvogel, B. (1985). *Anal. Biochem.* **148**, 182–189.
Linn, T. C., Pelley, J. W., Pettit, G. H., Hucho, F., Randall, D. D. and Reed, L. J. (1972). *Arch. Biochem. Biophys.* **188**, 327–342.
Ludlow, J. W., Guikema, J. A. and Consigli, R. A. (1986). *Anal. Biochem.* **154**, 104–109.
Matthews, J. and Reed, L. J. (1963). *J. Biol. Chem.* **238**, 1869–1876.
Miernyk, J. A. and Randall, D. D. (1987a). *Plant Physiol.* **83**, 306–310.
Miernyk, J. A. and Randall, D. D. (1987b). *Plant Physiol.* **83**, 311–315.
Miernyk, J. A. and Randall, D. D. (1987c). In "Plant Mitochondria" (A. L. Moore and R. B. Beechey, eds.), pp. 223–226, Plenum, London.
Miernyk, J. A. and Randall, D. D. (1988). *J. Cell Biol.* **107**, 279a.
Miernyk, J. A. and Randall, D. D. (1989). *J. Biol. Chem.* **264**, 9141–9144.
Miernyk, J. A., Camp, P. J. and Randall, D. D. (1985). *Curr. Top. Plant Biochem. Physiol.* **4**, 175–190.
Miernyk, J. A., Fang, T. K. and Randall, D. D. (1987a). *J. Biol. Chem.* **262**, 15338–15340.
Miernyk, J. A., Rapp, B. J., David, N. R. and Randall, D. D. (1987b). In "Plant Mitochondria" (A. L. Moore and R. B. Beechey, eds.), pp. 189–197, Plenum, London.
Mullinax, T. R., Stepp, L. R., Brown, J. R. and Reed, L. J. (1985). *Arch. Biochem. Biophys.* **243**, 655–659.
Neuburger, M., Journet, E-P., Bligny, R., Carde, J.-P. and Douce, R. (1982). *Arch. Biochem. Biophys.* **217**, 312–323.
Nash, D. and Wiskich, J. T. (1983). *Plant Physiol.* **71**, 627–634.
Pettit, F. H., Humphreys, J. and Reed, L. J. (1982). *Proc. Natl. Acad. Sci. USA* **79**, 3945–3958.
Pettit, F. H., Pelley, J. W. and Reed, L. J. (1975). *Biochem. Biophys. Res. Commun.* **65**, 575–582.
Poulsen, L. L. and Wedding, R. T. (1970). *J. Biol. Chem.* **245**, 5709–5717.
Randall, D. D. and Rubin, P. M. (1977). *Plant Physiol.* **59**, 1–3.
Randall, D. D., Rubin P. M. and Fenko, M. (1977). *Biochim. Biophys. Acta.* **485**, 336–349.
Randall, D. D., Williams, M. and Rapp, B. J. (1981). *Arch. Biochem. Biophys.* **207**, 437–444.
Rao, K. P. and Randall, D. D. (1978). *Arch. Biochem. Biophys.* **200**, 461–466.
Rapp, B. J., Miernyk, J. A. and Randall, D. D. (1987). *J. Plant Physiol.* **127**, 293–306.
Reed, L. J. (1974). *Acc. Chem. Res.* **7**, 40–56.
Reed, L. J., Koike, M., Levitch, M. E. and Leach, F. R. (1958). *J. Biol. Chem.* **232**, 143–158.
Reed, L. J. and Willms, C. R. (1966). In "Methods in Enzymology", Vol. IX (W. A. Wood, ed.) pp. 247–257. Academic Press, New York.
Reed, L. J., Damuni, Z. and Merryfield, M. L. (1985). *Curr. Top. Cell Regul.* **27**, 41–49.
Reid, E. E., Thompson, P., Lyttle, C. R. and Dennis, D. T. (1977). *Plant Physiol.* **59**, 842–848.
Rubin, P. M. and Randall, D. D. (1977a). *Plant Physiol.* **60**, 34–39.
Rubin, P. M. and Randall, D. D. (1977b). *Arch. Biochem. Biophys.* **178**, 342–349.
Schuller, K. A. and Randall, D. D. (1989). *Plant Physiol.* **89**, 1207–1212.
Schwartz, E. R. and Reed, L. J. (1969). *J. Biol. Chem.* **244**, 6074–6079.
Schwartz, E. R. and Reed, L. J. (1970). *Biochemistry* **9**, 1434–1439.
Sokatch, J. R. and Burns, G. (1984). *Arch. Biochem. Biophys.* **188**, 70–77.
Stepp, L. R., Pettit, F. H., Yeaman, S. J. and Reed, L. J. (1983). *J. Biol. Chem.* **258**, 9454–9458.
Szutowicz, A., Stepien, M. and Piec, B. (1981). *Anal. Biochem.* **115**, 81–87.
Teague, W. M., Pettit, F. H., Wu, T.-L., Silberman, S. R. and Reed, L. J. (1982). *Biochemistry* **21**, 5585–5592.
Thompson, P., Reid, E. E., Lyttle, C. R. and Dennis, D. T. (1977). *Plant Physiol.* **59**, 849–853.
Walker, J. L. and Oliver, D. J. (1986). *J. Biol. Chem.* **261**, 2214–2221.
Williams, M. and Randall, D. D. (1979). *Plant Physiol.* **64**, 1099–1103.
Yanagawa, H. and Egami, F. (1976). *J. Biol. Chem.* **251**, 3637–3644.
Yeaman, S. J. (1986). *Trends Biochem. Sci.* **11**, 293–296.
Yeaman, S. J., Hutcheson, E. T., Roche, T. E., Pettit, F. H., Brown, J. R., Reed, L. J., Watson, D. C. and Dixon, G. H. (1978). *Biochemistry* **17**, 2364–2370.

11 Enzymes of Fatty Acid Synthesis

JOHN L. HARWOOD, M. C. WALSH and
KEVIN A. WALKER*

*Department of Biochemistry, University of Wales College of Cardiff,
P.O. Box 903, Cardiff CF1 1ST, UK*

I.	Introduction	194
II.	Acetyl-CoA carboxylase	195
	A. Assay systems	196
	B. Purification	197
	C. Localisation studies	197
III.	Fatty acid synthetase	197
	A. Acyl carrier protein	198
	B. Priming reaction	201
	C. Malonyl-CoA:ACP transacylase	202
	D. Condensing enzymes	203
	E. β-Ketoacyl-ACP reductase	205
	F. β-Hydroxyacyl-ACP dehydrase	206
	G. Enoyl-ACP reductases	207
	H. Overall fatty acid synthesis	208
IV.	Desaturation	208
	A. Stearoyl-ACP Δ9-desaturase	208
	B. Δ 12-Desaturation using 1-acyl-2-oleoyl-phosphatidylcholine	210
	C. Linolenate formation	210
V.	Elongation of fatty acids	212
	A. Systems useful for the study of fatty acid elongation	212
	B. Purification of elongases	213
VI.	Synthesis of other plant fatty acids	213
	A. α-Linolenate formation	213
	B. Ricinoleic acid formation	214
	References	214

* Current address: Schering Agrochemicals Ltd., Chesterford Park Research Station, Saffron Walden, Essex CB10 1Xl, UK.

METHODS IN PLANT BIOCHEMISTRY Vol. 3
ISBN 0-12-461013-7

Copyright © 1990 Academic Press Limited
All rights of reproduction in any form reserved

I. INTRODUCTION

In this chapter we will deal with those enzymes involved in the *de novo* formation and later modification of plant fatty acids. *De novo* synthesis is catalysed by the combined action of acetyl-CoA carboxylase and fatty acid synthetase. Over the last few years most of the component proteins for these activities have been purified. Much less, however, is known of the (mainly) membrane-bound enzymes which catalyse the elongation and desaturation of fatty acids. The assay systems for, and some information on, the properties of these enzymes are also included, since the prevalent fatty acids of plant tissues are made by further reactions in addition to *de novo* synthesis (see Harwood, 1980). An outline of the pathways is shown in Fig. 11.1 and the reactions catalysed by the enzymes under discussion are shown in Table 11.1.

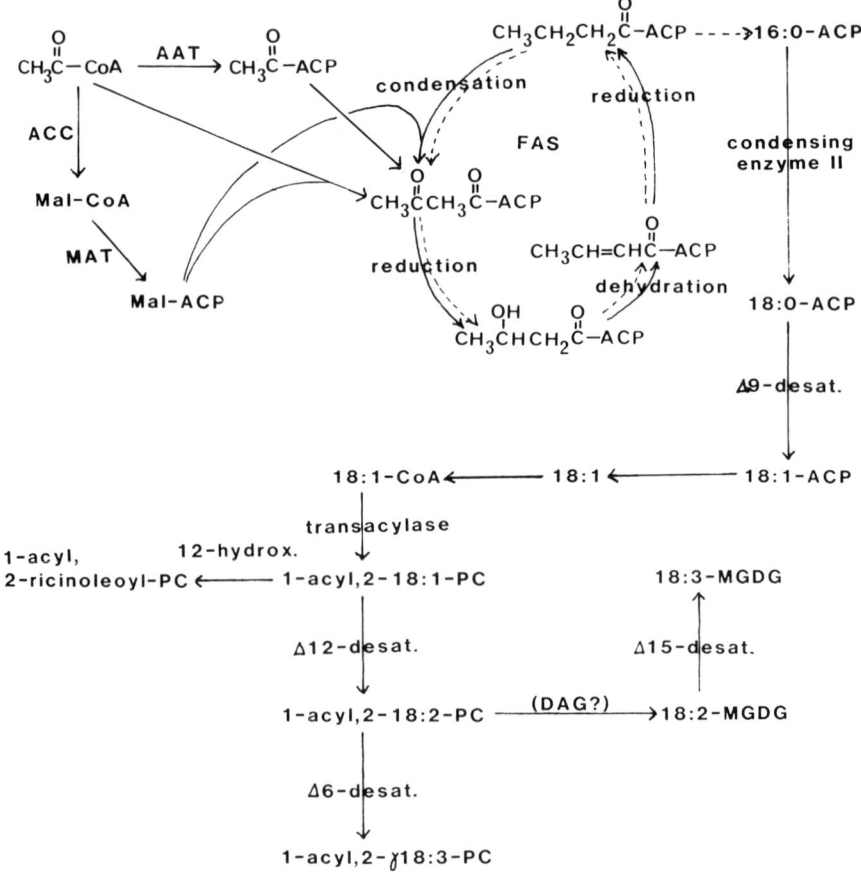

FIG. 11.1. Summary of pathways for fatty acid synthesis in plants showing individual enzyme steps.

TABLE 11.1. Assay principles used for measuring the partial reactions of fatty acid synthetase.

Enzyme	Method
1. Acetyl-CoA:ACP transacylase [^{14}C]Acetyl-CoA + ACP $\rightleftharpoons$ [^{14}C]Acetyl-ACP + CoA	Precipitate acetyl-ACP Count.
2. Malonyl-CoA:ACP transacylase [^{14}C]Malonyl-CoA + ACP $\rightleftharpoons$ [^{14}C]Malonyl-ACP + CoA	Precipitate malonyl-ACP Count.
3. Acetoacetyl-ACP synthetase [^{14}C]Acetyl-CoA + Malonyl-ACP $\rightleftharpoons$ [^{14}C]Acetoacetyl-ACP + CoA (+cerulenin)	Precipitate acetoacetyl-ACP Count. (Reaction conditions not fully defined.)
4. β-Ketoacyl-ACP synthetase I Acetyl-ACP + Malonyl-ACP $\rightleftharpoons$ Acetoacetyl-ACP + CO_2 + ACP	Measure absorbance at 303 nm (acetoacetate formation).
or nAcyl-ACP + Malonyl-ACP $\rightleftharpoons$ (n + 2)Acyl-ACP + ACP + CO_2	In presence of NADPH and other FAS enzymes, measure absorbance at 340 nm. With [^{14}C]malonyl-CoA measure counts in acyl chains. Use of NaH[^{14}C]CO_2 permits a CO_2-exchange assay.
5. β-Ketoacyl-ACP synthetase II Palmitoyl-ACP + [2-^{14}C]Malonyl-ACP $\rightleftharpoons$ [2-^{14}C]β-ketooctadecanoyl-ACP + ACP + CO_2	Reaction products reduced and counts in acyl chain measured.
6. β-Ketoacyl-ACP reductase β-Ketoacyl-ACP + NAD(P)H $\rightleftharpoons$ β-Hydroxyacyl-ACP + NAD(P)	Change in absorbance at 340 nm.
7. β-Hydroxyacyl-ACP dehydrase β-Hydroxylacyl-ACP $\rightleftharpoons$ Enoyl-ACP + H_2O	Back reaction followed with crotonyl-ACP and decrease in absorbance at 263 nm.
8. Enoyl-ACP Enoyl-ACP + NAD(P)H $\rightleftharpoons$ Acyl-ACP + NAD(P)	Change in absorbance at 340 nm.

II. ACETYL-CoA CARBOXYLASE (EC 6.4.1.2)

Acetyl-CoA carboxylase catalyses the ATP-dependent formation of malonyl-CoA from acetyl-CoA and bicarbonate. This reaction can be regarded as the first committed step for *de novo* fatty acid synthesis. The product of the reaction, malonyl-CoA, is also used for fatty acid elongation. The source of the acetyl-CoA substrate is somewhat controversial—generation through the activities of a plastid pyruvate decarboxylase/dehydrogenase or acetyl-CoA synthetase being proposed (see Harwood, 1988).

Until fairly recently it was thought that plant acetyl-CoA carboxylase was similar to that from *Escherichia coli* in that the enzyme consists of three separable proteins—biotin carboxyl carrier protein (BCCP), biotin carboxylase and BCCP:acetyl-CoA transcarboxylase. Work over the last few years, however, has pointed to plant acetyl-CoA carboxylase being a high molecular weight multifunctional protein like that from mammals (Harwood, 1988). We shall concentrate more on these later reports.

A. Assay Systems

Acetyl-CoA carboxylase is usually assayed by the acetyl-CoA-dependent incorporation of ^{14}C-labelled bicarbonate into acid-stable malonyl-CoA. A typical procedure would be as follows:

The incubation system (200 µl) contains 0.1 M Tricine-KOH (pH 8.0) 1 mM ATP, 2.5 mM MgCl$_2$, 50 mM KCl, 30 mM NaH^{14}CO$_2$ (1.17 µCi), 1 mM dithiothreitol, 0.3 mM acetyl-CoA and up to 200 µg of crude fraction (e.g. stroma) protein. The reaction is initiated by the addition of acetyl-CoA and incubated with shaking at 30°C. Linearity varies but is generally seen for about 5 min. The reaction is terminated by the addition of 50 µl 6 M HCl and aliquots (e.g. 25 µl) can then be removed, dried on 2 cm disks of Whatman 3MM paper (crude assays may require heating to remove acid stable volatiles) and counted in a scintillation counter.

For individual enzyme preparations it is most important to evaluate assay conditions for optimal component concentrations and linearity. Reference to individual papers (listed later) will show that the concentrations of substrates and cofactors which have been used vary considerably—although it is not always clear whether these represent optimal requirements. In some cases, additional components are added. For example, the enzyme from castor bean (*Ricinus communis* L.) endosperm was assayed in the presence of 0.1% (v/v) Triton X-100 (Finlayson and Dennis, 1983) and that from oil seed rape (*Brassica napus* L.) with bovine serum albumin (Slabas and Hellyer, 1985).

Since the assay depends on acetyl-CoA-dependent ^{14}CO$_2$ fixation, the presence of other carboxylases in crude fractions may lead to unacceptably high blank values. These activities can be minimised by a simple desalting procedure on a Sephadex G-25 (1 × 5 cm) column to remove other potential substrates. Such methods, in our hands, lead to blank values of about 5% of acetyl-CoA carboxylase activity in crude soluble leaf extracts.

For other special purposes (e.g. examination of regulatory properties) it may be desirable to alter some incubation conditions—such as those for adenylate nucleotides (Eastwell and Stumpf, 1983) as well as Mg^{2+} (Nikolau and Hawke, 1984). However, as discussed by Harwood (1988), there is no evidence for the stimulation of plant acetyl-CoA carboxylase by tricarboxylic acids or phosphorylation/dephosphorylation—in contrast to mammalian enzymes.

Individual assay systems for a number of plant acetyl-CoA carboxylases are referenced in Table 11.2. Even when assaying a plant species which is listed in Table 11.2, the reader is strongly advised to check assay conditions since different cultivars and developmental age can alter the specific activity of the acetyl-CoA carboxylase preparations obtained.

TABLE 11.2. Reference list for acetyl-CoA carboxylase studies.

Plant preparations	Reference
Avocado mesocarp	Mohan and Kekwick (1980)
Maize leaf extract	Nikolau et al. (1981)
Maize leaf enzyme	Nikolau and Hawke (1984)
Parsley cells, enzyme	Egin-Buhler and Ebel (1983)
Wheat leaf extract	Eastwell and Stumpf (1983)
Spinach chloroplast stroma	Laing and Roughan (1982)
Oil seed rape enzyme	Slabas and Hellyer (1985)
Soybean seed enzyme	Charles and Cherry (1986)
Barley, pea leaf extracts	Walker et al. (1988)
Barley, maize, mung bean, spinach, wheat enzyme	Rendina et al. (1988)

B. Purification

Purification of plant acetyl-CoA carboxylase is notoriously difficult. The multifunctional protein seems to be very susceptible to inactivation, possibly by proteolysis. To date the only purifications of active enzymes to essential homogeneity are from seed tissues (Slabas and Hellyer, 1985; Charles and Cherry, 1986). Two key aspects of the purification procedure are the inclusion of proteinase inhibitors such as phenylmethylsulphonyl fluoride (PMSF) (Egin-Buhler et al., 1980) and the use of an affinity step on avidin-monomer-Sepharose 4B (Egin-Buhler and Ebel, 1983). The latter technique allows the rapid purification of the enzyme which appears to be important for its stability. SDS-PAGE analysis of purified protein generally shows that its molecular mass is in the range 220–240 kDa (Hellyer et al., 1986), although the native enzyme probably has double that mass (e.g. Egin-Buhler and Ebel, 1983).

C. Localisation Studies

A method used to study the distribution of biotinyl proteins in various plant tissues has been applied to studies on acetyl-CoA carboxylase. The technique relies on the rapid preparation of tissue extracts (following an isolation of mesophyll, epidermal or bundle sheath cells if required) and the separation of proteins by SDS-PAGE. After such separation, the proteins are transferred from the gel onto nitrocellulose paper electrophoretically by Western blotting. The transferred proteins are then probed with [^{125}I]streptavidin to reveal those bands containing biotin (Nikolau et al., 1984). Although such a technique is not specific for acetyl-CoA carboxylase, it can give valuable information about its possible molecular weight, especially in cases where the number of streptavidin-binding bands are limited. An important caveat, however, is that adequate transfer of all protein bands should occur—this may not be the case, especially for higher molecular weight components.

III. FATTY ACID SYNTHETASE

When acyl carrier protein (ACP) was isolated from plants (Simoni et al., 1967), it was clear that plant fatty acid synthetase was a Type II dissociable enzyme complex. In

1982, three laboratories, working independently, published papers describing the partial or complete purification of individual component enzymes from barley leaves (Hoj and Mikkelson, 1982), avocado mesocarp (Caughey and Kekwick, 1982) and safflower seeds or spinach leaves (see Stumpf, 1984). Since that time further work has provided more details on these individual proteins with the recent inclusion of cloning experiments.

A. Acyl Carrier Protein

Acyl carrier protein (ACP) is a low molecular weight acidic protein which acts as a cofactor for over a dozen enzymes in plant lipid metabolism. The plant ACPs appear to be very similar to the more extensively studied *Escherichia coli* ACP (Ohlrogge, 1987) and, indeed, it is usual to use *E. coli* ACP in assays of plant fatty acid synthesis and for the synthesis of acyl-ACP substrates. Care should be exercised when drawing conclusions as to the physiological significance of such results because different ACPs have been shown to give different activities with various enzymes of plant lipid metabolism (see e.g. Guerra *et al.*, 1986).

1. Measurement of ACP

ACP can be quantified in a number of ways:

(a) Early studies made use of the ability of ACP to participate in malonyl-CoA/$^{14}CO_2$ exchange. The assay depends on the simultaneous presence of β-ketoacyl-ACP synthase, malonyl-CoA:ACP transacylase and acetyl-CoA:ACP transacylase (Majerus *et al.*, 1969). The method is sensitive but suffers from several methodological complications (Ohlrogge, 1987).

(b) An easier alternative uses malonyl-CoA:ACP transacylase to catalyse the transfer of [^{14}C]malonate onto malonyl-ACP (Ohlrogge *et al.*, 1979; Hoj and Svendsen, 1983).

(c) Alternatively, the *E. coli* acyl-ACP synthetase (Rock and Cronan, 1979) can be used to measure nanogram amounts of plant ACP in crude extracts (Kuo and Ohlrogge, 1984a). The *E. coli* enzyme is also used for the preparation of acyl-ACP substrates.

(d) A radioimmunoassay can also be used. Originally, ^{125}I was utilised in a competitive binding assay with antibodies to spinach ACP (Ohlrogge *et al.*, 1979). Recently, the assay has made use of *E. coli* acyl-ACP synthetase to form [^{3}H]palmitoyl-ACP of high specific radioactivity in crude plant extracts. This allows the assay of ACP in a variety of plants without significant interference (Kuo and Ohlrogge, 1984b).

2. Measurement of acyl-ACP

The first useful method for the estimation of plant acyl-ACPs was that developed by Mancha *et al.* (1975). This was later improved (Sanchez and Mancha, 1980) and has been used in a number of studies for examination of the pattern of acyl-ACPs made by different plant preparations (e.g. Sanchez and Mancha, 1980). The technique involves the extraction of free fatty acids and oxygen esters into petroleum ether from aqueous isopropanol. Acyl-ACPs are then precipitated in the presence of $(NH_4)_2SO_4$ and chloroform–methanol, while acyl-CoAs remain in the supernatant.

A typical extraction would involve homogenisation of 300 g of tissue in a similar volume of isopropanol–0.1 M potassium phosphate buffer (pH 7) (1:1) followed by filtration and centrifugation. The supernatant is extracted three times with petroleum ether saturated with 50% aqueous isopropanol. Glacial acetic acid (12 µl ml^{-1}), ammonium sulphate (25 µl saturated solution ml^{-1}) and chloroform–methanol (1:2; 4 ml ml^{-1}) are then added to the aqueous phase which is mixed, stood for 20 min and then centrifuged. The acyl-ACPs are present in the pellet while the acyl-CoAs are in the supernatant. The isolated acyl-ACPs and acyl-CoAs can then be saponified and assayed for radioactivity. Alternatively, they can be saponified, methylated and analysed by gas–liquid chromatography (GLC) or radio-GLC.

A variation of the Mancha procedure was developed by Soll and Roughan (1982) for measuring acyl-ACP pools during fatty acid synthesis by isolated spinach chloroplasts. They considered that non-esterified fatty acids could contaminate significantly the acyl-ACP fraction. Therefore, they used sodium methoxide to transesterify the acyl-ACPs. Argentation thin layer chromatography can be used to separate different classes of methyl esters from the non-esterified fatty acids, which are not methylated. Alternatively, for simple estimation of the pool size, the radioactively labelled non-esterified fatty acids can be extracted into petroleum ether before sodium methoxide treatment. Extraction of non-esterified fatty acids plus methyl esters, after treatment, then allows the acyl-ACP pool to be estimated by difference.

3. Purification of ACP

Although ACP has been isolated from seed tissues (Slabas *et al.*, 1986), it has proved much easier to purify it from leaves. The first detailed study on plant ACPs (Simoni *et al.*, 1967) described the purification of ACP from spinach leaves and avocado mesocarp. These purified proteins exhibited cross-reactivity for fatty acid synthesis, although it was noted that plant ACP gave rise to a different reaction product balance when compared to that produced by *E. coli* ACP, especially when the *E. coli* synthetase system was used. A more noticeable feature is that total plant fatty acid synthesis was more active with *E. coli* ACP, something which has been confirmed more recently (Hoj and Svendsen, 1983). Apart from the convenience of using *E. coli* ACP for plant fatty acid synthesis assays, the extra activity represents another desirable characteristic.

ACP has a number of unusual features which are used to aid its purification. It is stable to heat treatment which can be used to inactivate lipid metabolising enzymes—although nowadays it is often desirable to use the non-ACP protein fractions for other purposes. A particularly useful aspect of the charged nature of the protein is its high solubility. Thus, a 70% $(NH_4)_2SO_4$ cut can be used to produce a soluble fraction containing ACP. In addition, the low isoelectric point of ACP (3.9–4.2) allows its tight binding to anion-exchange resins such as DEAE-cellulose and subsequent purification by salt gradient elution. We have also used FPLC on Mono-Q columns for similar purposes. Moreover, gel filtration on Sephacryl S-300 has been employed for barley ACPs (Hoj and Svendsen, 1983). Finally, it is possible to use an immunoaffinity purification, which was originally developed for *Euglena* ACP (Ernst-Fonberg *et al.*, 1977), for plant ACP. The latter procedure allows a rapid purification of ACP from crude preparations, although the capacity of the affinity column is, of course, limited (Ohlrogge and Kuo, 1985).

4. Properties of ACP

ACP is a highly conserved protein exhibiting a high degree of sequence homology between *E. coli* and plant ACPs. Spinach and *E. coli* ACP show 40% homology while barley and spinach ACPs are 70% homologous. It is likely, therefore, that many of the structural features described for *E. coli* ACP also apply to barley or spinach ACP (Ohlrogge, 1987).

Matsumura and Stumpf (1968) examined the amino acid sequence around the active site and in other peptide fragments from spinach ACP. They obtained evidence for heterogeneity and, indeed, isoforms of plant ACP have been found in barley (Hoj and Svendsen, 1984), spinach (Kuo and Ohlrogge, 1984c), castor oil seed and soybean (Ohlrogge and Kuo, 1985). Amino terminal sequence data has been provided for the barley ACP I and II (Hoj and Svendsen, 1984) while a complete sequence for spinach ACP I is available (Kuo and Ohlrogge, 1984c). The different isoforms of ACP seem to exhibit different tissue expression (e.g. Ohlrogge and Kuo, 1985) and this may be related to a role in controlling lipid metabolism (see Harwood, 1988).

Availability of the primary structure of spinach ACP, together with antibodies to the protein, has allowed experiments on gene cloning. A synthetic gene of 268 base pairs which encoded spinach ACP I was constructed from two gene fragments, encoding the amino terminal and carboxy terminal portions. These gene fragments were assembled according to the amino acid sequence data for spinach ACP I into phage M13mp19. The partial gene constructs were joined and inserted into a plasmid and the synthetic gene subcloned into the *E. coli* expression vector pkk233-2. Using a *trc* promotor, yields of 6 mg l^{-1} of the spinach ACP I were obtained. The protein was enzymatically active because *E. coli* was able to attach the phosphopantetheine prosthetic group to the synthetic plant gene product (Beremand et al., 1987). Sequence data on the first 30 amino acid residues of the recombinant ACP showed that it was immunologically identical to authentic spinach ACP I (Guerra et al., 1988). This new source of plant ACP in large quantities should prove useful for future studies on the functional role of ACP in lipid metabolism.

5. Formation of holo-ACP

In *E. coli* it is well known that during the biosynthetis of ACP, the prosthetic group is transferred from CoA to apo-ACP (see e.g. Alberts and Vagelos, 1966). Recent studies have revealed that a similar holo-(ACP) synthase is present in the cytosol from spinach leaves and castor oil seeds (Elhussein et al., 1988). The enzyme has been partially purified from spinach by a combination of $(NH_4)_2SO_4$ fractionation and anion-exchange and gel-permeation chromatography. The localisation of this synthase in the cytosol means that the precursor protein (5–6 kDa transit sequence) is taken up by plastids and directly processed proteolytically to the mature holo-ACP (Elhussein et al., 1988).

B. Priming Reaction

Traditionally, acetyl-CoA:ACP transacylase (EC 2.3.1.28) has been described as the first component of Type II fatty acid synthetases and the enzyme which catalyses the priming reaction. Particular interest has been placed on this transacylase since there is evidence that it may be rate-limiting for overall fatty acid synthesis in plants (see Stumpf, 1987). However, a recent report by Jackowski and Rock (1987) described a special short-chain (acetoacetyl-ACP synthase) condensing enzyme from *E. coli* which not only seems to be the slowest step in fatty acid biosynthesis, but will also catalyse malonyl-ACP decarboxylation and acetyl-CoA:ACP transacylation. These authors suggest: 'it is possible that acetyltransacylase is not a distinct enzyme but a partial reaction of acetoacetyl-ACP synthase'. Experiments with spinach leaves provide evidence that in plants also a specific acetoacetyl-ACP synthase is present which, by using acetyl-CoA substrate, can bypass the so-called 'rate-limiting' acetyl transacylase reaction (Jaworski *et al.*, 1989). It may be that this acetoacetyl-ACP synthase also catalyses acetyl-CoA:ACP transacylase and that the enzymes are identical.

Acetyl-CoA:ACP transacylase can be partly separated from other fatty acid synthetase enzymes by PEG fractionation of crude extracts from developing safflower (*Carthamus tinctorius* L.) seeds (Shimakata and Stumpf, 1982a). For barley chloroplast stroma separations, partial resolution of four component enzymes was obtained by gel filtration on a Sephacryl S-300 column. Calibration of the column with molecular weight markers revealed a molecular mass of about 87 kDa for acetyl-CoA:ACP transacylase (Hoj and Mikkelson, 1982).

The acetyl transacylase has been purified some 180-fold from spinach leaf extracts (Shimakata and Stumpf, 1983a). Following $(NH_4)_2SO_4$ precipitation, the enzyme was partly purified by gel filtration on Sephacryl S-300, affinity chromatography on Affi-Gel Blue and ion-exchange chromatography on DEAE-cellulose. Although the final preparation showed similar affinities for acetyl-CoA, butyryl-CoA and hexanoyl-CoA (K_m = 8.0, 8.3, and 8.6 µM, respectively), the V_{max} was much higher for the two carbon substrate (Shimakata and Stumpf, 1983a). The acetyltransacylase was sensitive to SH-reagents including sodium arsenite. When a number of plant extracts were assayed for the various component reactions of fatty acid synthetase, acetyl transacylase (along with β-ketoacyl-ACP synthetase I and II) had the lowest activity. This led to the suggestion that it was probably rate-limiting for overall fatty acid synthesis (Shimakata and Stumpf, 1983a). In agreement with this, addition of acetyl-ACP to fatty acid synthesising systems usually stimulates overall activity (Walker *et al.*, 1988).

1. Assay of acetyl-CoA:ACP transacylation

The usual assay system is based on that developed by Hendren and Block (1980) for *E. coli*. It uses [^{14}C]acetyl-CoA substrate and relies on the precipitation of [^{14}C]acetyl-ACP product by acid at the end of the reaction. Dithiothreitol (DTT) is included (presumably to protect vicinal SH groups) in the different plant assay systems (Hoj and Mikkelson, 1982; Shimakata and Stumpf, 1982a, 1983a).

C. Malonyl-CoA:ACP Transacylase (EC 2.3.1.39)

Malonyl-CoA:ACP transacylases have been purified from a number of plant tissues including avocado (Caughey and Kekwick, 1982), barley (Hoj and Mikkelson, 1982), spinach (Stapleton and Jaworski, 1984), soybean (Guerra and Ohlrogge, 1986) and leek (Lessire and Stumpf, 1983). None of the above preparations was homogeneous. However, the avocado malonyl-CoA:ACP transacylase has been reported to have been purified to homogeneity (Hilt, 1984) and that from barley essentially to homogeneity (Hoj and Svendsen, 1983). Hilt (1984) reported that the purified enzyme from avocado had a molecular mass of 42.5 kDa in good agreement with values of about 41 kDa obtained by Caughey and Kekwick (1982). The barley enzyme (Hoj and Svendsen, 1983) had a molecular mass of 34.5 kDa on SDS-PAGE but 41 kDa on gel filtration. In soybean leaves two isoforms are found and these can be resolved by blue dye affinity chromatography (Guerra and Ohlrogge, 1986). In contrast to leaf tissue, soybean seeds only contained one isoform. That tissue-specific malonyl-CoA:ACP transacylases occur in plants is confirmed by the observation by Lessire and Stumpf (1983) that the enzymes from epidermal and parenchymal cells have molecular masses of 38 and 45 kDa, respectively. Some properties of the various plant malonyl-CoA:ACP transacylases are listed in Table 11.3.

TABLE 11.3. Comparative aspects of malonyl-CoA:ACP transacylases (MCT) from different plant tissues.

Source	Molecular mass (kDa)	K_m(malonyl-CoA)	K_m(ACP)	Remarks
Avocado fruit	40.5	3.3 µM	42 µM	
Barley chloroplasts	41			
Spinach leaves	31	0.5 mM	0.4 mM	
Anabaena variabilis	36	0.3 mM	0.4 mM	
Soybean	43	9.4(MCT$_1$) µM		Both forms present in leaves, MCT$_1$ predominant in seeds.
	43	15.0(MCT$_2$) µM		
Leek leaves	38	13.7 µM		Epidermal cells.
	45	21.7 µM		Parenchymal cells.

Taken from Harwood (1988) with permission.

1. Assay

Malonyl-CoA:ACP transacylase is assayed in a manner similar to that for acetyl-CoA:ACP transacylase except that [^{14}C]malonyl-CoA is used as labelled substrate (Table 11.1). Hoj and Mikkelson (1982) added bovine serum albumin (BSA) and relied on the excess precipitated protein to ensure quantitative recovery of [^{14}C]malonyl-ACP. Others have used a filtration procedure to remove product and Guerra and Ohlrogge (1986) compared filter papers or glass fibre disks for recovery and non-specific binding.

They concluded that comparable results were obtained by both techniques but that disks allowed rapid and reproducible processing of a greater number of enzyme assays. An alternative assay was used by Caughey and Kekwick (1982), where the formation of malonylpantetheine and CoA was coupled to the formation of acetyl-CoA from acetyl phosphate by phosphotransacetylase; the reaction was monitored by the increase in absorbance at 323 nm (Wieland *et al.*, 1979).

D. Condensing Enzymes (EC 1.3.1.41 and 2.3.1.41)

Two condensing enzymes have been separated from spinach leaves and these have been termed β-ketoacyl-ACP synthase I and II, respectively (Stumpf, 1987). By analogy with *E. coli* fatty acid synthetase, one might, in fact, have predicted a second long chain condensing enzyme such as that revealed in *fab* F mutants (Garwin *et al.*, 1980a). In addition, as mentioned in Section III.B, it seems likely that a third condensing enzyme, responsible for acetoacetyl-ACP formation, is also present in plant tissues (Jaworski *et al.*, 1989). This acetoacetyl-ACP synthase has been reported in spinach leaves. It is cerulenin-insensitive, uses acetyl-CoA rather than acetyl-ACP, and can also form a limited amount of 3-ketohexanoyl-ACP.

β-Ketoacyl-ACP synthase activity has been separated from other component FAS enzymes from plants other than spinach. In barley, the enzyme was assayed by following acetoacetyl-ACP formation from acetyl-ACP and malonyl-ACP and had a molecular mass of 92 kDa on gel filtration (Hoj and Mikkelson, 1982). Schuz *et al.* (1982) used cell suspension cultures of parsley (*Petroselinum hortense* L.) and separated the activity from acetyl and malonyl transacylases by $(NH_4)_2SO_4$ precipitation and ion-exchange and hydroxyapatite chromatography. On Sephacryl S-300 filtration, the parsley enzyme had a molecular mass of about 70 kDa. The condensing enzyme was also partly purified from photoautotrophic *Euglena gracilis* cultures (Hendren and Bloch, 1980).

As mentioned above, two condensing enzymes have been separated from each other using spinach extracts. These β-ketoacyl-ACP synthases show different sensitivities towards cerulenin and arsenite and also have characteristic substrate chain-length specificities. Although other plants, generally, have not been assayed specifically for two condensing enzymes, the ability of arsenite to inhibit palmitate conversion to stearate in a whole variety of plants (see Harwood and Stumpf, 1971; Harwood, 1988) presumably means that both enzymes are present.

β-Ketoacyl-ACP synthase I was purified 180-fold from spinach leaves using PEG precipitation and DEAE-cellulose, blue-agarose, hydroxyapatite and cellulose phosphate chromatography. It was completely free from contamination with other FAS enzymes and had a molecular mass of 56 kDa by gel filtration. It was active with acyl-ACP substrates in the C_2–C_{14} range with hexanoyl-ACP being most effective (Table 11.4). Cerulenin caused strong inhibition but arsenite was relatively ineffective (Shimakata and Stumpf, 1983b).

β-Ketoacyl-ACP synthase II was purified 295-fold from spinach leaves using $(NH_4)_2SO_4$ precipitation, Sephacryl S-300 gel filtration and Cibacron blue-agarose and cellulose phosphate chromatography (Shimakata and Stumpf, 1982c). Its molecular mass, by gel filtration, was only slightly more than synthase I, at 57.5 kDa. However, β-ketoacyl-ACP synthase II was not inhibited by cerulenin, was very sensitive to arsenite and showed a very different substrate specificity compared to β-ketoacyl-ACP synthase I (Table 11.4). Thus, β-ketoacyl-ACP synthase II is clearly responsible for chain lengthening of palmitate during *de novo* fatty acid synthesis (Harwood, 1988).

TABLE 11.4. Substrate specificities of β-ketoacyl-ACP synthesis from spinach leaves.

Substrate	β-ketoacyl-ACP synthetase I		β-ketoacyl-ACP synthetase II		
	$K_m{}^a$	$V_{max}{}^a$	K_m	V_{max}	
Acetyl-ACP	6.3	437.9	—	0	
Hexanoyl-ACP	7.5	589.6	—	—	Assay (b)[b]
Decanoyl-ACP	5.1	278.0	—	—	
Decanoyl-ACP	9.6	104	13.3	65	
Myristoyl-ACP	9.1	102	13.9	388	Assay (b)[b]
Palmitoyl-ACP	—	6	3.6	215	
Stearoyl-ACP	—	1	—	5	

Results taken from Shimakata and Stumpf (1982c) with permission.
[a] Units: $K_m = \mu M$, V_{max} = nmol $(10 \min)^{-1}$ (mg protein)$^{-1}$.
[b] Assay (a) was spectrophotometric while Assay (b) used the incorporation of radioactivity from [2-^{14}C] malonyl-ACP into fatty acids.

The β-ketoacyl-ACP synthase I from developing oilseed rape seeds has been purified to homogeneity very recently. Simultaneously, the β-ketoacyl-ACP synthase II was separated (Mackintosh *et al.*, 1989). The synthase I was purified some 12 000-fold but because of its low prevalence a final yield of only 8 μg from 300 g seeds was achieved. As noted consistently before, work with developing seeds requires using material of the correct age—in the case of studies with lipid synthesis during the early to mid-stages of fat deposition. The β-ketoacyl-ACP synthase I from rape had a molecular mass of 43 kDa on SDS-PAGE but appeared to run as a homodimer on native gels (Mackintosh *et al.*, 1989).

1. Assay systems

Several methods for assaying β-ketoacyl-ACP synthase have been used:

(a) A CO_2-exchange assay using malonyl-CoA, ACP, *E. coli* malonyl transacylase and $NaH^{14}CO_3$ (Schuz *et al.*, 1982). This method is based on the procedure of Majerus *et al.* (1969) and is rapid but not specific for a particular condensing enzyme.
(b) An assay using acetyl-CoA, ACP, *E. coli* malonyl transacylase and [2-^{14}C]malonyl-CoA was also used with a partially purified synthetase preparation to allow product identification (Schuz *et al.*, 1982). The reaction was stopped with NaOH, the products extracted into ethyl acetate and identified by TLC and HPLC.
(c) Hoj and Mikkelson (1982) used acetyl-ACP and malonyl-ACP as substrates and measured the increase in absorbance at 303 nm due to the formation of acetoacetyl-ACP. The two substrates could be made enzymatically using *E. coli* ACP and the acetyl and malonyl transacylases purified from barley (Hoj and Mikkelson, 1982). This assay is easy but is not very sensitive and requires the preparation of acyl-ACP substrates (see Method (f) below).

(d) Shimakata and Stumpf (1982c, 1983b) used C_2 to C_{10} ACP substrates (for preparation see Shimakata and Stumpf, 1982a), malonyl-CoA, ACP, malonyl transacylase, NADPH, EDTA and DTT. For maximal activity a pre-incubation of malonyl-CoA with the malonyl transacylase was needed. Activity was assessed by following the difference in absorbance at 340 nm.

(e) For longer chain acyl-ACP substrates, Shimakata and Stumpf (1982c, 1983b) used the radioactive method described by Garwin et al. (1980b). The assay system contained acyl-ACP, ACP, [2-^{14}C]malonyl-CoA and malonyl transacylase. After incubation the reaction was terminated by the addition of reducing agent. Following reduction, toluene was added and an aliquot of the organic phase taken for counting (Garwin et al., 1980b). For specific measurement of β-ketoacyl-ACP synthase II, palmitoyl-ACP was used as substrate (Shimakata and Stumpf, 1982c).

(f) Methods (c)–(e) described above require the preparation of acyl-ACP substrates and the availability of transacylase(s). In order to circumvent the inconvenience of these requirements, Mackintosh et al. (1989) have developed a sensitive assay which does not need so many specialised reagents. The assay relies on a cerulenin-treated E. coli extract which contains active enzymes to convert malonyl-CoA and acyl-CoA (C_6–C_{18}) to their respective ACP derivatives. The reaction is then followed as in Method (a) by using the back-reaction to incorporate radioactivity from NaH^{14}CO$_3$ into [^{14}C]malonyl-CoA.

(g) The acetoacetyl-ACP synthase can be specifically measured by incubating fractions with [^{14}C]acetyl-CoA and ACP under conditions needed to measure acetyl transacylase (Section III.B.1). Cerulenin is included to inhibit β-ketoacyl-ACP synthase I and the radioactive acetoacetyl-ACP can be precipitated out for counting or for further analysis on native PAGE (Jaworski et al., 1989).

E. β-Ketoacyl-ACP Reductase (EC 1.1.1.100)

This enzyme has been studied in several plant extracts and purified to homogeneity from spinach leaves, avocado mesocarp and oilseed rape. Hoj and Mikkelson (1982) partly separated β-ketoacyl-ACP reductase from other components of the FAS system by $(NH_4)_2SO_4$ fractionation and gel filtration on two Sephacryl S-300 columns. Likewise, Shimakata and Stumpf (1982a) used PEG fractionation of safflower seed extracts followed by Sephadex G-200 gel filtration and obtained a fraction containing β-ketoacyl-ACP reductase together with enoyl-ACP reductase I. These two reductases were separated from each other by further chromatography on hydroxyapatite.

In both of the above separations, NADPH was used as a cofactor for assay of the β-ketoacyl-ACP reductase. However, when studying avocado, Caughey and Kekwick (1982) found, surprisingly, that a second (NADH-requiring) form of the enzyme was also present. This NADH-form appeared to be present in the avocado plastid fraction and was not due to microbody contamination or β-oxidation enzymes. However, it proved difficult to release from plastids and further work proceeded with the NADPH-form. Early experiments resulted in about a 100-fold purification from avocado (Caughey and Kekwick, 1982). More recently, the enzyme has been purified to homogeneity from avocado and rape seeds (Sheldon et al., 1988). Three steps are involved after $(NH_4)_2SO_4$ fractionation—chromatography on Procion Red H-E3B Sepharose, gel filtration on Ultrogel AcA34 or Superose 12, and hydroxylapatite chromatography. The enzyme from avocado had a molecular mass of 28 kDa on SDS-PAGE while the purified enzyme from rape seed appeared as five components in the molecular mass range 28–30 kDa. Cleveland mapping, however, suggested that all five components were related. Indeed, *in vitro* translation of rape embryo poly(A^+) RNA and detection with reductase antiserum showed a single polypeptide of mass about 30 kDa suggesting that the multiple bands seen on SDS-PAGE were due to proteolysis during purification. Gel filtration suggested that the native reductase is tetrameric (Sheldon et al., 1988).

The β-ketoacyl-ACP reductase from spinach leaves has been purified to homogeneity (Shimakata and Stumpf, 1982b) by PEG fractionation, blue-agarose chromatography, Sephadex G-200 gel filtration, hydroxyapatite chromatography and Sephacryl S-300 gel filtration. The enzyme had a subunit molecular mass of 24.2 kDa but ran as a tetramer of 97 kDa on gel filtration. Although NADH could be used by the purified β-ketoacyl-ACP reductase, NADPH was more effective (Shimakata and Stumpf, 1982b).

1. Assay methods

(a) Caughey and Kekwick (1982) assayed the avocado reductases spectrophotometrically by following the change in absorbance at 340 nm due to the NADH- or NADPH-catalysed reduction of *S*-acetoacetyl-*N*-acetylcysteamine substrate.

(b) Acetoacetyl-ACP was used as substrate by Shimakata and Stumpf (1982b) and by Hoj and Mikkelson (1982), who also used acetoacetyl pantetheine. Acetoacetyl-ACP itself can be prepared by reacting reduced ACP with diketene (Vaglos et al., 1969; Hendren and Block, 1980). Again, change in absorbance at 340 nm is then used to measure β-ketoacyl-ACP reductase activity.

F. β-Hydroxyacyl-ACP Dehydrase

The dehydrase has been partly purified from developing safflower seeds (Shimakata and Stumpf, 1982a) and purified to homogeneity from spinach leaves (Shimakata and Stumpf, 1982b). The latter purification was achieved using blue-agarose chromatography, Sephadex G-200 filtration and hydroxyapatite chromatography. The purified dehydrase had a subunit molecular weight of 19 kDa on SDS-PAGE but migrated as a tetramer of 85 kDa on gel filtration. It showed a stereospecificity for D-β-hydroxybutyryl-ACP substrate, had activity towards substrates in the C_4–C_{16} range, and was strongly inhibited by *p*-chloromercuribenzoic acid (*p*-CMB) but not by N-ethylmaleimide (NEM) or arsenite.

1. Assay method

The enzyme is measured by the back-reaction, i.e. the hydration of enoyl-ACP substrate. Crotonyl-ACP is made by reacting reduced ACP with crotonic anhydride (Weeks and Wakil, 1969). The rate of hydration of the crotonyl-ACP substrate is then followed by a decrease in absorbance at 236 nm with an $E_{263} = 6.7 \times 10^3$ mol^{-1} cm (Shimakata and Stumpf, 1982a).

G. Enoyl-ACP Reductases (EC 1.3.1.9 and 1.3.1.10)

Two forms of enoyl-ACP reductase have been detected. Type I is NADH-specific (EC 1.3.1.9) while Type II uses NADPH (EC 1.3.1.10) in preference to NADH (Harwood, 1988). Both types were present in safflower, castor bean and rape seeds (Slabas et al., 1984) but only Type I seems to be present in leaf tissues or avocado. The Type I enoyl-ACP reductase has been partly purified from avocado mesocarp. It had a molecular weight of about 64.5 kDa on gel filtration and a very high K_m for crotonyl-ACP (Caughey and Kekwick, 1982). In developing safflower seeds, enoyl-ACP reductases I and II were distinguished from each other by PEG fractionation, different pH optima and different hydrogen donor and substrate specificity. In the latter connnection, safflower enoyl-ACP reductase II could not use crotonyl-ACP as substrate but utilised 2-decenyl-ACP effectively. Moreover, while crotonyl-CoA served as a poor substrate for enoyl-ACP reductase I (K_m 0.67 mM), it was not used at all by reductase II (Shimakata and Stumpf, 1982a). The avocado enzyme would use neither crotonyl-CoA nor crotonylpantetheine (Caughey and Kekwick, 1982).

The only enoyl-ACP reductase detected in spinach leaves is the NADH-specific form (I). The enzyme has been purified to homogeneity by PEG fractionation, blue-agarose chromatography, Sephadex G-200 filtration, hydroxyapatite chromatography and DEAE ion-exchange chromatography (Shimakata and Stumpf, 1982b). The purified enzyme was absolutely specific for NADH and, like the spinach β-hydroxyl-ACP dehydrase, was inhibited by p-CMB but not by NEM or arsenite. The enoyl reductase had a subunit molecular mass of 32.5 kDa on SDS-PAGE but migrated as a tetramer of 115 kDa on gel filtration.

The rape seed enoyl-ACP reductase I was purified to homogeneity by a combination of affinity chromatography using blue sepharose, ion exchange HPLC and hydrophobic interaction HPLC. It migrated with a molecular mass of 140 kDa on gel filtration but careful SDS-PAGE revealed two dissimilar subunits with masses of 33.6 and 34.8 kDa (Slabas et al., 1986). The amino acid composition of the rape seed enzyme was very similar to that of the spinach enzyme. It was interesting that rape seed enoyl-ACP reductase I had no activity towards C_{18} substrate, suggesting that the NADPH-requiring reductase II had to be used for stearate formation (Slabas et al., 1984). Since the Type I enzyme was the only one detected in spinach leaves, one presumes that it must have activity with C_{18} substrate (Harwood, 1988).

1. Assay methods

(a) Caughey and Kekwick (1982) used an assay with S-crotonyl-N-acetyl cysteamine and NADH which was based on the method of Muessing and Porter (1975). Activity was followed by the absorbance change at 340 nm and crotonyl substrate was prepared by the procedure of Simon and Shimin (1953).

(b) Shimakata and Stumpf (1982a,b) used a similar procedure but with crotonyl-ACP and NADH as substrates for enoyl-ACP reductase I and 2-decenyl-ACP and NADPH as substrates for enoyl-ACP reductase II. The substrates were generated by reacting reduced ACP with crotonic anhydride or the mixed anhydride of 2-decenoic acid, respectively (Shimakata and Stumpf, 1982a).

H. Overall Fatty Acid Synthesis

For many purposes, especially when studying such aspects as chain length terminations, it is necessary to measure overall fatty acid synthetase activity. Many individual incubation systems have been developed but they all have the same basic characteristics. Typically, NADPH, NADH, *E. coli* ACP and [2-^{14}C] malonyl-CoA are needed. In crude systems, it may not be necessary to add acetyl-CoA (or acetyl-ACP) primer due to decarboxylation of malonyl-CoA. Following the incubation period, total counts in fatty acids can be analysed by extracting into a suitable organic solvent and measuring radioactivity. In order to include radioactivity present as acyl-thioesters or acyl-lipids, sodium hydroxide hydrolysis must be used before extraction. In addition, because the pattern of [^{14}C]fatty acyl products is usually of interest, some method of analysing these is needed. Although some progress has been made recently with radio-HPLC systems (see Christie, 1987), gas chromatographic separation coupled to a gas-flow proportional counter is the only suitable analytical method presently available. A typical procedure using [^{14}C]malonyl-CoA is given in Bolton and Harwood (1977) or Walker and Harwood (1985).

If no radio-GLC is available, then it is possible to fractionate fatty acids by reversed phase TLC (Christie, 1987), but if only saturated fatty acids are being examined then radio-HPLC is suitable. (The main problem with HPLC separations of fatty acids is the poor resolution of palmitate from oleate.) Alternatively, fractions can be collected manually using a splitter on a GLC (see Lem and Williams, 1981) but this is very tedious and may give rise to errors due to inefficient collection.

IV. DESATURATION

A. Stearoyl-ACP Δ9-Desaturase (EC 1.14.99.6)

Nagai and Bloch (1968) were the first to demonstrate that oleate was formed in plants (*Chlorella* and spinach leaves) by a stearoyl-ACP desaturase. Subsequently the desaturase was studied in developing safflower seeds (Jaworski and Stumpf, 1974) and has been purified from the same source (McKeon and Stumpf, 1982).

The stearoyl-ACP desaturase is presumed to be localised in plastids, since that is where ACP is concentrated (Ohlrogge et al., 1979) and, in consequence, saturated fatty acid synthesis is normally coupled to Δ9-desaturation. Therefore, when isolated chloroplasts (or other plastids) are incubated with suitable precursors (usually [^{14}C]acetate) they synthesise [^{14}C]palmitate and [^{14}C]oleate as their principal products (e.g. Walker and Harwood, 1985). However, the Δ9-desaturase is generally much less active in soluble preparations from chloroplasts (see Walker and Harwood, 1985) either because the enzyme is (loosely) bound to membranes or because it needs thylakoid electron transport for its reductant.

In contrast to the higher plant chloroplast system, cyanobacteria may synthesise oleate via complex lipid substrates. Several workers have failed to detect stearoyl-ACP desaturase in these organisms and newly synthesised stearate is, in fact, transferred rapidly to monoglucosyldiacylglycerol before being desaturated (see Jaworski, 1987).

The soluble nature of the safflower seed stearoyl-ACP desaturase permitted its purification with relative ease compared to other desaturases. McKeon and Stumpf (1982) used a simple three-step method with $(NH_4)_2SO_4$ precipitation, DEAE-ion-exchange chromatography and affinity chromatography on an ACP-linked column. The enzyme was purified some 200-fold and existed in its native form as a dimer of molecular mass 68 kDa. The safflower desaturase is highly specific for stearoyl-ACP (Table 11.5). Both chain length and thiol content seem to control activity. In contrast, while crude extracts of *Euglena gracilis* showed a similar specificity to the safflower desaturase, the partially purified enzyme had equal activity with stearoyl-ACP and stearoyl-CoA (Nagai and Bloch, 1968).

TABLE 11.5. Substrate specificity of steoroyl-ACP Δ9-desaturase from safflower.

Substrate	Relative activity (%)	K_m (μM)	V_{max} (nmol min^{-1} (mg protein)$^{-1}$)
Palmitoyl-ACP	1	0.51	0.96
Stearoyl-ACP	100	0.38	106
Stearoyl-CoA	5	8.30	106

Data taken from McKeon and Stumpf (1982) with permission.

The source of electrons for stearoyl-ACP desaturation is not entirely clear. Although oleate formation in crude extracts is inhibited by CN^- (Jaworski and Stumpf, 1974), there is no evidence for a cyanide-sensitive component in the purified safflower enzyme (see Stumpf, 1984). Reduced ferredoxin has been repeatedly implicated as the source of electrons to the desaturase (Jaworski, 1987) and the *Euglena* enzyme is capable of distinguishing ferredoxins from various sources (Nagai and Bloch, 1968). Usually in artificial systems a NADPH–ferredoxin oxidoreductase system is employed, although thylakoid membranes plus a suitable electron donor plus light are more effective even in non-photosynthetic systems (see Jaworski, 1987). However, the participation of ferredoxin *in vivo* has still to be confirmed. Oxygen is also required and the K_m for the safflower desaturase is 56 μM—well below oxygen solubility at room temperature. Therefore, it has been concluded that variations in dissolved oxygen due to temperature are unlikely to affect stearate desaturation in safflower (McKeon and Stumpf, 1982; Stumpf, 1984).

1. Assay of stearoyl-ACP desaturase

The major difficulty in the enzyme's assay probably lies in the preparation of sufficient substrate. The method used is essentially that of Jaworski and Stumpf (1974) modified by McKeon and Stumpf (1982). Stearoyl-ACP desaturase is assayed by the production of [1-^{14}C]oleate from [1-^{14}C]stearoyl-ACP. The reaction requires NADPH, NADPH: ferredoxin reductase, ferredoxin and molecular oxygen. Additionally, catalase is included in the assay medium as it stimulates desaturase activity; DTT and BSA are also included (McKeon and Stumpf, 1982). The reaction is stopped by addition of KOH and the mixture is saponified, extracted and methylated. The methyl esters are then analysed by silver nitrate TLC and liquid scintillation counting or by radio-GLC.

B. Δ12-Desaturation using 1-Acyl-2-oleoyl-phosphatidylcholine (EC 1.3.1.35)

As discussed extensively recently (see Stumpf, 1984; Jaworski, 1987; Harwood, 1988) the desaturation of oleate to linoleate occurs after the acid has been attached to phosphatidylcholine (PC). This desaturation takes place on the endoplasmic reticulum and, therefore, requires that the plastid-synthesised oleate be transferred out of the chloroplast. This involves additional enzyme reactions—certainly oleoyl-ACP thioesterase, oleoyl-CoA synthetase and lyso PC:oleoyl-CoA acyl transferase. In some plants ('16:3-plants') a different pathway for polyunsaturated fatty acid formation is used. Therefore, other substrates for Δ12 desaturation are almost certainly involved. Although there is no direct evidence in higher plants, monogalactosyldiacylglycerol (MGDG) and phosphatidylglycerol (PG) are prime candidates (see Jaworski, 1987) and, indeed, MGDG is likely to be a major substrate for all desaturations in Cyanobacteria (Murata and Nishida, 1987; Harwood and Jones, 1989). In safflower microsomes, phosphatidylethanolamine has also been implicated in Δ12-desaturation (Sanchez and Stumpf, 1984).

1. Measurement of oleate desaturation

Since the Δ12-desaturase has not even been successfully solubilised (Murphy et al., 1983), still less purified, assay systems all use crude membrane preparations. Typically [^{14}C]oleate is supplied as [^{14}C]oleoyl-CoA and reliance is made on the associated acyl-CoA:lysoPC acyltransferase. NADH is the required electron donor as well as O_2. Typical assay details will be found in Slack et al. (1979), Stymne and Glad (1981), Slack et al. (1976) or Gennity and Stumpf (1985). The products can be analysed by argentation-TLC or, better, by radio-GLC. When acyl-CoAs and lipids need to be analysed, reverse-phase C_{18} columns can be used (Stymne and Glad, 1981).

C. Linolenate Formation

1. In vivo studies

The major (probably exclusive) route for α-linolenate formation is via sequential desaturation of stearate through oleate and linoleate. Whereas PC is believed to be the main substrate for oleate desaturation to linoleate, the conversion of the latter probably involves MGDG in all photosynthetic tissues. Measurement of linolenate formation in leaves must of necessity use young developing tissues. Studies with maize (Slack et al., 1977; Hawke and Stumpf, 1980), broad bean (Heinz and Harwood, 1977), barley, pea and wheat (Wharfe and Harwood, 1978) and oat leaves (Ohnishi and Yamada, 1980a,b), all implicated MGDG rather than PC as the main substrate for linoleate desaturation. However, in some plants (or under some conditions) PC may play an additional role as a substrate for linoleate desaturation (see e.g. Williams and Khan, 1982). In fact, the overall pathways for α-linolenate synthesis can be categorised as typically 'prokaryotic' (as in 16:3-plants or cyanobacteria) or 'eukaryotic' (as in 18:3-plants). In the latter case, desaturation of linoleoyl-PC on the endoplasmic reticulum is intermediate in the formation of dilinolenoyl-MGDG.

For a fuller discussion of these aspects of fatty acid synthesis refer to Heinz and Roughan (1983), Jaworski (1987), Roughan (1987) and Harwood (1988).

2. Study of linoleate desaturation in vitro

Very few experiments on α-linolenate synthesis have been conducted *in vitro*—mainly due to the great difficulty of obtaining active preparations. Roughan *et al.* (1979), using chloroplasts from spinach, were able to show a sequential appearance of radiolabel in oleate, then linoleate, then linolenate of the MGDG fraction when [^{14}C]acetate was the precursor. In a more direct test, chloroplasts were prepared from young, expanding leaves from lettuce or pea and incubated with [^{14}C]linoleoyl-MGDG and a crude fraction containing lipid exchange proteins. Direct desaturation was detected (Jones and Harwood, 1980). Linoleoyl-PC could also serve as a substrate (Table 11.6), most likely because the linoleate was transferred to MGDG before desaturation (Ohnishi and Yamada, 1980b, 1982).

TABLE 11.6. Desaturation of [^{14}C]linoleoyl-lipids by isolated chloroplasts from lettuce.

Lipid substrate	Additions	% Distribution of [^{14}C]fatty acids			
		16:0	18:1	18:2	18:3
MGDG[a]	—	tr.	n.d.	61	39
MGDG[a]	Chloroplasts + S.F.	n.d.	n.d.	46	54
MGDG	—	25	48	27	tr.
MGDG	Chloroplasts	25	47	28	tr.
MGDG	Chloroplasts + S.F.	24	34	25	17
PC[a]	—	n.d.	n.d.	100	n.d.
PC[a]	Chloroplasts + S.F.	3	tr.	97	n.d.
PC[a]	Chloroplasts + S.F. + UDP-gal + G 3-P	tr.	tr.	77	23

Results taken from Jones and Harwood (1980) with permission and where further details will be found.
tr., trace; n.d., not determined.
[a] Lipids labelled *in vivo* from [^{14}C]linoleic acid. Other samples labelled from [^{14}C]acetate. Abbreviations: MGDG, monogalactosyldiacylglycerol; PC, phosphatidylcholine; S.F., soluble fractions containing lipid exchange proteins; G 3-P, glycerol 3-phosphate; UDP-gal., UDP-galactose.

Since the Δ15-desaturase has not been isolated, measurement of its activity has been indirect, using either chloroplasts or homogenates:

(a) Homogenates of developing soybean cotyledons are capable of sequentially desaturating oleate (from oleoyl-CoA) to linolenate, apparently using PC as the substrate (Stymne and Appelqvist, 1980).
(b) Isolated chloroplasts from spinach are incubated for different periods with [^{14}C]acetate and the fatty acyl pattern of the MGDG fraction analysed by AgNO$_3$-TLC. Again sequential appearance of label in oleate, linoleate and then α-linolenate was seen (Roughan *et al.*, 1979), this time associated with MGDG.
(c) [^{14}C]Linoleoyl-MGDG and [^{14}C]linoleoyl-PC were isolated from pre-labelled leaf tissue and incubated with isolated chloroplasts in the presence of lipid

exchange protein, UDP-galactose and glycerol 3-phosphate (Jones and Harwood, 1980). Desaturation was analysed by radio-GLC.

(d) Ohnishi and Yamada (1982) used [^{14}C]linoleoyl-PC also but in the presence of a specific PC-exchange protein. Linolenate formation was associated with the MGDG of oat chloroplasts.

V. ELONGATION OF FATTY ACIDS

Fatty acids are elongated by endoplasmic reticulum-located enzyme systems using malonyl-CoA as the two-carbon donor (Harwood, 1988). ACP does not appear to be involved. NADPH is usually used as reductant, though in some cases NADH is also effective (e.g. Agrawal et al., 1984). Evidence from a number of laboratories using different systems points to several different elongation systems being involved in total very long chain fatty acid synthesis (see Harwood, 1988). All plants need to form saturated very long chain fatty acids for formation of surface coverings (cutin, suberin, waxes) but, in addition, some seeds (such as *Brassica juncea*) accumulate very long chain monounsaturated acids in their seed oil triacylglycerol. Such plants have enzymes which elongate monounsaturated-CoA (Agrawal and Stumpf, 1985a,b).

A. Systems Useful for the Study of Fatty Acid Elongation

1. Monocotyledon leaf tissue

Hawke and Stumpf (1980) first used leaves from young barley or oat tissues and showed that they synthesised a very high proportion of radiolabelled very long chain fatty acids. Etiolated and green tissues are both effective. However, microsomal fractions from such leaves were not active (K. A. Abulnaja and J. L. Harwood, unpubl. res.).

2. Germinating peas

Germinating peas only form saturated fatty acids, including very long chain compounds, during the first 24 h of germination. Microsomal fractions from such tissues contain active elongation systems which can be studied with [^{14}C]malonyl-CoA (e.g. Harwood and Stumpf, 1971; Jordan and Harwood, 1980).

3. Aged potato slices

This tissue is extremely active at forming fatty acids, including very long chains, from [^{14}C]acetate. Induction of the elongation reactions has been studied in detail and indicates that at least three systems are involved (Walker and Harwood, 1986). The elongases are sensitive to thiocarbamate herbicides (Bolton and Harwood, 1976) as are those from other plant tissues (e.g. Harwood and Stumpf, 1971).

4. Leek epidermal cells

Microsomal fractions synthesise very long chain fatty acids from saturated acyl components of the endogenous lipid pool with malonyl CoA as the C_2 donor. This system has an absolute requirement for ATP and is inhibited by acetyl-ACP (Agrawal *et al.*, 1984). Exogenous saturated acyl CoAs can be elongated in the absence of ATP but oleoyl CoA is ineffective as a substrate. Characterisation of this system indicates that there may be two elongases, one converting stearoyl-CoA to arachidoyl-CoA and another elongating arachidoyl-CoA (Agrawal and Stumpf, 1985a,b; Lessire *et al.*, 1985). A fairly comprehensive survey of the leek system has been made (Cassagne *et al.*, 1987).

5. Developing oil seeds

Various oil seeds which accumulate very long chain fatty acids have been used to study elongation. These studies include the use of microsomal fractions with acyl-CoA substrates (see Harwood, 1988; Pollard and Singh, 1987).

B. Purification of Elongases

Attempts to purify plant elongases have been made with the leek microsomal system. Agrawal and Stumpf (1985a) solubilised the leek elongase system using Triton X-100 or octyl-glucoside at concentrations above their critical micellar concentrations and at detergent:protein ratios of 2. Deoxycholate inhibited the activity even at lower concentrations. Further purification of the Triton X-100 solubilised activity, by gel filtration or anion-exchange or hydroxylapatite chromatography, resolved a single peak of elongase activity. The authors suggested all the elongation components may therefore be located on a single protein. However, it is more likely that poor solubilisation produced a protein aggregate containing all the constituent elongation activities. This is a common problem when solubilising membrane proteins with non-ionic detergents.

A second study by Lessire *et al.* (1985) used the same detergents but at lower concentrations and detergent:protein ratios and similar activity profiles were achieved. Further purification of the Triton X-100 solubilised activity by sucrose density gradient centrifugation and gel filtration resolved two separate elongase activities with molecular weights of 650 and 350 kDa for the C_{20}-CoA and C_{18}-CoA elongases, respectively. Further investigation also resolved these two activities by anion-exchange chromatography and the polypeptide profiles of these revealed no high molecular weight components but six or seven polypeptides were enriched in each case, suggesting a prokaryotic type multi-subunit complex as the structure of the elongase (Lessire *et al.*, 1987).

VI. SYNTHESIS OF OTHER PLANT FATTY ACIDS

Two other systems for fatty acid synthesis which have been studied in some detail are those for γ-linolenic acid and ricinoleic acid. The formation of the hundreds of other unusual plant fatty acids remains a largely unexplored area, deserving further attention.

A. γ-Linolenate Formation

γ-Linolenate (octadeca-6,9,12-trienoate), a key intermediate in the conversion of linoleate to arachidonate by mammals, is particularly enriched in the triacylglycerol stores of the common borage (*Borago officinalis*). The Δ6-desaturase responsible for synthesising γ-linolenate from linoleate is present in the microsomal fraction from maturing (15–20 days after pollination) cotyledons. It appears to use a phosphatidylcholine substrate in which linoleate is present at the *sn*-2 position (Griffiths *et al.*, 1988).

1. Measurement of the Δ6-desaturase

Microsomal ($9.45 \times 10^6 g$ min pellet) assays were carried out at 25°C with constant shaking and in complete darkness. [^{14}C]Oleoyl-phosphatidylcholine (prepared using the borage preparation with [^{14}C]oleoyl-CoA), NADH, catalase and bovine serum albumin and a reaction pH of 7.2 seemed necessary (Griffiths *et al.*, 1988). After incubation, the total lipids could be extracted, separated by TLC and analysed by radio-GLC or by positional degradation as required. Measurements of desaturation of endogenous linoleate in microsomal lipids were also made as detailed by Griffiths *et al.* (1988).

B. Ricinoleic Acid Formation (oleate Δ^{12}-hydroxylase; EC 1.14.13.26)

Ricinoleic acid (D(+)-12-hydroxy-*cis*-9-octadecenoic acid) is the major fatty acyl component of castor bean oil. The *in vitro* formation of this acid was partially characterised by Galliard and Stumpf (1966). In the original experiments oleoyl-CoA was used as substrate with O_2 and NADH as cofactors. Although there is now some later evidence that oleoyl-phosphatidylcholine may be the true substrate, oleoyl-CoA is a much more effective exogenous substrate than oleoyl-phosphatidylcholine (Moreau and Stumpf, 1981). The hydroxylase is not inhibited by high concentrations of CO or cyanide. Therefore, although it is a mixed-function oxidase, neither cytochrome P450 nor similar systems seem to be involved (Moreau and Stumpf, 1981).

1. Measurement of Δ^{12}-hydroxylase activity

The routine assay used in the above studies used [^{14}C]oleoyl-CoA, NADH and a microsomal ($6.1 \times 10^6 g$ min pellet obtained from developing castor seeds when endosperm filled all but the outer 1 mm of the volume of the seed) fraction at a pH of 7.0. After assay (30 min, 30°C) samples were saponified, acidified and the fatty acids extracted. They were analysed by radio-GLC and the conversion of substrate to product estimated (Moreau and Stumpf, 1981).

REFERENCES

Agrawal, V. P. and Stumpf, P. K. (1985a). *Lipids* **20**, 361–366.
Agrawal, V. P. and Stumpf, P. K. (1985b). *Arch. Biochem. Biophys.* **240**, 154–165.
Agrawal, V. P., Lessire, R. and Stumpf, P. K. (1984). *Arch. Biochem. Biophys.* **230**, 580–589.
Alberts, A. W. and Vagelos, P. R. (1966). *J. Biol. Chem.* **241**, 5201–5204.
Beremand, P. D., Hannapel, D. J., Guerra, D. J., Kuhn, D. N. and Ohlrogge, J. B. (1987). *Arch. Biochem. Biophys.* **256**, 90–100.
Bolton, P. and Harwood, J. L. (1976). *Phytochemistry* **15**, 1507–1509.
Bolton, P. and Harwood, J. L. (1977). *Biochim. Biophys. Acta* **489**, 15–24.

Cassagne, C., Lessire, R., Bessoule, J. J. and Moreau, P. (1987). *In* "Biochemistry, Structure and Function of Plant Lipids" (P. K. Stumpf, J. B. Mudd and W. D. Nes, eds), pp. 481–488. Plenum Press, New York.
Caughey, I. and Kekwick, R. G. O. (1982). *Eur. J. Biochem.* **123**, 553–561.
Charles, D. J. and Cherry, J. H. (1986). *Phytochemistry* **25**, 1067–1071.
Christie, W. W. (1987). "High Performance Liquid Chromatography of Lipids." Pergamon Press, Oxford.
Eastwell, K. C. and Stumpf, P. K. (1983). *Plant Physiol.* **72**, 50–55.
Egin-Buhler, B. and Ebel, J. (1983). *Eur. J. Biochem.* **133**, 335–339.
Egin-Buhler, B., Loyal, R. and Ebel, J. (1980). *Arch. Biochem. Biophys.* **203**, 90–100.
El Hussein, S. A., Miernyk, J. A. and Ohlrogge, J. B. (1988). *Biochem. J.* **252**, 39–45.
Ernst-Fonberg, M. L., Schongalla, A. M. and Walker, T. M. (1977). *Arch. Biochem. Biophys.* **178**, 166–173.
Finlayson, S. A. and Dennis, D. T. (1983). *Arch. Biochem. Biophys.* **225**, 576–585.
Galliard, T. and Stumpf, P. K. (1966). *J. Biol. Chem.* **241**, 5806–5812.
Garwin, J. L., Klages, A. L. and Cronan, J. E., Jr. (1980a). *J. Biol. Chem.* **255**, 3263–3265.
Garwin, J. L., Klages, A. L. and Cronan, J. E., Jr. (1980b). *J. Biol. Chem.* **255**, 11949–11956.
Gennity, J. M. and Stumpf, P. K. (1985). *Arch. Biochem. Biophys.* **239**, 444–454.
Griffiths, G., Stobart, A. S. and Stymne, S. (1988). *Biochem. J.* **252**, 641–647.
Guerra, D. J. and Ohlrogge, J. B. (1986). *Arch. Biochem. Biophys.* **246**, 274–285.
Guerra, D. J., Ohlrogge, J. B. and Frentzen, H. (1986). *Plant Physiol.* **82**, 448–453.
Guerra, D. J., Dziewanowska, K., Ohlrogge, J. B. and Beremand, P. (1988). *J. Biol. Chem.* **263**, 4386–4391.
Harwood, J. L. (1980). *In* "The Biochemistry of Plants" (P. K. Stumpf and E. E. Conn, eds), Vol. 4, pp. 1–55. Academic Press, New York.
Harwood, J. L. (1988). *Ann. Rev. Plant. Physiol.* **39**, 101–138.
Harwood, J. L. and Jones, A. L. (1989). *Adv. Botanical Res.* **16**, 1–54.
Harwood, J. L. and Stumpf, P. K. (1971). *Arch. Biochem. Biophys.* **142**, 281–291.
Harwood, J. L. and Stumpf, P. K. (1972). *Arch. Biochem. Biophys.* **148**, 282–290.
Hawke, J. C. and Stumpf, P. K. (1980). *Arch. Biochem. Biophys.* **203**, 296–306.
Heinz, E. and Harwood, J. L. (1977). *Hoppe-Seyler's Z. Physiol. Chem.* **358**, 897–908.
Heinz, E. and Roughan, P. G. (1983). *Plant Physiol.* **72**, 273–279.
Hellyer, A., Bambridge, H. E. and Slabas, A. R. (1986). *Biochem. Soc. Trans.* **14**, 565–568.
Hendren, R. W. and Bloch, K. (1980). *J. Biol. Chem.* **255**, 1504–1508.
Hilt, K. L. (1984). Ph.D. Thesis, University of California, Davis, CA.
Hoj, P. B. and Mikkelson, J. D. (1982). *Carlsberg Res. Commun.* **47**, 119–141.
Hoj, P. B. and Svendsen, I. (1983). *Carlsberg Res. Commun.* **48**, 285–306.
Hoj, P. B. and Svendsen, I. (1984). *Carlsberg Res. Commun.* **49**, 483–492.
Jackowski, S. and Rock, C. O. (1987). *J. Biol. Chem.* **262**, 7927–7931.
Jaworski, J. G. (1987). *In* "The Biochemistry of Plants" (P. K. Stumpf and E. E. Conn, eds), Vol. 9, pp. 159–174. Academic Press, New York.
Jaworski, J. G. and Stumpf, P. K. (1974). *Arch. Biochem. Biophys.* **162**, 158–165.
Jaworski, J. G., Clough, R. C. and Barnum, S. R. (1989). *Plant Physiol.* **90**, 41–44.
Jones, A. V. M. and Harwood, J. L. (1980). *Biochem. J.* **190**, 851–854.
Jordan, B. R. and Harwood, J. L. (1980). *Biochem. J.* **191**, 791–797.
Kuo, T. M. and Ohlrogge, J. B. (1984a). *Arch. Biochem. Biophys.* **230**, 110–116.
Kuo, T. M. and Ohlrogge, J. B. (1984b). *Anal. Biochem.* **136**, 497–502.
Kuo, T. M. and Ohlrogge, J. B. (1984c). *Arch. Biochem. Biophys.* **234**, 290–296.
Laing, W. A. and Roughan, P. G. (1982). *FEBS Lett.* **144**, 341–344.
Lem, N. W. and Williams, J. P. (1981). *Plant Physiol.* **68**, 944–949.
Lessire, R. and Stumpf, P. K. (1983). *Plant Physiol.* **69**, 897–903.
Lessire, R., Juguelin, H. and Cassagne, C. (1985). *FEBS Lett.* **187**, 314–320.
Lessire, R., Besoule, J. J. and Cassagne, C. (1987). *In* "The Metabolism, Structure and Function of Plant Lipids" (P. K. Stumpf, J. B. Mudd and W. D. Nes, eds), pp. 525–527. Plenum Press, New York and London.
McKeon, T. A. and Stumpf, P. K. (1982). *J. Biol. Chem.* **257**, 12141–12147.

Mackintosh, R. W., Hardie, D. G. and Slabas, A. R. (1989). *Biochim. Biophys. Acta.* **1002**, 114–124.
Majerus, P. W., Alberts, A. W. and Vagelos, P. R. (1969). *In* "Methods in Enzymology" (J. H. Lowenstein, ed.), Vol. 14, pp. 43–50. Academic Press, New York.
Mancha, M., Stokes, G. B. and Stumpf, P. K. (1975). *Anal. Biochem.* **68**, 600–608.
Matsumura, S. and Stumpf, P. K. (1968). *Arch. Biochem. Biophys.* **125**, 932–941.
Mohan, S. B. and Kekwick, R. G. O. (1980) *Biochem. J.* **187**, 667–676.
Moreau, R. A. and Stumpf, P. K. (1981). *Plant Physiol.* **67**, 672–676.
Muessing, R. A. and Porter, J. M. (1975). *Methods Enzymol.* **3**, 45–59.
Murata, N. and Nishida, I. (1987). *In* "The Biochemistry of Plants" (P. K. Stumpf and E. E. Conn, eds), Vol. 9, pp. 315–347. Academic Press, New York.
Murphy, D. J., Woodrow, I. E., Latzko, E. and Mukherjee, K. D. (1983). *FEBS Lett.* **162**, 442–446.
Nagai, J. and Bloch, K. (1968). *J. Biol. Chem.* **243**, 4626–4633.
Nikolau, B. J. and Hawke, J. C. (1984). *Arch. Biochem. Biophys.* **228**, 86–96.
Nikolau, B. J., Hawke, J. C. and Stumpf, P. K. (1981). *Arch. Biochem. Biophys.* **211**, 605–612.
Nickolau, B. J., Wurtels, E. S. and Stumpf, P. K. (1984). *Plant Physiol.* **75**, 895–901.
Ohlrogge, J. B. (1987). *In* "The Biochemistry of Plants" (P. K. Stumpf and E. E. Conn, eds), Vol. 9, pp. 137–157. Academic Press, New York.
Ohlrogge, J. B. and Kuo, T. M. (1985). *J. Biol. Chem.* **260**, 8032–8037.
Ohlrogge, J. B., Kuhn, D. N. and Stumpf, P. K. (1979). *Proc. Natl. Acad. Sci. USA* **76**, 1194–1198.
Ohnishi, J. and Yamada, M. (1980a). *Plant Cell Physiol.* **21**, 1595–1606.
Ohnishi, J. and Yamada, M. (1980b). *Plant Cell Physiol.* **21**, 1607–1618.
Ohnishi, J. and Yamada, M. (1982). *Plant Cell Physiol.* **23**, 767–773.
Pollard, M. R. and Singh, S. S. (1987). *In* "The Metabolism, Structure and Function of Plant Lipids" (P. K. Stumpf, J. B. Mudd and W. D. Nes, eds), pp. 455–463. Plenum Press, New York.
Rendina, A. R., Felts, J. M. Beaudoin, J. D., Craig-Kennard, A. C., Look, L. L., Paraskos, S. L. and Hagenah, J. A. (1988). *Arch. Biochem. Biophys.* **265**, 219–225.
Rock, C. O. and Cronan, J. E. (1979). *J. Biol. Chem.* **254**, 9778–9785.
Roughan, P. G. (1987). *In* "The Metabolism, Structure and Function of Plant Lipids" (P. K. Stumpf, J. B. Mudd and W. D. Nes, eds), 247–254. Plenum Press, New York.
Roughan, P. G., Mudd, J. B., McManus, T. T. and Slack, C. R. (1979). *Biochem. J.* **18**, 571–574.
Sanchez, J. and Mancha, M. (1980). *Phytochemistry* **19**, 817–820.
Sanchez, J. and Stumpf, P. K. (1984). *Arch. Biochem. Biophys.* **228**, 185–196.
Schuz, R., Ebel, J. and Hahlbrock, K. (1982). *FEBS Lett.* **140**, 207–209.
Sheldon, P. S., Safford, R., Slabas, A. R. and Kekwick, R. G. O. (1988). *Biochem. Soc. Trans.* **16**, 392–393.
Shimakata, T. and Stumpf, P. K. (1982a). *Arch. Biochem. Biophys.* **217**, 144–154.
Shimakata, T. and Stumpf, P. K. (1982b). *Arch. Biochem. Biophys.* **218**, 77–91.
Shimakata, T. and Stumpf, P. K. (1982c). *Proc. Natl. Acad. Sci. USA* **79**, 5808–5812.
Shimakata, T. and Stumpf, P. K. (1983a). *J. Biol. Chem.* **258**, 3592–3598.
Shimakata, T. and Stumpf, P. K. (1983b). *Arch. Biochem. Biophys.* **220**, 39–45.
Simon, R. J. and Shemin, D. (1953). *J. Am. Chem. Soc.* **75**, 2250.
Simoni, R. D., Criddle, R. S. and Stumpf, P. K. (1967). *J. Biol. Chem.* **242**, 573–581.
Slabas, A. R. and Hellyer, A. (1985). *Plant Sci.* **39**, 177–182.
Slabas, A. R., Harding, J., Hellyer, A., Sidebottom, C., Gwynne, H., Kessell, R. and Tombs, M. P. (1984). *In* "Structure, Function and Metabolism of Plant Lipids" (P. A. Siegenthaler and W. Eichenberger, eds), pp. 3–10. Elsevier, New York.
Slabas, A. R., Sidebottom, C. M., Hellyer, A., Kessell, R. M. J. and Tombs, M. P. (1986). *Biochim. Biophys. Acta.* **877**, 271–280.
Slack, C. R., Roughan, P. G. and Terpstra, J. (1976). *Biochem. J.* **155**, 710–780.
Slack, C. R., Roughan, P. G. and Balsingham, N. (1977). *Biochem. J.* **162**, 289–296.
Slack, C. R., Roughan, P. G. and Browse, J. (1979). *Biochem. J.* **179**, 649–656.
Soll, J. and Roughan, P. G. (1982). *FEBS Lett.* **146**, 189–192.

Stapleton, S. R. and Jaworski, J. G. (1984). *Biochim. Biophys. Acta.* **794**, 240–248.
Stumpf, P. K. (1984). *In* "Fatty Acid Metabolism and its Regulation" (S. Numa, ed.), pp. 155–199. Elsevier, Amsterdam.
Stumpf, P. K. (1987). *In* "The Biochemistry of Plants" (P. K. Stumpf and E. E. Conn, eds), Vol. 9, pp. 121–136. Academic Press, New York.
Stymne, S. and Appelqvist, L. A. (1980). *Plant Sci. Lett.* **17**, 287–292.
Stymne, S. and Glad, G. (1981). *Lipids* **16**, 298–305.
Vagelos, P. R., Alberts, A. W. and Majerus, P. W. (1969). *Methods Enzymol.* **14**, 61–63.
Walker, K. A. and Harwood, J. L. (1985). *Biochem. J.* **226**, 551–556.
Walker, K. A. and Harwood, J. L. (1986). *Biochem. J.* **237**, 41–46.
Walker, K. A., Ridley, S. M., Lewis, T. and Harwood, J. L. (1988). *Biochem. J.* **254**, 307–310.
Weeks, G. and Wakil, S. J. (1969). *Methods Enzymol.* **14**, 66–73.
Wharfe, J. and Harwood, J. L. (1978). *Biochem. J.* **174**, 163–169.
Wieland, F., Renner, L., Verfurth, C. and Lynen, F. (1979). *Eur. J. Biochem.* **94**, 189–197.
Williams, J. P. and Khan, M. U. (1982). *Biochim. Biophys. Acta* **713**, 177–184.

12 Enzymes of Lipid Degradation

ANTHONY H. C. HUANG

Department of Botany and Plant Sciences, University of California, Riverside, CA 92521, USA

I.	Introduction	219
II.	Lipases	220
	A. Introduction	220
	B. Technical difficulties in the assay of lipase activities	221
	C. Lipase assay	222
III.	Acyl hydrolase	224
	A. Introduction	224
	B. Acyl hydrolase assay	224
IV.	Phospholipase D	225
	A. Introduction	225
	B. Phospholipase D assay	225
	References	226

I. INTRODUCTION

Information on lipolytic enzymes in higher plants is important in understanding their physiological roles, their action in agricultural products during storage, and their potential use in biochemistry and industry. However, our knowledge of these enzymes is still very limited. The nomenclature for plant lipolytic enzymes, especially with reference to substrate specificity, is a common source of confusion. This confusion, together with technical difficulties in the handling of water-insoluble substrates and products, has discouraged researchers. Nevertheless, many of the problems have been recognised and overcome recently. Recent reviews on plant lipolytic enzymes include those by Galliard (1980, 1983) and Huang (1984, 1985, 1987).

Plant lipid degrading enzymes which have been studied extensively can be divided into two groups according to their mode of action (Table 12.1). Group 1 consists of enzymes which can release primary products such as fatty acid, glycerol, choline, phosphocholine, etc. from the various lipids. Group 2 includes enzymes that act on the primary products released by enzymes from the first group.

TABLE 12.1. Active lipid degrading enzymes in plant tissues.

Group 1 (releasing primary products)	Group 2 (acting on the primary or secondary products)
Lipase (EC 3.1.1.3)	Acting on fatty acids:
Acyl hydrolase, possessing activities of	α-oxidation
Phospholipase A1 (EC 3.1.1.32)	β-oxidation
Phospholipase A2 (EC 3.1.1.4)	ω-oxidation
Phospholipase B (EC 3.1.1.5)	lipoxygenase
Glycolipase	
Sulpholipase	Acting on acetate produced
Monoacylglycerol lipase	by β-oxidation:
Phospholipase D (EC 3.1.4.4)	glyoxylate cycle

Naturally, it is impossible to cover the methodology of all the lipid degrading enzymes in this short article. Therefore, I will only describe the three enzymes listed in Group 1. These three enzymes have not only very high *in vitro* activities, but also wide distributions among plant species. However, only lipase has a clearly defined metabolic function during seed germination. The physiological roles of acyl hydrolase and phospholipase D are still unclear. Each of the three enzymes has been purified from many sources, and exhibits diverse, species- and tissue-dependent differences in its properties. The procedure for purification of the enzyme from one source probably will not be applicable to that from another source. Because of this non-applicability, and because of the limitation in the length of this article, I will not describe the purification procedure. Instead, the reports of enzyme purification are cited.

The readers are referred to published reviews on enzymes in Group 2. Fatty acid oxidation enzymes have been dealt with partly, or completely, by Kindl (1987), Galliard (1980), and Vick and Zimmerman (1987). The glyoxylate cycle enzymes have been reviewed by Beevers (1980) and Kindl (1987).

II. LIPASES (EC 3.1.1.3)

A. Introduction

True lipases, which attack the fatty acyl linkage of water-insoluble triacylglycerols, are present in high activities in oilseeds (Galliard, 1980; Huang, 1987). They catalyse the hydrolysis of reserve triacylglycerols in post-germinative growth of seedlings. A common feature among lipases from diverse seeds, is that the enzyme activities are absent in ungerminated seeds and increase post-germination. Other than this similarity, oilseed lipases from diverse species exhibit differences in their properties. These differences include substrate specificity, pH for optimal activity, reactivity toward sulphydryl reagents, hydrophobicity of the molecule, and subcellular location. The enzymes from maize lipid bodies (Lin and Huang, 1984) and castor bean glyoxysomes (Maeshima and Beevers, 1985) have been purified to apparent homogeneity.

B. Technical Difficulties in the Assay of Lipase Activities

In the study of lipases, technical difficulties in the assay of the enzyme activities need to be overcome. Some of the difficulties are not unusual, and are common among enzyme assays dealing with water-insoluble substrates. In the assay of lipase activity, the substrate emulsions are generally prepared by sonicating triacylglycerol in the presence of an emulsifying agent, such as Acacia (gum arabic), or a detergent such as Triton X-100. The fatty acid released is quantitated by a pH stat continuously, or at time intervals by colorimetry or radioactive analysis. The assay is time-consuming. The substrate emulsion is unstable, and the sizes of the emulsions are non-uniform and are not totally reproducible from experiment to experiment. Therefore, some researchers use artificial triacylglycerols containing short acyl moieties so that the emulsion is much more stable and uniform. Some other researchers measure the esterase activity of the lipase by using an artificial substrate which upon enzymatic hydrolysis generates a fluorescent product. As will be explained, these short-cuts actually generate more problems than those they are intended to solve. An account of the technical problems in measuring lipase activities is itemised below, and some of them are especially critical when dealing with plant tissues (Huang, 1987):

(1) A lipase from a certain seed species is likely to be relatively specific on the native storage triacylglycerols, or artificial triacylglycerols containing the major fatty acid components of the storage triacylglycerols in that same species. Thus, when assaying seed lipase activity, a suitable triacylglycerol should be used.

(2) Highly active non-specific acyl hydrolases are present in many plant tissues (see Section III). These hydrolases are active on various acyl lipids other than triacylglycerols. Under ideal assay conditions, their activities are several orders of magnitude higher than the lipase activity in the same tissues. Thus, using an artificial ester substrate (generally a monoester of a fatty acid and a fluorescent moiety) to study seed lipases generates uncertain results. The problem may be relatively less important in the study of lipases in other organisms, but is very critical in plant tissues due to the presence of the ubiquitous and highly active acyl hydrolases.

(3) The problem described in (2) is compounded by the general unawareness of the purity of commercial triacylglycerol preparations used as substrates. Impurity of a small amount of monoacylglycerols or other acyl lipids would lead to the assay of the dominating acyl hydrolases mentioned above. It has been reported that commercially available tributyrin preparation contains monobutyrin as a contaminant (Brockerhoff and Jensen, 1974), and trilinolein preparation, labelled as 99% pure, contains a small amount of monoacylglycerols (from one company) or fatty acids (from another company) (Lin and Huang, 1983). The impurities can be removed easily by a preparative thin layer chromatography (to be described) or silica gel column chromatography (Christie, 1982).

(4) The optimal pH for lipase activity on the native triacylglycerols may be quite different from that on an artificial substrate. For example, rapeseed lipase has an optimal activity at pH 8.5 on N-methylindoxylmyristate, but at pH 6.5 on native triacylglycerols, trilinolein or trierucin. It is inappropriate to assay the optimal pH for activity using an artificial substrate (or impure substrate), and then use this pH to study substrate specificity and other enzyme properties.

(5) Lipase inhibitors are present in some seeds before and after germination. These inhibitors are not the classical enzyme inhibitors which bind to or act on the enzyme molecule. Instead, the lipase inhibitors are proteins that bind to the surface of the substrate micelles in an *in vitro* enzyme assay. The binding prevents the normal functioning of the lipase which acts on the interfacial area between the aqueous medium and the emulsion surface. If the seed extract does not contain an overwhelming amount of the protein inhibitors, as is the case with peanut cotyledons, lipase activity can be detected and measured by simply adding more substrate micelles to the assay system. However, in other seed tissues, such as soybean cotyledons, the amount of protein inhibitors is in great excess and a simple increase in the substrate micelles is not sufficient to overcome the inhibition. Methods should be designed to remove these proteins prior to assay of lipase activities. The methods of removal could take advantage of the fact that these protein inhibitors are probably of amphipathic nature.

In an overall assessment, items (1)–(4) are minor technical difficulties that can be overcome easily. The substrate triacylglycerol should be the native triacylglycerol extracted from the ungerminated seed, or artificial triacylglycerols containing the major storage fatty acid moieties. The release of fatty acid is monitored by a pH stat continuously or at time intervals by colorimetry or radioactive assay (if the triacylglycerol is available in radioactive form). The triacylglycerol obtained commercially, irrespective of the purity claimed by the chemical company, should be checked for purity and, if necessary, purified. Artificial substrates such as those used in fluorescence measurement of acyl hydrolase activity could be employed in routine enzyme assays such as in enzyme purification, but should be used only after the lipase has been fairly well studied and separated from the non-specific acyl hydrolases. Overcoming the lipase inhibitors described in (5) will have to be designed and is likely to be species specific. So far, no such procedure has been published.

C. Lipase Assay

Lipase from an individual seed species exhibits high activities on the native substrates which are present in the same seed species. Many seeds contain oleic and linoleic acids as the major acyl components in the triacylglycerols, and the lipases from these seeds are generally active on triolein or trilinolein. Therefore, either of the two substrates can be used for the lipase assay. The substrate should be free of contaminating monoacylglycerols. The substrate preparation obtained commercially should be checked by thin layer chromatography for purity irrespective of the claims of purity made by the manufacturers. To do this, the substrate is dissolved in chloroform (5–10 µg/10 µl for each spot) and spotted onto a TLC plate coated with 250 µm of Silica Gel G (Brinkman Instruments, Inc., Westbury, NY, USA). The plate is developed in 50:50:1 (v/v/v) hexane:diethyl ether:acetic acid, and allowed to react with iodine vapour (put some crystals of iodine in a TLC tank). Standards of tri-, di- and mono-olein/linolein and the free acid (5–10 μg/10 μl for each spot) are run on the TLC plate. The mobilities of the components in descending order are triacylglycerol, fatty acid, 1,3-diacylglycerol, 1,2-diacylglycerol and monoacylglycerols. A small percentage of free fatty acid in the triacylglycerol preparation will not interfere with the assay. If monoacylglycerol is present, the triacylglycerol preparation is purified on a similar TLC plate. The triacylglycerol preparation is applied as a line across the whole plate (200 μg per one 20 cm × 20 cm plate). A similar but smaller plate (20 cm × 5 cm) is run in the same tank as a marker. After development, the small plate is allowed to react with iodine vapour. By making a comparison between the two plates, the position of triacylglycerol in the large plate can be identified. The silica gel containing the triacylglycerol line is scraped off and extracted with chloroform. After centrifugation to remove the silica gel, the chloroform supernatant is obtained, and the chloroform is evaporated.

The activity of true lipase is measured by a colorimetric method. Although the released fatty acids can be measured more rapidly using an automatic titrator, the colorimetric method has the advantage of requiring no specific equipment or set-up. Furthermore, if many enzyme samples are to be assayed, the colorimetric method probably consumes about the same amount of time per assay, and all the assays can be performed simultaneously for a more uniform quantitation. In the colorimetric assay, the fatty acids produced are converted to copper soaps and measured using 2,2'-diphenylcarbazide. The reaction is performed at room temperature in a 5 ml tube. The 1 ml reaction mixture contains 0.1 M Tris-HCl (pH 7.5) (depending on specific enzyme), 5 mM dithiothreitol, 5 mM substrate, and enzyme preparation. Triolein or trilinolein (50 mM) is first emulsified in 2 ml of 5% gum arabic for 1 min at low speed with a Bronwill Biosonic IV ultrasonic generator fitted with a microprobe. The reaction is stopped at time intervals (ranging from 5 min to 2 h, depending on the amount of enzyme used) to ensure that proper kinetics are observed. Each 0.1 ml aliquot of the reaction mixture is put in a 7 ml screw top tube (pre-washed with chromic sulphuric acid) and boiled in a boiling water bath for 5 min. After cooling to room temperature, 4 ml of chloroform–heptane–methanol (4:3:2; v/v/v) is added. The tube is closed with a Teflon screw cap, and shaken horizontally for 15 min. Two millilitres of 0.1 M sodium phosphate (pH 2.5) is added and the tube is shaken horizontally for 3 min. After centrifugation in a table-top centrifuge, the upper layer of methanol–water is pipetted out and discarded. One millilitre of 0.01 M HCl is added, and the tube is shaken horizontally for 3 min. After centrifugation, the upper layer of HCl solution is pipetted out and discarded. Then 1.5 ml of copper reagent (0.1 M $Cu(NO_3)_2$, 0.2 M triethanolamine, 0.06 N NaOH, and 6 M NaCl) is added. The tube is closed with a Teflon screw cap, and shaken horizontally for 30 min. After centrifugation in a table-top centrifuge, 2 ml of the chloroform layer is transferred to a tube, and 0.1 ml of colour reagent (10 ml freshly prepared 0.4%, 2,2'-diphenylcarbazide solution in 100% ethanol, plus 0.1 ml 1 M triethanolamine added immediately before use) is added. After 5 min or more, the absorbance is read at 550 nm. Oleic acid or linoleic acid is used to produce a standard curve that is linear up to a concentration of 0.05 μmol per 2 ml chloroform.

The seed of jojoba contains intracellular wax ester instead of triacylglycerols as food reserve for post-germinative growth. The localisation of wax esters in the subcellular lipid bodies and the appearance of a lipid hydrolase for lipid degradation in post-germinative growth are similar to those in triacylglycerol-storing oilseeds. The enzyme can be assayed by the lipase assay procedure using wax esters instead of triacylglycerols as the substrate. Alternatively, if the enzyme preparation is known not to contain the non-specific acyl hydrolase, such as in the latter steps of enzyme purification, the simpler assay for acyl hydrolase activity (Section III) can be employed.

III. ACYL HYDROLASE

A. Introduction

An enzyme that hydrolyses acyl groups from several classes of lipids, including glycolipids, phospholipids, sulpholipids, and mono- and diacylglycerols, but is inactive on triacylglycerols, is present in many plant tissues (Galliard, 1980; Huang, 1987). Acyl hydrolase releases both fatty acids from diacyl glycerolipids and, in many cases, there is no preference for either the *sn* 1- or 2-position of the acyl ester linkage. Thus, the enzyme possesses a combined catalytic capacity of phospholipase A_1, A_2 and B, as well as glycolipase, sulpholipases and monoacylglycerol lipase. Similarities of the enzymes from various tissues include the following: (1) they exert a similar pattern of substrate specificity as described above; (2) they occur as isozymes in each tissue and they have fairly similar patterns of substrate specificity; (3) they have similar pH for optimal activity on a particular substrate and generally this pH shifts to a more alkaline value in the presence of detergent; and (4) they catalyse acyl transferase reactions.

The hydrolytic activities on various classes of acyl lipids are apparently carried out by a single protein. The following evidence with occasional exceptions has been obtained using purified or partially purified enzymes on, usually but not exclusively, galactolipids and phosphatidylcholine.

(1) The activity ratio of the enzyme preparation on galactolipid and phospholipid remains fairly constant throughout an enzyme purification procedure.
(2) The activity ratio is also similar after treatment of the enzyme with high or low temperature and chemical-modifying reagents.
(3) The enzyme carries out acyltransferase reactions with each of the substrates.
(4) Each substrate inhibits activity at a similar high concentration.
(5) The optimal activity is at an acidic pH and shifts to a high pH in the presence of detergent.
(6) Competition of the two substrates for enzyme activity exists, suggesting that the activities reside not only in a single protein, but also within the same active site.
(7) Potato tuber of a special variety in which the acyl hydrolase activity is very low contains a proportional reduction of activity on each of the substrates (Galliard, 1980).

The enzymes from potato tubers, leaves, and rice bran have been purified to apparent homogeneity (see review by Huang, 1987).

B. Acyl Hydrolase Assay

Acyl hydrolase activity is measured at room temperature continuously with a fluorometer using an artificial substrate. The reaction mixture of 4 ml contains 0.1 M Tris-HCl buffer (pH 7.5) (depending on specific enzyme), 2 mM dithiothreitol, and enzyme preparation. The reaction is initiated by the addition of 0.1 ml of 33 mM (final concentration, 0.83 mM) N-methylindoxylmyristate (from US Biochem. Corp.) dissolved in ethylene glycol monomethyl ether. Fluorescence measurements are made with a Turner Model III fluorometer with excitation filter No. 405 (405 nm maxima) and emission filter No. 2A-12 (>510 nm) attached to a recorder. The reaction is usually linear for the first 10 min. Although unnecessary, a more sophisticated fluorometer can be used, and the volume of the reaction mixture can be reduced to 1 ml. The activity is expressed on a relative basis of ΔF unit per unit time.

The commercial supply of N-methylindoxylmyristate may be unreliable. If it cannot be obtained, fatty esters of 4-methylumbelliferone are used (Hasson and Laties, 1976). Many fatty derivatives of 4-methylumbelliferone are available commercially (from Sigma Corp., St Louis, MO, USA; and several other biochemical companies). Acyl hydrolase from a particular tissue or species may exhibit some degree of specificity towards the acyl derivatives. Nevertheless, 4-methylumbelliferyl laurate (inexpensive and obtainable from U.S. Biochem. Corp.) is probably an active substrate of acyl hydrolase from most if not all sources. In the assay, 1 mM of 4-methylumbelliferone laurate is used instead of 0.83 mM N-methylindoxylmyristate, and excitation filter No. 7-60 (365 nm maxima) is used instead of No. 405. If desirable, the actual activity in nmol min^{-1} can be calculated from a standard curve of fluorescence units versus 4-methylumbelliferone (stable and commercially available) concentrations in the assay system minus enzyme at the same pH. The assay using 4-methylumbelliferone is about an order of magnitude more sensitive than the assay using N-methylindoxylmyristate.

IV. PHOSPHOLIPASE D (EC 3.1.4.4)

A. Introduction

The enzyme is present in high activity in diverse plant tissues and species (Heller, 1978; Galliard, 1980). It hydrolyses a phosphatide to phosphatidic acid and a nitrogenous base. It has a broad substrate specificity, hydrolysing phosphatidylcholine, phosphatidylethanolamine, phosphatidylserine, and lysophosphatides. The acyl moieties of the phosphatides can be of short or native (C_{16} and C_{18}) chain length. The enzyme also catalyses the transphosphatidylation between a phosphatide and an alcohol such as methanol or glycerol. Organic solvent such as diethyl ether, or detergent such as SDS, is a required activator, as is Ca^{2+}. Depending on the tissues and species, the enzyme may be recovered either in the soluble fraction or in the particulate fraction from tissue homogenates. The enzymes from leaves and seeds of many species (see review by Galliard, 1980) and citrus callus (Witt et al., 1987) have been purified. In spite of the fairly extensive studies on the enzymes from plants, animals, and bacteria, its physiological role remains unclear.

B. Phospholipase D Assay

The enzyme activity can be assayed colorimetrically (by monitoring the release of choline) or radiometrically (by monitoring the release of either choline or phosphatidic acid) (Heller, 1978; Taki and Kanfer, 1981; Grossman et al., 1974). The radiometric assay is much more sensitive than the colorimetric assay. The enzyme activity can also be assayed using *p*-nitrophenylphosphocholine as an artificial substrate by monitoring the release of *p*-nitrophenol spectrophotometrically after treatment of the enzymatic product (*p*-nitrophenylphosphate) with a phosphatase (Gupta and Wold, 1980). This assay has not been used extensively, and it works only if the enzyme preparation does not contain phospholipase C activity. In the radiometric assay, the availability of specific labelled [^{14}C]phosphatidylcholine commercially in recent years offers many options. Water-soluble labelled choline released from either uniformly labelled or choline-specific labelled phosphatidylcholine can be monitored. Alternatively, labelled phosphatidic acid released from either uniformly labelled or acyl-labelled phosphatidylcholine can be assayed. In addition to specific positional labelling, labelled phosphatidylcholine with defined acyl moieties is also available. The monitoring of the release of water-soluble radioactive choline is comparatively easy, but less reliable. At the start workers should ascertain the enzymatic activity by identifying phosphatidic acid as the reaction product. The following is a procedure recommended for both initial, and routine, assays.

To a 7 ml screw top tube, 50 µl of chloroform containing 1 µC of 2 µmol of [^{14}C]U-phosphatidylcholine (dioleoyl) is added. (The radioactive and non-radioactive phosphatidylcholines are obtainable from New England Nuclear Corp. and Sigma Corp., respectively.) The chloroform is evaporated to dryness under a stream of nitrogen. A volume of 100 µl 10 mM SDS is added to resuspend the phosphatidylcholine. The tube is vortexed briefly until a fine suspension forms. Afterward, 500 µl water, 100 µl 1 M Na acetate (pH 5.7) 100 µl 0.5 M $CaCl_2$ are added successively. Additional water and an appropriate amount of enzyme are added to make a final volume of 1 ml. The tube is incubated at 30°C for 10 min in a shaking water bath. The reaction is terminated with 2.5 ml chloroform–methanol (1:1; v/v). To the mixture, a solution of 1.1 ml 1 M KCl and 0.2 M H_3PO_4 is added. The tube is closed with a Teflon screw cap, and shaken horizontally for 15 min. After centrifugation in a table-top centrifuge, the lower chloroform phase is obtained. The upper phase is re-extracted with 1 ml chloroform. The two chloroform phases are combined, and evaporated to dryness under a stream of nitrogen. The residue is resuspended in 50 µl of chloroform, and spotted onto a silica gel TLC plate. The plate is developed in chloroform–methanol–acetic acid–water (85:15:10:3; v/v/v/v). In this system, phosphatidic acid moves twice as fast as phosphatidylcholine, and can be visualised by either reacting with iodine or autoradiography. The silica gel is scraped off and counted for radioactivity.

REFERENCES

Beevers, H. (1980). *In* "The Biochemistry of Plants" (P. K. Stumpf and E. E. Conn, eds), Vol. 4, pp. 117–130. Academic Press, New York.
Brockerhoff, H. and Jensen, R. G. (1974). "Lipolytic Enzymes". Academic Press, New York.
Christie, W. W. (1982). "Lipid Analysis," 2nd edn. Pergamon, Oxford.
Galliard, T. (1980). *In* "The Biochemistry of Plants" (P. K. Stumpf and E. E. Conn, eds), Vol. 4, pp. 85–116. Academic Press, New York.
Galliard, T. (1983). *In* "Lipids in Cereal Technology" (P. J. Barnes, ed.), pp. 111–147. Academic Press, New York.
Grossman, S., Oestreicher, G., and Singer, T. P. (1974). *Meth. Biochem. Anal.* **22**, 177–204.

Gupta, M. N. and Wold, F. (1980). *Lipids* **15**, 594–596.
Hasson, E. P. and Laties, G. G. (1976). *Plant Physiol.* **57**, 142–147.
Heller, M. (1978). *Adv. Lipid Res.* **16**, 267–326.
Huang, A. H. C. (1984). *In* "Lipases" (B. Borgström and H. L. Brockman, eds), pp. 419–442. Elsevier, Amsterdam.
Huang, A. H. C. (1985). *In* "Modern Methods of Plant Analyses" (J. F. Jackson and H. F. Linskens, eds), pp. 145–151. Springer, Berlin and New York.
Huang, A. H. C. (1987). *In* "The Biochemistry of Plants" (P. K. Stumpf and E. E. Conn, eds), Vol. 9, pp. 91–119. Academic Press, New York.
Kindl, H. (1987). *In* "The Biochemistry of Plants" (P. K. Stumpf and E. E. Conn, eds), Vol. 9, pp. 31–52. Academic Press, New York.
Lin, Y. H. and Huang, A. H. C. (1983). *Arch. Biochem. Biophys.* **225**, 360–369.
Lin, Y. H. and Huang, A. H. C. (1984). *Plant Physiol.* **76**, 719–722.
Maeshima, M. and Beevers, H. (1985). *Plant Physiol.* **79**, 489–493.
Taki, T. and Kanfer, J. N. (1981). *Methods Enzymol.* **71**, 746–750.
Vick, B. A. and Zimmerman, D. C. (1987). *In* "The Biochemistry of Plants" (P. K. Stumpf and E. E. Conn, eds), Vol. 9, pp. 53–90. Academic Press, New York.
Witt, W., Yelenosky, G. and Mayer, R. T. (1987). *Arch. Biochem. Biophys.* **259**, 164–170.

13 Enzymes of Phospholipid Synthesis

THOMAS. S. MOORE, Jr.
Louisiana State University, Baton Rouge, LA 70803-1705, USA

I.	Introduction	230
II.	General characteristics	232
III.	Category I enzymes	232
	A. Cholinephosphotransferase	232
	B. Ethanolaminephosphotransferase	233
	C. CTP cytidyltransferase	234
IV.	Category II enzymes	234
	A. Serine phosphatidyltransferase	234
	B. Inositol phosphatidyltransferase	235
	C. Glycerol-P phosphatidyltransferase	235
V.	Category III enzymes	236
	A. Phosphatidate phosphatase	236
	B. Phosphatidylglycerol-phosphate phosphatase	236
VI.	Category IV enzymes	236
	A. Ethanolamine exchange	236
	B. Serine exchange	236
	C. Inositol exchange	237
VII.	Category V enzymes	237
	A. PtdEtn methyltransferases	237
	B. Cardiolipin synthesis	238
	C. PtdSer decarboxylase	238
VIII.	Summary	238
	Acknowledgements	238
	References	238

I. INTRODUCTION

Synthesis of phospholipids requires the coordination of several metabolic pathways to provide the glycerol backbone, the acyl units and the headgroups. Pathways synthesising fatty acids, glycerol-P, serine, inositol, choline and ethanolamine are involved (Moore, 1982). Integration of activities of at least the plastids, mitochondria, cytoplasm, endoplasmic reticulum and Golgi apparatus is required (Kinney and Moore, 1987b). Yet, despite these complexities, the concentrations of phospholipids in cells or organelles of a particular type and stage of development are remarkably constant. Phospholipid composition appears to be a highly conserved character.

How is this control achieved? Synthesis and degradation rates of the enzymes, compartmentalisation, and more direct regulation of enzyme activity all appear to be involved. However, the full story is far from being understood and much research remains to be done. It is hoped that this chapter will provide a basis for more research in this area, and thereby a fuller understanding of this important field of metabolism.

It is obvious that this chapter cannot cover all enzymes related to phospholipid metabolism. Therefore, the methods described will be restricted to those enzymes involved in headgroup additions and modifications. The specific reactions to be discussed are outlined in the metabolic pathways of Fig.13.1, and classified in Table 13.1.

TABLE 13.1. Official and common names, and EC numbers for the enzymes described in this chapter. Reaction numbers refer to the reactions outlined in Fig. 13.1.

Reaction Name	EC number	Short name
1 Phosphatidate phosphohydrolase		Phosphatidate phosphatase
2 CDPcholine:1,2-diacylglycerol cholinephosphotransferase	2.7.8.2	Cholinephosphotransferase
3 CDPethanolamine:1,2-diacylglycerol ethanolaminephosphotransferase	2.7.8.1	Ethanolaminephosphotransferase
4 S-Adenosyl-L-methionine: phosphatidylethanolamine N-methyltransferase	2.1.1.17	PtdEtn methyltransferase
5 CTP: phosphatidate cytidylyltransferase	2.7.7.41	CTP cytidylyltransferase
6 CDPdiacylglycerol: L-serine 3-phosphatidyltransferase	2.7.8.8	PtdSer synthase
7 CDPdiacylglycerol: myo-inositol 3-phosphatidyltransferase	2.7.8.11	PtdIns synthase
8 CDPdiacylglycerol: sn-glycerol-3-phosphate 3-phosphatidyltransferase	2.7.8.5	PtdGro-P synthase
9 Phosphatidylglycerophosphate phosphohydrolase	3.1.3.27	PtdGro-P phosphatase
10 Not described in plants		
11 Phosphatidylserine carboxylyase	4.1.1.65	PtdSer decarboxylase
12 Phosphatidyl-X: ethanolamine phosphatidyltransferase		Ethanolamine exchange enzyme
13 Phosphatidyl-X: L-serine phosphatidyltransferase		Serine exchange enzyme
14 Phosphatidylinositol: myo-inositol phosphatidyltransferase		Inositol exchange enzyme

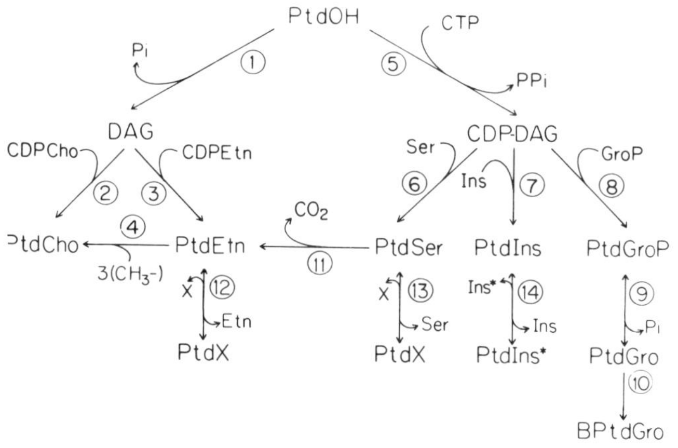

13. ENZYMES OF PHOSPHOLIPID SYNTHESIS

FIG. 13.1. Pathways of phospholipid headgroup additions and modifications in plants. The numbers are correlated with the specific enzymes involved in Table 13.1 and with some general *in vitro* requirements of the reactions in Table 13.2. The asterisks indicate a labelled *myo*-inositol. Abbreviations: BPtdGro, *bis*-phosphatidylglycerol; CDP, cytidine diphosphate; CDPCho, cytidine diphosphocholine; CDP-DAG, CDP diacylglycerol; CDPEtn, CDP ethanolamine; CTP, cytidine triphosphate; DAG, diacylglycerol; Etn, ethanolamine; GroP, glycerol phosphate; Ins, inositol; PPi, pyrophosphate; PtdCho, phosphatidyl choline; PtdEtn, phosphatidylethanolamine; PtdGro, phosphatidylglycerol; PtdGroP, phosphatidylglycerol-phosphate; PtdIns, phosphatidylinositol; PtdOH, phosphatidic acid; PtdSer, phosphatidylserine; PtdX, phosphatidyl unit with unknown headgroup.

TABLE 13.2. General requirements of the enzymes catalysing addition or modification of phospholipid headgroups in castor bean endosperm. The cations and their concentrations are those which give maximum *in vitro* enzyme activities. The reaction numbers correspond to the numbers assigned in Fig. 13.1 and the enzymes listed in Table 13.1.

Reaction	Organelle	Cation	(mM)	pH	Substrate K_m (μM) Aqueous	Lipid
1	ER	?		?	?	?
2	ER	Mg^{2+}	(3.0)	7.5	10	?
	Mitochondria	Mg^{2+}	(10.0)	7.5	8	?
3	ER	Mg^{2+}	(3.0)	6.5	8	?
4	ER	None		9.0	31	?
	Mitochondria	None		?	?	?
5	ER	Mn^{2+}	(7.5)	6.5	17	94
	Mitochondria	Mn^{2+}	(1.5)		67	132
6	ER	Mn^{2+}	(2.0)	8.0	35	?
7	ER	Mn^{2+}	(1.5)	8.5	300	1350
8/9	ER	Mn^{2+}	(5.0)	7.3	50	3
	Mitochondria	Mn^{2+}	(5.0)	7.3	50	2
10	?	?		?	?	?
11	ER	?		?	?	?
12	ER	Ca^{2+}	(2.0)	7.8	5	?
13	ER	Ca^{2+}	(2.0)	7.8	?	?
14	ER	Mn^{2+}	(15.0)	8.0	26	?

Plant phospholipid synthesising enzymes have been discussed in several review articles (Kates and Marshall, 1975; Mudd, 1980; Moore, 1982; Kinney and Moore, 1987b), and the techniques for assaying some are contained in a recent review (Moore, 1987). Methods for separating and identifying the products and some precursors of these phospholipids are reviewed in a separate volume of this series dealing with lipids (edited by J. L. Harwood).

II. GENERAL CHARACTERISTICS

A summary of the general requirements of the enzymes for headgroup addition and modification in castor bean endosperm is contained in Table 13.2. The reactions can be subdivided into three categories according to their characteristics. These are (with reaction numbers from Fig. 13.1 given in parentheses):

I. Utilising a water-soluble CDP derivative and generally requiring Mg^{2+} (Reactions 2,3,5).
II. Utilising a lipid-soluble CDP derivative and frequently preferring Mn^{2+} (Reactions 6,7,8).
III. Phosphatases (Reactions 1,9).
IV. Headgroup exchangers requiring Ca^{2+} (Reactions 12,13) or Mn^{2+} (Reaction 14).
V. Others (Reactions 4,10,11).

The reactions of Categories I and II could be considered to be the fundamental headgroup addition enzymes, while the other three categories (III, IV and V) would be headgroup modifiers. These reactions are discussed below in order of the above five categories.

III. CATEGORY I ENZYMES

Two of these enzymes are thought to be involved in synthesis of the two major phospholipids of most cellular membranes, phosphatidylcholine (PtdCho) and phosphatidylethanolamine (PtdEtn). Their activities increase early in organ development (Johnson and Kende, 1971; Kinney and Moore, 1987a) and presumably are needed to provide membrane for newly synthesised organelles (Kagawa *et al.*, 1973; Moore and Troyer, 1983). The cholinephosphotransferase activity has been examined in more detail than has ethanolamine incorporation.

A. Cholinephosphotransferase

This enzyme is responsible for the final step in what appears to be the major pathway leading to PtdCho synthesis. It often is the most active phospholipid synthesising enzyme (headgroup addition) in cells and is readily measured *in vitro*. It is quite active in fractions containing endoplasmic reticulum (ER) membranes (Devor and Mudd, 1971; Moore, 1976), but also occurs with high activity in Golgi membranes (Montague and

Ray, 1977; Sauer and Robinson, 1985). In addition, it has been found in the outer mitochondrial membrane, in which case it requires a higher concentration of Mg^{2+} than the ER activity (Sparace and Moore, 1981).

The assay conditions are straightforward and general requirements of the enzyme from both ER (Moore, 1976) and mitochondria (Sparace and Moore, 1981) have been reported. The ER activity of castor bean was measured at 30°C in the presence of 50 mM Tris-HCl buffer (pH 7.5), 1 mM dithiothreitol, 2–3 mM $MgCl_2$ and 0.2 mM CDP[1,2-^{14}C]choline (5 mCi mmol^{-1}). The reaction is started by the addition of the CDPcholine and may be terminated by addition of 3.3 ml of chloroform–methanol–H_2O (1:2:0.3; v/v). The radiolabelled PtdCho is extracted by adding 1 ml of chloroform to the above mixture, followed by 2–3 ml of 1 M KCl and mixing vigorously. The upper aqueous phase is removed by inserting a Pasteur pipette to just above the interface and carefully applying suction (aspiration may be used). The chloroform layer is washed two to three times with 1 M KCl. Following this procedure, the chloroform layer may be concentrated for chromatography or measurement of radioactivity.

The K_ms reported for cytidinediphosphocholine (CDPcholine) using plant enzyme are around 10 μM (Devor and Mudd, 1971; Moore, 1976). The K_ms for diacylglycerol (DAG) have been more difficult to measure, since the enzyme is tightly bound to the membranes where it is found. The use of indirect methods (measuring the ester fatty acid composition of PtdCho produced from membrane diacylglycerol obtained by phospholipase C treatment of the membranes) has led to the conclusion that in potato tuber the cholinephosphotransferase is non-specific (Jolliot et al., 1982), but in pea leaf microsomes it prefers the 1–16:0, 2–18:1 substrate (Demandre et al., 1986). Assays with a partially solubilised (with 20 mM β-octylglucopyranoside) enzyme from castor bean endosperm resulted in a preference similar to the latter case (Moore, 1986). The optimal pH is in the range 7.5–8.0 and Mg^{2+} is preferred by the castor bean endosperm, but Mn^{2+} has been reported to be equal to Mg^{2+} in stimulating the enzyme from spinach; Ca^{2+} inhibits the activity (reviewed in Moore, 1982, 1987). Assays for mitochondrial activity may require an increase in the Mg^{2+} concentration to 10 mM. Dithiothreitol and sulphydryl reagents have had variable effects (Moore, 1982, 1987).

B. Ethanolaminephosphotransferase

This reaction is similar to that catalysed by cholinephosphotransferase, and initially the two were thought to be catalysed by the same enzyme. Although this is unlikely (Moore, 1982), the two enzymes have not been separated from plants as they have from mammalian cells (Bell and Coleman, 1980).

The reaction has been performed at 37°C (Sparace et al., 1981; 30°C should be fine) and contains in a volume of 0.5 ml: 10 mM 2-(-N-morpholino)ethanesulphonic acid (Mes) buffer (pH 6.5), 3 mM $MgCl_2$, 1 mM dithiothreitol and 7.5 μM CDP-[1,2-^{14}C]-ethanolamine (3.0 mCi mmol^{-1}) and enzyme. The reaction is stopped and the product extracted as with cholinephosphotransferase (Section III.A).

Reported K_ms for CDPcholine are 20 μM for spinach microsomes (Macher and Mudd, 1974) and 6 μM for castor bean endosperm (Sparace et al., 1981). CDPcholine is highly competitive in both cases. The castor bean activity strongly prefers Mg^{2+}, while the spinach enzyme works best with Mn^{2+}; the pH optima range from 6.5 to 8 (Macher and Mudd, 1974; Sparace et al., 1981).

C. CTP Cytidyltransferase

This enzyme catalyses a branch point reaction which ultimately leads to formation of the four phospholipids present in the lowest concentration in most plant membranes, phosphatidylserine (PtdSer), phosphatidylinositol (PtdIns), phosphatidylglycerol (PtdGro) and cardiolipin. The one exception is in the chloroplast, where PtdGro is the predominant phospholipid (Mudd and deZacks, 1981). Obviously this enzyme, which has the potential to regulate the availability of substrate (CDPdiacylglycerol; CDP-DAG) to these reactions, is highly significant. However, the enzyme has received little attention.

The reaction has been assayed in ER, mitochondrial inner, and plastid envelope fractions (Moore, 1984). The assay (Kleppinger-Sparace and Moore, 1985) is conducted (with castor bean endosperm ER) at 30°C in a volume of 0.5 ml which contains: 10 mM Mes (pH 6.5), 7.5 mM $MnCl_2$, 2.5 mM phosphatidate (derived from egg lipids), 0.12 mM [^{3}H]CTP (32.2 mCi mmol^{-1}) and enzyme. The reaction is terminated and radiolabelled CDP-DAG extracted as described in Section III.A.

The K_m for CTP in castor bean enzyme was estimated to be 17 μM (Kleppinger-Sparace and Moore, 1985), but the saturation kinetics were complex. This value is somewhat below previously reported values (Bahl et al., 1970). Mn^{2+} is preferred over Mg^{2+} and a slightly acid pH is optimal.

IV. CATEGORY II ENZYMES

The enzymes of Category II are similar to each other in requirements. All utilise CDP-DAG and an unbound molecule for the headgroup. All three enzymes generally demonstrate a preference for Mn^{2+}.

A. Serine Phosphatidyltransferase

This activity was first reported in plants by Marshall and Kates (1974). These researchers assayed the activity in a 1200–100 000 × g pellet from spinach leaves, in a reaction mixture which contained in a final volume of 1.0 ml: 7.5 μM [U-^{14}C]serine (2.1 × 10^6 dpm), 0.13 mM CDPdiglyceride, 50 mM sodium phosphate buffer (pH 8.0), 1.5 mM β-mercaptoethanol, and enzyme.

We were initially unable to obtain this activity in castor bean endosperm and argued that all PtdSer synthesis in these cells occurred by the exchange reaction (see Section VI.C). More recently, however, we have re-examined this situation and found that the activity does occur. The discrepancy appears to have arisen from the fact that Mn^{2+} is not included in the assay of Marshall and Kates, and is absolutely required by the castor bean endosperm. The standard conditions we use for this assay are, in a final volume of 0.5 ml (Feild and Moore, 1988): 50 mM Tricine (pH 8.0), 0.25 mM L-[^{14}C]-L-serine (1 mCi mmol^{-1}), 1.5 mM 2-mercaptoethanol, 2 mM $MnCl_2$ and 0.8 mM CDP-DAG. The activity is present in fractions containing the endoplasmic reticulum.

The occurrence of this enzyme in plants has some evolutionary significance. The activity is thought not to play a role in animals (Bell and Coleman, 1980), but has been described from bacteria (Patterson and Lennarz, 1971).

B. Inositol Phosphatidyltransferase

This enzyme was first described from plants by Sumida and Mudd (1970) and has subsequently been studied in several plant species. The stimulation of its activity by low concentrations of some detergents has led to solubilisation of the enzyme from soybean cotyledons (Robinson and Carman, 1982), but rigorous purification has not been pursued. Perhaps recent interest in PtdIns and its phosphorylated derivatives will lead to a change in this situation.

The activity is readily measured, although it may be low and radiolabelled *myo*-inositol is still somewhat expensive. The assay may be conducted at 30–37°C for 30 min in a final volume of 0.5 ml. The assay mixture contains: 50 mM Tris-HCl (pH 8.5), 1.5 mM $MnCl_2$, and a 1 mM suspension of CDPdipalmitoylglyceride. Enzyme is added and the reaction started by the addition of 1.2 mM *myo*-[2-^{3}H]inositol (1.79 mCi $mmol^{-1}$). The reaction is terminated and the radiolabelled PtdIns extracted by the procedure described for cholinephosphotransferase (Section III.A).

The *in vitro* activity has exhibited apparent K_ms of 0.045–0.3 mM for CDP-DAG and from 0.027–1.35 mM for *myo*-inositol (Sumida and Mudd, 1970; Sexton and Moore, 1978). Mn^{2+} is strongly preferred to fulfil a cation requirement. The activity may be stimulated by low concentrations of some detergents, but higher detergent and CDP-DAG concentrations are inhibitory, presumably by disruption of the membranes (Sexton and Moore, 1978).

C. Glycerol-P Phosphatidyltransferase

This enzyme has not been examined independently of Reaction 8 (see Table 13.1), the phosphatidylglycerol phosphatase. Under the conditions of the reaction, the phosphatase seems to react immediately with the substrate, and so the product of the combined reactions is primarily PtdGro (although a small quantity of the phosphorylated substrate may be found in some situations: Marshall and Kates, 1972). Therefore, the phosphatidyltransferase appears to be rate-limiting and so the measured requirements reflect the transferase rather than the phosphatase. The reactions occur in both the endoplasmic reticulum and mitochondria (Douce and Dupont, 1969; Marshall and Kates, 1972; Moore, 1974) and also in the chloroplast (Mudd and deZacks, 1981). It is one of the few phospholipid synthesis reactions stimulated by the addition of detergent (Marshall and Kates, 1972; Moore, 1974), but detergents inhibit the chloroplast activity (Mudd and deZacks, 1981).

The assay (Moore, 1974) is performed at 30°C in a final volume of 0.25 ml containing: 0.4 mM Tris-HCl (pH 7.3), 8 mM $MnCl_2$, 0.15% (w/w) Triton X-100, CDP-DAG (0.8 mM), 0.5 mM *sn*-[$U^{14}C$]glycerophosphate (1.9 mCi $mmol^{-1}$), and enzyme. The reaction may be stopped by the addition of 1 ml methanol. The assay tubes are then incubated on ice, 3 ml of chloroform added and mixed, and the emulsion broken by centrifugation. The aqueous layer is removed and the chloroform washed twice with 1 M KCl.

Michaelis constants for glycerol-P in the ER and mitochondria have been reported to be 50 and 250 μM (Marshall and Kates, 1972; Moore, 1974), and those for CDP-DAG about 2–3 μM for castor bean (Moore, 1974) and 40 μM from spinach (Marshall and Kates, 1972). Mn^{2+} stimulates best at 5 mM while Mg^{2+} is a poor substitute. Triton X-100 stimulates up to about 0.075% (w/w) (Moore, 1974).

The chloroplast activity has been measured at 25°C with illumination using an organelle suspension in 0.3 M Sorbitol, 33 mM Tricine-NaOH (pH 7.9), 2 mM $MgCl_2$, 2 mM $MnCl_2$, 1.5 mM ATP, 0.15 mM sodium acetate, 10 mM $KHCO_3$, 0.2 mM K_2HPO_4, 0.2 mM CoA, 0.5 mM dithiothreitol and 0.4 mM DL-glycerol-3-P, including 523 000 dpm of L-[U-^{14}C]glycerol 3-P in a final volume of 1 ml (Mudd and deZacks, 1981; Sparace and Mudd, 1982a,b). This complex mixture allows for synthesis of PtdGro from newly formed lipid substrate as well as added headgroup precursor.

V. CATEGORY III ENZYMES

A. Phosphatidate Phosphatase

This central enzyme has proven difficult to assay accurately *in vitro*, and thus is poorly understood. A primary problem is provision of the lipid substrate to the enzyme. Some success has been achieved by treating membranes with phospholipase D, followed by testing for conversion of the resultant phosphatidic acid (PtdOH) to diacylglycerol (DAG) (Block *et al.*, 1983).

B. Phosphatidylglycerol-phosphate Phosphatase

Little is known about this enzyme in plants. The enzyme was reported to be inhibited with $HgCl_2$ by Douce and Dupont (1969), but Marshall and Kates (1972) did not find this at concentrations up to 10 mM with enzyme from spinach and obtained only 21% inhibition with *p*-chloromercuribenzoate.

VI. CATEGORY IV ENZYMES

A. Ethanolamine Exchange

An ethanolamine Ca^{2+}-stimulated exchange activity was first explored in plants by Vandor and Richardson (1968) and has been re-examined recently by Shin and Moore (1988). This latter study was stimulated by data which suggested that PtdEtn formed by this reaction could be utilised as the substrate for methylation to provide PtdCho (Kinney and Moore, 1987a). The reaction is typical of exchange reactions in requiring Ca^{2+}, and has been investigated extensively in mammalian tissues (Bell and Coleman, 1980). The *in situ* substrate is unknown.

The reaction may be assayed by measuring incorporation of radiolabelled ethanolamine into a chloroform-soluble product (Shin and Moore, 1988). The assay is performed in a final volume of 0.5 ml containing: 200 mM Hepes buffer (pH 7.8), 2 mM $CaCl_2$, enzyme, and 30 μM [1,2-^{14}C]-ethanolamine-HCl (mCi mmol^{-1}). The reaction is started by the addition of the ethanolamine, followed by termination after one hour and product extraction as described for the cholinephosphotransferase assay (Section III.A).

B. Serine Exchange

The serine exchange reaction was first reported in 1975 and was thought to represent the sole means of synthesis of PtdSer in castor bean endosperm (Moore, 1975). This is not the case, however, since the reaction utilising CDP-DAG has now been found in this tissue (Feild and Moore, 1988). Thus, unlike mammalian and bacterial organisms, both reactions occur in plants. The reason for having both remains to be determined.

The enzyme activity is measured by reacting radiolabelled serine with an unknown lipid acceptor (Moore, 1975; the acceptor is endogenous to the added membrane) in a final volume of 0.5 ml. The assay contains: 200 mM Hepes buffer (pH 7.8), 10 mM $CaCl_2$, enzyme and L-[3-^{14}C]serine (2.5 mCi mmol^{-1}). The reaction is maintained at 30°C for 60 min, followed by stopping and extracting as described in Section III.A.

The K_m of this reaction for serine is about 20 μM. Ethanolamine inhibits the reaction, but not competitively (Moore, 1975). The pH optimum shifts to a lower pH with increases in Ca^{2+} concentration.

C. Inositol Exchange

This enzyme appears to catalyse a futile reaction, and its role is not understood. It was first described from plants by Sexton and Moore (1981) and more recently studied by Morré and coworkers (Morré et al., 1984; Sandelius and Morré, 1987). This enzyme is unique from the other two enzymes in this category in that it requires Mn^{2+} rather than Ca^{2+}, and the latter cation will not substitute.

The assay is designed to measure the quantity of radiolabelled myo-inositol incorporated into a chloroform-soluble compound which has been identified as PtdIns (Sexton and Moore, 1981). This reaction is performed in the absence of CDP-DAG in order to eliminate activity of the transferase, since conditions are otherwise similar. The assay may be conducted by incubating a mixture of 50 mM Hepes buffer (pH 8.0), 15 mM $MnCl_2$, enzyme and myo-[2-^{3}H]inositol (1.79 mCi mmol^{-1}) for 1 h at 37°C. The mixture without the inositol is incubated for 5 min to equilibrate the mixture to the temperature, and the reaction started by addition of the label. The reaction is stopped and extracted as in the cholinephosphotransferase assay (Section III.A).

One central problem for such reactions is determination of the identity of the lipid substrate. Sandelius and Morré (1987) have demonstrated by pre-labelling experiments that newly synthesised PtdIns will serve as a substrate, and so the reaction can indeed behave as diagrammed (Fig. 13.1, Reaction 14). In these experiments, they incubated the enzyme under assay conditions, then washed out excess precursor and reincubated with unlabelled myo-inositol. The loss of radiolabel from the PtdIns fraction was followed.

VII. CATEGORY V ENZYMES

A. PtdEtn Methyltransferases

The final product of this set of three methyl additions is PtdCho. The methylation pathway, or at least parts of it, has been demonstrated in several plant species (Moore, 1987). The reactions have been characterised as a total methyl incorporation, and it remains uncertain as to how many enzymes are involved (Moore, 1976) or even what the role of the pathway may be (Mudd and Datko, 1986).

The assay measures the incorporation of methyl groups from radiolabelled S-adenosyl-L-methionine into a chloroform-soluble product. The assay is conducted at 37°C for 30 min in a final volume of 0.5 ml containing: 50 mM Tris-HCl buffer (pH 9.0), 1 mM PtdEtn, 1.0 mM S-adenosyl-L-[methyl-^{14}C]methionine (1 mCi mmol^{-1}) and enzyme (Moore, 1976). The reaction is started by addition of the enzyme and stopped and extracted as described in Section III.A.

Both spinach (Marshall and Kates, 1974) and castor bean (Moore, 1976) enzymes have pH optima for the total reaction of 8–9. An apparent K_m of 31 µM has been determined (Moore, 1976) and added PtdEtn stimulates the castor bean about 33%, but has no effect on the spinach reaction. The single and doubly methylated intermediates stimulate both reactions. The final product is PtdCho, with varying amounts of mono- and dimethylated intermediates, depending on the tissue utilised (Moore, 1976).

B. Cardiolipin Synthesis

Despite the fact that cardiolipin is abundant in plant mitochondria (Harwood, 1980) and small amounts may be synthesised under conditions leading to PtdGro, synthesis, this enzyme has not been characterised for plant cells.

C. PtdSer Decarboxylase

Experiments with both *in vitro* and *in vivo* (Kates and Marshall, 1975; Moore, 1975; Kinney and Moore, 1987a) measurements have provided evidence that this reaction exists in plants. However, it has not been directly measured or characterised.

VIII. SUMMARY

Most of the enzymes engaged in headgroup additions and modifications for phospholipids probably are now known and some initial characterisation exists for most of them. Yet, we still understand little about details of the syntheses, incorporation into the membranes, and regulation of the activity of any of these enzymes in any biological system, particularly plants. Achieving an understanding of these areas will be difficult, but remains an exciting challenge for the future, perhaps not so distant.

ACKNOWLEDGEMENTS

Preparation of this manuscript and much of the research by the author were supported by grants from the US National Science Foundation, most recently DCB-8703739.

REFERENCES

Bahl, J., Guillot-Salomon, T. and Douce, R. (1970). *Physiol. Veg.* **8**, 55–74.
Bell, R. M. and Coleman, R. A. (1980). *Ann. Rev. Biochem.* **49**, 459–487.
Block, M. A., Dorne, A.-J., Joyard, J. and Douce, R. (1983). *FEBS Lett.* **169**, 111–115.

Demandre, C., Justin, A. M., Naguyen, X. V., Gawer, M., Tremolieres, A. and Mazliak, P. (1986). *In* "The Metabolism, Structure, and Function of Plant Lipids" (P. K. Stumpf, J. B. Mudd and W. D. Nes, eds), pp. 273–282. Plenum Press, New York.
Devor, K. A. and Mudd, J. B. (1971). *J. Lipid Res.* **12**, 403–411.
Douce, R. and Dupont, J. (1969). *C. R. Acad. Sci. Ser. D.* **268**, 1657–1660.
Feild, M. and Moore, T. S. (1988). *Plant Physiol.* **86**, 83S.
Harwood, J. (1980). *In* "The Biochemistry of Plants. A Comprehensive Treatise, Lipids: Structure and Function" (P. K. Stumpf, ed.), Vol. 4, pp. 1–55. Academic Press, New York.
Johnson, K. D. and Kende, H. (1971). *Proc. Natl. Acad. Sci. USA* **68**, 2674–2677.
Jolliot, A., Justin, A. M., Bimont, E. and Mazliak, P. (1982). *Plant Physiol.* **70**, 206–210.
Kagawa, T., Lord, J. M. and Beevers, H. (1973). *Plant Physiol.* **51**, 61–65.
Kates, M. and Marshall, M. O. (1975). *In* "Recent Advances in the Chemistry and Biochemistry of Plant Lipids" (T. Galliard and E. I. Mercer, eds), pp. 115–159. Academic Press, New York.
Kinney, A. J. and Moore, T. S. (1987a). *Plant Physiol.* **84**, 78–81.
Kinney, A. J. and Moore, T. S. (1987b). *In* "Models in Plant Physiology and Biochemistry" (D. Newman and K. G. Wilson, eds), pp. 131–133. CRC Press, Boca Raton.
Kleppinger-Sparace, K. F. and Moore, T. S. (1985). *Plant Physiol.* **77**, 12–15.
Macher, B. A. and Mudd, J. B. (1974). *Plant Physiol.* **53**, 171–175.
Marshall, M. O. and Kates, M. (1972). *Biochim. Biophys. Acta* **260**, 558–570.
Marshall, M. O. and Kates, M. (1974). *Can. J. Biochem.* **52**, 469–482.
Montague, M. J. and Ray, P. M. (1977). *Plant Physiol.* **59**, 225–230.
Moore, T. S. (1974). *Plant Physiol.* **54**, 164–168.
Moore, T. S. (1975). *Plant Physiol.* **56**, 177–180.
Moore, T. S. (1976). *Plant Physiol.* **57**, 383–386.
Moore, T. S. (1982). *Ann. Rev. Plant Physiol.* **33**, 235–259.
Moore, T. S. (1984). *In* "Structure, Function and Metabolism of Plant Lipids" (P.-A. Siegenthaler and W. Eichenberger, eds), pp. 83–91. Elsevier, New York.
Moore, T. S. (1986). *In* "The Metabolism, Structure, and Function of Plant Lipids" (P. K. Stumpf, J. B. Mudd and W. D. Nes, eds), pp. 265–272. Plenum Press, New York.
Moore, T. S. (1987). *Methods Enzymol.* **148**, 585–596.
Moore, T. S. and Troyer, G. D. (1983). *In* "Biosynthesis and Function of Plant Lipids," pp. 16–27. American Society for Plant Physiology, Rockville, MD.
Morré, D. J., Gripshover, B., Monroe, A. and Morré, J. T. (1984). *J. Biol. Chem.* **259**, 15 364–15 368.
Mudd, J. B. (1980). *In* "The Biochemistry of Plants, Lipids: Structure and Function" (P. K. Stumpf, ed.), Vol. 4, pp. 250–282. Academic Press, New York.
Mudd, J. B. and de Zacks, R. (1981). *Arch. Biochem. Biophys.* **209**, 584–591.
Mudd, S. H. and Datko, A. H. (1986). *Plant Physiol.* **82**, 126–135.
Patterson, P. H. and Lennarz, W. J. (1971). *J. Biol. Chem.* **246**, 1062–1072.
Robinson, M. L. and Carman, G. M. (1982). *Plant Physiol.* **69**, 146–149.
Sandelius, A. S. and Morré, D. J. (1987). *Plant Physiol.* **84**, 1022–1027.
Sauer, A. and Robinson, D. G. (1985). *J. Exp. Bot.* **36**, 1257–1266.
Sexton, J. C. and Moore, T. S. (1978). *Plant Physiol.* **62**, 978–980.
Sexton, J. C. and Moore, T. S. (1981). *Plant Physiol.* **68**, 18–22.
Shin, S.-H. and Moore, T. S. (1988). *Plant Physiol.* **86**, 83S.
Sparace, S. A. and Moore, T. S. (1981). *Plant Physiol.* **67**, 261–265.
Sparace, S. A. and Mudd, J. B. (1982a). *Plant Physiol.* **70**, 1260–1264.
Sparace, S. A. and Mudd, J. B. (1982b). *In* "Biochemistry and Metabolism of Plant Lipids" (J. F. G. M. Wintermans and P. J. C. Kuiper, eds), pp. 111–119. Elsevier, Amsterdam.
Sparace, S. A., Wagner, L. K. and Moore, T. S. (1981). *Plant Physiol.* **67**, 922–925.
Sumida, S. and Mudd, J. B. (1970). *Plant Physiol.* **45**, 712–718.
Vandor, S. L. and Richardson, K. E. (1968). *Can. J. Biochem.* **46**, 1309–1315.

14 Nitrate Reductase and Nitrite Reductase

JOHN L. WRAY[1] and ROGER J. FIDO[2]

[1]*Plant Molecular Genetics Unit, Sir Harold Mitchell Building, University of St. Andrews, St. Andrews, Fife KY16 9TH, UK*

[2]*Department of Agricultural Sciences, University of Bristol, AFRC Institute of Arable Crops Research, Long Ashton Research Station, Long Ashton, Bristol BS18 9AF, UK*

I.	Nitrate Reductase	241
	A. Historical	241
	B. Introduction	242
	C. Nitrite reductase purification	244
	D. Immunochemistry of nitrate reductase	247
II.	Nitrite reductase	249
	A. Historical	249
	B. Introduction	250
	C. Nitrite reductase purification	251
	D. Immunochemistry of nitrite reductase	253
	Acknowledgements	253
	References	253

I. NITRATE REDUCTASE

A. Historical

Evans and Nason (1953) first reported the occurrence of nitrate reductase in the roots and leaves of higher plants and its partial purification and characterisation from the leaves of soybean. Since that time nitrate reductase (NR) has been extensively studied and its properties have been the subject of a number of reviews (Hageman and Reed, 1980; Hewitt and Notton, 1980; Guerrero et al., 1981; Campbell and Smarrelli, 1986; Rajasekhar and Oelmüller, 1987; Campbell, 1988).

B. Introduction

In higher plants the leaves are the major site of nitrate assimilation (Hageman, 1979), and nitrate reduction, nitrite reduction, and the subsequent assimilation of ammonium into glutamate are essentially photosynthetic reactions (Abrol et al., 1983). However, roots have also been shown to contribute significantly to the total nitrate-reducing capacity of some plants (Oaks and Hirel, 1985).

The NR enzymes catalyse the first enzymatic, and possibly rate-controlling (Beevers and Hageman, 1969), step of the nitrate assimilation pathway, involving the two-electron reduction of nitrate to nitrite. Almost all plants examined possess the NADH-linked enzyme (EC 1.6.6.1) which has a pH optimum of 7.5. A second, bispecific, NR (EC 1.6.6.2) which uses NADH or NADPH as electron donor has been described for some plant species, for example, rice (Shen et al., 1976), maize (Campbell, 1978) and soybean (Orihuel-Iranzo and Campbell, 1980). More recently a third type of NR has been identified in soybean. This is a constitutive NADH-NR (EC number not yet assigned) with a pH optimum of 6.5 (Streit et al., 1987). NADH-NR (EC 1.6.6.1), which is substrate inducible and undergoes rapid turnover *in vivo*, is the best characterised form of NR and is discussed below.

1. Structure and function of NADH:nitrate reductase

NADH:nitrate reductase (EC 1.6.6.1) is a large homodimer with one of each of the prosthetic groups FAD, haem (cytochrome b_{557}) and molybdenum cofactor per 100–115 kDa subunit (Redinbaugh and Campbell, 1985). The molybdenum cofactor, a complex between molybdenum and a phosphorylated pterin, molybdopterin (Kramer et al., 1987), is common to all molybdoenzymes with the exception of nitrogenase. Electron flow from NADH is via the flavin and haem redox centres to the molybdenum cofactor redox centre, which acts as the terminal electron donor to nitrate (reviewed in Hewitt and Notton, 1980).

In addition to the overall physiological reaction (NADH-dependent nitrate reduction) the enzyme carries so-called partial activities which are considered to be catalysed by specific regions (domains) of the enzyme molecule. These partial activities are of two broad types: those (NADH:dehydrogenase) activities which are independent of the nitrate binding site [reduction by NADH of ferricyanide (NADH:ferricyanide reductase), cytochrome c (NADH:cytochrome c reductase) or dichlorophenol indophenol (NADH:dichlorophenol indophenol reductase)] and, secondly, those activities which are independent of the NADH binding site and where reduction of nitrate is mediated by electrons donated from dithionite-reduced flavin nucleotide ($FADH_2$: or $FMNH_2$: nitrate reductase) via the haem redox centre or by electrons donated from dithionite-reduced methyl viologen (MV:nitrate reductase) or bromophenol blue (BPB:nitrate reductase) via the molybdenum redox centre.

The partial functions are especially useful in the characterisation of nitrate reductase mutants (see for example Müller and Mendel, 1989) and in the identification and study of the functional domains of the enzyme in association with limited proteolysis studies (Solomonson et al., 1986; Poulle et al., 1987).

With respect to the latter, limited proteolysis of the tetrameric 375 kDa *Chlorella* NR demonstrated the presence of a small (30 kDa) monomeric fragment which contained FAD and the NADH-binding site and possessed NADH-ferricyanide activity and a large (280 kDa) homotetrameric fragment which contained haem and molybdenum but no FAD and possessed reduced methylviologen nitrate reductase activity. These results are consistent with a structure–function model of NR in which FAD/NADH-binding domains are exposed on the surface of the molecule, a protease-sensitive hinge region connects the nitrate-reducing and NADH-dehydrogenase moieties and the quaternary structure is maintained via association sites on the haem/molybdenum domain (Solomonson et al., 1986).

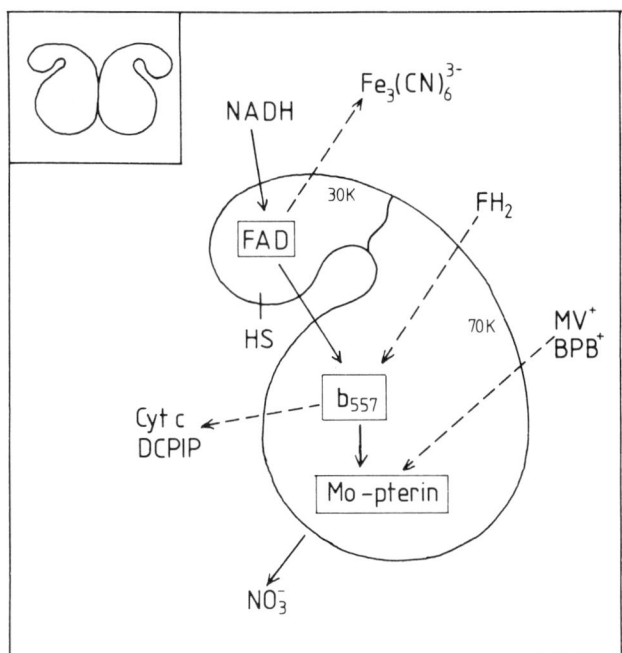

FIG. 14.1. Structure–function model of higher plant nitrate reductase by analogy with the *Chlorella vulgaris* enzyme. Features include two identical subunits, each composed of domains connected by protease-sensitive regions. Solid arrows denote physiological electron transfer. Dashed arrows denote artificial electron transfer (partial reactions).

Equivalent fragments have been cleaved from the dimeric spinach NR by limited proteolysis (Notton et al., 1989) and a structure–function model of the higher plant NR is shown in Fig. 14.1. Some properties of spinach NADH-NR are shown in Table 14.1.

TABLE 14.1. Properties of the NADH-nitrate reductase of spinach *Spinacea oleracea* L.

Molecular weight (kDa)	210–230
Subunit number	2
$S_{20,w}$ (s)	8.1
Stokes radius (nm)	6.0
Amino acid residues	ND
pH optimum	7.5
Specific activity	103
Molecular activity (μmol 2e min^{-1} nmol^{-1} haem^{-1})	
NADH-NR	9
NADH-FR	224
NADH-CR	83
FADH$_2$-NR	12
MV-NR	32
K_m NO$_3^-$ (μM)	13
K_m NADH (μM)	7
Prosthetic groups (mol mol^{-1} enzyme)	
FAD	2
haem	2
Mo	2
Midpoint potential (mV)	
FAD	ND
haem	-60
Mo	Mo(vi)/Mo(v) -8
	Mo(v)/Mo(iv) -42
Spectral properties	
Absorption maximum (nm)	
Oxidised form	280, 412 (Soret)
Reduced form	423, 525(β), 557(α)
Extinction coefficients (mM^{-1} cm^{-1})	
412 (ox) nm	127
423, 525, 557 (red.) nm	172, 17.5, 29

Data compiled from: Barber *et al.* (1987); M. J. Barber and B. A. Notton, unpublished; Fido and Notton (1984); Fido *et al.* (1979).
Note that the prosthetic group stoichiometry has not been determined for the spinach enzyme but the values quoted are compatible with other higher plant nitrate reductases.
ND, not determined.

C. Nitrate Reductase Purification

The many difficulties encountered when extracting and purifying higher plant NR has slowed progress in examination of the structure and function of the enzyme. The low abundance of the enzyme in plant tissues (<5 mg kg^{-1} fresh weight) combined with susceptibility to proteolysis, as well as the loss of FAD and molybdenum components during extraction/purification, all lead to poor recovery of active enzyme.

1. Extraction and in vitro stability

Higher plant NR is inherently unstable *in vivo* and *in vitro*, and the complex nature of this instability is reflected in the many different classes of compounds added to extraction buffer. It is most probable that many types of inactivating mechanism simultaneously affect the enzyme.

The *in vitro* stability of barley NR was reported to be a function of seedling age, growth temperature and extraction buffer, NR being most stable when extracted from young seedlings (Brown *et al.*, 1981; Kuo *et al.*, 1982a). Multiple additions of protectants are used, especially when extracting NR from cereal species, and include EDTA, FAD and sulphydryl compounds such as cysteine and DTT which prevent the oxidation of essential sulphydryl groups on the enzyme. Exogenous protein such as casein or bovine serum albumin (BSA) is also routinely added to stabilise NR. Proteinase inhibitors, including leupeptin, pepstatin and PMSF, combined with high pH buffers (pH 8.2–8.5) have been used to inhibit proteolysis during purification procedures (Wray and Kirk, 1981; Kuo *et al.*, 1982a; Campbell and Wray, 1983).

2. Purification

In order to minimise the time manipulating extracts during NR purification, methods have necessarily become shorter and more specific. Probably the single major advance in the purification of NR came from the work of Solomonson (1975) using the green alga *Chlorella vulgaris*. *Chlorella* NR was shown to possess a dinucleotide fold (NADH-binding site) and could be retained by, and specifically eluted from, the affinity-medium Blue dextran-Sepharose by micromolar concentrations of NADH. Work with higher plant NR, however, proved less successful and demonstrated the binding of NR to be less specific, requiring high salt concentrations to effect elution (Campbell, 1976).

The affinity-medium Blue Sepharose was subsequently developed which showed higher binding capacity for NR, and both media were used successfully in columns and in batch procedures to purify NR from many higher plant species including soybean (Campbell, 1976; Streit *et al.*, 1987), wheat (Sherrard and Dalling, 1979), barley (Kuo *et al.*, 1982b; Campbell and Wray, 1983), and spinach (Fido and Notton, 1984).

Other media used include hydroxylapatite (Fido and Notton, 1984), butyl Toyopearl 650-M (Nakagawa *et al.*, 1985), NADH-Sepharose (Heimer *et al.*, 1976), zinc chelate (Redinbaugh and Campbell, 1983) and AMP-Sepharose (Fido and Notton, 1984).

Purified NR has been shown to be extremely stable for many months when stored in the presence of 50% glycerol at $-20°C$ (Solomonson *et al.*, 1975).

3. Enzyme assay methods

The ability of NR to utilise alternative electron acceptors and donors has made it possible to assay the various partial activities of the enzyme as well as the overall physiological reaction.

(a) *NADH nitrate reductase.* The overall NADH-NR activity is usually determined by the reduction of nitrate to nitrite and subsequent colorimetric measurement of the nitrite produced. The assay mixture contains 50 mM potassium phosphate buffer (pH

7.5), 10 mM KNO_3, 0.1 mM NADH and enzyme (usually 0.1–0.2 ml) in a final volume of 1 ml (Wray and Filner, 1970). The reaction is started by addition of enzyme. After incubation for the desired period (15–20 min) at 25°C the reaction is stopped by addition of 1 ml of each of the diazo-coupling reagents 1% (w/v) in sulphanilamide 3 N HCl and 0.02% (w/v) N-(1-naphthyl)-ethylenediamine dihydrochloride (Snell and Snell, 1949). The tube contents are mixed and, where appropriate, precipitated protein sedimented in a bench-top centrifuge. After 15 min the absorbance of the pink diazo dye compound is measured at 540 nm and readings converted to amounts of nitrite by reference to a previously established standard plot (0–100 nmol of nitrite). Enzyme is added to the control tube after addition of sulphanilamide.

Where appropriate, residual NADH, which has been reported to interfere with dye formation, may be removed by treatment of the reaction mixture with zinc acetate and phenazine methosulphate prior to addition of the diazo-coupling reagents (Scholl et al., 1974).

An alternative assay method follows the nitrate-dependent oxidation of NADH by measuring the rate of change in absorbance at 340 nm, with stoichiometry found between amount of NADH consumed and nitrite formed (Eaglesham and Hewitt, 1975).

(b) *$FMNH_2$:nitrate reductase.* The assay mixture contains 50 mM potassium phosphate buffer (pH 7.5), 0.6 mM FMN, 10 mM KNO_3 2.9 mM $Na_2S_2O_4$ and an appropriate volume of enzyme (usually 0.1–0.2 ml) (Wray and Filner, 1970). The reaction is started by the addition of 0.05 ml of freshly prepared $Na_2S_2O_4$ (10 mg ml^{-1}) in 95 mM $NaHCO_3$ to give a final volume of 1 ml. The $Na_2S_2O_4$ bleaches the bright yellow FMN to a pale straw colour. After incubation for the desired period (15–20 min) at 25°C the reaction is stopped by vigorously mixing the tube contents to reoxidise the $FMNH_2$ to FMN. The controls lack $Na_2S_2O_4$. Nitrite formation is measured as described above. Where appropriate, interference from substances derived from $Na_2S_2O_4$ may be minimised by treatment of the assay mixture with formaldehyde prior to addition of the diazo-coupling reagents (Senn et al., 1976).

(c) *Reduced methyl viologen:nitrate reductase.* The assay mixture contains 25 mM potassium phosphate buffer (pH 7.5), 10 mM KNO_3, 0.2 mM methyl viologen dye, 3.2 mM $Na_2S_2O_4$ and enzyme (Dailey et al., 1982). After addition of enzyme (usually 0.1–0.2 ml) and incubation at 30°C for 1 min the reaction is started by the addition of 0.1 ml of freshly prepared $Na_2S_2O_4$ (14 mg ml^{-1} in 25 mM phosphate buffer, pH 8.2). After incubation at 30°C for 30 min the reaction is stopped by vigorously mixing the tube contents to reoxidise the reduced methyl viologen dye (tube contents turn colourless). Controls, containing all the reaction components and enzymes, are stopped immediately after addition of the $Na_2S_2O_4$. Nitrite measurement and, where appropriate, formaldehyde treatment are performed as described above.

(d) *Reduced bromophenol blue:nitrate reductase.* A further NR activity more recently reported (Hoarau et al., 1986) uses bromophenol blue (BPB) as electron donor and has an activity some 10–15 times greater than NADH-NR activity. BPB appears not to donate electrons to the NADH site of NR and behaves more like the electron donor, methyl viologen. However, evidence indicates that these two artificial electron donors interact with the enzyme at different sites (Campbell, 1986).

(e) *NADH:cytochrome c reductase.* The assay mixture contains 50 mM potassium phosphate buffer (pH 7.5), 0.1% (w/v) equine heart cytochrome c, 0.2 mM NADH and enzyme (usually 10–100 µl) in a final volume of 0.4 ml (Wray and Filner, 1970). The reaction is started by the addition of enzyme and the reduction of cytochrome c is measured at 25°C by following the rate of increase in absorbance at 550 nm. NADH is omitted from the control. The millimolar extinction coefficient of reduced cytochrome c (equine heart) at 550 nm is 29.5.

(f) *NADH:ferricyanide reductase.* The assay mixture contains 50 mM potassium phosphate buffer (pH 7.5), 0.1 mM NADH, 0.5 mM potassium ferricyanide and suitably diluted enzyme in a final volume of 1 ml. The reaction is started by addition of enzyme and oxidation of NADH monitored at 340 nm (Kay and Barber, 1986).

(g) *NAD(P)H:nitrate reductase.* Unlike NADH:nitrate reductase (EC 1.6.6.1), which has been detected in almost all plants examined, the bispecific NAD(P)H:nitrate reductase (EC 1.6.6.2) has been identified in only a few plant species. Assays for NAD(P)H:nitrate reductase from maize (Campbell, 1978), rice (Shen *et al.*, 1976), soybean (Campbell, 1976; Streit *et al.*, 1985), *Erythrina senegalensis* (Stewart and Orebamjo, 1979) and the *nar*-1 mutant of barley (Kleinhofs *et al.*, 1986) have been described.

(h) *Protein.* Protein measurements, used to determine the specific activity of the enzyme, are commonly made using the Coomassie Blue dye-binding method of Bradford (1976).

D. Immunochemistry of Nitrate Reductase

One method to study chemical similarity and structural homology of proteins is through immunological cross-reactivities. Nitrate reductase has proved to be a good immunogen requiring only microgram quantities of protein to elicit an immune response. The difficulty in obtaining highly purified enzyme for use as antigen may be overcome, in some instances, by excising NR directly from native polyacrylamide gels (Smarrelli and Campbell, 1981; Somers *et al.*, 1983). Antiserum prepared in this way has been used to distinguish between NR-deficient mutants of barley and test for commonality of structure between NR protein of mutants and wild-type plants.

Immunological comparisons of NR from a number of different plant species have been made using either the immunodiffusion technique of Ouchterlony, rocket immunoelectrophoresis, or enzyme inhibition. Smarrelli and Campbell (1981), using immunodiffusion and enzyme inhibition, demonstrated that all NRs had antigenic determinants in common, but showed a degree of reactivity which varied with the different species. Similarly Snapp *et al.* (1984) used monospecific antiserum raised to the barley enzyme against NR from nine higher plant species. Immunoelectrophoresis and enzyme inhibition studies showed NRs from monocotyledonous species were antigenically similar to, but dicotyledonous species were antigenically different from, barley NR. However, inactivation of NR from all species by the barley antiserum demonstrated the presence of conserved antigenic sites within the protein structure.

Comprehensive reviews covering many aspects of the immunology of NR are available (Kleinhofs et al., 1986; Notton, 1989).

1. Monoclonal antibodies

Monoclonal antibodies (McAb) have been raised to higher plant NR by a number of research groups. Cherel et al. (1985) raised McAbs to semi-purified maize leaf NADH-NR. Antibodies were selected by enzyme inhibition and Western blotting. Six were subsequently used to compare NR from different plant species using a two-site enzyme-linked immunosorbent assay (ELISA) (Cherel et al., 1986). The assay consisted of an initial coating of McAb which bound NR from extracts, followed by a polyclonal antiserum prepared against maize leaf NR. Detection was made by alkaline phosphatase-labelled antispecies antibody. McAbs raised to squash and corn NR (Campbell, 1987) have been shown to differentiate between electron donor sites on the enzyme.

The McAbs raised (in rats) against spinach NR purified by the method of Fido and Notton (1984) were selected by ELISA against the same enzyme and by enzyme inhibition. Four antibodies (MAC 74, 75, 76 and 77) inhibited nitrate-reducing activities ($FMNH_2$-NR and MV-NR) and two others (MAC 78 and 79) partially inhibited the dehydrogenase activity (Notton et al., 1985). When the spinach enzyme was subjected to SDS-PAGE (Laemmli, 1970), followed by electrophoretic transfer to nitrocellulose (Western blotting) essentially as described by Towbin et al. (1979), and challenged by each McAb, only two (MAC 78 and 79) recognised the SDS-treated enzyme. When the McAbs were tested for enzyme inhibition of NR activity from crude extracts of a large number of other plant species, including both monocotyledonous and dicotyledonous species, one McAb (MAC 74) was especially effective in inhibiting the activity from every species tested. Two further McAbs recently obtained (MAC 231 and 232) recognise both spinach and *Chlorella* NR, but do not inhibit enzyme activity (Notton et al., 1988).

2. Immunoaffinity purification

In an attempt to improve the speed and specificity of NR purification one McAb (MAC 74), chosen because of its known ability to cross-react with NR from a number of higher plant species, was used to prepare an immunoaffinity medium (Fido, 1987). The McAb was purified and coupled to Sepharose 4B. The NR of a crude extract from 500 g of spinach leaves was concentrated by ammonium sulphate precipitation. After re-solution, the NR was passed from a pre-column of rat γ-globulin bound to Sepharose 4B (which removed some non-specifically binding proteins) directly onto the immuno-column. After thorough washing of the immunocolumn the enzyme was eluted with 1 M KNO_3. The recovery of active enzyme was approximately 60%, with a purification of greater than 1500-fold.

Analysis of the eluted protein by SDS-PAGE revealed a doublet at 110–115 kDa as well as other antigenically positive fragments, including a clearly discernible band at 50–55 kDa. These various fragments were subsequently shown to remain associated with NR during native PAGE, which revealed only a single band, as well as during molecular sieving upon Biogel or a FPLC Superose 6 column. Affinity chromatography upon 5′AMP Sepharose (Fido and Notton, 1984) removed these fragments, but recovery of active enzyme was poor, with active enzyme retained on the medium.

Using MAC 74, immunoaffinity purification of NR has been successfully applied to a number of other higher plant species (R. J. Fido, unpubl. res.).

3. Quantification

Early attempts to estimate the amount of soluble NR cross-reacting material (CRM) in higher plants was made by Graf *et al.* (1975) with antiserum to spinach NR. Enzymically inactive NR was quantified by determining the protection it afforded to active NR from inactivation by antibodies. In a similar way Kuo *et al.* (1981) compared CRM in barley extracts of induced NR with that of NR-deficient mutants.

Improved and increased sensitivity of detection of NR-CRM in crude extracts has been made by preparing monospecific antiserum for use in a sandwich-type ELISA. Nitrate reductase in extracts of squash (Campbell and Ripp, 1984), spinach (Maki *et al.*, 1986) and corn (Campbell and Remmler, 1986) were detectable at the nanogram level. The amount of NR present was estimated by peroxidase activity of labelled second antibody bound to antigen NR antibodies. Using ELISA, the leaves of corn plants were estimated to contain 4–5 µg and roots 0.24 µg NR protein per g tissue.

An indirect sandwich ELISA was developed by Whitford *et al.* (1987) using especially one McAb (MAC 74) raised to spinach NR. The initial coating onto 96-well polystyrene plates was rabbit polyclonal antibody to NR, which bound antigen protein in crude extracts. Detection was made by measuring the amount of MAC 74 which bound to NR using a peroxidase-labelled sheep anti-rat antibody. The assay was able to detect nanogram amounts of NR, and revealed changes in the amount of NR-CRM in spinach leaves upon removal and re-supply of nitrate to the nutrient medium. Both NADH-NR activity and ELISA signal decreased when nitrate was removed, but the ELISA signal was restored prior to NR activity upon re-supply of nitrate. This suggested that, as well as *de novo* synthesis of NR in response to nitrate, a conformational change was required before the enzyme could express activity.

II. NITRITE REDUCTASE

A. Historical

In 1961 Huzisige and Satoh demonstrated that light stimulated the disappearance of nitrite in the presence of chloroplast grana and a soluble fraction derived from spinach leaves and ascribed this loss to a photochemical reduction process. Subsequently, the presence of nitrite reductase activity in higher plants was reported by Hageman *et al.* (1962) who showed that incubation of leaf extracts of marrow with reduced benzyl viologen allowed the stoichiometric conversion of nitrite to ammonium ions. Paneque *et al.* (1963) reported that nitrite can be reduced in the light by spinach chloroplast grana in the presence of chloroplast extract, suggested that ferredoxin is the primary physiological electron donor replaced by benzyl viologen in the experiments of Hageman *et al* (1962), and concluded that 'nitrite reduction can be considered as one of the most simple and typical examples of photosynthesis...'. Subsequently, they demonstrated that reduced ferredoxin is indeed the natural electron donor to nitrite reductase (Losada *et al.*, 1963). Independent evidence for the participation of ferredoxin in the photochemical reduction of nitrite was presented by Huzisige *et al.* (1963) and Hewitt and Betts (1963). In 1966 Ramirez et al. (1966) and Joy and Hageman (1966) reported the purification of nitrite reductase from spinach and maize, respectively, and confirmed its ability to use reduced ferredoxin as electron donor in the reduction of nitrite to ammonium ions.

Over the last twenty or so years the enzyme has been intensively studied at the physiological, biochemical and molecular levels and has been the subject of a number of reviews (Beevers and Hageman, 1980; Guerrero et al., 1981; Huffaker, 1982; Rajasekhar and Oelmüller, 1987; Siegel and Wilkerson, 1989; Wray, 1989; Back et al., 1989).

B. Introduction

Nitrite reductase (ferredoxin nitrite oxidoreductase EC 1.7.7.1; NiR) of leaves is located within the chloroplast (Dalling et al., 1972; Miflin, 1974) and in C_4 plants is found predominantly in the mesophyll cells (Neyra and Hageman, 1978). NiR of non-green tissue, such as scutellum and root, is localised in plastids (Dalling et al., 1972; Emes and Fowler, 1979) and in vivo electron donation is probably via a ferredoxin-like protein (Ninomiya and Sato, 1984; Suzuki et al., 1985; Wada et al., 1986) which is reduced by a pyridine nucleotide reductase with electrons probably derived from the pentose phosphate pathway (Emes and Fowler, 1983). Isoforms of NiR have been identified in a number of plants (Kutscherra et al., 1987). In wild oats inheritance patterns showed that two of the isoforms were governed by a single Mendelian locus with codominant alleles (Heath-Pagliuso et al., 1984), suggesting that the NiR apoprotein is probably encoded in the nuclear DNA.

1. Structure and function of nitrite reductase

The NiR enzymes are usually isolated as monomeric polypeptides of c. 63 kDa with very similar amino acid composition [for example, spinach (Vega and Kamin, 1977; Ida and Mikami, 1986); *Cucurbita pepo* (Hucklesby et al., 1976); barley (Serra et al., 1982; Ip et al., 1989); wheat (Small and Gray 1984) and pea (Bowsher et al., 1988)]. All enzymes examined by chemical analysis and by visible and EPR spectroscopy have been shown to contain one sirohaem, an iron tetrahydroporphyrin of the isobacteriochlorin type (Murphy et al., 1974), and one 4Fe/4S centre per enzyme molecule (Ida and Mikami, 1986). Evidence from Mössbauer, ENDOR and EPR studies, discussed in some detail in Siegel and Wilkerson (1989), indicate that there is electronic overlap between the 4Fe/4S centre and the sirohaem, and that the active centre of the enzyme comprises these two prosthetic groups bridged together through a cysteine S atom. The molecular cloning of the spinach NiR structural gene has recently allowed the position of the cysteine within the protein to be deduced (Back et al., 1988).

A form of NiR having a molecular weight of 86 kDa has been isolated from spinach leaves (Hirasawa et al., 1987) and one with a molecular weight of 100 kDa from etiolated shoots of the bean, *Phaseolus angularis* (Ishiyama and Tamura, 1985). These two forms of the enzyme are reported to have a dissociable subunit, of 24 or 35 kDa respectively, in addition to the 63 kDa polypeptide reported by others. This enzyme from spinach is reported to have a greater specificity for reduced ferredoxin (as opposed to reduced methyl viologen) as electron donor than does the 63 kDa form of the enzyme. However, the 63 kDa form of the spinach enzyme is equally and highly active catalytically with either reduced ferredoxin or reduced methyl viologen as electron donor (Ida and Mikami, 1986). Preliminary evidence indicates that the 86 kDa form of the spinach enzyme may contain two moles of sirohaem per mole of enzyme (Hirasawa et al., 1987).

The properties of the 63 kDa form of spinach NiR are shown in Table 14.2.

TABLE 14.2. Properties of the nitrite reductase of spinach, *Spinacea oleracea* L.

Molecular weight (kDa)	63
Subunit number	1
$S_{20,w}$ (s)	4.38
Stokes radius (nm)	3.3
Amino acid residues	562
pH optimum	7.5
Specific activity (μmol min^{-1} mg^{-1})	207
Molecular activity (10^3 mol min^{-1} mol^{-1} enzyme)	
Ferredoxin-linked	12.3
Methyl viologen-linked	12.7
K_m NO$_2^-$ (μM)	360
K_m ferredoxin (μM)	6
K_m methyl viologen (μM)	120
Prosthetic groups (mol mol^{-1} enzyme)	
(4 Fe-4S)	1
Sirohaem	1
Midpoint potential (mV)	
(4 Fe-4S)	-550
Sirohaem	-50
Spectral properties	
Absorption maximum (nm)	
Oxidised form	280, 389(Soret), 574(α), 691
Reduced form	280, 400, 551, 585
Extinction coefficient (mM^{-1} cm^{-1})	
389 nm	60.9

Data compiled from: Back et al. (1988); Ida and Mikami (1986); Lancaster et al. (1979); Stoller et al. (1977); Vega and Kamin (1977).
Note that Hirasawa et al. (1987) have reported the isolation of an 87 kDa form of this enzyme which contains subunits of 63 and 24 kDa.

C. Nitrite reductase purification

1. Purification

Nitrite reductase has been purified to apparent electrophoretic homogeneity from leaves [spinach (Ida et al., 1976; Vega and Kamin, 1977; Ida and Mikami, 1986; Hirasawa et al., 1987); marrow, *Curcurbita pepo* (Hucklesby et al., 1976); barley (Serra et al., 1982; Ip et al., 1989); wheat (Small and Gray, 1984); and bean, *Phaseolus angularis* (Ishiyama et al., 1985)] and roots [for example barley (Ida et al., 1974); bean, *Phaseolus angularis* (Nagaoka et al., 1984); and pea (Bowsher et al., 1988)] of several plant species.

A number of different, multi-step, purification procedures yielding apparently electrophoretically homogeneous enzyme have been described. Many employ affinity chromatography on ferredoxin-Sepharose, introduced by Ida et al. (1976), as a final or penultimate step, in combination with ammonium sulphate/acetone fractionation and one or more of ion-exchange chromatography, gel filtration and chromatography on butyl Toyopearl or hydroxylapatite. Ida and Mikami (1986) have described in detail a purification procedure for spinach nitrite reductase which involves acetone and ammonium sulphate fractionation, hydrophobic chromatography on benzyl and phenyl-Sepharose, and affinity chromatography on Blue- and ferredoxin-Sepharose. This is the most extensive purification procedure so far published and gives enzyme with the highest specific activity (207 µmol NO_2^- reduced per min per mg protein) so far reported, a purification factor of 1900, a recovery of 26% and a yield of 26 mg protein from 18 kg of spinach leaves. The enzyme is homogeneous as judged by gel electrophoresis, sedimentation behaviour and chemical analysis.

2. *Enzyme assay methods*

Enzyme assays, in general, utilise either dithionite-reduced methyl viologen dye or dithionite-reduced ferredoxin as electron donor and measure the rate of loss of nitrite colorimetrically.

(a) *Dithionite reduced methyl viologen:nitrite reductase.* The assay mixture contains 33 mM potassium phosphate buffer (pH 7.5), 2 mM potassium nitrite, 1 mM methyl viologen, 11.6 mM $Na_2S_2O_4$ and enzyme in a final volume of 1 ml (Wray and Filner, 1970). The reaction is started by the addition of 0.2 ml of $Na_2S_2O_4$ (10 mg ml^{-1}) in 0.29 M sodium bicarbonate. After incubation for the desired period (10–20 min) at 25°C in open tubes the reaction is stopped by vigorous mixing of the tube contents on a Whirlmixer until the dithionite and reduced methyl viologen are oxidised, as indicated by the loss of the blue colour of the reduced dye. A 0.1 ml aliquot of the reaction mixture is diluted to 3 ml with water and 1 ml of each of the diazo-coupling reagents 1% (w/v) sulphanilamide in 3 M HCl and 0.02% (w/v) *N*-(1-naphthyl)ethylenediamine dihydrochloride added (Snell and Snell, 1949). After 20 min the absorbance of the pink diazo dye compound is measured at 540 nm and readings converted to amounts of nitrite by reference to a previously established standard plot (0–100 nmol nitrite). In control tubes the enzyme is added to the incubation subsequent to the oxidation of the dye.

A method which assays NiR activity by measuring the rate of nitrite-dependent oxidation of reduced methyl viologen dye at 604 nm has also been described (Asada et al., 1969).

(b) *Dithionite reduced ferredoxin:nitrite reductase.* The assay is performed as described for dithionite reduced methyl viologen:nitrite reductase (above), but with 0.5 mM ferredoxin in place of the methyl viologen dye. The preparation of spinach ferredoxin has been described (Ida et al., 1976).

Many minor variations of these procedures have been published (for example Ramirez et al., 1966: Ida and Morita, 1973; Vega and Kamin, 1977) and some involve incubation in closed tubes under nitrogen (see for example, Fry et al., 1982). Usually, however, assays are performed in open tubes.

D. Immunochemistry of Nitrite Reductase

Production of polyclonal antibodies to the NiR from pea (Gupta et al., 1984), wheat (Small and Gray, 1984), spinach (Hirasawa et al., 1984; Ida, 1987), bean, *Phaseolus angularis* (Ishiyama et al., 1985), barley (Ip et al., 1987) and marrow, *Cucurbita pepo* (Bowsher et al., 1988) has been reported. The use of antibodies to NiR in various immunological procedures such as Ouchterlony double diffusion analysis (Gupta et al., 1984; Hirasawa et al., 1984; Ishiyama et al., 1985; Ida, 1987; Bowsher et al., 1988), rocket immunoelectrophoresis (Gupta and Beevers, 1984; Gupta et al., 1984: Ida, 1987), immunoprecipitation (Hirasawa et al., 1984; Small and Gray, 1984; Ishiyama et al., 1985; Ogawa and Ida, 1987; Ida, 1987) and Western blotting (Small and Gray, 1984) has been described. Antibodies have been used to quantitate the level of NiR protein in cell-free extracts (Gupta and Beevers, 1984), to demonstrate that nitrite reductase is synthesised *de novo* in response to nitrate and light (Gupta and Beevers, 1984; Small and Gray, 1984; Ogawa and Ida, 1987), to demonstrate, by immunoprecipitation of *in vitro* translation products, that nitrite reductase is synthesised as a precursor (Small and Gray, 1984; Gupta and Beevers, 1985, 1987; Ogawa and Ida, 1987), to assess phylogenetic relatedness (Hirasawa et al., 1984; Ida, 1987), and to clone the nitrite reductase structural gene by immunoscreening of an expression library (Back et al., 1988). Production of monoclonal antibodies has not yet been reported. Methods of immunochemical analysis in higher plants are discussed in Linskens and Jackson (1986).

ACKNOWLEDGEMENTS

Long Ashton Research Station is funded through the Agricultural and Food Research Council. We thank Brian Notton for his helpful discussion and Michael Barber for permission to reproduce Fig. 14.1.

REFERENCES

Abrol, Y. P., Sawhney, S. K. and Naik, M. S. (1983). *Plant Cell Environ.* **6**, 595–599.
Asada, K., Tamura, G. and Bandurski, R. S. (1969). *J. Biol. Chem.* **244**, 4904–4915.
Back, E., Burkhart, W., Moyer, M., Privalle, L. and Rothstein, S. (1988). *Mol. Gen. Genet.* **212**, 20–26.
Back, E., Burkhart, W., Moyer, M., Privalle, L. and Rothstein, S. (1989). *In* "Molecular and Genetic Aspects of Nitrate Assimilation" (J. L. Wray and J. R. Kinghorn, eds), pp. 284–296. Oxford University Press, Oxford.
Barber, M. J., Notton, B. A. and Solomonson, L. P. (1987). *FEBS Lett.* **213**, 372–374.
Beevers, L. and Hageman, R. H. (1969). *Ann. Rev. Plant Physiol.* **20**, 495–522.
Beevers, L. and Hageman, R. H. (1980). *In* " The Biochemistry of Plants, Vol. 5, Amino acids and Derivatives" (B. J. Miflin, ed.), pp. 115–168. Academic Press, New York.
Bowsher, C. G., Emes, M. J., Cammack, R. and Huckleby, D. P. (1988). *Planta* **175**, 334–340.
Bradford, M. M. (1976). *Anal. Biochem.* **72**, 248–254.

Brown, J., Small, I. S. and Wray, J. L. (1981). *Phytochemistry* **20**, 389–398.
Campbell, J. McA and Wray, J. L. (1983). *Phytochemistry* **22**, 2375–2382.
Campbell, W. H. (1976). *Plant Sci. Lett.* **7**, 239–247.
Campbell, W. H. (1978). *Z. Pflanzenphysiol.* **88**, 357–361.
Campbell, W. H. (1986). *Plant Physiol.* **82**, 729–732.
Campbell, W. H. (1987). *In* "Second International Symposium on Nitrate Assimilation—Molecular and Genetic Aspects" L3. St. Andrews, Scotland.
Campbell, W. H. (1988). *Physiol. Plant.* **74**, 214–219.
Campbell, W. H. and Ripp, K. G. (1984). *Ann. New York Acad. Sci.* **435**, 123–125.
Campbell, W. H. and Remmler, J. L. (1986). *Plant Physiol.* **80**, 435–441.
Campbell, W. H. and Smarrelli, J. (1986). *In* "Biochemical Basis of Plant Breeding" (C. A. Neyra, ed.), Vol. II, pp. 1–39. CRC Press, Boca Raton, FL, USA.
Cherel, I., Grosclaude, J. and Rouzé, P. (1985). *Biochem. Biophys. Res. Commun.* **129**, 686–693.
Cherel, I., Marion-Poll, A., Mayer, C. and Rouzé, P. (1986). *Plant Physiol.* **81**, 376–378.
Dailey, F. A., Warner, R. L., Somers, D. A. and Kleinhofs, A. (1982). *Plant Physiol.* **69**, 1200–1204.
Dalling, M. J., Tolbert, N. E. and Hageman, R. H. (1972). *Biochim. Biophys. Acta* **283**, 505–512.
Eaglesham, A. R. J. and Hewitt, E. J. (1975). *Plant Cell Physiol.* **16**, 1137–1149.
Emes, M. J. and Fowler, M. W. (1979). *Planta* **144**, 249–253.
Emes, M. J. and Fowler, M. W. (1983). *Planta* **158**, 97–102.
Evans, H. J. and Nason, A. (1953). *Plant Physiol.* **28**, 233–254.
Fido, R. J. (1987). *Plant Sci.* **50**, 111–115.
Fido, R. J. and Notton, B. A. (1984). *Plant Sci. Lett.* **37**, 87–91.
Fido, R. J., Hewitt, E. J., Notton, B. A., Jones, O. T. G. and Nasrulhaq-Boyce, A. (1979). *FEBS Lett.* **99**, 180–182.
Fry, I. V., Cammack, R., Hucklesby, D. P. and Hewitt, E. J. (1982). *Biochem. J.* **205**, 235–238.
Graf, L., Notton, B. A. and Hewitt, E. J. (1975). *Phytochemistry* **14**, 1241–1243.
Guerrero, M. G., Vega, J. M. and Losada, M. (1981). *Ann. Rev. Plant Physiol.* **32**, 169–204.
Gupta, S. C. and Beevers, L. (1984). *Plant Physiol.* **75**, 251–252.
Gupta, S. C. and Beevers, L. (1985). *Planta* **166**, 89–95.
Gupta, S. C. and Beevers, L. (1987). *Plant Physiol.* **83**, 750–754.
Gupta, S. C., Fletcher, J. and Beevers, L. (1984). *Z. Pflanzenphysiol.* **114**, 321–329.
Hageman, R. H. (1979). *In* "Nitrogen Assimilation of Plants" (E. J. Hewitt and C. V. Cutting, eds), pp. 591–612. Academic Press, London.
Hageman, R. H. and Reed, A. J. (1980). *In* "Methods in Enzymology" (A. San Pietro, ed.), Vol. 69, pp. 270–280. Academic Press, New York.
Hageman, R. H., Cresswell, C. F. and Hewitt, E. J. (1962). *Nature* **193**, 247–250.
Heath-Pagliuso, S., Huffaker, R. C. and Allard, R. W. (1984). *Plant Physiol.* **76**, 353–358.
Heimer, Y. M., Krasmin, S. and Riklis, E. (1976). *FEBS Lett.* **62**, 30–32.
Hewitt, E. J. and Betts, G. F. (1963). *Biochem. J.* **89**, 20P.
Hewitt, E. J. and Notton, B. A. (1980). *In* "Molybdenum and Molybdenum-Containing Enzymes" (M. P. Coughlan, ed.), pp. 273–325. Pergamon Press.
Hirasawa, M., Fukushima, K., Tamura, G. and Knaff, D. B. (1984). *Biochim. Biophys. Acta.* **791**, 145–154.
Hirasawa, M., Shaw, R. W., Palmer, G. and Knaff, D. B. (1987). *J. Biol. Chem.* **262**, 12428–12433.
Hoarau, J., Hirel, B. and Nato, A. (1986). *Plant Physiol.* **80**, 946–949.
Hucklesby, D. P., James, P. M., Banwell, M. J. and Hewitt, E. J. (1976). *Phytochemistry* **15**, 599–603.
Huffaker, R. C. (1982). *In* "Encyclopaedia of Plant Physiology, New Series" (D. Boulter and B. Parthier, eds), Vol. 14A, pp. 370–400. Springer, Berlin.
Huzisige, H. and Satoh, K. (1961). *Bot. Mag., Tokyo* **74**, 178–185.
Huzisige, H., Satoh, K., Tanaka, K. and Hayasida, T. (1963). *Plant Cell Physiol.* **4**, 307–322.
Ida, S. (1987). *Plant Sci.* **49**, 111–116.
Ida, S. and Morita, Y. (1973). *Plant Cell Physiol.* **14**, 661–671.
Ida, S. and Mikami, B. (1986). *Biochim. Biophys. Acta* **871**, 167–176.

Ida, S., Mori, E. and Morita, Y. (1974). *Planta* **121**, 213–224.
Ida, S., Kobayakawa, K. and Morita, Y. (1976). *FEBS Lett.* **65**, 305–308.
Ip, S. M., Kerr, J. and Wray, J. L. (1987). *In* "Abstracts Second International Symposium on Nitrate Assimilation—Molecular and Genetic Aspects" B21. St. Andrews, Scotland.
Ip, S. M., Kerr, J., Ingledew, W. J. and Wray, J. L. (1989). *Plant Sci*, in press.
Ishiyama, Y. and Tamura, G. (1985). *Plant Sci. Lett.* **37**, 251–256.
Ishiyama, Y., Shinoda, I., Fukushima, K. and Tamura, G. (1985). *Plant Sci.* **39**, 89–95.
Joy, K. W. and Hageman, R. H. (1966). *Biochem. J.* **100**, 263–273.
Kay, C. J. and Barber, M. J. (1986). *J. Biol. Chem.* **261**, 14125–14129.
Kleinhofs, A., Narayanan, K. R., Somers, D. A., Kuo, T. M. and Warner, R. L. (1986). *In* "Immunology in Plant Sciences" (H. F. Linskens and J. F. Jackson, eds), pp. 190–211. Springer, Berlin.
Kramer, S. P., Johnson, J. L., Ribeiro, A. A., Millington, D. S. and Rajagopalan, K. V. (1987). *J. Biol. Chem.* **262**, 16357–16363.
Kuo, T. M., Kleinhofs, A., Somers, D. A. and Warner, R. L. (1981). *Mol. Gen. Genet.* **181**, 20–23.
Kuo, T. M., Warner, R. L. and Kleinhofs, A. (1982a). *Phytochemistry* **21**, 531–533.
Kuo, T. M., Somers, D. A., Kleinhofs, A. and Warner, R. L. (1982b). *Biochim. Biophys. Acta* **708**, 75–81.
Kutscherra, M., Jost, W. and Schlee, D. (1987). *J. Plant Physiol.* **129**, 383–393.
Laemmli, U. K. (1970). *Nature* **227**, 680–685.
Lancaster, J. R., Vega, J. M., Kamin, H., Orme-Johnson, N. R., Orme-Johnson, W. H., Krueger, R. J. and Siegel, L. M. (1979). *J. Biol. Chem.* **254**, 1268–1272.
Linskens, H. F. and Jackson, J. F. (1986). "Immunology in Plant Sciences". Modern Methods of Plant Analysis, New Series, Vol. 4. Springer, Berlin.
Losada, M., Paneque, A., Ramirez, J. M. and del Campo, F. F. (1963). *Biochem. Biophys. Res. Commun.* **10**, 298–303.
Maki, H., Yanagishi, K., Sato, K., Ogura, N. and Nakagawa, H. (1986). *Plant Physiol.* **82**, 739–741.
Miflin, B. J. (1974). *Plant Physiol.* **54**, 550–555.
Müller, A. J. and Mendel, R. R. (1989). *In* "Molecular and Genetics Aspect of Nitrate Assimilation" (J. L. Wray and J. R. Kinghorn, eds), pp. 166–185. Oxford University Press, Oxford.
Murphy, M. J., Siegel, T. M., Tove, S. R. and Kamin, H. (1974). *Proc. Natl. Acad. Sci. USA* **71**, 612–616.
Nagaoka, S., Hirasawa, M., Fukushima, K. and Tamura, G. (1984). *Agric. Biol. Chem.* **48**, 1179–1188.
Nakagawa, H., Yonemura, Y., Tamamoto, H., Sato, T., Ogura, N. and Sato, R. (1985). *Plant Physiol.* **77**, 124–128.
Neyra, C. A. and Hageman, R. H. (1978). *Plant Physiol.* **62**, 618–621.
Ninomiya, Y. and Sato, S. (1984). *Plant Cell Physiol.* **25**, 453–458.
Notton, B. A. (1989). *In* "Molecular and Genetic Aspects of Nitrate Assimilation" (J. L. Wray and J. R. Kinghorn, eds), pp. 155–165. Oxford University Press, Oxford.
Notton, B. A., Fido, R. J. and Galfre, G. (1985). *Planta* **165**, 114–119.
Notton, B. A., Barber, M. J., Fido, R. J., Whitford, P. N. and Solomonson, L. P. (1988). *Phytochemistry* **27**, 1965–1968.
Notton, B. A., Fido, R. J., Whitford, P. N. and Barber, M. J. (1989). *Phytochemistry* **28**, 2261–2266.
Oaks, A. and Hirel, B. (1985). *Ann. Rev. Plant Physiol.* **36**, 345–365.
Ogawa, M. and Ida, S. (1987). *Plant Cell Physiol.* **28**, 1501–1508.
Orihuel-Iranzo, B. and Campbell, W. H. (1980). *Plant Physiol.* **65**, 595–599.
Paneque, A., Del Campo, F. F. and Losada, M. (1963). *Nature* **198**, 90–91.
Poulle, M., Oaks, A., Bzonek, P., Goodfellow, V. J. and Solomonson, L. P. (1987). *Plant Physiol.* **85**, 375–378.
Rajasekhar, V. K. and Oelmüller, R. (1987). *Physiol. Plant* **71**, 517–520.
Ramirez, J. M., Del Campo, F. F., Paneque, A. and Losada, M. (1966). *Biochim. Biophys. Acta* **118**, 58–71.

Redinbaugh, M. G. and Campbell, W. H. (1983). *Plant Physiol.* **71**, 205–207.
Redinbaugh, M. G. and Campbell, W. H. (1985). *J. Biol. Chem.* **260**, 3380–3385.
Scholl, R. L., Harper, J. E. and Hageman, R. H. (1974). *Plant Physiol.* **53**, 825–828.
Senn, D. R., Carr, P. W. and Klatt, L. N. (1976). *Anal. Biochem.* **75**, 464–471.
Serra, J. L., Ibarlucea, J. M., Arizmendi, J. M. and Llama, M. J. (1982). *Biochem. J.* **201**, 167–170.
Shen, T. C., Funkhauser, E. A. and Guerrero, M. G. (1976). *Plant Physiol.* **58**, 292–297.
Sherrard, J. H. and Dalling, M. J. (1979). *Plant Physiol.* **63**, 346–353.
Siegel, L. M. and Wilkerson, J. O. (1989). *In* "Molecular and Genetic Aspects of Nitrate Assimilation" (J. L. Wray and J. R. Kinghorn, eds), pp. 263–283. Oxford University Press, Oxford.
Small, I. S. and Gray, J. C. (1984). *Eur. J. Biochem.* **145**, 291–297.
Smarrelli, J. and Campbell, W. H. (1981). *Plant Physiol.* **68**, 1226–1230.
Snapp, S., Somers, D. A., Warner, R. L. and Kleinhofs, A. (1984). *Plant Sci. Lett.* **36**, 13–18.
Snell, F. D. and Snell, C. T. (1949). *In* "Colorimetric Methods of Analysis", Vol II, pp. 804–805. Van Nostrand, New York.
Solomonson, L. P. (1975). *Plant Physiol.* **56**, 853–855.
Solomonson, L. P., Lorimer, G. H., Hall, R. L., Borchers, R. and Bailey, J. L. (1975). *J. Biol. Chem.* **250**, 4120–4127.
Solomonson, L. P., Barber, M. J., Robbins, A. P. and Oaks, A. (1986). *J. Biol. Chem.* **261**, 11290–11294.
Somers, D. A., Kuo, T. M., Kleinhofs, A. and Warner, R. L. (1983). *Plant Physiol.* **71**, 145–149.
Stewart, G. R. and Orebamjo, T. O. (1979). *New Phytol.* **83**, 311–319.
Stoller, M. L., Malkin, R. and Knaff, D. M. (1977). *FEBS Lett.* **81**, 271–274.
Streit, L., Nelson, R. S. and Harper, J. E. (1985). *Plant Physiol.* **78**, 80–84.
Streit, L., Martin, B. A. and Harper, J. E. (1987). *Plant Physiol.* **84**, 654–657.
Suzuki, A., Oaks, A., Jacquot, J. P., Vidal, J. and Gadal, P. (1985). *Plant Physiol.* **78**, 374–378.
Towbin, H., Staehelin, T. and Gordon, J. (1979). *Proc. Natl. Acad. Sci. USA.* **76**, 4350–4354.
Vega, J. M. and Kamin, H. (1977). *J. Biol. Chem.* **252**, 896–909.
Wada, K., Onda, M. and Matsubara, H. (1986). *Plant Cell Physiol.* **27**, 407–415.
Whitford, P. N., Fido, R. J. and Notton, B. A. (1987). *Phytochemistry* **26**, 2467–2470.
Wray, J. L. (1989). *In* "Molecular and Genetic Aspects of Nitrate Assimilation" (J. L. Wray and J. R. Kinghorn, eds), pp. 244–262. Oxford University Press, Oxford.
Wray, J. L. and Filner, P. (1970). *Biochem J.* **119**, 715–725.
Wray, J. L. and Kirk, D. W. (1981). *Plant Sci. Lett.* **23**, 207–213.

15 Enzymes of Ammonia Assimilation

PETER J. LEA,[1] RAY D. BLACKWELL,[1] FENG-LING CHEN[2] and URSULA HECHT[3]

[1]*Division of Biological Sciences, University of Lancaster, Lancaster LA1 4YQ, UK*

[2]*Department of Biological Sciences, University of Warwick, Coventry CV4 7AL, UK*

[3]*Biologisches Institut II der Universität, Schänzlestrasse 1, D-7800 Freiburg i. Br., FRG*

I.	Introduction	258
	A. Glutamine synthetase	259
	B. NADH- and ferredoxin-GOGAT	259
II.	Glutamine synthetase	260
	A. Extraction	260
	B. Assay	261
	C. Purification	264
	D. Properties and structure	265
III.	Ferredoxin- and NADH-GOGAT	267
	A. Extraction	267
	B. Ferredoxin-GOGAT: Assay	268
	C. Ferredoxin-GOGAT: Purification	269
	D. Ferredoxin-GOGAT: Structure	270
	E. NADH-GOGAT: Assay	270
	F. NADH-GOGAT: Purification	273
	G. NADH-GOGAT: Structure	273
	Acknowledgement	273
	References	273

I. INTRODUCTION

Prior to 1970 it was assumed that ammonia was directly incorporated into glutamate via the operation of one single enzyme glutamate dehydrogenase (EC 1.4.1.1). However, Tempest *et al.* (1970a) described an enzyme glutamate synthase (GOGAT: glutamine: 2-oxoglutarate aminotransferase) that was capable of converting glutamine and 2-oxoglutarate to two molecules of glutamate. Coupled to glutamine synthetase they demonstrated a two-step mechanism whereby bacteria could assimilate ammonia into amino acids when grown on limiting quantities of inorganic nitrogen (Tempest *et al.*, 1970b).

Reaction 1 Glutamine synthetase (GS; EC 6.3.1.2)

$$\text{L-Glutamate} + \text{ATP} + \text{NH}_3 \longrightarrow \text{L-Glutamine} + \text{ADP} + \text{Pi} + \text{H}_2\text{O}$$

Reaction 2a Glutamate synthase (NAD(P)H-GOGAT; EC 1.4.1.13)

$$\text{2-Oxoglutarate} + \text{L-Glutamine} + \text{NAD(P)H}_2 \longrightarrow \text{2 L-Glutamate} + \text{NAD(P)}$$

In 1974, Lea and Miflin presented evidence to show that chloroplasts of higher plants contained a ferredoxin-dependent glutamate synthase (Fd-GOGAT) that catalysed a reaction similar to Reaction 2a:

Reaction 2b Glutamate synthase (Ferredoxin GOGAT; EC 1.4.7.1)

$$\text{2-Oxoglutarate} + \text{L-Glutamine} + \text{reduced ferredoxin} \longrightarrow \text{2 L-Glutamate} + \text{oxidised ferredoxin}$$

The combined action of the two enzymes (GS and GOGAT) is frequently referred to as the glutamate synthase cycle.

Despite a considerable amount of controversy in the literature, there is now sufficient evidence from studies using ^{13}N, ^{15}N, inhibitors and more recently mutants, to clearly indicate that the glutamate synthase cycle is the sole mechanism of ammonia assimilation in higher plants (for reviews see Miflin and Lea, 1976, 1982; Lea and Miflin, 1979; Joy, 1988; Lea *et al.*, 1989). Although there has recently been a revival of interest in the role of glutamate dehydrogenase in nitrogen-fixing species of *Bacillus* (Kanamori *et al.*, 1988), there is no evidence to suggest that the enzyme operates in the assimilation of ammonia in higher plants (Rhodes *et al.*, 1989).

Ammonia within the plant cell may be generated by a number of different mechanisms. In the root, ammonium ions may be taken in directly under certain soil conditions, or produced by the reduction of nitrate (see Wray and Fido, Chapter 14, this volume). In legume root nodules, ammonia is generated directly by the nitrogenase enzyme inside the *Rhizobium* bacteria and then excreted into the plant (Vance *et al.*, 1988). Within the leaf ammonia may again be generated by the reduction of nitrate (Andrews, 1986), by the conversion of glycine to serine in photorespiration (Keys *et al.*, 1978; Givan *et al.*, 1988; Lea *et al.*, 1989) and by the metabolism of transport compounds, for example asparagine (Lea and Miflin, 1980; Sieciechowicz *et al.*, 1988) and ureides (Schubert,

1986). In developing seeds ammonia is again generated by the breakdown of the nitrogen transport compounds, arginine, asparagine and ureides. The metabolism of amino acids liberated during proteolysis, possibly via glutamate dehydrogenase, may also produce ammonia in germinating seeds (Lea and Joy, 1983; Srivastava and Singh, 1987). Small amounts of ammonia may also be formed in other aspects of metabolism, in particular the operation of threonine dehydratase in the synthesis of isoleucine and phenylalanine ammonia lyase in the formation of lignins and flavonoids. These minor routes of ammonia synthesis have been discussed in detail by Joy (1988).

A. Glutamine Synthetase

Glutamine synthetase (GS) appears to be distributed throughout the plant, although levels of GS activity may differ markedly from one tissue to another (Lee and Stewart, 1978; Stewart and Rhodes, 1978). In most cases, the activity of the enzyme is higher in the leaf than in the root, although if a large proportion of nitrogen is assimilated in the root the situation may be reversed (Lee and Stewart, 1978).

Studies with ion-exchange chromatography have shown that two separate forms of GS exist in green tissues of a wide variety of species including soybean hypocotyl (Stasiewiz and Durham, 1979), rice (Guiz et al., 1979) and pea leaves (Evstigneeva et al., 1977). Furthermore, in most leaves tested, two distinct isoenzymes, revealed by staining for GS activity, can be separated by electrophoresis on starch gels (Barratt, 1980). In leaves of a wide range of higher plants, McNally and colleagues observed four different patterns of GS isoforms (McNally et al., 1983). The groups were characterised as having: (1) only cytosolic GS; (2) only chloroplastic GS; (3) cytosolic GS as a minor component of total GS, with chloroplastic GS making up the rest; and (4) approximately equal activities of cytosolic and chloroplastic GS.

The majority of studies of GS activity in roots have suggested that it is predominantly located in the cytosol (Suzuki et al., 1981) with a small amount being present in plastids (Miflin, 1974; Emes and Fowler, 1979). However, in a recent detailed study of root GS, Vezina et al. (1987) were able to calculate that up to 60% of alfalfa and pea GS was present in plastids in roots grown on nitrate, the remainder being in the cytosol. Vézina et al. (1987) were able to separate the two forms by DEAE-Sephacel ion-exchange chromatography.

The nitrogen-fixing root nodules of legumes contain high levels of GS activity which may account for up to 2% of the total soluble protein (McParland et al., 1976). There are large increases in GS activity during the development of the nodule which coincide with the appearance of nitrogenase and leghemoglobin (Robertson et al., 1975; Cullimore and Bennett, 1988). In *Phaseolus vulgaris*, a novel nodule specific GS is synthesised in the cytosol during nodule development (Cullimore et al., 1983; Lara et al., 1983, 1984).

B. NADH- and Ferredoxin-GOGAT

Two types of enzyme activity dependent on ferredoxin or pyridine nucleotide, were first detected in various plant tissues (Dougall, 1974; Fowler et al., 1974; Lea and Miflin, 1974). However, it was not immediately obvious whether they were catalysed by the same or by different protein molecules. NADH-specific and Fd-dependent forms have been clearly separated by gel filtration (Cullimore and Sims, 1981a) and ion-exchange chromatography (Márquez et al., 1984) of *Chlamydomonas* extracts and by ion-exchange chromatography of extracts of mustard cotyledons (Hecht et al., 1988).

Antibodies raised against Fd-GOGAT in rice cross-reacted with the enzyme in the leaves and roots. However, there was no evidence of cross-reaction with the NADH-dependent enzyme (Suzuki et al., 1982, 1987), thus confirming that the two enzyme activities are catalysed by distinct proteins. Absolute proof that the NADH- and ferredoxin-dependent GOGAT activities are under separate genetic control was obtained by the isolation of mutant lines deficient in Fd-GOGAT. Such mutants which have been obtained in *Arabidopsis* (Somerville and Ogren, 1980) and in barley (Wallsgrove et al., 1986; Kendall et al., 1986; Blackwell et al., 1988) contain normal levels of NADH-GOGAT activity.

There have been some early reports that the NADH-dependent form of GOGAT may also use NADPH as a coenzyme (Beevers and Storey, 1976; Kang and Titus, 1981; Chiu and Shargool, 1979). Although there is good evidence that a NADPH-GOGAT operated in bacteria (Tempest et al., 1970a,b; Yelton and Yoch, 1981), the precise role of the enzyme in plant tissues is not clear at the present time.

In green plant tissue the ferredoxin-dependent form of the enzyme predominates. During the greening process Fd-GOGAT activity increases rapidly, whilst NADH-GOGAT remains constant or declines (Matoh and Takahashi, 1981; Wallsgrove et al., 1982; Hecht et al., 1988). Fd-GOGAT activity is found exclusively in the leaf chloroplast (Wallsgrove et al., 1980), a result confirmed by immunogold antibody staining of leaf mesophyll chloroplasts (Botella et al., 1988).

In the non-green tissues of roots (Suzuki et al., 1981; Matoh and Takahashi, 1982) and developing legume cotyledons (Matoh et al., 1979; Beevers and Storey, 1976) the two forms of glutamate synthase appear to be present. In developing nitrogen-fixing root nodules there is rapid increase in NADH-GOGAT activity (Robertson et al., 1975; Awonaike et al., 1981). The enzyme has recently been studied in more detail (Chen and Cullimore, 1988; Anderson et al., 1989) and two specific forms of NADH-GOGAT have been detected in *Phaseolus* root nodules (Chen and Cullimore, 1988), both of which appear to be located in the plastids (Chen and Cullimore, 1989).

II. GLUTAMINE SYNTHETASE

A. Extraction

In our own hands the best method to isolate GS from barley leaves has been to grind 200 mg of liquid nitrogen frozen material in 2 ml of buffer containing 50 mM Tris-HCl (pH 7.8), 1 mM EDTA, 1 mM DTT, 10 mM $MgSO_4$, 5 mM sodium glutamate and 10% (v/v) ethanediol. Following centrifugation the supernatant may be assayed directly or desalted on a small Sephadex G-25 column.

Imidazole was at one time a popular buffer for the isolation of GS. Stewart and Rhodes (1977) utilised a 50 mM imidazole buffer (pH 7.2) containing 2 mM $MnCl_2$, 0.5 mM EDTA, 1 mM DTT, 20% (v/v) ethylene glycol and 5% (w/v) Polyclar AT for the extraction of GS from *Lemna* fronds. In a comparative study of GS from the roots and leaves of rice, Hirel and Gadal (1980) employed a 100 mM Tris-HCl (pH 7.6) buffer containing 1 mM MgCl$_2$, 1 mM EDTA and 10 mM 2-mercaptoethanol. In order to maintain the integrity of organelles in roots, Vézina et al. (1987) used a high (1.2 M) sucrose-containing buffer of 100 mM Tricine (pH 7.8), 0.1% BSA, 1 mM MgCl$_2$, 1% (w/v) soluble PVP and 6 mM 2-mercaptoethanol. In extracts of Phaseolus root nodules, sulphydryl reagents do not stabilise GS and have in fact been shown to promote the almost complete deactivation of both isoenzymic forms without the addition of other protectants. A suitable extraction method is to grind 10 g of frozen nodules with 20% (w/v) Polyclar AT and 10 ml of Tris-HCl buffer (pH 7.8) containing 5 mM sodium glutamate, 10 mM MgSO$_4$ and 10% glycerol. The brei should be filtered through four layers of muslin prior to centrifugation (Cullimore et al., 1983).

B. Assay

1. Synthetase

Due to difficulties in obtaining a rapid separation of glutamine from the substrate glutamate, hydroxylamine (NH$_2$OH) has been used as an alternative substrate for ammonia:

$$\text{Glutamate} + \text{NH}_2\text{OH} + \text{ATP} \xrightarrow{\text{Mg}^{2+}} \gamma\text{-Glutamylhydroxamate} + \text{ADP} + \text{Pi} + \text{H}_2\text{O}$$

The γ-glutamylhydroxamate may be rapidly quantified by a simple reaction with acidified ferric chloride to yield a brown colour that may be determined spectrophotometrically at 540 nm. For the determination of GS activity in barley leaves the 500 µl mixture for the 'synthetase' assay contains: 50 mM glutamate, 5 mM hydroxylamine hydrochloride, 50 mM MgSO$_4$, 20 mM ATP in 100 mM Tris-HCl (pH 7.8). The pH values of the glutamate and hydroxylamine solutions should be carefully adjusted to 7.8. The ATP solution should be made up fresh before each assay and pH adjusted to just above 7.0.

The reaction is started by the addition of 200 µl of enzyme extract, and incubated at 30°C. The reaction is terminated by the addition of 700 µl of ferric chloride reagent (0.67 M ferric chloride, 0.37 M HCl and 20% v/v TCA). Control tubes should also be incubated comprising the enzyme extract and all the reagents except ATP. For precipitation of protein the tubes should be centrifuged at 10 000 × g for 5 min. The absorbance of the supernatant is determined at 540 nm. The time-course is normally linear for at least 20 min. It is essential to construct a standard curve using authentic γ-glutamylhydroxamate to obtain an accurate quantification of GS activity.

The precise concentrations of the reactants used will depend upon the source of the GS activity. Rhodes et al. (1976) using Lemna fronds maintained a reaction medium of 2 ml containing 18 mM ATP, 45 mM MgSO$_4$, 6 mM hydroxylamine, 90 mM glutamate and 50 mM imidazole buffer at pH 7.2. O'Neal and Joy (1973), as a basis of a series of elegant papers on pea leaf GS, used a reaction medium containing 80 mM glutamate, 20 mM MgSO$_4$, 8 mM ATP, 6 mM hydroxylamine and 1 mM EDTA in 0.1 M Tricine with the final pH adjusted to 7.8 with NaOH.

2. Transferase

A second assay based on a further reaction catalysed by GS also yields the hydroxamate product that reacts with acidified ferric chloride.

$$\text{Glutamine} + \text{NH}_2\text{OH} \xrightarrow[\text{ADP, Mn}^{2+}]{\text{Arsenate}} \gamma\text{-Glutamylhydroxamate} + \text{NH}_3$$

The reaction is normally termed the transferase assay and has been used widely in studies on bacterial GS, for determining the adenylation state of the enzyme (Shapiro and Stadtman, 1970). As there is no evidence that higher plant GS is subject to regulation by adenylation/deadenylation, we do not recommend that the assay is employed for physiological studies. The assay does have a major advantage in that much higher 'apparent' rates of activity are detected. The assay therefore has particular value for the rapid determination of GS activity in column eluates. Cullimore et al. (1983) demonstrated that in *Phaseolus* root nodules the ratio of transferase:synthetase activities varied for different isoenzymic forms between 35 and 100, and could be used as a diagnostic tool for the different forms. In *Lemna* this figure was shown to be closer to 20 (Rhodes et al., 1976) and there was little variation in this ratio when the plant was grown on a range of nitrogen sources.

A typical reaction mixture contains in 1.0 ml: 100 mM Tris-acetate buffer (pH 6.4), 100 mM L-glutamine, 60 mM hydroxylamine, 0.5 mM ADP, 1.0 mM $MnCl_2$ and 20 mM sodium arsenate (Cullimore and Sims, 1980). Rhodes et al. (1976) utilised a 2.0 ml reaction medium containing 0.17 mM ADP, 1.5 mM $MnCl_2$, 17 mM hydroxylamine, 65 mM L-glutamine, 33 mM sodium arsenate and 100 mM Tris-acetate buffer (pH 6.4).

Over 90% of the experiments published on plant GS have made use of the fact that hydroxylamine acts as an alternative substrate. Unfortunately, the product measured is not glutamine. We would urge workers always to check their assay under more physiological conditions using one of the methods described below.

3. Biosynthetic

(a) *Determination of [^{14}C]Glutamine.* The method employs [^{14}C]glutamate as a substrate. Cullimore and Sims (1981b) described an assay used for determining GS in *Chlamydomonas* cells. The incubation mixture (1.5 ml) contained 50 mM Imidazole acetate buffer (pH 7.2), 0.55 mM EDTA, 4 mM ATP, 40 mM $MgCl_2$, 4 mM NH_4Cl, and 3.0 mM [^{14}C]glutamate (7.4 kBq). The centrifuged medium was then applied to a Dowex-1-acetate column (0.5 × 3 cm) and the [^{14}C]glutamine eluted by washing with 3 ml of 50 mM acetic acid. The method was based on that described by Prusiner and Milner (1970). However, it is possible that plant cells may also contain glutamate decarboxylase activity that would convert glutamate to γ-aminobutyrate. This ^{14}C-labelled product would not be separated from [^{14}C]glutamine in the above system. Pishak and Phillips (1979), utilising GS isolated from a range of animal tissues, devised a modification of the above method that would take into account the action of glutamate decarboxylase. The eluate from the Dowex-1-acetate column should be passed directly onto a separate 5 cm long Amberlite CG-50 (H^+) column which will remove any [^{14}C]γ-aminobutyrate. As before, the radioactivity in the eluate can be determined directly by counting in an aqueous liquid scintillation medium. It is recommended that with any previously untested plant extracts that the two-column and one-column methods are compared to ascertain whether there is any significant glutamate decarboxylase activity present.

(b) *Determination of Glutamine by HPLC.* The synthesis of glutamine may be determined directly by a method described by Martin *et al.* (1982a). The contents of the assay medium are reacted with *o*-phthaldialdehyde and separated in 4 min on a µBondapak C_{18}-column. The rates of GS activity determined were shown to be comparable to the other assay methods. A more detailed account of the determination of free amino acids in plant tissue has been described by Martin *et al.* (1982b).

Using a similar method GS activity has also been determined by Vézina *et al.* (1987) in legume roots. The enzyme extract (250 µl), was incubated in 500 µl assay mixture containing 100 mM imidazole, 100 mM $MgCl_2$, 24 mM ATP, 24 mM NH_4Cl and 160 mM glutamate at a pH of 8.0. The reaction was terminated by the addition of 100 µl sulphosalicylic acid (0.3 g ml^{-1}) followed by centrifugation and neutralisation with NaOH. The sample was diluted with 10% (v/v) methanol–H_2O, reacted with *o*-phthaldialdehyde and the glutamine derivative separated on a C_{18} column by HPLC essentially as described by Fleury and Ashley (1983).

(c) *Coupled spectrophotometric assay of GS.* One of the arguments frequently used to establish that GS is the primary enzyme involved in the assimilation of ammonia is that it has a very high affinity for the substrate with a K_m of the order 10^{-4} to 10^{-5} M. The observant reader will have noted that whilst being very useful for measuring maximum activities of GS, none of the methods previously described allow for the measurement of initial rates using low concentrations of ammonia as a substrate.

However, a method is available which links the ADP formed in the reaction to the conversion of phosphoenolpyruvate (PEP) to pyruvate by pyruvate kinase. The pyruvate may then be determined by the oxidation of NADH in a spectrophotometer by lactate dehydrogenase. The method has been used for the assay of GS from animals (Wellner *et al.*, 1966), bacteria (Kingdom *et al.*, 1970) and plants (O'Neal and Joy, 1973). Stewart and Rhodes (1977) described the following method for the assay of GS from *Lemna*: the reaction mixture contained in a volume of 2 ml: 5 mM ATP, 20 mM $MgSO_4$, 1 mM EDTA, 50 mM KCl, 0.2 mM NADH, 1 mM PEP, 50 mM sodium glutamate, 50 mM Imidazole-HCl, 1 unit lactate dehydrogenase and 3.2 units pyruvate kinase at a final pH of 7.2. The reaction was initiated by the addition of varying concentrations of ammonium acetate, after pre-incubation of the enzyme in the reaction medium to remove all traces of ammonia. Activity was determined by following the rate of NADH oxidation at 340 nm. The coupling enzymes were desalted on a column of Sephadex G-25, equilibrated with 50 mM Imidazole-HCl (pH 7.2) prior to use.

4. Detection of GS activity following gel electrophoresis

Barratt (1980) devised a simple method for the detection of GS activity following starch gel electrophoresis (Siciliano and Shaw, 1968). The method involves the incubation of the gel slice in the optimum reaction mixture required for either the transferase or synthetase assay. The gel should be fully covered by the assay mixture and left for varying periods of up to three hours. After this time the medium is decanted, the gel blotted and the acidified ferric chloride reagent added gently to the gel with a pasteur pipette. Dark brown bands of γ-glutamylhydroxamate appear almost immediately and the gel should be washed gently with distilled water to reduce the background yellow colour. The method was used very successfully to demonstrate two isoenzymic forms of GS in a range of leaf extracts (Barratt, 1980). The detection method has also been used following polyacrylamide gel electrophoresis of root nodule extracts (Cullimore et al., 1983; Robert and Wong, 1986).

A major drawback to the method described by Barratt is that the activity bands rapidly diffuse and great skill is required to photograph the gels at the correct time. The liberation of phosphate during the true biosynthetic reaction may be used as an alternative assay for GS following gel electrophoresis. Hirel and Gadal (1980) briefly described a method in which gels were incubated in full reaction medium containing ammonia as a substrate, and the phosphate released from ATP detected by the addition of Fiske and SubbaRow (1926) reagent or 0.2 M $CaCl_2$.

Since that time more sensitive methods for the detection of phosphate by precipitation with calcium (Nimmo and Nimmo, 1982) or reaction with malachite green (Zlotnick and Gottlieb, 1986) have been described. These methods have been used very successfully by Relton et al. (1988) for the detection of the low activity enzyme, aspartate kinase, in plant extracts. The methods are described in full in this volume by Bonner and Lea (Chapter 18). We see no reason, provided suitable controls are used (i.e. lack of substrates or addition of methionine sulphoximine) to exclude the presence of non-specific ATPases, that this method cannot be modified for the detection of GS.

C. Purification

GS has been purified and characterised from a large number of plant tissues including pea leaves (O'Neal and Joy, 1973), *Lemna* (Stewart and Rhodes, 1977), rice roots and leaves (Hirel and Gadal, 1980, 1981), spinach leaves (Hirel et al., 1982), tobacco leaves (Hirel et al., 1984), *Phaseolus* roots, nodules and leaves (Cullimore et al., 1983; Lara et al., 1984), to name but a few. Readers should refer to recent review articles for full documentation (Stewart et al., 1980; Cullimore and Bennett, 1988; Vance et al., 1988).

Hirel and Gadal (1980) published a simple method of purifying three different forms of GS from rice. GS_1 (the cytosolic leaf form) was purified from 200 g of etiolated leaves, GS_2 (the leaf chloroplastic form) from non-aqueously isolated chloroplasts, and GS_R (the root form) from 200 g of roots. Table 15.1 shows the purification obtained for the three different forms. Using a similar purification procedure, where a phenyl-Sepharose and a Sephacryl S-300 chromatography step were included, Cullimore et al., (1983) purified the two isoenzymes of GS located in *Phaseolus* root nodules. The purification procedure is shown in Table 15.2.

Lara et al. (1984) adopted the use of an inhibitor affinity resin to purify GS from *Phaseolus*. The partially purified extract was loaded onto a column of Sepharose-anthranilic acid (Palacios, 1976) and eluted with AMP. The method was an extremely powerful last step in the purification of GS from non-photosynthetic tissue, but was not so successful for the enzyme isolated from green leaves.

TABLE 15.1. Purification of GS_1, GS_2, and GS_R from rice plants.

Purification step	Total protein (mg)	Total units (μmol min^{-1})	Specific activity (μmol min^{-1} mg protein^{-1})	Fold purification
GS_1				
Etiolated leaf crude extract	1.456	145.6	0.1	1
$(NH_4)_2SO_4$ (40–60%), DEAE-Sephacel	64.3	26	0.4	4
Sephacryl S-300	0.4	9.3	23.2	232
Hydroxyapatite	0.090	3	33.3	333
GS_2				
Non-aqueous isolated chloroplasts	208	33.6	0.16	1
DEAE-Sephacel	33.6	8	0.23	1.4
Sephacryl S-300	0.7	2.8	4	25
Hydroxyapatite	0.060	1.86	31	193
GS_R				
Root crude extract	301	137.5	0.45	1
$(NH_4)_2SO_4$ (40–60%), DEAE-Sephacel	13.7	16	1.17	2.6
Sephacryl S-300	0.140	2.6	18.5	41.1
Hydroxyapatite	0.028	1	35.7	79.3

From Hirel and Gadal (1980).

D. Properties and Structure

The kinetic properties of plant GS have been reviewed in detail by Stewart *et al.* (1980). The enzyme is notable for having a high affinity for ammonia but a relatively low affinity for glutamate. Whilst there is considerable evidence that GS activity is regulated *in vivo* by a range of metabolites in bacteria (Ginsburg and Stadtman, 1973), there have been few reports to suggest this in higher plants. In our opinion the role of GS is to assimilate ammonia as rapidly as possible and therefore the only regulation is through the availability of substrates.

In *Chlorella*, GS may be activated by a light-dependent thioredoxin-related mechanism (Tischner and Hüttermann, 1980), although in higher plants this process could not be established (W. R. Mills, unpubl. res.). Light does, however, play a major role in stimulating the synthesis of GS in leaves (Guiz *et al.*, 1979; Hirel *et al.*, 1982; Tobin *et al.*, 1985; Edwards and Coruzzi, 1989; Schmidt and Mohr, 1989). Studies with mutants of barley, containing various levels of GS activity, suggest that the enzyme is present in at least a two-fold excess to allow for the rate of ammonia production (Blackwell *et al.*, 1987).

It has been described previously that GS isoenzymes can be separated by ion-exchange chromatography and native gel electrophoresis. However, the molecular

TABLE 15.2. Purification of two forms of glutamine synthetase from *Phaseolus* root nodules.

	Synthetase activity (µmol min^{-1})	Total protein (mg)	Specific activity (µmol min^{-1} mg protein^{-1})	Fold purification	Recovery (%)	Transferase: synthetase ratio
Crude extract	76.9	977	0.08	1.0	100	57
Protamine-sulphate supernatant	80.0	777	0.10	1.3	104	54
35–55% saturated (NH$_4$)$_2$SO$_4$ precipitate	—	227	—	—	—	—
Sephacryl S-300	70.0	90.0	0.78	9.9	91	62
GS$_1$ DEAE-Sephacel	34.6	4.4	7.9	100	45	30
Phenyl-Sepharose	32.6	2.7	11.9	152	43	34
Hydroxyapatite	19.1	1.7	11.3	144	25	36
GS$_2$ DEAE-Sephacel	9.2	12.2	0.75	9.6	12	67
Phenyl-Sepharose	5.3	1.4	3.7	48	7	87
Hydroxyapatite	3.7	1.0	3.6	46	5	88

From Cullimore et al. (1983).

weights of the enzymes isolated from a range of sources vary between 300 and 370 kDa and comprise eight subunits. GS in *Phaseolus* occurs as five forms: two leaf, one root and two nodule (Lara *et al.*, 1983, 1984; Cullimore and Bennett, 1988). The leaf chloroplast GS is composed of 43 kDa polypeptides encoded by a single gene, while the cytosolic forms of roots and leaves are composed of two 39 kDa polypeptides, designated α- and β-subunits. In addition there is a third cytosolic 39 kDa polypeptide which is most prevalent in nodules designated γ. The α-, β- and γ-polypeptides are all encoded by separate genes.

In pea plants, five distinct GS polypeptides are differentially expressed in roots and nodules (Tingey *et al.*, 1987). The predominant leaf GS is composed of a 44 kDa polypeptide localised in the chloroplast stroma. Leaves also express at a much lower level a cytosolic 38 kDa polypeptide, which is the major form of GS in the root cytosol. Roots and leaves also express, at a very reduced level, three 37 kDa GS polypeptides. The high GS activity that appears as root nodules develop is accompanied by striking increases in the three 37 kDa polypeptides. Nodules also express at low levels the 44 and 38 kDa polypeptides, but there appears to be no strongly nodule-enhanced specific GS form as there is in *Phaseolus*.

In the mature leaves of *Nicotiana* only one single GS polypeptide of 44 kDa was detected that was localised in the stroma of intact chloroplasts. In the roots, two GS polypeptides of 38 kDa molecular weight were detected (Tingey and Coruzzi, 1987). A previous report that the chloroplast GS of *Nicotiana* was glycosylated (Nato *et al.*, 1984) was not confirmed (Tingey and Coruzzi, 1987).

III. FERREDOXIN- AND NADH-GOGAT

A. Extraction

For the extraction of both NADH- and Fd-GOGAT, for *in vitro* enzyme activity measurements in crude extracts, phosphate buffer (pH 7.5) is normally used (Wallsgrove *et al.*, 1982; Suzuki *et al.*, 1987; Matoh *et al.*, 1980; Hecht *et al.*, 1988). More recently in bean nodules, Hepes has been the preferred buffer for NADH-GOGAT (Chen and Cullimore, 1988).

Protection against the oxidation of SH-groups by the addition of 2-mercaptoethanol or DTT is essential. Enzyme activity can be stabilised (especially for long-lasting purification procedures) by the addition of 2-oxoglutarate, KCl, ethanediol or PMSF (Wallsgrove *et al.*, 1977; Marquez *et al.*, 1988). As Fd-GOGAT is located in the plastids, it might be necessary in some cases to add Triton X-100, to ensure complete extraction of the enzyme. In addition, NADH-GOGAT appears to be sensitive to low temperatures and it is advisable not to freeze the enzyme during the extraction and purification procedures.

Extraction buffer for green leaves (Marquez *et al.*, 1988): 50 mM KH_2PO_4-KOH (pH 7.5), 100 mM KCl, 5 mM EDTA, 12.5 mM 2-mercaptoethanol, 1 mM PMSF, 2 mM 2-oxoglutarate, 20% (v/v) ethanediol, 0.05% (v/v) Triton X-100.

Extraction buffer for cotyledons (Hecht *et al.*, 1988): 100 mM KH_2PO_4-KOH (pH 7.5), 0.5 mM EDTA, 100 mM KCl, 0.1% (v/v) 2-mercaptoethanol, 0.5% (v/v) Triton X-100.

Extraction buffer for root nodules (Chen and Cullimore, 1988): 50 mM Hepes (pH 7.5), 0.5 M sucrose, 10 mM DTT, 1 mM EDTA, 1 mM PMSF.

B. Ferredoxin-GOGAT: Assay

The electron donor reduced ferredoxin can be replaced in the *in vitro* assay by reduced methyl viologen as described for other ferredoxin-dependent enzymes (e.g. nitrite reductase). Provided the methyl viologen was present in saturating amounts, no significant difference in activity with ferredoxin or methyl viologen could be found for the barley leaf enzymes (Marquez *et al.*, 1988).

For the determination of Fd-GOGAT activity (e.g. in mustard cotyledons: Hecht *et al.*, 1988) the reaction mixture normally consists of 100 µl 100 mM glutamine in 500 mM KH_2PO_4-KOH pH 7.5, 100 µl 100 mM oxoglutarate in 500 mM KH_2PO_4-KOH pH 7.5, 100 µl 150 mM methyl viologen and 400 µl crude extract. After pre-incubation at 30°C, the reaction was started by addition of 100 µl reductant (47 mg $Na_2S_2O_4$, 50 mg $NaHCO_3$ dissolved in 1 ml of distilled water). After 20 min incubation at 30°C, the reaction was terminated by adding 1 ml of ethanol.

The reaction mixture for Fd-GOGAT determination in other plant tissues differs only slightly (Suzuki *et al.*, 1982; Galvan *et al.*, 1984). Some authors stop the reaction by boiling the mixture (Wallsgrove *et al.*, 1977; Galvan *et al.*, 1984). The transaminase inhibitor aminooxyacetate should be added if there is evidence of reductant-independent glutamate synthesis (Wallsgrove *et al.*, 1977; Galvan *et al.*, 1984).

Fd-GOGAT activity is determined by the quantitative measurement of glutamate, the product in the reaction between glutamine and 2-oxoglutarate. Glutamate, therefore, has to be separated from the mixture after the termination of the enzymatic reaction.

Five different procedures have been applied:

(1) Paper chromatography (Wallsgrove *et al.*, 1977). An aliquot of the reaction mixture is spotted onto Whatman No. 4 chromatography paper and the separation performed in 25% (w/v) phenol in the presence of ammonia vapour.

(2) TLC (Blackwell, 1988). An aliquot of the reaction mixture is spotted onto Whatman Chromedia CC41 cellulose plates and chromatographed using ethanol–ammonia–water (32 : 8 : 4; v/v/v).

(3) Paper electrophoresis (Wallsgrove *et al.*, 1982; Chen and Cullimore, 1988). The electrophoresis can be performed on Whatman 3 MM paper in 45 mM Na acetate buffer (pH 5.1), at 3 kV for 30 min or at 45 V for 3 h.

(4) HPLC (Martin *et al.*, 1982a; Suzuki *et al.*, 1987; Avila *et al.*, 1987). The amino acids formed after the enzymatic reaction are derivatised with OPA and applied to a µBondapak C_{18}-column (reverse phase) and eluted by isocratic elution with 20 mM sodium phosphate buffer (pH 6.8)–methanol (67 : 36; v/v) (Martin *et al.*, 1982a).

(5) Anion-exchange chromatography on Dowex 1 × 8 resin (Matoh *et al.*, 1980; Suzuki and Gadal, 1982; Galvan *et al.*, 1984; Hecht *et al.*, 1988). The assay mixture is transferred completely to a Dowex-acetate column (1 × 8, 200–400 mesh size, 12 mm diameter, 35 mm length). Glutamine is eluted from the column with 15 ml of distilled water. So as to remove residual water, the column is centrifuged for 2 min at 1500 rpm. Glutamate is then eluted from the column with 5 ml 3 M acetic acid and the centrifugation step is repeated. An aliquot of 5 ml of acetic acid was found to be sufficient to elute more than 95% of the glutamate.

Three different methods have been employed to quantify the glutamate that has been separated from the reaction mixture:

(1) The use of [^{14}C]glutamine as substrate and detection of [^{14}C]glutamate by scintillation counting (Wallsgrove et al., 1982; Chen and Cullimore, 1988).
(2) Derivatisation with o-phthaldialdehyde (OPA) and 2-mercaptoethanol (Suzuki et al., 1987, Martin et al., 1982a) or dansyl-chloride (Avila et al., 1987) and subsequent elution from reverse phase HPLC columns (see previously). For the OPA derivatives a simple UV detector set at 340 nm may be employed to assay the column eluate.
(3) Quantification of glutamate by reaction with ninhydrin (Wallsgrove et al., 1977; Matoh et al., 1980; Hecht et al., 1988). Paper chromatograms or TLC plates should be sprayed with a standard ninhydrin reagent. The glutamate spots are then removed and placed in test tubes and the colour eluted with a mixture of 600 ml acetone, 600 ml ethyl acetate, 600 ml water, 600 ml methanol, 18 g cadmium acetate, 18 ml glacial acetic acid and its extinction at 500 nm determined (Wallsgrove et al., 1977). For the eluates of Dowex-acetate columns the concentration of glutamate in the eluate can also be determined by a ninhydrin assay. An aliquot of 500 µl of the eluate is added to 1 ml of a ninhydrin solution (0.4 g ninhydrin, 80 ml 95% ethanol, 1 g $CdCl_2$, 10 ml acetic acid, 20 ml H_2O). After incubation at 80°C in a water bath for precisely 10 min, the samples are cooled and absorbance is measured at 506 nm.

C. Ferredoxin-GOGAT: Purification

Purification of Fd-GOGAT has been carried out for the enzyme from bean (Wallsgrove et al., 1977), maize (Matoh et al., 1980), rice leaves and roots (Suzuki and Gadal, 1982), Chlamydomonas (Galvan et al., 1984), spinach (Hirasawa and Tamura, 1984), tomato (Avila et al., 1987), and barley (Marquez et al., 1988).

The decisive step in purification of Fd-GOGAT was achieved by utilising a Fd-Sepharose column for affinity chromatography. The recently elaborated isolation procedure for barley leaves (Marquez et al., 1988) was performed using $(NH_4)_2SO_4$-fractionation, elution from Sephadex G-100, chromatography on DEAE Sephacel and finally affinity chromatography on Fd-Sepharose (Table 15.3). Fd-GOGAT protein obtained by this method represented 1% of total soluble protein in green barley leaves.

TABLE 15.3. Purification of ferredoxin-glutamate synthase from barley leaves.

Purification step	Volume (ml)	Total protein (mg)	Total activity (µkat)	Specific activity (nkat mg^{-1} protein)	Yield (%)	Fold purification
Crude extract	135	527	1.35	2.5	(100)	1.0
30–60% $(NH_4)_2SO_4$	7	210	0.8	3.8	59	1.5
Sephadex G-100	29	145	0.47	3.2	35	1.3
DEAE-Sephacel	90	13	0.38	29	28	11.8
Ferredoxin-Sepharose	1.4	0.9	0.23	255	17	103

From Marquez et al. (1988).

D. Ferredoxin-GOGAT: Structure

Fd-GOGAT is a monomeric iron–sulphur flavoprotein in spinach (Hirasawa and Tamura, 1984) and *Chlamydomonas* (Galvan *et al.*, 1984; Marquez *et al.*, 1984) of M_r about 140 kDa. However, in green leaves of rice, Fd-GOGAT appears to have an M_r of 250 kDa consisting of two identical polypeptides of M_r of 125 kDa (Suzuki and Gadal, 1982).

Fd-GOGAT is mainly located in the chloroplasts of green leaves and in plastids in other tissues, with a pH optimum of 6.9–7.5. The K_m for ferredoxin ranges between 2 and 5.5 µM, for glutamine between 100 and 1000 µM and for α-oxoglutarate between 7 and 70 µM. Additionally to ferredoxin, FAD acts as a cofactor. A comparison of the properties of Fd-GOGAT from a range of different sources as originally presented by Marquez *et al.* (1988) is shown in Table 15.4.

E. NADH-GOGAT: Assay

There are two common methods of determining NADH-GOGAT activity in crude extracts:

(a) Following the consumption of NADH spectrophotometrically at 340 nm.
(b) Determining the formation of glutamate in the reaction of 2-oxoglutarate with glutamine.

The advantage of Method (a) is its easy performance, though it requires simultaneous controls with one of either of the substrates to ensure that NADH consumption is specific for NADH-GOGAT and not due to non-specific side reactions catalysed by other enzyme activities. The advantage of Method (b) is that it follows the same procedure used for Fd-GOGAT determination, where the consumption of reduced ferredoxin or methyl viologen cannot be followed directly. A major problem in green tissue is that NADH-GOGAT activity is relatively low compared to the ferredoxin-dependent activity (Wallsgrove *et al.*, 1982; Suzuki *et al.*, 1987). In mustard cotyledons the NADH-dependent activity decreases rapidly as soon as plastidogenesis begins (2–2.5 days after sowing: Hecht *et al.*, 1988). However, maximum level of activity determined, expressed on a protein basis (0.2 nkat mg^{-1} protein) is equivalent to the activity determined in bean nodules (Chen and Cullimore, 1988).

Method (a). For measuring the oxidation of NADH, the assay mixture contains NADH (0.1 mM), glutamine (10 mM), 2-oxoglutarate (10 mM), 100 mM KH_2PO_4-KOH (pH 7.5) and 200 µl crude extract. Two controls must be performed for each measurement with glutamine and 2-oxoglutarate being left out of the reaction medium in turn. The reaction is followed at 340 nm in a spectrophotometer for about 10 min at 30°C. The method described above was optimised for the assay of NADH-GOGAT in mustard cotyledons (Hecht *et al.*, 1988) but it has also been used with slight variations for leaves (Matoh *et al.*, 1982; Avila *et al.*, 1987) and for root nodules (Chen and Cullimore, 1988).

Method (b). A radioactive assay employing [^{14}C]glutamine has been used for NADH-GOGAT determination in barley and pea leaves (Wallsgrove *et al.*, 1982) and bean root nodules (Chen and Cullimore, 1988). The reaction mixture is the same as

TABLE 15.4. Properties of ferredoxin-glutamate synthase from barley, and other higher plants.

Properties	Barley	Rice[a]	Spinach[b]	Maize[c]	Chlamydomonas[d]
Molecular weight (kDa)					
Subunit(s)	154 ± 3	115		ND	151
Native protein	177 ± 14[e]		170		144
	199 ± 30[f]				
Stokes' radius (nm)	4.9				4.1
Sedimentation coefficient (s)	8.8 × 10⁻¹³	230[g]	140	160	8.3 × 10⁻¹³
Frictional ratio	1.3	ND	ND	ND	1.2
Specific activity (nkat mg⁻¹ protein)	258	ND	ND	ND	173
Percentage of soluble protein in green tissue	0.2–1.0	598	ND	235	0.1–1.0
Catalytic activity in green tissue (nkat g⁻¹ fresh weight)	30	0.1–0.5	ND	ND	27
$K_m^{(app)}$ methyl viologen (mM)	1.9	42	ND	5	4.7
Absorption maxima (nm)	278,355,372,437	ND	279,360,438	ND	278,377,437
A_{278}/A_{437}	8.2	278	5.0	280	7.0

[a] Suzuki and Gadal (1982); [b] Hirasawa and Tamura (1984); [c] Matoh et al. (1980); [d] Galván et al. (1984); [e] from Stokes' radius and sedimentation coefficient; [f] electrophoresis in gradient gels; [g] from gel filtration; ND, not determined. From Marquez et al. (1988).

TABLE 15.5. Purification of two isoenzymes of NADH-GOGAT from root nodules.

Purification step	Total activity (μmol min^{-1})	Total protein (mg)	Specific activity (μmol min^{-1} mg protein^{-1})	Fold purification[a]	Recovery (%)
Crude extract	21.50	1248	0.017	1	100
Sephacryl S-300 column	12.33	144	0.084	4.9	57.3
Blue Sepharose column	10.58	11.8	0.87	51.4	49.2
HPLC ion-exchange column					
NADH-GOGAT I	1.37	1.48	0.92	216	6.3
NADH-GOGAT II	3.36	1.94	1.73	136	15.6
HPLC gel filtration column					
NADH-GOGAT I	0.47	0.13	3.77	887	2.1
NADH-GOGAT II	1.26	0.16	8.05	631	5.8

[a] The degree of purification of each isoenzyme was based on a determined activity ratio of NADH-GOGAT I:NADH-GOGAT II of 1:3 in crude nodule extracts. From Chen and Cullimore (1988).

for the Fd-GOGAT measurement except that methyl viologen or ferredoxin is replaced by 0.6 mM NADH (see Section III. B). Carbon-14 labelled glutamate is separated from the reaction mixture by paper electrophoresis and the radioactivity measured in a liquid scintillation counter (see Fd-GOGAT assay).

F. NADH-GOGAT: Purification

NADH-GOGAT has been purified from lupin root nodules (Boland and Benny, 1977), soybean cell cultures (Chiu and Shargool, 1979), etiolated pea shoots (Matoh *et al.*, 1980), *Chlamydomonas* (Galvan *et al.*, 1984; Marquez *et al.*, 1984), tomato (Avila *et al.*, 1987), bean root nodule (Chen and Cullimore, 1988) and alfalfa root nodule (Anderson *et al.*, 1989). The purification of bean root nodule NADH-GOGAT was achieved by $(NH_4)_2SO_4$ precipitation, gel filtration on Sephacryl S-300, chromatography on Blue Sepharose columns, chromatography on HPLC ion-exchange column and chromatography on HPLC gel filtration columns (Chen and Cullimore, 1988: Table 15.5).

G. NADH-GOGAT: Structure

NADH-GOGAT appears to be an iron–sulphur flavoprotein with a single subunit of M_r about 230 kDa and is specific for NADH as reductant, showing little or no activity with NADPH or ferredoxin. No significant amount of NADPH-GOGAT activity was found in soybean root nodules (Suzuki *et al.*, 1984), in bean root nodules (Chen and Cullimore, 1988) or in mustard cotyledons (Hecht *et al.*, 1988), whereas in pea cotyledons (Beevers and Storey, 1976) soybean cells (Chiu and Shargool, 1979) and *Chlamydomonas* (Marquez *et al.*, 1984) a considerable amount of NADPH-GOGAT activity was detected in crude extract. Avila *et al.*, (1987) showed that the K_m for NADPH in tomato leaves was higher (6 μM) than for NADH (1.7 μM). As purified NADH-GOGAT from higher plants does not function with NADPH, it is likely that NADPH-GOGAT activity is due to phosphorylase activity in the crude extracts. NADH-GOGAT is mainly located in non-green tissue such as nodules, roots, developing cotyledons with a pH-optimum range from 7.5 to 8.5. The K_m for NADH is between 4 and 13 μM, for glutamine between 400 and 1000 μM and for 2-oxoglutarate between 39 and 960 μM. FAD/FMN non-haem iron with acid-labile sulphur act as cofactor. In bean root nodules two different NADH-GOGAT forms could be discerned which differ in biochemical properties and in their physiological role in the root nodule (Chen and Cullimore, 1988).

ACKNOWLEDGEMENT

We are indebted to Dr Julie Cullimore for helpful advice and criticism of the manuscript.

REFERENCES

Anderson, M. P., Vance, C. P., Heichel, G. H. and Miller, S. S. (1989). *Plant Physiol.* **90**, 351–358.
Andrews, M. (1986). *Plant Cell Environm.* **9**, 511–519.
Avila, C., Botella, J. R., Canovas, F. M., de Castro, I. N. and Valpuesta, V. (1987). *Plant Physiol.* **85**, 1036–1039.

Awonaike, K. O., Lea, P. J. and Miflin, B. J. (1981). *Plant Sci. Lett.* **23**, 189–195.
Barratt, D. H. P. (1980). *Plant Sci. Lett.* **18**, 249–255.
Beevers, L. and Storey, R. (1976) *Plant Physiol.* **57**, 862–866.
Blackwell, R. D. (1988). "Isolation and Characterisation of Mutants of Higher Plants unable to carry out Photorespiration" Ph.D. thesis, University of Lancaster, UK.
Blackwell, R. D., Murray, A. J. S. and Lea, P. J. (1987). *J. Exp. Bot.* **38**, 1799–1809.
Blackwell, R. D., Murray, A. J. S. and Lea, P. J. (1988). *J. Exp. Bot.* **39**, 845–858.
Boland, M. J. and Benny, A. G. (1977). *Eur. J. Biochem.* **99**, 531–535.
Botella, J. R., Verbelen, J. P. and Valpuesta, V. (1988). *Plant Physiol.* **87**, 255–257.
Chen, F.-L. and Cullimore, J. V. (1988). *Plant Physiol.* **88**, 1411–1417.
Chen, F.-L. and Cullimore, J. V. (1989). *Planta* **178**, in press.
Chiu, J. Y. and Shargool, P. D. (1979). *Plant Physiol.* **63**, 409–415.
Cullimore, J. V. and Bennett, M. J. (1988). *J. Plant Physiol.* **132**, 387–393.
Cullimore, J. V. and Sims, A. P. (1980). *Planta* **150**, 392–396.
Cullimore, J. V. and Sims, A. P. (1981a). *Phytochemistry* **20**, 597–600.
Cullimore, J. V. and Sims, A. P. (1981b). *Phytochemistry* **20**, 933–940.
Cullimore, J. V., Lea, P. J. and Miflin, B. J. (1982). *Israel J. Bot.* **31**, 155–162.
Cullimore, J. V., Lara, M., Lea, P. J. and Miflin, B. J. (1983). *Planta* **157**, 245–253.
Dougall, D. K. (1974). *Biochem. Biophys. Res. Commun.* **58**, 639–646.
Edwards, J. W. and Coruzzi, G. M. (1989). *The Plant Cell* **1**, 241–248.
Emes, M. and Fowler, M. W. (1979). *Planta* **145**, 287–292.
Evstigneeva, Z. G., Pushkin, A. V., Radyukina, N. A. and Kretovich, V. L. (1977). *Doklady Akademii Nauk SSR* **237**, 962–964.
Fiske, C. H. and SubbaRow, Y. (1926) *J. Biol. Chem.* **66**, 375–400.
Fleury, M. O. and Ashley, D. V. (1983). *Anal. Biochem.* **133**, 330–335.
Fowler, M. W., Jessup, W. and Sarkissian, G. S. (1974). *FEBS Lett.* **46**, 340–342.
Galvan, F., Marquez, A. J. and Vega, J. M. (1984). *Planta* **162**, 180–187.
Ginsburg, A. and Stadtman, E. R. (1973). *In* "The Enzymes of Glutamine Metabolism" (S. Prusiner and E. R. Stadtman, eds), pp. 9–44. Academic Press, London and New York.
Givan, C. V., Joy, K. W. and Kleczkowski, L. A. (1988). *TIBS* **13**, 433–437.
Guiz, C., Hirel, B., Shedlofsky, G. and Gadal, P. (1979). *Plant Sci. Lett.* **15**, 271–278.
Harel, E., Lea, P. J. and Miflin, B. J. (1977). *Planta* **134**, 195–200.
Hecht, U., Oelmüller, R., Schmidt, S. and Mohr, H. (1988). *Planta* **175**, 130–138.
Hirel, B. and Gadal, P. (1980). *Plant Physiol.* **66**, 619–623.
Hirel, B. and Gadal, P. (1981). *Z. Pflanzenphysiol.* **102**, 315–319.
Hirel, B., Vidal, J. and Gadal, P. (1982). *Planta* **155**, 17–23.
Hirel, B., Weatherley, C., Cretin, C., Bergoonioux, C. and Gadal, P. (1984). *Plant Physiol.* **74**, 448–450.
Hirasawa, M. and Tamura, G. (1984). *J. Biochem.* **96**, 983–994.
Joy, K. W. (1988). *Can. J. Bot.* **66**, 2103–2109.
Kanamori, K., Weiss, R. L. and Roberts, J. D. (1988). *J. Biol. Chem.* **263**, 2817–2823.
Kang, S.-M. and Titus, J. S. (1981). *J. Am. Soc. Hort. Sci.* **106**, 765.
Kendall, A. C., Wallsgrove, R. M., Hall, N. P., Turner, J. C. and Lea, P. J. (1986). *Planta* **168**, 316–323.
Keys, A. J., Bird, I. F., Cornelius, M. J., Lea, P. J., Wallsgrove, R. M. and Miflin, B. J. (1978). *Nature* **275**, 741–743.
Kingdom, H., Hubbard, J. S. and Stadtman, E. R. (1970). *Biochemistry* **7**, 2136–2142.
Lara, M., Cullimore, J. V., Lea, P. J., Miflin, B. J., Johnston, A. W. B. and Lamb, J. W. (1983). *Planta* **157**, 254–258.
Lara, M., Porta, H., Padilla, J., Folch, J. and Sanchez, F. (1984). *Plant Physiol.* **76**, 1019–1023.
Lea, P. J. and Joy, K. W. (1983). *In* "Metabolisation of Reserves in Germinating Seeds" (C. Nozzolillo, P. J. Lea and F. A. Loewus, eds), pp. 77–109. Plenum, New York.
Lea, P. J. and Miflin, B. J. (1974). *Nature* **251**, 614–616.
Lea, P. J. and Miflin, B. J. (1979). *In* "Encyclopedia of Plant Physiology" (M. Gibbs and E. Latzko, eds), Vol. VI, pp. 445–455. Springer, Berlin, Heidelberg and New York.

Lea, P. J. and Miflin, B. J. (1980). *In* "The Biochemistry of Plants" (B. J. Miflin, ed.), Vol. 5, pp. 569–607. Academic Press, London.
Lea, P. J., Blackwell, R. D., Murray, A. J. S. and Joy, K. W. (1989). *In* Recent Advances in Phytochemistry" (E. E. Conn, ed.), Vol. 23, pp. 157–189. Plenum, New York.
Lee, J. A. and Stewart, G. R. (1978). *Adv. Bot. Res.* **6**, 1–43.
Marquez, A. J., Galvan, F. and Vega, J. M. (1984). *Plant Sci. Lett.* **34**, 305–314.
Marquez, A. J., Avila, C., Forde, B. G. and Wallsgrove, R. M. (1988). *Plant Physiol. Biochem.* **26**, 645–651.
Martin, F., Suzuki, A. and Hirel, B. (1982a). *Anal. Biochem.* **125**, 24–29.
Martin, F., Maudinas, B. and Gadal, P. (1982b). *Ann. Bot.* **50**, 401–406.
Matoh, T. and Takahashi, E. (1981). *Plant Cell Physiol.* **22**, 727–731.
Matoh, T. and Takahashi, E. (1982). *Planta* **154**, 289–294.
Matoh, T., Takahashi, E. and Ida, S. (1979). *Plant Cell Physiol.* **20**, 1455–1459.
Matoh, T., Ida, S. and Takahashi, E. (1980). *Bull. Res. Inst. Food Sci. Kyoto Univ.* **43**, 1–6.
McNally, S. F., Hirel, B., Gadal, P., Mann, A. F. and Stewart, G. R. (1983). *Plant Physiol.* **72**, 22–25.
McParland, R. H., Guevara, J. G., Becker, R. R. and Evans, H. J. (1976). *Biochem. J.* **153**, 597–606.
Miflin, B. J. (1974). *Plant Physiol.* **54**, 550–555.
Miflin, B. J. and Lea, P. J. (1976). *Phytochemistry* **15**, 873–885.
Miflin, B. J. and Lea, P. J. (1980). *In* "Biochemistry of Plants" (B. J. Miflin, ed.) Vol. 5, pp. 65–113. Academic Press, New York.
Miflin, B. J. and Lea, P. J. (1982). *In* "Encyclopedia of Plant Physiology" (D. Boulter and B. Parthier, eds), Vol. 14A, pp. 5–64. Springer, Berlin, Heidelberg and New York.
Nato, F., Hirel, B., Nato, A. and Gadal, P. (1984). *FEBS Lett.* **175**, 443–446.
Nimmo, H. G. and Nimmo, G. A. (1982). *Anal. Biochem.* **121**, 17–22.
O'Neal, T. D. and Joy, K. W. (1973). *Arch. Biochem. Biophys.* **159**, 113–122.
Palacios, R. (1976). *J. Biol. Chem.* **251**, 4787–4791.
Pishak, M. R. and Phillips, A. T. (1979). *Anal. Biochem.* **94**, 82–88.
Prusiner, S. and Milner, L. (1970). *Anal. Biochem.* **37**, 429–438.
Relton, J. M., Bonner, P. L. R., Wallsgrove, R. M. and Lea, P. J. (1988). *Biochim. Biophys. Acta* **953**, 48–60.
Rhodes, D., Rendon, G. A. and Stewart, G. R. (1976). *Planta* **129**, 203–210.
Rhodes, D., Brunk, D. G. and Magalhaes, J. R. (1989). *In* "Recent Advances in Phytochemistry" (E. E. Conn, ed.), Vol. 23, pp. 191–226. Plenum, New York.
Robert, F. M. and Wang, P. P. (1986). *Plant Physiol.* **81**, 142–148.
Robertson, J. B., Warbarton, M. P. and Farnden, K. J. F. (1975). *Plant Physiol.* **55**, 255–260.
Schmidt, S. and Mohr, H. (1989). *Planta* **177**, 526–534.
Schubert, K. R. (1986). *Ann. Rev. Plant Physiol.* **37**, 539–574.
Shapiro, B. M. and Stadtman, E. R. (1970). *Ann. Rev. Microbiol.* **24**, 501–524.
Siciliano, M. J. and Shaw, C. R. (1968). *In* "Chromatography and Electrophoresis", 2nd edn (I. Smith, ed.), p. 185. Interscience Publishers, New York.
Sieciechowicz, K. A., Joy, K. W. and Ireland, R. J. (1988). *Phytochemistry* **27**, 663–671.
Somerville, C. R. and Ogren, W. L. (1980). *Nature* **286**, 257–259.
Srivastawa, H. S. and Singh, R. P. (1987). *Phytochemistry* **26**, 597–610.
Stasiewiz, S. and Durham, V. L. (1979). *Biochem. Biophys. Res. Commun.* **87**, 627–634.
Stewart, G. R. and Rhodes, D. (1976). *FEBS Lett.* **64**, 296–299.
Stewart, G. R. and Rhodes, D. (1977). *New Phytol.* **79**, 257–268.
Stewart, G. R. and Rhodes, D. (1978). *New Phytol.* **80**, 307–316.
Stewart, G. R., Mann, A. F. and Fentem, P. A. (1980). *In* "The Biochemistry of Plants" (B. J. Miflin, ed.), Vol. 5, pp. 271–327. Academic Press, London and New York.
Suzuki, A. and Gadal, P. (1982). *Plant Physiol.* **69**, 848–852.
Suzuki, A., Gadal, P. and Oaks, A. (1981). *Planta* **151**, 457–461.
Suzuki, A., Vidal, J. and Gadal, P. (1982). *Plant Physiol.* **70**, 827–832.
Suzuki, A., Vidal, J., Nguyen, J. and Gadal, P. (1984). *FEBS Lett.* **173**, 204–208.

Suzuki, A., Audet, C. and Oaks, A. (1987). *Plant Physiol.* **84**, 578–581.
Tempest, D. W., Meers, J. L. and Brown, C. M. (1970a). *Biochem. J.* **117**, 405–407.
Tempest, D. W., Meers, J. L. and Brown, C. M. (1970b). *J. Gen. Microbiol.* **64**, 187–194.
Tingey, S. V. and Coruzzi, G. M. (1987). *Plant Physiol.* **84**, 366–373.
Tingey, S. V., Walker, E. L. and Coruzzi, G. M. (1987). *EMBO J.* **6**, 1–9.
Tischner, R. and Hüttermann, A. (1980). *Plant Physiol.* **66**, 805–808.
Tobin, A. K., Ridley, S. M. and Stewart, G. R. (1985). *Planta* **163**, 544–548.
Vance, C. P., Egli, M. A., Griffith, S. M. and Miller, S. S. (1988). *Plant Cell Environm.* **11**, 413–427.
Vézina, L., Hope, H. J. and Joy, K. W. (1987). *Plant Physiol.* **83**, 58–62.
Wallsgrove, R. M., Harel, E., Lea, P. J. and Miflin, B. J. (1977). *J. Exp. Bot.* **28**, 588–596.
Wallsgrove, R. M., Keys, A. J., Bird, I. F., Cornelius, M. J., Lea, P. J. and Miflin, B. J. (1980). *J. Exp. Bot.* **31**, 1005–1017.
Wallsgrove, R. M., Lea, P. J. and Miflin, B. J. (1982). *Planta* **154**, 473–476.
Wallsgrove, R. M., Kendall, A. C., Hall, N. P., Turner, J. C. and Lea, P. J. (1986). *Planta* **168**, 324–329.
Wellner, V. P., Zoukis, M. and Meister, A. (1966). *Biochemistry* **5**, 3509–3514.
Yelton, M. M. and Yoch, D. C. (1981). *J. Gen. Microbiol.* **123**, 335–342.
Zlotnick, G. W. and Gottlieb, M. (1986). *Anal. Biochem.* **153**, 121–125.

16 Aminotransferases

ROBERT J. IRELAND[1] and KENNETH W. JOY[2]

[1]Department of Biology, Mount Allison University, Sackville, NB, E0A 3C0, Canada

[2]Department of Biology, Carleton University, Ottawa, Ontario, K1S 5B6, Canada

I.	Introduction	277
II.	Assays	278
III.	Aspartate aminotransferase	280
	A. Assay by OAA determination	280
	B. Assay by coupling to MDH	281
	C. Assay by coupling to GDH	281
	D. Extraction and purification	281
	E. Properties	282
IV.	Alanine : 2'-oxoglutarate aminotransferase	282
	A. Assay	283
	B. Extraction and purification	283
	C. Properties	283
V.	Glyoxylate aminotransferases	283
	A. Assay	284
	B. Extraction and purification	284
	C. Properties	285
VI.	Other aminotransferases	285
	References	286

I. INTRODUCTION

Following the assimilation of nitrogen into glutamate by the GS-GOGAT system, the glutamate can then act as a nitrogen source for the synthesis of most of the other amino acids. Nitrogen transfer from glutamate and other amino acids frequently involves the activity of aminotransferases, which are also known as transaminases. All primary amino acids can be acted upon by aminotransferases, and since these enzymes are reversible, they can act either in the synthesis or catabolism of a given amino acid. The general aminotransferase reaction is:

$$\begin{array}{c} R_1 \\ | \\ HCNH_2 \\ | \\ COOH \\ \text{amino acid 1} \end{array} + \begin{array}{c} R_2 \\ | \\ C{=}O \\ | \\ COOH \\ \text{keto acid 2} \end{array} \rightleftharpoons \begin{array}{c} R_1 \\ | \\ C{=}O \\ | \\ COOH \\ \text{keto acid 1} \end{array} + \begin{array}{c} R_2 \\ | \\ HCNH_2 \\ | \\ COOH \\ \text{amino acid 2} \end{array}$$

where an amino group is transferred from the α-carbon of amino acid 1 to the α-carbon of keto acid 2, leaving the keto acid analogue (keto acid 1) of amino acid 1, and producing amino acid 2. Pyridoxal 5'-phosphate (PLP) is a coenzyme in this reaction, but in plants is very tightly bound to the enzyme, obviating the need to add it to most assay mixtures. Aminotransferases do not appear to be regulated by other metabolites or ions; their activities are largely controlled by substrate and product concentrations, or perhaps by local pH. Aminotransferases are found in most subcellular compartments, such as the cytosol, mitochondrion, chloroplast and peroxisome, where they play roles not only in nitrogen redistribution, but also in more specialised activities such as photorespiration, carbon and hydrogen shuttles, and the synthesis of secondary metabolites.

Plants were first shown to possess aminotransferase activity in 1938 by Virtanen and Laine. Since that time many plant aminotransferases have been discovered and characterised, but few have been extensively purified; characterisation of these enzymes has largely been done on partially purified preparations.

Since the aminotransferases are so similar in many of their properties and in the reactions they catalyse, the emphasis in this review will be placed on those enzymes which have received the most attention in the literature. In most cases, the methods and comments relating to these enzymes can also be applied to other aminotransferases.

Because of the reversible nature of aminotransferases, the same enzyme may be given different names according to the direction being studied at the time. A prime example of this is EC 2.6.1.1, known variously as glutamate:oxaloacetate aminotransferase (GOT, more common in the animal literature), asparate:2'-oxoglutarate aminotransferase, or as is usual when the enzyme is from a plant source, simply aspartate aminotransferase. The names used in this review are those commonly found in the plant literature, but the inclusion of EC numbers will help reduce confusion.

Several comprehensive reviews on plant aminotransferases have been published in the last ten years (Wightman and Forest, 1978; Givan, 1980; Ireland and Joy, 1985). Reviews on assay procedures are also available (Bergmeyer and Bernt, 1974; Cooper, 1985; see also Ashton *et al.*, Chapter 3, this volume).

II. ASSAYS

Assay methods usually involve measurement of the amino acid product by chromatography (e.g. HPLC, TLC or paper), or the keto acid product by coupling to another reaction, usually one that involves NAD^+ or $NADP^+$, which permits assay by monitoring change in absorbance at 340 nm. The keto acid product can also be determined by radioactive means or by derivatisation with dinitrophenylhydrazone (DNP) or other reagents. The authors prefer methods which quantify all of the amino acids present at the beginning and end of the incubation period, such as amino acid analysis by HPLC of o-phthaldialdehyde (OPA) derivatives or traditional ion-exchange. These methods can be expensive and time-consuming, but have the considerable advantage of revealing the presence of contaminating amino acids in substrates and the presence of proteases (which can produce unwanted amino acids that may interfere with the reaction) in the extract being assayed (Streeter, 1977).

Similar assay procedures can be used for numerous aminotransferases: for example, thin layer or other chromatographic separations of amino acid products can be used in the assay of nearly all aminotransferases. Assays are usually specific for individual products, not individual enzymes, so many of the assays described can be used, with only slight modification, for other aminotransferases. The following is a simple general procedure which we have found useful for assaying several aminotransferase activities in crude extracts and partially purified preparations.

The assay mixture (1.5 ml), containing Tris-HCl buffer (50 mM; pH 7.8), amino acid (20 mM), oxo acid (20 mM), and enzyme (used to start the reaction), is incubated at 30°C for up to 60 min. At various times (e.g. 0, 10, 20, 30 min), aliquots (200 μl) are removed and placed into 1.5 ml microcentrifuge tubes containing a 'stop solution', the nature of which will depend on the subsequent chromatographic procedure to be used. The amino acid product can be determined by amino acid analysis or by thin layer or paper chromatography followed by elution and spectrophotometric quantification of the relevant spots. If amino acid analysis is to be used, the stop solution consists of 100 μl sulphosalicylic acid (250 mg ml^{-1}), which stops the reaction and precipitates the proteins. After centrifugation for 3 min at $12\,000 \times g$, an aliquot of the supernatant is removed, adjusted to the required pH, then stored frozen until analysis. If paper (large sheets of Whatman No. 4) or thin layer (20 cm silica gel or cellulose) chromatography is to be used, the stop solution is 200 μl of 95% ethanol. After centrifugation the supernatant is removed and applied to the paper or plate. The solvent used will depend on the substrate/product to be separated, e.g. butanol–acetic acid–water (10:2:5 by volume) gives a good separation of aspartate, glutamate and alanine. After chromatography the papers/plates are dried and then sprayed with ninhydrin (0.5 g ninhydrin, 0.05 g cadmium acetate, 1.0 ml acetic acid, 5.0 ml water, 50 ml acetone). The chromatograms are left in an enclosed, darkened container in the presence of sulphuric acid vapour for 12 h, then the spots cut out/scraped off and the colour eluted into 5 ml methanol for 2 h. After centrifugation, the absorbance of these solutions is determined at 500 nm. By chromatographing and eluting standards of known concentration, the amount of amino acid in each spot can be determined: the relationship between absorbance and the amount of each amino acid present in the spot is linear over the range 1–50 μg (true for all of the common or 'protein' amino acids). (See Forest and Wightman, 1971; Lea, 1982.)

Notes:

(a) Crude extracts should be desalted prior to assay by dialysis or by passage through a small (9×55 mm) Sephadex G-25 column.

(b) When using glyoxylate as the oxo acid, use 10 mM or less, since higher concentrations frequently inhibit the reaction.
(c) Most aminotransferases have broad pH optima, and are usually active at pH 7.8. This may need to be adjusted in some cases: we have found peroxisomal and chloroplastic aminotransferases tend to have optima around pH 8–8.4, whereas their cytosolic counterparts have optima in the range pH 7.2–8.0.
(d) The addition of PLP to the assay mixture is not usually necessary: in highly purified extracts or in unusual cases the addition of 50–100 μM PLP may be required.
(e) Control assays should be done without amino acid and without oxo acid. When glyoxylate is being used, a boiled enzyme control should also be done since non-enzymic transfer of amino groups to glyoxylate can occur.
(f) A useful inhibitor, fairly specific for aminotransferases, is aminooxyacetic acid, which usually completely inhibits aminotransferase activity at 1 mM in assays. Isonicotinic acid and hydroxylamine are also used but these are less specific and less reliable (see Givan, 1980).

III. ASPARTATE AMINOTRANSFERASE (EC 2.6.1.1)

Aspartate + 2′-oxoglutarate ⇌ Oxaloacetate + glutamate

This is certainly the most well studied of the plant aminotransferases: it is present in all tissues examined thus far, and is frequently present as organelle-specific isoenzymes. Huang et al. (1976) report the presence of four distinct isozymes of aspartate aminotransferase in spinach leaves, each in a separate subcellular compartment—cytosol, chloroplast, mitochondrion and peroxisome. In C_4 plants, different forms of aspartate aminotransferase are found in high levels in bundle sheath and mesophyll cells (e.g. Hatch, 1973; Hatch and Mau, 1973).

Assays can involve chromatographic separation and quantification of the amino acid product (see above), direct spectrophotometric measurement of oxaloacetate (OAA), or coupling to malate dehydrogenase (MDH) or glutamate dehydrogenase (GDH). Radioactive assays have also been described (e.g. Wong and Cossins, 1969; Reynolds et al., 1981). Some recipes include PLP, others do not: the PLP requirement should be checked for the particular enzyme under investigation, but it is unlikely that PLP will be required in assays of unpurified preparations.

A. Assay by OAA Determination

This procedure is subject to interference from other reactions and is often unsuitable for crude extracts, especially those that contain large amounts of chlorophyll. However, it offers the convenience of being done in the spectrophotometer, and provides a direct, real-time measurement of one of the products. It is useful when characterizing enzymes that have been at least partially purified. The assay (total volume, 3 ml) is performed in a cuvette containing aspartate (10–50 mM), 2′-oxoglutarate (5–25 mM), PLP (50 μM), Tris-HCl buffer (50 mM, pH 8.0), and enzyme, at 25–35°C. The production of OAA is followed by monitoring the increase in absorbance at 280 nm for 3–5 min (Ellis and Davies, 1961; Wong and Cossins, 1969; Kanamori and Matsumoto, 1974).

Notes:

(a) The reaction is usually started by the addition of 2′-oxoglutarate, after the other components have been pre-incubated for 5–10 min.
(b) Controls are run containing either no 2′-oxoglutarate or no aspartate
(c) Other buffers and pH values have been used, e.g. phosphate (0.2 M, pH 7–8.5).

B. Assay by Coupling to MDH

This procedure uses MDH to reduce the OAA product to malate, with the concomitant oxidation of NADH to NAD^+, which can be followed spectrophotometrically. It has the advantage of being simple and rapid, but close attention must be paid to the controls. The reaction mixture (1 ml) contains Hepes buffer (50 mM, pH 7.5), aspartate (25 mM), NADH (0.2 mM), 2′-oxoglutarate (25 mM), MDH (5 units), and enzyme, which is added to start the reaction. Activity is monitored by following the decrease in absorbance at 340 nm (e.g. Hatch, 1973; Reed and Hess, 1975; Huang *et al.*, 1976; Balkow and Wildner, 1982).

Notes:

(a) Controls done with no aspartate or no 2′-oxoglutarate.
(b) Other buffers include Tris-HCl and potassium phosphate (0.06–0.1 M, pH 7.5–8).

C. Assay by Coupling to GDH

This is a variation on the above procedure: 2′-oxoglutarate produced in the reverse reaction is measured by using GDH, which again results in the oxidation of NADH (e.g. Huang *et al.*, 1976). The reaction mixture (1 ml) contains potassium phosphate buffer (60 mM, pH 7.5), glutamate (10 mM), NAD(P)H (0.14 mM), GDH (3 units), NH_4Cl (3 mM) and OAA (10 mM). The assay is started by the addition of enzyme and absorbance at 340 nm is recorded.

D. Extraction and Purification

There are few reports of this enzyme being extensively purified from plants, but many of partial purification from a variety of plant tissues, usually involving ammonium sulphate precipitation followed by ion-exchange and gel permeation chromatography. The enzyme is fairly stable during extraction (all operations should be done at 0–5°C) and on subsequent storage (5°C or −20°C). Tissue is usually disrupted (using a pestle and mortar or a blender) in four volumes of buffer, commonly Tris or phosphate (0.05–0.2 M, pH 7–7.5, 5°C), sometimes containing additives such as glycerol (20%), EDTA (2 mM) or isoascorbate (2 mM). The homogenate is filtered (e.g. through eight layers of cheesecloth) and clarified by centrifugation (12–20 000 × g for 10–20 min) before precipitation and chromatography.

Reynolds et al. (1981) described such a procedure for aspartate aminotransferase from lupin nodules, in which the enzyme is precipitated between 40 and 75% ammonium sulphate and then centrifuged as above. The pellet is resuspended in about 5 ml of Tris-HCl (10 mM, pH 7.5, containing 20 µg ml^{-1} PLP) then applied to a Sephadex G-100 column (40 × 2 cm). The column is equilibrated and eluted with Tris-HCl (10 mM, pH 7.5, containing 20 µg ml^{-1} PLP). Active fractions are pooled and applied to a DEAE-cellulose column (25 × 1.5 cm, equilibrated with 10 mM Tris-HCl, pH 7.5). The enzyme is eluted (15–20 ml h^{-1}) from the column with a gradient (300 ml) of 0–0.25 M NaCl (in 10 mM Tris-HCl, pH 7.5). This resolves two forms of aspartate aminotransferase, one eluting at 75 mM NaCl (form 1), the other at 160 mM (form 2). Analysis by PAGE showed form 1 to contain only one major protein, which corresponded with an aspartate aminotransferase activity stain. Further purification of form 2 involves desalting by passage through Sephadex G-25 (40 × 2 cm, equilibrated with 10 mM Tris-HCl, pH 7.5), followed by chromatography on DEAE-Sephacel (17.5 × 1.0 cm, equilibrated with 10 mM Tris-HCl, pH 7.5). Elution is with a linear gradient (200 ml) of 0–0.5 M NaCl (in 10 mM Tris-HCl, pH 7.5), at a flow rate of 15 to 20 ml h^{-1}. The enzyme elutes at 275 mM NaCl, and produced only one band on PAGE. Both forms of the enzyme are stabilised by the addition of glycerol (35–43%) and PLP (100 µg ml^{-1}), and can be stored for several weeks at $-20°C$. For all chromatographic procedures, fractions are collected into tubes that contain 100 µl of PLP (500 µg ml^{-1}).

Reed and Hess (1975) described a similar procedure for aspartate aminotransferase from oat leaves, which also resolved two forms of the enzyme. Following ammonium sulphate precipitation and desalting on Sephadex G-15, the enzyme was applied to a DEAE cellulose column and two peaks of activity were eluted by use of a 0–0.2 M NaCl gradient. The enzyme eluting at low ionic strength was further purified on DEAE-Sephadex, and the higher ionic strength form by ammonium sulphate precipitation (50–55%) followed by preparative PAGE.

E. Properties

As with all aminotransferases, Michaelis–Menten kinetics are observed: the K_m for aspartate is usually between 1 and 4 mM; for oxoglutarate, 0.2 to 1.0 mM; for OAA, 0.02 to 0.1 mM, but for glutamate is quite variable, from 4 to 36 mM (e.g. Wong and Cossins, 1969; Reed and Hess, 1975). Most reports describe the enzyme to be highly specific for these four substrates (e.g. Wong and Cossins, 1969; Reynolds et al., 1981). Optimal pH is between 7.5 and 8.9 (e.g. Wong and Cossins, 1969; Forest and Wightman, 1972). The molecular weight is reported to be between 75 (Verjee and Evered, 1969) and 130 kDa (Reed and Hess, 1975; Forest and Wightman, 1972), but most descriptions place it at about 100 kDa consisting of two identical subunits (e.g. Reynolds et al., 1981).

IV. ALANINE : 2'-OXOGLUTARATE AMINOTRANSFERASE (EC 2.6.1.2)

$$\text{Alanine} + 2'\text{-oxoglutarate} \rightleftharpoons \text{Pyruvate} + \text{glutamate}$$

This enzyme, referred to as GPT (glutamate:pyruvate transaminase) in the animal literature, has been extracted from various plant tissues, and has been found in mitochondria (Yu and Spencer, 1970), plastids (Thomas and Stoddart, 1974), peroxisomes (Noguchi and Hayashi, 1981) and cytosol (Yu and Spencer, 1970; Thomas and Stoddart, 1974). As with aspartate aminotransferase, high levels are found in C_4 plants: different forms are found in mesophyll and bundle sheath cells (Hatch and Mau, 1973).

A. Assay

In either direction, the amino acid product can be assayed chromatographically as above, or, in the oxoglutarate-producing direction, the reaction can be coupled to GDH, as described for aspartate aminotransferase (in Section III.C, replace the OAA with pyruvate; see Rech and Crouzet, 1974). In the other direction, pyruvate production can be followed by coupling the reaction with lactate dehydrogenase (LDH), and monitoring the associated NADH consumption (Rech and Crouzet, 1974; Splittstoesser et al., 1976). This reaction mixture (3 ml) contains 50 mM Tris (pH 7.25), 50 mM alanine, LDH (5 units), 0.1 mM NADH, and enzyme. Following 5 min pre-incubation, the reaction is started by the addition of 3 mM oxoglutarate, and A_{340} is monitored. A control containing no oxoglutarate is also run.

Other assays include measurement of radioactive oxoglutarate produced from the transamination of labelled glutamate (e.g. Hatch and Mau, 1973), and in the other direction, the spectrophotometric (546 nm) determination of the DNP derivative of the pyruvate produced (e.g. Lea, 1982).

B. Extraction and Purification

Similar procedures to those used for aspartate aminotransferase have been used. Rech and Crouzet (1974) obtained a 660-fold purification of alanine aminotransferase from tomato by ammonium sulphate precipitation followed by chromatography on DEAE-Sephadex, then twice on Sephadex G-200. Noguchi and Hayashi (1981) obtained a homogeneous preparation of alanine aminotransferase from cytosolic and peroxisomal fractions of spinach leaves: extracts were chromatographed on DEAE-cellulose, then subjected to isoeletric focusing followed by Sephacryl S-200 and HA chromatography.

C. Properties

The enzyme has a pH optimum between 8 and 8.5, and its molecular weight is 100 kDa (Rech and Crouzet, 1974; Noguchi and Hayashi, 1981). The Michaelis constant for alanine is 2–3 mM; for oxoglutarate, 0.1–0.9 mM; for glutamate, 1–5 mM, and for pyruvate, 0.02– 0.09 mM (Hatch, 1973; Rech and Crouzet, 1974). Highly purified alanine aminotransferase from spinach leaf peroxisomes also catalysed the transamination of glutamate with glyoxylate (Noguchi and Hayashi, 1981).

V. GLYOXYLATE AMINOTRANSFERASES

$$\text{Serine} + \text{glyoxylate} \rightleftharpoons \text{OH-pyruvate} + \text{glycine}$$
$$\text{Glutamate} + \text{glyoxylate} \rightleftharpoons 2'\text{-Oxoglutarate} + \text{glycine}$$
$$\text{Alanine} + \text{glyoxylate} \rightleftharpoons \text{Pyruvate} + \text{glycine}$$

The principal glyoxylate aminotransferases in plant cells are serine:glyoxylate aminotransferase (EC 2.6.1.45) and glutamate:glyoxylate aminotransferase (EC 2.6.1.4), which are reported also to catalyse alanine:glyoxylate transamination (e.g. Rehfield and Tolbert, 1972). They are located in the peroxisome where they participate in photorespiration. Ireland and Joy (1983a,b) found that the transamination of asparagine with glyoxylate or pyruvate was also catalysed by serine glyoxylate aminotransferase. These enzymes are also described by Blackwell et al., Chapter 8 this volume.

A. Assay

Glyoxylate aminotransferases can be assayed by measuring glycine production chromatographically (using 80% phenol, containing 0.004% 8'-hydroxyquinoline as the solvent: King and Waygood, 1968). For the serine enzyme, the hydroxypyruvate produced can be determined using hydroxypyruvate reductase. The reaction mixture (1.0 ml) contains 50 mM Hepes (pH 7.0), 0.15 mM NADH, 1.0 mM glyoxylate, 20 mM serine, hydroxypyruvate reductase (0.05 units) and enzyme (Nakamura and Tolbert, 1983). Absorption at 340 nm is monitored, and controls containing no serine are also run.

Alanine:glyoxylate aminotransferase can be assayed by coupling the pyruvate production with LDH (as in Section IV.A; replace OG with glyoxylate, see Nakamura and Tolbert, 1983).

Radioactive assays are usually used to determine glutamate:glyoxylate aminotransferase activity: the assay mixture (0.5 ml) contains 50 mM Hepes (pH 7.1), 1.0 mM [1-^{14}C]glyoxylate (0.03 µCi), 20 mM glutamate, and the reaction started by the addition of enzyme. After 20 min, 0.2 ml H_2O_2 (18%) is added to oxidise the glyoxylate to CO_2 and then 0.2 ml of 2 M HCl is added to drive off the CO_2. After leaving to stand for 60 min, 10 ml of scintillation fluid is added and the [1-^{14}C]glycine produced is counted in a scintillation counter (Nakamura and Tolbert, 1983). For other radioactive assays, see Rehfield and Tolbert (1972), and Walton and Butt (1981).

Asparagine aminotransferase activity can be measured in purified or partially purified preparations by spectrophotometric determination of the product, 2'-oxosuccinamate, which has an absorption maximum at 289 nm at alkaline pH. The assay mixture (1.5 ml) contains 20 mM asparagine, 8 mM glyoxylate (or 20 mM pyruvate), 50 mM Tris-HCl (pH 8.1), 20 µM PLP (for purified enzyme only) and is started by the addition of enzyme. The mixture is incubated at 30°C for up to 60 min, and aliquots (0.3 ml) are removed at intervals, mixed with 0.133 M NaOH (0.9 ml), and their A_{289} measured. Controls without enzyme should also be run.

B. Extraction and Purification

The following procedure was employed (Ireland and Joy, 1983a) to produce a 90% pure (as determined by PAGE) preparation of serine:glyoxylate aminotransferase from pea leaves. Tissue (25 g) is homogenised in 5 vols of 50 mM Tris-HCl (pH 7.5, containing 2 mM dithiothreitol), filtered through four layers of cheesecloth and the pH adjusted to 5.4 with HCl. Following centrifugation (20 000 × g, 10 min) the supernatant is adjusted to pH 8.0 and PEG 6000 (50% w/w in 50 mM Tris-HCl, pH 8.0) added slowly to give a final concentration of 9% (w/w). After centrifugation, the PEG concentration is raised to 17% (w/w) and the suspension centrifuged again. The pellets are resuspended in

10 ml imidazole buffer (10 mM, pH 7.1) and pumped into the bottom of a column (60 × 2.5 cm) of DEAE-Sephadex A-25. The proteins are eluted off the top of the column by ion-filtration chromatography (see Kirkegaard, 1973), the gradient being developed by pumping 0.5 M KCl (in 10 mM Imidazole buffer, pH 7.1) up through the column at 1.5 ml min^{-1}. Active fractions are pooled and concentrated by precipitation with 17% PEG as above. Pellets are resuspended in 5 ml Tris-HCl (50 mM, pH 7.0) and applied to a column of Sephacryl S-300 (85 × 2.5 cm) equilibrated and eluted with the same buffer at 1 ml min^{-1}. Active fractions are pooled and applied to DEAE-Sephacel (20 × 1.6 cm), then eluted with a 0–0.4 M gradient of KCl (in 50 mM Tris-HCl, pH 7.0). Active fractions are pooled and glycerol added to give a final concentration of 15% (v/v). The enzyme loses less than 20% activity over 5 days at 5°C.

C. Properties

As with other aminotransferases, the molecular weights of glyoxylate aminotransferases are in the range 100–105 kDa (Noguchi and Fujiwara, 1982; Ireland and Joy, 1983a). For serine glyoxylate aminotransferase, pH optima have been reported between 7.0 (spinach leaf: Nakamura and Tolbert, 1983) and 8.2 (wheat leaves: King and Waygood, 1968). The K_m values are usually reported to be 0.4–0.9 mM for serine, 0.2–0.6 mM for glyoxylate (King and Waygood, 1968; Smith, 1973), 4 mM for asparagine and 2–5 mM for pyruvate, which can substitute for glyoxylate in the transamination of serine (e.g. Rehfield and Tolbert, 1972) and asparagine (Ireland and Joy, 1983a). The apoenzyme of pea leaf serine:(asparagine):glyoxylate aminotransferase showed an absolute requirement for PLP (Ireland and Joy, 1983a): the K_m was 2.5 µM. Unlike most aminotransferases, serine:glyoxylate aminotransferase does not appear to be freely reversible, requiring high substrate concentrations to drive the reaction in the direction of glyoxylate production (see Nakamura and Tolbert, 1983; Carpe and Smith, 1974).

Glutamate:glyoxylate aminotransferase has a pH optimum at 8.0–8.5, with K_m values of 1.7–3.8 mM for glutamate and 0.2–2.4 mM for glyoxylate. For alanine:glyoxylate aminotransferase activity, K_m values of 1.6–3.1 mM for alanine and 0.15 mM for glyoxylate have been reported (Noguchi and Fujiwara, 1982; Nakamura and Tolbert, 1983).

VI. OTHER AMINOTRANSFERASES

For descriptions of other plant aminotransferases the reader is referred to the reviews cited in the introduction. Purification, assays and properties are generally similar to those described above. For example, an enzyme that catalysed the transamination of branched-chain amino acids was purified from barley leaves by ammonium sulphate precipitation, gel permeation and ion-exchange chromatography, and had a pH optimum of 8.4, an M_r of 95 kDa and amino acid K_m values of 0.8–1.8 mM (Aarnes, 1981). There are some unique purification and assay procedures for individual aminotransferases, e.g. for arogenate and prephenate aminotransferases (Bonner and Jensen, 1987), and for aromatic amino acid aminotransferases (e.g. Gamborg and Wetter, 1963; Matheron and Moore, 1973). De-Eknamkul and Ellis (1987) used a combination of ion-exchange and chromatofocusing to purify three forms of tyrosine aminotransferase from *Anchusa officinalis*, which had M_r values ranging from 180–220 kDa and pH optima from 8.8–9.6.

REFERENCES

Aarnes, H. (1981). *Z. Pflanzenphysiol.* **102**, 81–89.
Balkow, C. and Wildner, G. F. (1982). *Planta* **154**, 477–484.
Bergmeyer, H. U. and Bernt, E. (1974). *In* "Methods of Enzymatic Analysis", 2nd edn (H. U. Bergmeyer, ed.), pp. 727–767. Academic Press, New York.
Bonner, C. and Jensen, R. (1987). *In* "Methods in Enzymology" (S. Kaufman, ed.), Vol. 142, pp. 479–494. Academic Press, New York.
Carpe, A. I. and Smith, I. K. (1974). *Biochim. Biophys. Acta* **370**, 96–101.
Cooper, A. J. L. (1985). *In* "Methods in Enzymology" (A. Meister, ed.), Vol. 113, pp. 66–82. Academic Press, New York.
De-Eknamkul, W. and Ellis, B. E. (1987). *Arch. Biochem. Biophys.* **257**, 430–438.
Ellis, R. J. and Davies, D. D. (1961). *Biochem. J.* **78**, 615–623.
Forest, J. C. and Wightman, F. (1971). *Can. J. Biochem.* **49**, 709–720.
Forest, J. C. and Wightman, F. (1972). *Can. J. Biochem.* **50**, 813–829.
Gamborg, O. L. and Wetter, L. R. (1963). *Can. J. Biochem. Physiol.* **41**, 1733–1740.
Givan, C. V. (1980). *In* "The Biochemistry of Plants" (B. J. Miflin, ed.), Vol. 5, pp. 329–357. Academic Press, New York.
Hatch, M. D. (1973). *Arch. Biochem. Biophys.* **156**, 207–214.
Hatch, M. D. and Mau, S.-L. (1973). *Arch. Biochem. Biophys.* **156**, 195–206.
Huang, H. C., Liu, K. D. F. and Youle, R. J. (1976). *Plant Physiol.* **58**, 110–113.
Ireland, R. J. and Joy, K. W. (1983a). *Arch. Biochem. Biophys.* **223**, 291–296.
Ireland, R. J. and Joy, K. W. (1983b). *Plant Physiol.* **72**, 1127–1129.
Ireland, R. J. and Joy, K. W. (1985). *In* "Transaminases" (P. Christen and D. E. Metzler, eds), pp. 376–384. John Wiley & Sons, New York.
Kanamori, T. and Matsumoto, H. (1974). *Plant Sci. Lett.* **3**, 431–436.
King, J. and Waygood, E. R. (1968). *Can. J. Biochem.* **46**, 771–779.
Kirkegaard, L. E. (1973). *Biochemistry* **12**, 3627–3632.
Lea, P. J. (1982). *In* "Techniques in Bioproductivity and Photosynthesis" (J. Coombs and D. O. Hall, eds), p. 135. Pergamon Press, Oxford.
Matheron, M. E. and Moore, T. C. (1973). *Plant Physiol.* **52**, 63–67.
Nakamura, Y. and Tolbert, N. E. (1983). *J. Biol. Chem.* **258**, 7631–7638.
Noguchi, T. and Hayashi, S. (1981). *Biochem. J.* **195**, 235–239.
Noguchi, T. and Fujiwara, S. (1982). *Biochem. J.* **201**, 209–214.
Rech, J. and Crouzet, J. (1974). *Biochim. Biophys. Acta* **350**, 392–399.
Reed, R. E. and Hess, J. L. (1975). *J. Biol. Chem.* **250**, 4456–4461.
Rehfield, D. W. and Tolbert, N.E. (1972). *J. Biol. Chem.* **247**, 4803–4811.
Reynolds, P. H. S., Boland, M. J. and Farnden, K. J. F. (1981). *Arch. Biochem. Biophys.* **209**, 524–533.
Smith, I. K. (1973). *Biochim. Biophys. Acta* **321**, 156–164.
Splittstoesser, W. E., Chu, M. C., Stewart, S. A. and Splittstoesser, S. A. (1976). *Plant Cell Physiol.* **17**, 83–89.
Streeter, J. G. (1977). *Plant Physiol.* **60**, 235–239.
Thomas, H. and Stoddart, J. L. (1974). *Phytochemistry* **13**, 1053–1058.
Verjee, Z. H. M. and Evered, D. F. (1969). *Biochim. Biophys. Acta* **185**, 103–110.
Virtanen, A. I. and Laine, T. (1938). *Nature* **141**, 748–749.
Walton, N. J. and Butt, V. S. (1981). *Planta* **153**, 232–237.
Wightman, F. and Forest, J. C. (1978). *Phytochemistry* **17**, 1455–1471.
Wong, K. F. and Cossins, E. A. (1969). *Phytochemistry* **8**, 1327–1338.
Yu, M. H. and Spencer, M. (1970). *Phytochemistry* **9**, 341–343.

17 Enzymes of Asparagine Metabolism

KENNETH W. JOY[1] and ROBERT J. IRELAND[2]

[1]*Biology Department, Carleton University, Ottawa, Ontario, K1S 5B6, Canada*

[2]*Biology Department, Mount Allison University, Sackville, NB, E0A 3C0, Canada*

I.	Introduction	287
II.	Asparagine synthetase	288
	A. Assay	288
	B. Extraction and purification	290
	C. Properties of asparagine synthetase	291
III.	Asparaginase	291
	A. Assay	292
	B. Extraction and purification	293
	C. Properties of asparaginase	294
IV.	Methods for assay of other enzymes of asparagine metabolism	294
	A. Asparagine aminotransferases	294
	B. 3-Cyanoalanine metabolism	295
	References	295

I. INTRODUCTION

In addition to the requirement as a protein amino acid, asparagine has a secondary role in many plants as a nitrogenous transport and storage compound. It is frequently produced in germinating seedlings, and is also synthesised in roots and nitrogen-fixing nodules, leading to its appearance as a major component of the xylem sap transported from roots to shoots. Asparagine is also synthesised by leaves. In most instances, synthesis appears to be through the action of asparagine synthetase, which adds an amide group to aspartate, utilising the amide group of glutamine, or ammonia. In spite of the common presence of the amide, asparagine synthetase has been assayed in high activity in only a few tissues, notably germinating legume seeds and legume root nodules. In cyanogenic plants, asparagine may also be synthesised by the condensation of cyanide and cysteine, and hydrolysis of the resulting 3-cyanoalanine.

The distribution of enzymes which degrade asparagine is fairly limited, and this may be a reason for the value of asparagine as a stable storage and transport compound. Nevertheless, in rapidly growing tissues (developing leaves or seeds) there is active metabolism of asparagine to provide nitrogen for amino acid synthesis. Main routes for utilisation involve either deamidation (by asparaginase) to release ammonia and aspartate, or initial transamination followed by subsequent deamidation of the transamination products. Recent reviews on asparagine metabolism are available (Lea and Miflin, 1980, and Sieciechowicz et al., 1988).

II. ASPARAGINE SYNTHETASE (EC 6.3.5.4)

$$\text{Aspartate} + \text{glutamine} + \text{ATP} \longrightarrow \text{Asparagine} + \text{glutamate} + \text{AMP} + \text{PPi}$$

A. Assay

Direct assay of asparagine, by use of HPLC or an amino acid analyser, provides an unequivocal method of measurement, and a direct check for potential problems from proteolysis or transamination during the assay. Alternatively, the detection of labelled asparagine from [^{14}C]aspartate has been frequently used; careful separation of the small amount of labelled product from precursor substrate is needed, for example by ion-exchange (Prusiner and Milner, 1970; Streeter, 1973). As the differential binding of aspartate and asparagine is dependent on pH (Prusiner and Milner, 1970) and may be modified by changes of medium, a check of the method under the conditions of use is recommended. Asparagine synthetase will also utilise hydroxylamine, producing aspartyl hydroxamate (Rognes, 1975), but the identical reaction is also performed by aspartate kinase (Wong and Dennis, 1973; Aarnes and Rognes, 1974; see Bonner and Lea, Chapter 18 this volume), so this method is of doubtful value for asparagine synthetase.

Plant extracts can be assayed directly, or after preliminary clean-up by ammonium sulphate precipitation; removal of small molecular weight components (with Sephadex G-25 or similar material, equilibrated with extraction buffer, using a bed volume at least three- to four-fold the volume of enzyme) is desirable. The unpurified enzyme is quite unstable, and various stabilising components are required in the extraction (and assay) medium.

1. Procedure

The assay may be carried out in 1.5 ml microcentrifuge tubes. The reaction is started by adding enzyme to an assay mix to give the following final concentrations in a final volume of 0.5 ml: Tris-HCl buffer (pH 7.8), 50 mM; aspartate, 10 mM (with or without added label); glutamine, 5 mM; ATP, 5 mM; magnesium sulphate, 10 mM; DTT, 2 mM; and EDTA, 0.1 mM. Incubate at 30°C for 30–60 min.

2. Termination of reaction and estimation of product

(a) *Amino acid analysis.* Terminate the reaction by adding 0.1 ml sulphosalicylic acid solution (containing 200 mg ml^{-1}), centrifuge 3 min at $10\,000 \times g$ to remove protein. Remove an aliquot, then bring to the appropriate pH and dilution for the amino acid analysis system available. The analytical system must be able to resolve asparagine from glutamine and aspartate. Modification of a simple HPLC method used for glutamine assay may be suitable (Martin *et al.*, 1982). Note: with pre-column derivatisation methods, care must be taken that the derivatising agent is not exhausted by the high concentrations of aspartate and glutamine in the reaction mix; halved concentrations of these substrates may be preferable.

(b) *Detection of [^{14}C]asparagine.* [^{14}C]Aspartate (0.25 µCi or up to 10^6 cpm) must also be added to the assay mixture. For crude preparations, 4-[^{14}C]aspartate may be preferable to uniformly labelled substrate (Rognes, 1970).

The reaction is terminated by addition of 1 ml ethanol, followed by centrifugation. The supernatant is decanted, evaporated, and taken up in 0.7 ml of water, then added to a column (5 × 50 mm) of washed Dowex 1 (8%, 200–400 mesh, in the formate form). Asparagine does not bind, and the solution running through, together with a 1.5 ml water wash, is collected in a scintillation vial, and counted after addition of scintillant (Streeter, 1973, adds the assay mix directly to the column). Separation by electrophoresis (Rognes, 1970), or paper chromatography (Lea and Fowden, 1975) has also been used.

Notes

(a) Appropriate controls must be carried out, including termination at zero time, and minus glutamine.
(b) Aspartate and ATP require neutralisation; the final pH of the bulk assay mix should be checked.
(c) If ammonia is to be used as an alternative source of amide nitrogen, 10 mM ammonium sulphate should replace glutamine.
(d) The reaction requires Cl$^-$ ions (Rognes, 1980). Addition of chloride is necessary if buffers without chloride replace Tris-HCl.
(e) Ca^{2+} is inhibitory (Rognes, 1975; Joy, 1985). Addition of EGTA to extraction/assay media may increase detectable activity.
(f) Some workers (e.g. Oaks and Ross, 1984) have added 1 mM aminooxyacetate to the assay mix to inhibit transaminase reactions.
(g) The presence of asparaginase in crude preparations may prevent the detection of asparagine synthesis in the assay (Rognes, 1970; Huber and Streeter, 1985). In germinating seedlings, the developing apex may contain considerable asparaginase activity, and careful separation of tissue is recommended when working with cotyledons.

B. Extraction and Purification

Germinating legume cotyledons, e.g. lupins (Rognes, 1975; Lea and Fowden, 1975), provide most active preparations of asparagine synthetase. Moderate levels of activity have been obtained from root nodules (Huber and Streeter, 1985; Ta et al., 1988), but low levels are found in leaves and roots (see Joy et al., 1983). Asparagine synthetase is unstable during extraction, and additives such as glycerol (15–25%) and sulphydryl reagents (1–50 mM) are routinely added to extraction media, while magnesium chloride, ATP, aspartate and insoluble polyvinyl pyrrolidone (Polyclar AT) have also been used (e.g. Huber and Streeter, 1985). For alfalfa nodules, Ta et al. (1988) found that aspartate, glutamine or ATP improved enzyme recovery by about 20%, but inclusion of 5 mM EGTA gave an improvement of 70% or more, supporting the suggestion that inhibitory components of tissues (Joy et al., 1983) may include calcium (Joy, 1985).

1. Extraction

Tissue should be disrupted in 2.5 to 4 vols of chilled buffer, using a mortar for small samples, or a homogenising blender, or high speed disrupter (e.g. Polytron). A typical extraction buffer contains Tris-HCl (pH 8.5, 100 mM), 2-ME (50 mM), EDTA (0.1 mM), $MgCl_2$ (10 mM), glycerol (20%; v/v). The effect of other additives, such as ATP (0.1 mM), aspartate (2 mM), KCN (1 mM) should be evaluated for optimum recovery; a trial of EGTA (5 mM) is recommended in view of the inhibitory effect of calcium.

All operations should be carried out at 0–4°C. After filtration (cheesecloth) and centrifugation, the extract may be assayed directly (after treatment with Sephadex G-25), or an enzyme fraction may be precipitated by adding solid ammonium sulphate to give 45% (or higher) saturation, and resuspension of the protein in extraction buffer. For some tissues (e.g. fatty cotyledons) the protein precipitated by ammonium sulphate will float, rather than sedimenting on centrifugation. The precipitation process may cause some loss of activity, but the enzyme stability is improved.

2. Purification

The following procedure for lupin cotyledons (described fully by Rognes, 1975, 1980) has given up to 600-fold purification, with a yield of 5% or better.

(a) Tissue (500 g lupin cotyledons) is homogenised in 800 ml of buffer (Tris-HCl, pH 8.5, 100 mM; 2-ME, 42 mM; EDTA, 0.1 mM; KCN, 1 mM; glycerol, 25%). The extract is filtered through cheesecloth and clarified (30 min, 13 000 × g).

(b) The extract is brought to 42% saturation with solid ammonium sulphate, with addition of further 2-ME, and ammonium hydroxide to maintain pH 7.5; after stirring for 30 min, the precipitate is collected (20 min, 30 000 × g) and dissolved in buffer B2 (to give approximately 20 mg protein ml^{-1}). (Buffer B2: Tris-HCl, pH 7.5, 20 mM; 2-ME, 42 mM; DTT, 0.5 mM; EDTA, 0.1 mM; KCN, 1 mM; glycerol, 20%.)

(c) Inactive protein is removed by slow addition of protamine sulphate (2% in buffer B2) to the stirred solution to give a concentration of 0.16 mg protamine sulphate per mg protein. After centrifugation, the supernatant is retained.

(d) Further inactive protein is removed from the supernatant by addition of alumina C_γ gel (26 mg dry weight per ml) to give a gel:protein ratio of 2.5. After stirring for 30 min, the precipitate is removed by centrifugation, retaining the supernatant.

(e) Ammonium sulphate is added to 38% saturation; after centrifugation, the precipitated enzyme is dissolved in buffer D2, and desalted by passage through a column of Sephadex G-75, equilibrated in the same buffer. (Buffer D2: composition as for B2, with addition of ATP, 1 mM; $MgCl_2$, 5 mM; glutamine, 2 mM.)

(f) The desalted enzyme is added to a column of DEAE Sephadex A-50 (2 × 5 cm), equilibrated with buffer D2, and is eluted with 100 ml of a linear gradient of 0–0.6 M KCl in buffer D2, the peak of activity emerging at about 0.27 M KCl.

(g) Enzyme is precipitated from pooled active fractions, using 40% saturation of ammonium sulphate, and the redissolved protein (in buffer D2) is passed through a column of Sephadex G-200 (2 × 40 cm), equilibrated and eluted with D2.

(h) Active fractions from the gel filtration step are pooled, reprecipitated with ammonium sulphate, and dissolved in buffer B2. The enzyme solution is then subjected to a second gel filtration step on Sephadex G-200, in this case equilibrated and eluted with B2. The most active fractions are pooled, and are stable for several months when stored at $-30°C$.

Lea and Fowden (1975) obtained a preparation (100-fold purification, 30% yield) from lupin cotyledons, using ammonium sulphate precipitation and gel filtration on Sephadex G-75 and G-200 columns. Huber and Streeter (1985) found that the purification method for cotyledons was unsuitable for soybean nodules, and devised a three-step procedure using ammonium sulphate precipitation, alumina C_γ gel precipitation of inactive protein, and separation on a column of Reactive Blue 2 Agarose (e.g. Cibacron Blue 3GA-Agarose, Sigma Chemical Co., St. Louis, MO, USA; see also Hongo and Sato, 1981); a purification of 240-fold was achieved, with a yield of about 5%.

C. Properties of Asparagine Synthetase

The properties of enzyme from two types of lupin cotyledons (Rognes, 1975; Lea and Fowden, 1975) and from soybean nodules (Huber and Streeter, 1985) are quite similar. The following Michaelis constants were reported: aspartate, 0.8–1.4 mM; glutamine, 0.04–0.16 mM; (Mg)ATP, 0.076–0.15 mM. The pH optimum is broad, from 7.6 to 8.3. There is a requirement for Cl^-, and the enzyme shows glutaminase activity (Rognes, 1980). It is inhibited by the glutamine analogues azaserine and albizziine (Lea and Fowden, 1975) and by calcium and some other ions (Rognes, 1975; Lea and Fowden, 1975). In the absence of substrates, gel filtration indicates a molecular weight of 160 kDa, which associates to a dimer (320 kDa) in the presence of ATP (Rognes, 1975). Ammonia can be used in place of glutamine as the source of the amide nitrogen, but the reaction rate is usually only 30–40% of that with glutamine, and the affinity for ammonia is at least 10-fold lower than for glutamine.

III. ASPARAGINASE (EC 3.5.1.1)

$$\text{Asparagine} + H_2O \longrightarrow \text{Aspartate} + NH_3$$

Some of the inconsistencies in reports of the presence of asparaginase were resolved by the demonstration that many asparaginase preparations have a strong requirement for potassium ions (Sodek et al., 1980). There is a K^+ requirement for enzyme from *Pisum sativum*, *Vicia faba*, *Phaseolus multiflorus*, *Zea*, *Hordeum*, and some *Lupinus* varieties, while enzyme from seeds of *Lupinus polyphyllus* (Sodek et al., 1980), *L. arboreus* and *L. angustifolius* (Chang and Farnden, 1981) is potassium independent. In *Lupinus*, the K^+ requirement may differ between varieties of the same species (Lea et al., 1984).

A. Assay

Methods which are similar to those used for asparagine synthetase can be employed for asparaginase, by detection of the aspartate product by direct assay or as the labelled aspartate formed from [^{14}C]asparagine. Detection of the ammonia produced can also be employed.

1. Procedure

The assay can be carried out in 1.5 ml microcentrifuge tubes. Start the reaction by addition of enzyme to reaction mix to give the following final concentrations, in a volume of 0.3 ml: Tris-HCl buffer (pH 8.0), 20 mM; asparagine, 20 mM (with or without added label); KCl, 40 mM; DTT, 2 mM. Incubate at 34°C for 30 min.

2. Termination of reaction and estimation of product

(a) *Amino acid analysis*. Proceed as described above for asparagine synthetase, using sulphosalicylic acid, and carry out analysis for aspartate production.

(b) *Ammonia estimation*. Direct Nesslerisation has been used (e.g. Chang and Farnden, 1981), but there is frequently interference by components of the extract, particularly in crude preparations, and the following microdiffusion procedure (Shelp et al., 1985) has proved to be more useful. Remove 0.2 ml from the assay medium, and transfer to a scintillation vial containing 1 ml saturated potassium carbonate, and close immediately with a prepared stopper assembly. (A rubber stopper, supporting a glass rod with rounded tip. The tip is dipped into concentrated sulphuric acid and drained just before use.) Ammonia diffuses from the solution to the acid on the rod (which is supported about 2 cm above the carbonate surface). After 60 min at room temperature, the stopper is removed, and the rod is stirred in 1.5 ml of Nessler's solution (Vanselow, 1940)* and after 5 min the absorbance at 420 nm is read.

(c) *Detection of [^{14}C]aspartate*. [^{14}C]Asparagine must be added to the assay mixture (0.25–1 µCi). Aspartate can be separated by ion-exchange (Prusiner and Milner, 1970). The reaction is terminated by addition of 1 ml of cold buffer containing Imidazole (20 mM, pH 7.0) and 75 mM asparagine, and the mixture is immediately applied to a

---

* Mix 22.75 g mercuric iodide and 17.45 g potassium iodide in the minimum volume (about 12 ml) of water. A nearly saturated solution of potassium hydroxide (56 g in 70 ml water, then cooled) is added slowly with stirring, and volume is made to 500 ml. Allow to stand for several days, then decant and store in a dark bottle.

column of Dowex 1 (5 × 35 mm, chloride form). The column is washed with 6 ml of 10 mM Imidazole buffer (pH 7.0) containing 30 mM asparagine, then the aspartate is eluted with 2 ml HCl (30 mM), which is collected in a scintillation vial for counting. Labelled aspartate can also be separated from asparagine by paper chromatography (Lea et al., 1978), thin layer chromatography (Sodek et al., 1980), or electrophoresis (Rognes, 1970; Chang and Farnden, 1981).

Notes

(a) Appropriate, particularly zero time, controls should be included.
(b) Labelled and unlabelled asparagine frequently have aspartate as an impurity. This should be removed by passing, at pH 7–7.5, over a small column of Dowex 1 (chloride or formate form) (Prusiner and Milner, 1970).
(c) Various workers have used asparagine concentrations ranging from 100 mM to carrier-free label; 20 mM is about 2–4 times the K_m values reported for the purified enzyme.
(d) The enzyme appears to be quite resistant to denaturation by ethanol (to at least 60%), making this reagent unsuitable for termination of the assay.
(e) In analysis by pre-column derivatisation with OPA (e.g. Martin et al., 1982), sulphosalicylic acid produces a peak at the injection front; solvent conditions must be adjusted so that aspartate is clearly resolved from this.

B. Extraction and Purification

Asparaginase is not widely distributed, and is most commonly found in growing tissues such as developing (not germinating) seeds or leaves. The enzyme is quite unstable, and stabilising additions are needed during purification. In view of the dependence of enzyme from several sources, potassium should be included in extraction media. Several workers have used phosphate buffers.

1. Extraction

Tissue should be disrupted in 2.5 to 4 volumes of chilled buffer, containing Tris-HCl (50 mM, pH 8.0), DTT (2 mM), KCl (50 mM), and $CaCl_2$ (1 mM). After filtration through cheesecloth and centrifugation, the extract can be assayed directly after removal of low molecular weight components by passage over Sephadex G-25.

2. Purification

Asparaginase (potassium-independent) has been purified from maturing seeds of *L. arboreus* and *L. angustifolius* (up to 320-fold, with a yield of 8.5%) by Chang and Farnden (1981), whose method is summarised here:

(a) Tissue (200 g maturing seeds) is crushed in 300 ml of extraction buffer (sodium phosphate, 100 mM, pH 8.0; PMSF, 5 mM; EDTA, 1 mM; glycerol, 17% w/v). The extract is squeezed through cheesecloth, and centrifuged (40 000 × g, 30 min).
(b) The supernatant is stirred and 13 ml of 1 M $MnCl_2$ is slowly added. After an hour, the solution is centrifuged (20 000 × g, 10 min), retaining the supernatant.

(c) Solid ammonium sulphate is added to bring the solution to 40% saturation, and the inactive protein precipitate is removed, then the supernatant is brought to 60% saturation, and the precipitated protein is collected, and dissolved in 20 ml buffer A (sodium phosphate, 20 mM, pH 8.0; PMSF, 1 mM; EDTA, 1 mM; sucrose, 10%).
(d) The enzyme is loaded onto a column of Sephadex G-100 (3 × 100 cm), equilibrated and eluted (12 ml h^{-1}) with buffer A.
(e) Fractions containing asparaginase are pooled, and loaded onto a column of DEAE cellulose (3 × 20 cm), equilibrated with buffer B (Tris-HCl, 20 mM, pH 8.0; PMSF, 1 mM; EDTA, 1 mM; sucrose, 10%). The column is washed with buffer B, then enzyme is eluted with a gradient of 0–0.5 M sodium chloride in buffer B (total volume, 500 ml; 15 ml h^{-1}). Pooled active fractions are concentrated by ultrafiltration.
(f) The concentrated solution is subjected to a second purification on Sephadex G-100 (2 × 45 cm), eluted with buffer A. Active fractions are again pooled and concentrated.
(g) The enzyme preparation is applied to a column (1.2 × 15 cm) containing a mixture of equal amounts of hydroxylapatite and Sephadex G-25, equilibrated in buffer A, and eluted with a gradient of sodium phosphate in buffer A (20–100 mM, 200 ml). Peak fractions are pooled and concentrated.
(h) The concentrate is applied to a column of DEAE Sephadex A-25 (0.8 × 10 cm), equilibrated in buffer B, and eluted with a gradient of sodium chloride in buffer B (0–0.5 M, 150 ml).

Lea et al. (1978) purified enzyme from *L. polyphyllus* (470-fold, 60% yield) using ammonium sulphate precipitation, DEAE cellulose and Ultragel chromatography, obtaining considerable purification by use of preparative gel electrophoresis. The resistance to ethanol (60%) has been used as a purification step (Sodek et al., 1980) although the extent of purification was not reported.

C. Properties of Asparaginase

The properties of potassium-independent enzyme from different lupin species are quite similar (Lea et al., 1978; Chang and Farnden, 1981). The K_m for asparagine ranges from 7 to 12 mM while pH optimum is 8.0–8.5. The enzyme has quite a high specificity, and does not deamidate glutamine. A number of inhibitors are effective, notably 5-diazo-4-oxo-L-norvaline (DONV), and 3-cyanoalanine. The M_r is reported to be 72–75 kDa. Highly purified preparations of the potassium-dependent enzyme have not been described, although a partially purified preparation (Sodek et al., 1980) had a K_m for asparagine of about 3.2 mM, and molecular weight of 68 kDa.

IV. METHODS FOR ASSAY OF OTHER ENZYMES OF ASPARAGINE METABOLISM

A. Asparagine Aminotransferases

Several aminotransferases, particularly serine-glyoxylate aminotransferase (Ireland

and Joy, 1983), are able to utilise asparagine, and may have a role in asparagine metabolism (see Sieciechowicz et al., 1988). Methods for these enzymes are described in accompanying chapters (Blackwell et al., Chapter 8 and Ireland and Joy, Chapter 16, this volume). However, a method specific for asparagine aminotransferases is that of Cooper (1977), in which the absorbance of the transamination product 2-oxo succinamate is measured.

Release of ammonia from 2-oxosuccinamate (or its reduced product 2-hydroxysuccinamate) by deamidation must be regarded as a final step in the utilisation of asparagine nitrogen. Streeter (1977) has measured the production and deamidation of 2-oxosuccinamate.

B. 3-Cyanoalanine Metabolism

The cyanoalanine pathway can lead to synthesis of asparagine, although it is regarded largely as a mechanism for metabolism and detoxification of cyanide (Wurtele et al., 1985).

1. 3-Cyanoalanine synthase (EC 4.4.1.9)

$$\text{L-Cysteine} + \text{HCN} \longrightarrow \text{3-cyanoalanine} + \text{H}_2\text{S}$$

Hendrickson and Conn (1969) have purified the enzyme from lupin seedlings, using an assay based on colorimetric determination of H_2S. More recently, the enzyme has been purified from spinach by Ikegami et al. (1988), who measured 3-cyanoalanine after separation by thin layer chromatography and reaction with ninhydrin.

2. 3-Cyanoalanine hydrolase (hydratase) (EC 4.2.1.65)

$$\text{3-Cyanoalanine} + \text{H}_2\text{O} \longrightarrow \text{Asparagine}$$

This enzyme has been purified from lupin seedlings by Castric et al. (1972), who describe two methods for assay—either estimation of ammonia released (by added asparaginase) from the asparagine product, or estimation of transfer of label from [^{14}C]cyanoalanine to asparagine, separated by paper chromatography.

REFERENCES

Aarnes, H. and Rognes, S. E. (1974). *Phytochemistry* **13**, 2717–2724.
Castric, P. A., Farnden, K. J. F. and Conn, E. E. (1972). *Arch. Biochem. Biophys.* **152**, 62–69.
Chang, K. S. and Farnden, K. J. F. (1981). *Arch. Biochem. Biophys.* **208**, 49–58.
Cooper, A. J. L. (1977). *J. Biol. Chem.* **252**, 2032–2038.
Hendrickson, H. R. and Conn, E. E. (1969). *J. Biol. Chem.* **244**, 2632–2640.
Hongo, S. and Sato, T. (1981). *Anal. Biochem.* **114**, 163–166.
Huber, T. A. and Streeter, J. G. (1985). *Plant Sci.* **42**, 9–17.
Ikegami, F., Takayama, K., Tajima, C. and Murakoshi, I. (1988). *Phytochemistry* **27**, 2011–2016.
Ireland, R. J., and Joy, K. W. (1983). *Arch. Biochem. Biophys.* **223**, 291–296.
Joy, K. W. (1985). *Plant Physiol.* **77**, S-34.
Joy, K. W., Ireland, R. J. and Lea, P. J. (1983). *Plant Physiol.* **73**, 165–168.

Lea, P. J. and Fowden, L. (1975). *Proc. Roy. Soc. Lond.* **B192**, 13–26.
Lea, P. J. and Miflin, B. J. (1980). *In* "The Biochemistry of Plants" (B. J Miflin, ed.), Vol. 5, pp. 569–602. Academic Press, New York.
Lea, P. J., Fowden, L. and Miflin, B. J. (1978). *Phytochemistry* **17**, 217–222.
Lea, P. J., Festenstein, G. N., Hughes, J. S. and Miflin, B. J. (1984). *Phytochemistry* **23**, 511–514.
Martin, F., Suzuki, A. and Hirel, B. (1982). *Anal. Biochem.* **125**, 24–29.
Oaks, A. and Ross, D. W. (1984). *Can. J. Bot.* **62**, 68–73.
Prusiner, S. and Milner, L. (1970). *Anal. Biochem.* **37**, 429–438.
Rognes, S. E. (1970). *FEBS Lett.* **10**, 62–66.
Rognes, S. E. (1975). *Phytochemistry* **14**, 1975–1982.
Rognes, S. E. (1980). *Phytochemistry* **19**, 2287–2293.
Shelp, B. J., Sieciechowicz, K. A., Ireland, R. J. and Joy, K. W. (1985). *Can. J. Bot.* **63**, 1135–1140.
Sieciechowicz, K. A., Joy, K. W. and Ireland, R. J. (1988). *Phytochemistry* **27**, 663–671.
Sodek, L., Lea, P. J. and Miflin, B. J. (1980). *Plant Physiol.* **65**, 22–26.
Streeter, J. G. (1973). *Arch. Biochem. Biophys.* **157**, 613–624.
Streeter, J. G. (1977). *Plant Physiol.* **60**, 235–239.
Ta, T.-C., MacDowall, F. D. H., Faris, M. A. and Joy, K. W. (1988). *Biochem. Cell. Biol.* **66**, 1349–1354.
Vanselow, A. P. (1940). *Ind. Eng. Chem. Anal. Ed.* **12**, 516–517.
Wong, K. F. and Dennis, D. T. (1973). *Plant Physiol.* **51**, 322–326.
Wurtele, E. S., Nikolau, B. J. and Conn, E. E. (1985). *Plant Physiol.* **78**, 285–290.

18 Enzymes of Lysine Synthesis

P. L. R. BONNER and P. J. LEA

Division of Biological Sciences, University of Lancaster, Lancaster LA1 4YQ, UK

I.	Introduction	297
II.	Lysine-sensitive aspartate kinase	299
	A. Assay	299
	B. Purification	302
III.	Aspartate semialdehyde dehydrogenase	305
IV.	Dihydrodipicolinate synthase	306
	A. Assay	306
	B. Purification	306
V.	Dihydrodipicolinate reductase	308
	A. Assay	308
	B. Purification	308
VI.	Diaminopimelate epimerase	309
	A. Assay	309
	B. Isolation procedure	309
VII.	Diaminopimelate decarboxylase	310
	A. Assay	310
	B. Purification	311
VIII.	Summary	311
	References	311

I. INTRODUCTION

There are two distinct pathways for the synthesis of lysine. In fungi and euglenoids, lysine is synthesised via the intermediate aminoadipic acid derived from 2-oxoglutarate and acetyl CoA. In bacteria and higher plants the pathway is characterised by the condensation of aspartate semialdehyde and pyruvate and the formation of the intermediate 2,6,diaminopimelate (Vogel et al., 1970). The use of radioactive precursors of lysine and intermediates of the proposed pathway have confirmed the synthesis of lysine is via *meso*-2,6,diaminopimelate (Moller, 1974; Miflin and Lea, 1977; Mills et al., 1980).

The first two enzymes of the pathway, aspartate kinase (EC 2.7.2.4.) and aspartate semialdehyde dehydrogenase (EC 1.2.1.11), are common to the synthesis of threonine, isoleucine and methionine. Aspartate kinase has been detected in a variety of plant tissue including: maize (Bryan et al., 1970; Cheshire and Miflin, 1975), pea (Aarnes and Rognes, 1974), barley (Shewry and Miflin, 1977; Bright et al., 1982), wheat (Bright et al., 1978; Yamada et al., 1986), spinach (Kochhar et al., 1986) and carrot (Davies and Miflin, 1977; Relton et al., 1988). The role of aspartate kinase in *Lemna* has recently been reassessed by Giovanelli et al. (1989). The enzyme is present in plants in at least two forms, a threonine-sensitive and a lysine-sensitive isoenzyme, with the lysine form predominant in rapidly growing tissue (Bryan, 1980; Lea et al., 1979). The lysine-sensitive isoenzyme of aspartate kinase is subject to synergistic feedback inhibition by *S*-adenosylmethionine at low concentrations of lysine (Rognes et al., 1980). A detailed study of this regulatory mechanism has recently been carried out on the highly purified carrot tissue culture enzyme by P. L. R. Bonner (unpubl. res.). In barley leaves there are two forms of the lysine-sensitive isoenzyme of aspartate kinase that are under separate genetic control. Both forms are subject to synergistic inhibition by *S*-adenosylmethionine (Rognes et al., 1983; Arruda et al., 1984). The threonine-sensitive isoenzyme has been shown to be present in high amounts in germinating pea cotyledons (Aarnes and Rognes, 1974), *Raphanus sativus* and *Sinapsis alba* (Aarnes, 1974) and in various tissues of *Glycine max*. (Matthews and Widholm, 1979). Aspartate semialdehyde dehydrogenase has been shown to be present in pea (Sasaoka, 1961) and a range of maize tissues (Gengenbach et al., 1978).

The first enzyme unique to the synthesis of lysine is dihydropicolinate synthase (EC 4.2.1.52), which has been shown to be present in: maize (Cheshire and Miflin, 1975), wheat germ, potato tuber (Mazelis et al., 1977), carrot (Matthews and Widholm, 1978), spinach (Wallsgrove and Mazelis, 1981) and wheat suspension culture (Kumpaisal et al., 1987). The enzyme has been shown to be subject to feedback inhibition by very low concentrations of lysine (Wallsgrove and Mazelis, 1980, 1981; Kumpaisal et al., 1987). Dihydrodipicolinate reductase (EC 1.3.1.26) has been isolated from maize; the activity was not shown to be subject to end product inhibition by lysine (Tygai et al., 1983).

The three enzymes following dihydrodipicolinate reductase in the pathway, which have not been isolated from plants, include: Δ'-piperidine dicarboxylate acylase (EC 2.3.1–), acyl diaminopimelate aminotransferase (EC 2.6.1.17), and acyl diaminopimelate deacylase (EC 3.5.1.18). Diaminopimelate epimerase (EC 5.1.1.7) has recently been isolated from maize (Tyagi et al., 1982). It has been assumed that in plants the synthetic route to lysine is analogous to that in *Escherichia coli* (Bryan, 1980). However, reports by Misono and Soda (1980), suggest that certain bacteria can convert tetrahydropicolinate directly to *meso*-diaminopimelic acid.

The final enzyme in the pathway, diaminopimelate decarboxylase (EC 4.1.1.20), has been detected in: *Lemna perpusilla* and *Spirodela ogliorhiza* (Vogel and Hirvonen, 1971), maize (Sodek, 1978) and wheat germ (Mazelis and Crevelling, 1978; Kelland et al., 1985). An unusual enzyme, *meso*-diaminopimelate dehydrogenase, has been isolated from *Glycine max*. The precise role of this enzyme is not clear at the present time (Wenko *et al.*, 1985).

All of the enzymes of the pathway to lysine are located within the chloroplasts in leaves (Wallsgrove *et al.*, 1983), but are present in low abundance in plant tissue (aspartate kinase is present at <0.005% of extractable protein: Relton *et al.*, 1988). The best source of these enzymes is a fast growing cell suspension culture (e.g. carrot: Davies and Miflin, 1977; Bonner *et al.*, 1986; Relton, *et al.*, 1988; or wheat: Yamada *et al.*, 1986; Kumpaisal *et al.*, 1987). A growth curve of the culture will help establish the period of maximal activity of the enzyme of interest.

The enzymology, physiology, feedback regulation and genetics of the enzymes involved in the synthesis of lysine have been extensively reviewed (Miflin and Lea, 1977, 1982; Bryan, 1980; Bright *et al.*, 1984); this chapter will concentrate on the practicalities of their purification and assay.

II. LYSINE-SENSITIVE ASPARTATE KINASE

A. Assay

The most widely used assay for aspartate kinase is the hydroxamate assay, which depends on the conversion of the relatively unstable β-aspartylphosphate to the stable β-aspartylhydroxamate during the course of the assay and the development of colour by the addition of acidified ferric chloride. It should be noted that the assays described below can equally well apply to the threonine-sensitive enzyme (see Rognes, Chapter 19, this volume.).

1. Hydroxamate

The assay components are: 100 µl 250 mM aspartic acid (pH 7.4); 100 µl 100 mM Tris-HCl (pH 7.4) containing 1 mM dithiothreitol (DTT) and 20% (v/v) ethanediol; 50 µl 200 mM ATP (pH 7.4); 50 µl 250 mM MgSO$_4$; 50 µl 4M hydroxylamine (pH 7.4); and 50 µl H$_2$O. The assay is started by the addition of 100 µl of enzyme, incubated at 30°C for various times and terminated by the addition of 500 µl of 0.67 M FeCl$_3$ containing 0.37 M HCl and 20% (w/v) TCA. After centrifugation to remove precipitated protein, the absorbance of the supernatant is read at 505 nm; the extinction coefficient for aspartylhydroxamate at 505 nm is 750 cm^{-1} M^{-1}. There are problems with the assay, some of which have been discussed previously (Davies and Miflin, 1977). These problems are most noticeable when the enzyme is extracted from leaf tissue. A number of unidentified compounds may interact with the assay components to produce false values of activity; a low absorbance value (<0.05) should be regarded with caution. Buffer components, for example sodium azide, and high aspartic acid concentrations (>1 M), will also produce a false colour in the assay.

To ensure that the activity measured is derived from aspartate kinase, the assay solutions should be read against a blank containing all the components but minus the aspartate. Other controls include substituting the water component of the assay with 50 µl of 10 mM L-lysine, or the addition of 50 µl of 10 mM sodium pyrophosphate to inhibit contaminating aspartyl-tRNA synthetase activity. Only the activity which is aspartate-dependent, lysine-sensitive and uninhibited by pyrophosphate is derived from aspartate kinase. For the detection of the threonine-sensitive enzyme, threonine should be substituted for lysine, where appropriate.

2. Other

Other assays for aspartate kinase include a coupled assay which depends on the reduction of aspartyl phosphate to aspartate semialdehyde in the presence of yeast aspartate semialdehyde dehydrogenase (Cheshire and Miflin, 1975), and a radioactive assay which uses high voltage paper electrophoresis (PE) to separate out the [^{14}C]-aspartylhydroxamate (Rognes et al., 1980; Arruda et al., 1984). The latter assay is particularly useful for measuring aspartate kinase in small quantities of tissue or when the activity is very low, and was originally described by Aarnes and Rognes (1974) for the threonine-sensitive enzyme. [U-^{14}C]Aspartate is first purified over a Dowex 1-X8 column to remove any contaminating neutral substances. The reaction mixture in 0.1 ml contains: 12 mM L-[U-^{14}C]aspartate (26 MBq mmol^{-1}), 18 mM MgCl$_2$, 15 mM ATP, 400 mM hydroxylamine adjusted to pH 7.5 with KOH, 1 mM DTT and 10–100 µg enzyme protein. The reaction mixtures are incubated for 60–120 min at 30°C, and the reaction terminated by the addition of 25 µl of 10% (w/v) TCA and 10 µl of 30 mM β-aspartylhydroxamate. [^{14}C]Aspartate is separated from the [^{14}C]aspartylhydroxamate by PE at pH 5.0 (Rognes, 1970) and the radioactive product which runs close to the origin is determined by liquid scintillation counting. It is also possible to assay the phosphate released from aspartyl phosphate by a malachite green phosphate assay developed by Zlotnick and Gottlieb (1986) (see Section II.A.3 on gel electrophoresis). However, the enzyme preparation used in this assay system should be essentially free from contaminating ATPase activity to avoid high background values.

3. Gel electrophoresis

Non-dissociating gradient gel electrophoresis and an activity stain based on calcium phosphate precipitation (Nimmo and Nimmo, 1982) of the phosphate released from aspartyl phosphate is a useful method of detecting aspartate kinase during a purification protocol (Relton et al., 1988).

Electrophoresis is carried out at 4°C overnight in a Tris buffer system. Sections of the gel are placed in separate heat-sealed bags with 50 mM CaCl$_2$ included in the assay mixture containing: 50 mM Tris-HCl (pH 7.5), 50 mM MgCl$_2$, 1 mM DTT, 10 mM ATP, 50 mM aspartate and 10% (v/v) ethanediol. One bag contains the complete assay mixture plus 10 mM lysine and one contains the assay mixture minus aspartate. The band that is aspartate-dependent and lysine-sensitive is aspartate kinase (Fig. 18.1). If low activity levels of aspartate kinase are applied it should be possible to detect aspartate kinase by autoradiography. Inclusion of γ-labelled [^{32}P]ATP in the incubation mixture will result in a radioactive calcium phosphate precipitate, but the gel will require extensive washing with 50 mM Tris-HCl (pH 8.0) before development, to remove background activity.

An alternative staining procedure has also been developed, based on the malachite green method of the detection of inorganic phosphate (Zlotnick and Gottlieb, 1986). The gel slices after electrophoresis are transferred to the three different solutions as

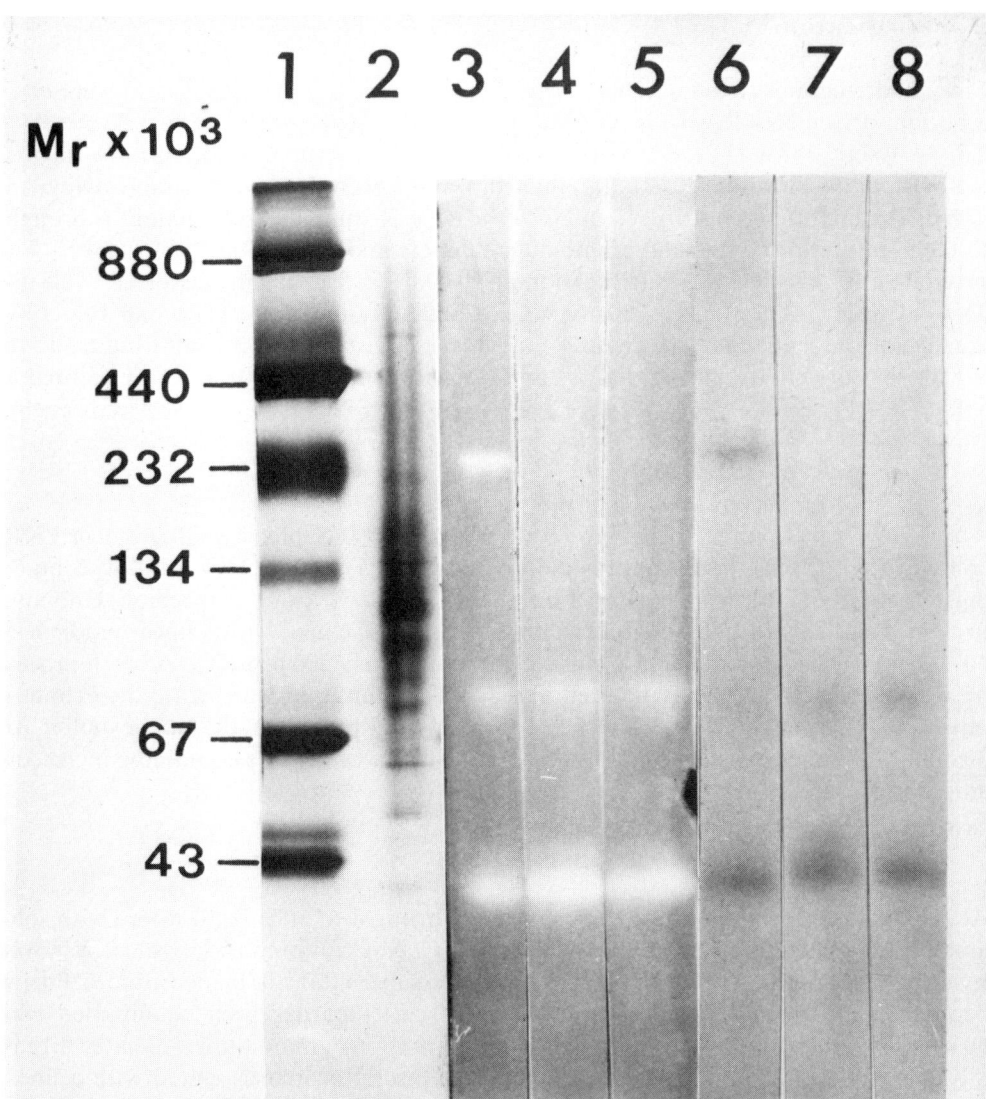

FIG. 18.1 Non-denaturing PAGE of partially purified aspartate kinase from carrot tissue culture. Tracks: 1, Calibration proteins. 2, Aspartate kinase preparation stained for protein with Coomassie Blue R-250. 3–5, Calcium phosphate activity stain (3, control; 4, minus aspartate; 5, plus lysine). 6–8, Malachite green activity stain (6, control; 7, minus aspartate; 8, plus lysine).

described before, but without the $CaCl_2$. Following incubation for 1–4 h, the gel slices are rinsed with water and incubated in a solution containing 0.033% (w/v) malachite green oxalate salt 1% (w/v) ammonium molybdate in 1 M HCl and 0.04% (v/v) Tween 80 for 30 min and washed with several changes of water. Phosphate-containing bands stain green and aspartate kinase activity can be detected by carrying out the appropriate controls as described previously and shown in Fig. 18.1.

The time taken for the band to develop will depend on the activity of the preparation applied.

B. Purification

Five hundred grams (fresh weight) of carrot suspension culture tissue is homogenised in a bottom-driven blender in 1 litre of 50 mM Tris-HCl (pH 7.4) containing: 50 mM KCl, 2 mM L-lysine, 1 mM DTT, 0.1 mM EDTA, 0.1 mM N-ethylmaleimide, 0.1 mM phenylmethylsulphonylfluoride, 20% (v/v) ethanediol and 5% (w/v) insoluble polyvinylpyrrolidone. The mixture is centrifuged at 20 000 × g for 30 min. The supernatant is brought to 60% (saturation) with solid ammonium sulphate and stirred for 30 min at 4°C. The precipitate is collected by centrifugation at 20 000 × g for 30 min, dissolved in 25 mM Tris-HCl (pH 7.4) containing: 50 mM KCl, 0.1 mM L-lysine, 1 mM DTT and 10% (v/v) ethanediol and dialysed overnight at 4°C against a large excess of the same buffer. There is little loss of activity during dialysis and after this stage the extract can be stored at −25°C.

1. Anion-exchange chromatography

After dialysis the extract is loaded onto a Fast Flow Q Sepharose (Pharmacia LKB) column (2.5 × 10 cm) that has been equilibrated with 25 mM Tris-HCl (pH 7.4) containing: 50 mM KCl, 0.1 mM L-lysine, 1 mM DTT and 10% (v/v) ethanediol. Unbound protein is eluted with the same buffer; bound protein is eluted with a linear gradient of 50–100 mM KCl; aspartate kinase elutes at approximately 200 mM KCl. A batch process on a sintered glass filter may be used instead of a column, in which case the extract is mixed with 100 ml of Fast Flow Q Sepharose equilibrated in the above buffer for 30 min. The resin is washed on a sintered glass filter, with buffers containing increasing amounts of KCl.

2. Phenyl sepharose

Active fractions from the ion-exchange step are brought to 30% (saturation) with solid ammonium sulphate and stirred for 30 min at 4°C. Any resulting precipitate is removed by centrifugation at 20 000 × g for 30 min. The supernatant is loaded onto a Phenyl Sepharose (Pharmacia LKB) column (2.5 × 10 cm) that has been equilibrated with 50 mM Tris-HCl (pH 7.4) containing: 30% (saturation) ammonium sulphate, 0.1 mM L-lysine, 1 mM DTT and 10% (v/v) ethanediol. Bound protein is eluted with a linear gradient of decreasing ammonium sulphate (30%–zero saturation). Active fractions are pooled and dialysed against 50 mM Tris-HCl (pH 7.4) containing: 500 mM KCl, 0.1 mM L-lysine, 1 mM DTT and 10% (v/v) ethanediol.

3. Metal chelating affinity chromatography (MCAC)

The dialysed extract is applied to a MCAC (Pharmacia LKB) column (2.5 × 10 cm) that has been 66% saturated with zinc ions and equilibrated with 50 mM Tris-HCl (pH 7.4) containing: 500 mM KCl, 0.1 mM L-lysine, 1 mM DTT and 10% (v/v) ethanediol. Unbound protein is eluted with a wash of the above buffer containing 2 M ammonium chloride. Aspartate kinase is eluted by washing the column with 50 mM Tris-HCl (pH 7.4) containing: 50 mM EDTA, 0.1 mM L-lysine, 1 mM DTT and 10% (v/v) ethandediol.

To decrease the volume of the eluting aspartate kinase peak, the wash containing

EDTA should be performed in the reverse direction to that of the sample application. This may be simply executed by turning the column upside down before the EDTA wash.

4. Gel filtration

Active fractions from the MCAC stage are immediately applied to a Sephacryl-S200 (Pharmacia LKB) column (2.6 × 100 cm) equilibrated with 25 mM Tris-HCl (pH 7.4) containing: 0.1 mM L-lysine, 1 mM DTT and 10% (v/v) ethanediol.

This step is primarily to remove contaminating zinc ions and EDTA before the sample is applied to a Mono Q column, so a fast flow rate can be used. This step in conjunction with the MCAC column will result in a preparation virtually free from contaminating ATPase activity.

5. Mono Q (HR5/5)

Active fractions from the previous step are loaded onto a Pharmacia Mono Q column via a 10 ml super loop. The column is equilibrated with 25 mM Tris-HCl (pH 7.4) containing: 0.1 mM L-lysine, 1 mM DTT and 10% (v/v) ethanediol. Active fractions are eluted with a linear salt gradient between 0 and 500 mM KCl.

6. Superose 6

The peak fractions of activity from the Mono Q stage are applied to a Superose 6 column equilibrated with 25 mM Tris-HCl (pH 7.4) containing: 0.1 mM L-lysine and 1 mM DTT. The fractions should be collected directly into ethanediol, to make a final concentration of 10% (v/v).

Table 18.1 summarises the overall purification that can be expected. For maximum yields, all steps should be performed at 4°C and all buffers should be made fresh and filtered through 0.2 μm membranes. The enzyme elutes as a single peak from Superose 6 with a molecular weight of 250 kDa but SDS-PAGE and silver staining of the peak fractions show more than five polypeptides present of varying molecular weights. This indicates that aspartate kinase is present in plants in lower abundance than predicted by Relton *et al.* (1988) and an additional step is required to reach homogeneity.

7. Hydroxyapatite

Hydroxyapatite Ultragel (Pharmacia LKB) can be inserted after the gel filtration step. The buffer for gel filtration should be changed to 2.5 mM potassium phosphate buffer (pH 7.4) containing: 0.1 mM L-lysine, 1 mM DTT and 10% (v/v) ethanediol. Problems with zinc phosphate precipitation on the column can be avoided by including a pre-column of Sephadex G-25 (Pharmacia LKB).

Active fractions from the gel filtration step can be applied to a hydroxyapatite column (2.5 × 5 cm) equilibrated with 2.5 mM potassium phosphate buffer (pH 7.4) containing: 0.1 mM L-lysine, 1 mM DTT and 10% (v/v) ethanediol. Bound protein is eluted with a linear gradient of potassium phosphate 2.5–30 mM (pH 7.4); the eluted enzyme may then be applied to the Mono Q column. The purist may insert a desalting

TABLE 18.1. Purification of aspartate kinase from (500 g) carrot suspension culture.

	Volume (ml)	Protein[a] (mg)	Specific activity (nmol min^{-1} mg^{-1})	Total activity (nmol min^{-1})	Yield (%)	Fold purification
Crude	4000	3.0	N.D.	—	—	—
0–60% (sat) Ammonium sulphate	550	18.0	1.4	17820	100	1
FFQ Sepharose	74	20.0	7.0	10360	58	5
Phenyl Sepharose	44	9.6	36	15206	85	26
MCAC	24	11.0	42	11088	62	30
S-200	31	2.4	187	13838	77	133
Mono Q (A)[b]	12	0.25	1560	4680	26	1110
(B)[c]	1.5	0.013	39400	768	4	28000
Superose 6	0.5	0.005	44300	110	0.6	32000

[a] Protein estimated using the method of Bradford (1977).
[b] Linear gradient of KCl from 0–500 mM KCl.
[c] Fractions reapplied to mono Q column and eluted with a linear gradient from 100–250 mM KCl.
N.D., not detectable.

step of Sephadex G-25 into the Tris buffer before application to the Mono Q column.

Since the first draft of this chapter we have become aware of work being carried out by S. B. Dotson at the University of Minnesota, USA. Dotson (1989) resolved two lysine-sensitive forms of aspartate kinase prepared from Black Mexican Sweet Corn suspension cultures. The two isoforms were purified >1200-fold to a minimum specific activity of 18 nkat mg^{-1} protein (Table 18.2). Additional purification steps were combined with preparative gel electrophoresis to obtain homogeneous enzyme. The aspartate kinase from maize suspension cells was shown to be a tetramer with a holoenzyme molecular weight of 254 kDa composed of 49 and 60 kDa molecular weight subunits. The tetramer appeared to dissociate during gel electrophoresis to 113 kDa molecular weight active species.

TABLE 18.2. Purification of early and late aspartate kinase (AK) isoforms from Black Mexican Sweet Corn (BMS) cell cultures. BMS suspension culture cells (80 g) were harvested 5 days after subculture (mid-log growth stage) for AK purification. Early and late isoforms resolved by the initial anion exchange chromatography were individually rechromatographed to further purify them, before the final specific activity was determined.

Step	Total protein (mg)	Total activity (nkat)	Specific activity (nkat mg^{-1})	Fold purification	Per cent recovery
Cell-free extract	1300	—	—	—	—
G-25	1500	21	0.014	1	100
Phenyl Sepharose	223	54	0.240	17	255
60% (NH$_4$)$_2$SO$_4$ ppt.	116	61	0.528	37	296
Gel filtration	15	45	3.06	216	224
Anion-exchange (pH 7.1)					
AK Early	0.36	6	18.0	1273	33
AK Late	0.62	12	18.9	1339	60

As part of a detailed kinetic analysis of maize aspartate kinase, Dotson (1989) employed a coupled assay measuring ADP production utilising pyruvate kinase and lactate dehydrogenase to monitor NADH oxidation at 340 nm in a spectrophotometer (McClure, 1969). Lea et al. (Chapter 15, this volume) have described in more detail the method for the measurement of glutamine synthetase which has been used previously for the assay of bacterial asparatate kinase (Wampler et al., 1970).

III. ASPARTATE SEMIALDEHYDE DEHYDROGENASE

This enzyme is poorly characterised in higher plants and has only been studied in any detail in maize tissues. Gengenbach et al. (1978) measured the activity in maize shoot, root, kernel, callus and suspension culture extracts. The activity in suspension culture (92.6 nmol min^{-1} mg^{-1} protein) was found to be particularly high. The activity was not inhibited to any great extent by 10 mM lysine, threonine or isoleucine. Methionine (10 mM) did have an inhibitory effect, in particular on the enzyme isolated from callus and suspension culture.

The activity was measured in the reverse direction by Gengenbach et al. (1978) by following the reduction of NADP at 340 nm in a spectrophotometer. The assay medium contained 30 mM Tricine (pH 9.0), 10 mM K_2HPO_4, 0.8 mM NADP and 5–50 µl enzyme extract in 1 ml. The reaction was started by the addition of 0.6 mM aspartate semi-aldehyde, prepared by the ozonolysis of allylglycine as described by Black and Wright (1955) and by Rognes (Chapter 19, this volume). The aspartate semialdehyde solution was neutralised immediately before use.

IV. DIHYDRODIPICOLINATE SYNTHASE

A. Assay

The measurement of the enzyme activity is dependent upon the reaction of dihydrodipicolinate with o-aminobenzaldehyde, as originally described by Yugari and Gilvarg (1965) and Stahly (1969). The substrate for the reaction, aspartate semialdehyde, is synthesised by the ozonolysis of allylglycine (Black and Wright, 1955; Rognes, Chapter 19, this volume) and stored at 4°C in 4 M HCl.

Wallsgrove and Mazelis (1980) modified the earlier assay of Mazelis et al. (1977) to include 100 mM Tris-HCl (pH 7.5), 100 mM sodium pyruvate, 5 mM aspartate semi-aldehyde (neutralised immediately before use) and enzyme in a volume of 0.5 ml. The reaction was stopped by the addition of 2 ml 0.22 M nitrate/0.55 M Na_2HPO_4 (pH 5.0) containing 0.5 mg o-aminobenzaldehyde dissolved in the minimum amount of ethanol. The colour was allowed to develop for 2 h and after clarification by centrifugation, the absorbance at 520 nm was determined. This assay has also been developed for detecting activity following gel electrophoresis (Wallsgrove and Mazelis, 1981). The use of the nitrate/phosphate buffer was found to be preferable to TCA as originally described (Mazelis et al., 1977), and prevented interference with the colour reaction by aspartate semialdehyde. A number of compounds including dithioerythritol and 2-mercaptoethanol were still found to inhibit colour formation.

Kumpaisal et al. (1987), working with wheat suspension culture enzyme, utilised a slight modification of the enzyme mixture which included 100 mM Tris-HCl (pH 8.0), 20 mM sodium pyruvate, 0.8 mM aspartate semialdehyde and 1 mg of o-aminobenzaldehyde dissolved in 30 µl of 98% (v/v) ethanol in a volume of 1 ml. The enzyme reaction was stopped by the addition of 0.2 ml of 12% (w/v) TCA and 3 M HCl (1:1; v/v). The colour reaction was allowed to develop the absorbance for 40 min at 30°C and measured at 540 nm following centrifugation. Care was taken to ensure that none of the inhibitors tested interfered with the formation of the purple colour.

B. Purification

1. Spinach

The original method described by Wallsgrove and Mazelis (1981) succeeded in producing a fraction from spinach leaves that was purified 87-fold and had a specific activity of 836 units mg^{-1} protein. Protamine sulphate, 35–55% saturation ammonium sulphate precipitation and ion-exchange chromatography on Cellex D were employed initially.

However, significant purification was only obtained following Octyl-Sepharose and Sephadex G-150 chromatography.

2. Wheat

Dihydrodipicolinate synthase has been purified over 5000-fold from wheat suspension cultures (Kumpaisal et al., 1987) to a specific activity of 170 200 units mg^{-1} protein. After the initial extraction of 1.65 kg of cells in 2.9 l of 100 mM potassium phosphate buffer (pH 7.5) containing: 2 mM EDTA, 3 mM DTT and 10 mM diethyldithiocarbamic acid, the supernatant was subjected to an ammonium sulphate cut between 30 and 60% (saturation). The precipitate was redissolved in buffer A—20 mM potassium phosphate buffer (pH 7.5) containing: 1 mM EDTA, 1 mM DTT and 10 mM pyruvate—and adjusted to 30% (saturation) with solid ammonium sulphate. The extract was applied to a hydrophobic column of Butyl-Toyopearl 650M (Toyo Soda Co.) and eluted step-wise by decreasing the ammonium sulphate concentration. Active fractions were pooled and concentrated on an Amicon YM-10 membrane and desalted on Sephadex G-25 into 20 mM potassium phosphate buffer (pH 7.5) containing: 100 mM KCl, 1 mM EDTA and 1 mM DTT. The sample was then applied to DEAE Sepharose CL-6B (Pharmacia LKB) column and eluted with a linear gradient of KCl (100–500 mM). Active fractions were precipitated by the addition of solid ammonium sulphate to 60% (saturation). The precipitate was redissolved in buffer A containing 30% saturation ammonium sulphate and applied to a Phenyl-Superose (Pharmacia LKB) column attached to an FPLC system and eluted with a linear gradient of 30% to 0% saturation ammonium sulphate. Two further FPLC columns were used—the ion-exchange resin Mono Q, eluted with a linear gradient of 0.1–0.5 M KCl, and gel filtration on Superose 12.

The final purification was 5131-fold (Table 18.3) and the enzyme was eluted as a single peak from the Superose 12 column (M_r 123 kDa). SDS-PAGE of the Superose 12 peak revealed additional protein bands. One band of activity was detected following non-dissociating PAGE as outlined by Wallsgrove and Mazelis (1981). Second dimension SDS-PAGE indicated that the subunit molecular weight of the enzyme was 32–35 kDa (Kumpaisal et al., 1987).

TABLE 18.3. Purification of dihydrodipicolinate synthase from wheat suspension cultures.

Purification step	Total protein (mg)	Specific activity (units mg^{-1})	Total activity (units)[a]	Yield[b] (%)	Fold purification[b]
Crude extract	16524	9.9	163187		
30–60% $(NH_4)_2SO_4$	16416	33	544512	100	1.0
Butyl-Toyopearl 650M	1314	246	323010	59.3	7.4
DEAE-Sepharose CL-6B	127	2683	341055	62.6	80.9
Phenyl-Superose	13	17871	237690	43.7	539
Mono Q	2	82425	181335	33.3	2485
Superose-12	0.7	170200	119140	21.9	5131

[a] A net absorbance increase of 0.001 min^{-1}.
[b] The ammonium sulphate precipitate was taken as the starting point for the calculation.
After Kumpaisal et al. (1987).

V. DIHYDRODIPICOLINATE REDUCTASE

A. Assay

The substrate dihydrodipicolinate was synthesised by the condensation of aspartate semialdehyde with oxaloacetate in alkali (Farkas et al., 1963), as further described by Tyagi et al. (1983). Three different assays for the reductase activity in maize extracts are described by the latter authors.

1. Dihydrodipicolinate removal

The reaction mixture contained 25 μmol Tris-HCl (pH 7.9), 2.9 μmol dihydrodipicolinate, and 0.33 μmol NADPH in 0.6 ml. The enzyme extract was incubated with the mixture for 1 h and 0.1 ml aliquot was added to 0.77 ml of 50% (v/v) aqueous acidic ethanol (0.13 N HCl) followed by 0.2 ml of o-aminobenzaldehyde (2.5% (w/v) in 50% (v/v) ethanol). The colour was allowed to develop for 40 min and the absorbance read at 560 nm. The assay measures the rate of disappearance of dihydrodipicolinate and is the converse of the synthase assay described previously.

2. Mutant complementation

The assay depends on the capacity of crude extracts of maize to restore the ability of extracts of the E. coli mutant CGSC/4549 deficient in dihydrodipicolinate synthase (obtained from the E. coli Genetic Stock Center, Yale University School of Medicine, USA) to synthesise diaminopimelate. The reaction mixture contained 82 μmol Imidazole buffer (pH 7.0), 13 μmol potassium aspartate, 1.3 μmol potassium glutamate, 6.5 μmol ATP, 6.5 μmol $MgSO_4$, 82 pmol NADP, 95 pmol NAD, 10 μmol pyruvate, 10 mg CGSC/4549 extract protein and maize extract in a final volume of 0.5 ml. The diaminopimelate synthesised after 60 min was determined by the acid ninhydrin method developed by Gilvarg (1958). A number of mutants of E. coli are available; a study of the review article by Bachmann and Low (1980) could provide similar methods of assay for the remaining enzymes in the lysine biosynthetic pathway.

3. NADPH oxidation

For partially purified extracts that contain little contaminating background NADPH oxidase activity, a straightforward spectrophotometric assay can be employed. The reaction mixture (1 ml) contains 32 μmol potassium phosphate buffer, 0.62 μmol dihydrodipicolinate, 0.6 μmol NADPH and 50 μg protein. The disappearance of NADPH is followed by measuring the absorbance at 340 nm.

B. Purification

The enzyme was isolated by Tyagi et al. (1983) from three-week-old maize kernels by homogenising 1.4 kg in 1 litre of 50 mM Tris-HCl buffer (pH 7.5) containing 7.3 mM 2-mercaptoethanol (buffer A) in the presence of 1 kg silica. The homogenate was filtered through cheesecloth and centrifuged at 27 500 × g for 15 min. The enzyme was precipitated by 60% saturation ammonium sulphate and dialysed overnight against buffer A. The extract was concentrated against polyethylene glycol and loaded onto a DEAE-cellulose column and the enzyme activity eluted with a linear gradient of 0–0.4 M KCl. A second DEAE-cellulose step involved loading the dialysed eluate onto a smaller column in buffer A and washing with 0.2 M KCl. Dihydrodipicolinate reductase activity was then eluted by washing with 0.2 M KCl again in buffer A. A subsequent Sephacryl-S200 step in buffer A containing 0.5 M KCl was employed to determine the molecular weight of the enzyme (84 kDa). A range of substrates was tested and the K_m for dihydrodipicolinate was determined as 4.3×10^{-4} M.

VI. DIAMINOPIMELATE EPIMERASE

The penultimate step in the synthesis of lysine in bacteria is the epimerisation of L-diaminopimelic acid (DAP) to *meso*-DAP (Antia *et al.*, 1957). In higher plants that do not require *meso*-Dap as a constituent of the cell wall there is no obvious reason for the conversion of L-DAP to *meso*-DAP prior to decarboxylation. However, Tyagi *et al.* (1982) have been able to demonstrate the presence of diaminopimelate epimerase (EC 5.1.1.7) in extracts of maize leaves.

A. Assay

The incubation medium contained 2.6 mg protein, 5 mM *meso*-DAP, 6.7 mM 2-mercaptoethanol, and 0.2 mM pyridoxal phosphate in 100 mM phosphate buffer (pH 7.4) in a final volume of 1 ml. As the reactions were carried out for a minimum of 5 h, the assay medium was sterilised by passage through a 0.45 μM filter. In order to prevent the decarboxylation of *meso*-DAP to lysine, 5 mM KCN was also added to the reaction medium. The reaction was terminated by the addition of 12 N HCl and the mixture placed in a boiling water bath for 15 min to remove precipitated protein. The supernatant was adjusted to 1 N HCl and any lysine formed was removed by passage through a Dowex 50 (H$^+$) column.

In order to determine the reverse conversion on *meso*-DAP to L-DAP, the eluate was spotted onto Whatman 3 MM paper and subjected to PC in methanol–pyridine–H$_2$O (80:10:20). L-DAP runs faster than *meso*-DAP in this system. The formation of L-DAP was either determined by the presence of a ninhydrin-positive spot or by counting radioactivity on the PC following the use of *meso*-[^{14}C]DAP.

DAP epimerase activity was also assayed using L-[^{14}C]DAP as a substrate, in a reaction coupled with *meso*-DAP decarboxylase. The assay method was similar to that described in Section VII. As would be expected, the evolution of $^{14}CO_2$ was inhibited by the addition of unlabelled *meso*-DAP.

B. Isolation Procedure

Leaves of two- to three-week-old maize (350 g) were homogenised in 400 ml of 100 mM potassium phosphate buffer (pH 7.4) containing 0.2 mM pyridoxal phosphate and 7.3 mM 2-mercaptoethanol. The centrifuged extract was concentrated with an Amicon PM10 ultrafilter under an atmosphere of N$_2$. Initially a 30–50% ammonium sulphate precipitate (dialysed against extraction buffer) was used as an enzyme source. However, later experiments suggested that a 40–60% ammonium sulphate cut would have been more appropriate. As far as we are aware the work described above by Tyagi *et al.* (1982) is the only report of the isolation of DAP epimerase in higher plants.

VII. DIAMINOPIMELATE DECARBOXYLASE

Meso-diaminopimelate (DAP) decarboxylase (EC 4.1.1.20) catalyses the final step in the synthesis of lysine by the pyridoxal phosphate-dependent decarboxylation of the R_D centre of DAP (Kelland *et al.*, 1985). The enzyme has been studied in bacteria (Asada *et al.*, 1981; Lakshman *et al.*, 1981) and in higher plants (Shimura and Vogel, 1966; Vogel and Hirvonen, 1971; Mazelis and Crevelling, 1978; Sodek, 1978), where it is localised solely in the chloroplasts (Mazelis *et al.*, 1976).

The molecular weight of the enzyme in higher plants varies between 75 kDa for wheat germ and 85 kDa for maize endosperm. Estimates of the K_m for *meso*-DAP vary between 0.1 and 0.3 mM (Mazelis and Crevelling, 1978; Sodek, 1978). The enzyme has the ability to bind pyridoxal phosphate, which increases its stability during storage. Lysine at concentrations up to 1 mM has little effect on enzyme activity; at 20 mM lysine inhibited the wheat germ DAP-decarboxylase by 60%. It is unlikely that inhibition at this high concentration would indicate any physiological feedback regulation.

DAP decarboxylase is the only pyridoxal-dependent α-decarboxylase known to act on a D-amino acid. Kelland *et al.* (1985) carried out a detailed analysis of the enzyme mechanism by two-dimensional ^{1}H-^{13}C heteronuclear NMR shift correlation spectroscopy with ^{2}H decoupling and compared it to the known mechanism operating in *Bacillus sphaericus* (Asada *et al.*, 1981).

A. Assay

The method is based on determining the $^{14}CO_2$ liberated by the decarboxylation of carboxyl-labelled *meso*-DAP. Mazelis and Crevelling (1978) utilised an assay medium containing either 80 mM potassium phosphate (pH 7.0) or Tricine (pH 8.0), 25 μM pyridoxal phosphate, 2 mM [1,7-^{14}C]DAP (0.1 μCi) in a final volume of 1 ml. The reaction was carried out in modified scintillation vials as described by Fox (1971) with 0.2 ml of hyamine hydroxide used to absorb the liberated $^{14}CO_2$. The reaction mixture was shaken at 30°C for 15 min and terminated by the addition of 0.2 ml 10% TCA, followed by additional shaking for a further 60 min. The radioactivity in the hyamine hydroxide was determined by scintillation counting. Kelland *et al.* (1985) adapted this method and included 1.2 mM EDTA and 3.5 mM dithioerythritol in the assay medium. They trapped the $^{14}CO_2$ released in a piece of 1.5 cm × 1.5 cm filter paper impregnated with 1 M hyamine hydroxide in the cap of the scintillation vial. Sodek (1978) again used a similar method involving Katz flasks which included in the main compartment 3 mM [1,7-^{14}C]*meso*-DAP, 0.06 mM pyridoxal phosphate, 100 mM potassium phosphate buffer (pH 7.0) in a final volume of 1 ml. After 1 h the reaction was terminated by the addition of 1 ml of 1 M HCl injected through the rubber septum and the flask was shaken for a further 30 min. The $^{14}CO_2$ was trapped in the centre well on a 1 cm^2 filter paper moistened with 50 μl of hyamine hydroxide.

B. Purification

Mazelis and Crevelling (1978) described a procedure for extracting the enzyme from a wheat germ acetone powder in 50 mM potassium phosphate buffer (pH 7.0) containing 50 µM pyridoxal phosphate and 0.05% (v/v) 2-mercaptoethanol. The purification procedure is shown in Table 18.4. The final preparation still contained a number of contaminating proteins as shown by PAGE. Kelland et al. (1985), using the same procedure, obtained a preparation of 283 pkat mg^{-1} protein.

Sodek (1978) used maize endosperm harvested 28 days after pollination as a particularly rich source of the enzyme. The purification procedure involved a 30–50% ammonium sulphate precipitation, step elution from DEAE-cellulose, Sephadex G-150 exclusion chromatography and a final gradient elution from DEAE-cellulose. The initial crude extract had a specific activity of 4 nkat mg^{-1} protein, and the final purified preparation a specific activity of 783 nkat mg^{-1} protein.

TABLE 18.4. Purification of diaminopimelate decarboxylase from wheat germ.

Fraction	Total protein (mg)	Total activity (nkat)	Specific activity (pkat mg^{-1})	Fold purification	Yield (%)
Original extract	6631	16.8	2.5	1.0	100
Protamine SO$_4$ supernate	6034	16.3	2.6	1.0	97
0–35% (NH$_4$)$_2$SO$_4$ ppt.	655	13.5	20	8.0	80
G-150 eluate	91	7.4	82	33	44
AlC$_\gamma$ eluate	14	5.9	433	173	35

After Mazelis and Crevelling (1977).

VIII. SUMMARY

The enzymes in the pathway of lysine synthesis are present in low abundance in plants—this makes the choice of starting material important. A fast-growing suspension culture of plant cells provides the best source of the enzymes and reduces interference in the subsequent enzyme assays. The use of fast flow resins and the FPLC system will increase the yields of the enzymes and with three enzymes still to be detected in plants, there is great scope for the plant biochemist in the pathway of lysine biosynthesis.

REFERENCES

Aarnes, H. (1974). *Physiol. Plant.* **32**, 400–402.
Aarnes, H. and Rognes, S. E. (1974). *Phytochemistry* **13**, 2717–2724.
Antia, M., Hoare, D. S. and Work, E. (1957). *Biochem. J.* **65**, 448–459.
Arruda, P., Bright, S. W. J., Kueh, J. S. H., Lea, P. J. and Rognes, S. E. (1984). *Plant Physiol.* **76**, 442–446.
Asada, Y., Tanizawa, K., Sawada, S., Suzuki, T., Misono, H., Soda, K. (1981). *Biochemistry* **20**, 6881–6886.

Bachman, B. J. and Low, K. B. (1980). *Microbiol. Rev.* **44**, 1–56.
Black, S. and Wright, N. G. (1955). *J. Biol. Chem.* **213**, 51–60.
Bonner, P. L. R., Hetherington, A. M. and Lea, P. J. (1986). *FEBS Lett.* **195**, 119–121.
Bradford, M. M. (1977). *Anal. Biochem.* **72**, 248–254.
Bright, S. W. J., Shewry, P. R. and Miflin, B. J. (1978). *Plant* **139**, 119–125.
Bright, S. W. J., Miflin, B. J. and Rognes, S. E. (1982). *Biochem. Genet.* **20**, 229–243.
Bright, S. W. J., Lea, P. J., Arruda, P., Hall, N. P., Kendall, A. C., Keys, A. J., Kueh, J. S. H., Parker, M. L., Rognes, S. E., Turner, J. C., Wallsgrove, R. M. and Miflin, B. J. (1984). *In* "The Genetic Manipulation of Plants and its Application to Agriculture" (P. J. Lea and G. R. Stewart, eds), pp. 141–169. Oxford University Press, Oxford.
Bryan, J. K. (1980). *In* "The Biochemistry of Plants: A Comprehensive Treatise" (B. J. Miflin, ed.), Vol. 5, pp. 403–452. Academic Press, New York and London.
Bryan, P. A., Crawley, R. D., Bruner, C. E. and Bryan J. K. (1970). *Biochem. Biophys. Res. Commun.* **41**, 1211–1217.
Cheshire, R. M. and Miflin, B. J. (1975). *Phytochemistry* **14**, 695–698.
Davies, H. M. and Miflin, B. J. (1977). *Plant Sci. Lett.* **9**, 323–332.
Dotson, S. B. (1989). "Purification and characterisation of aspartate kinase in wild-type corn and lysine plus threonine resistant mutants", Ph.D. Thesis, University of Minnesota, USA.
Farkas, W. R., Yugari, Y. and Gilvarg, C. (1963). *Fed. Proc.* **22**, 243.
Fox R. M. (1971). *Anal. Biochem.* **44**, 578–580.
Gegenbach, B. G., Walter, T. J., Green, C. E. and Hibberd, K. A. (1978). *Crop Sci.* **18**, 472–476.
Gilvarg, C. (1958). *J. Biol. Chem.* **233**, 1501–1504.
Giovanelli, J., Mudd, S. H. and Datko, A. H. (1989) *Plant Physiol.*, **90**, 1577–1583.
Kelland, J. G., Palcic, M. M., Pickard, M. A. and Verderas, J. C. (1985). *Biochemistry* **24**, 3263–3267.
Kochhar, S., Kochhar, V. K. and Sane, P. V. (1986). *Biochim. Biophys. Acta* **880**, 220–225.
Kumpaisal, R., Hashimoto, T. and Yamada, Y. (1987). Plant Physiol. **85**, 145–151.
Lakshman, M., Stenoy, B. C. and Rao, M. R. R. (1981). *J. Biosci.* **3**, 89–103.
Lea, P. J., Mills, W. R. and Miflin, B. J. (1979). *FEBS Lett.* **98**, 165–168.
Matthews, B. F. and Widholm, J. M. (1978). *Planta* **141**, 315–321.
Matthews, B. F. and Widholm, J. M. (1979). *Can. J. Bot.* **57**, 299–304.
Mazelis, M. and Crevelling, R. K. (1978). *J. Food Biochem.* **2**, 29–37.
Mazelis, M., Miflin, B. J. and Pratt, H. M. (1976). *FEBS Lett.* **64**, 197–200.
Mazelis, M., Whatley, F. R. and Whatley, J. (1977). *FEBS Lett.* **84**, 236–240.
McClure, W. R. (1969). *Biochemistry* **7**, 2782–2786.
Miflin, B. J. and Lea, P. J. (1977). *Ann. Rev. Plant Physiol.* **28**, 299–329.
Miflin, B. J. and Lea, P. J. (1982). *In* "The Encyclopaedia of Plant Physiology; Nucleic Acid and Proteins in Plants" (D. Boulter and B. Parthier, eds), Vol. 14, pp. 5–64. Springer, Berlin, Heidelberg and New York.
Mills, W. R., Lea, P. J. and Miflin, B. J. (1980). *Plant Physiol.* **65**, 1166–1172.
Misono, H. and Soda, K. (1980). *J. Biol. Chem.* **225**, 10599–10605.
Moller, B. L. (1974). *Plant Physiol.* **54**, 638–643.
Nimmo, H. G. and Nimmo, G. A. (1982). *Anal. Biochem.* **121**, 17–22.
Relton, J. M., Bonner, P. L. R., Wallsgrove, R. M. and Lea, P. J. (1988). *Biochim. Biophys. Acta* **953**, 48–60.
Rognes, S. E. (1970). *FEBS Lett.* **10**, 62–66.
Rognes, S. E., Lea, P. J. and Miflin, B. J. (1980). *Nature* **287**, 357–359.
Rognes, S. E., Bright, S. W. J. and Miflin, B. J. (1983). *Planta* **157**, 32–38.
Sasaoka, K. (1961). *Plant Cell Physiol.* **2**, 231–242.
Shewry, P. R. and Miflin, B. J. (1977). *Plant Physiol.* **59**, 69–73.
Shimura, Y. and Vogel, H. J. (1966). *Biochim. Biophys. Acta* **118**, 396–404.
Sodek, L. (1976). *Phytochemistry* **15**, 1903–1906.
Sodek, L. (1978). *Rev. Brazil. Bot.* **1**, 65–69.
Stahly, D. P. (1969). *Biochim. Biophys. Acta* **191**, 439–444.
Tyagi, V. V. S., Henke, R. R. and Farkas, W. R. (1982). *Biochim. Biophys. Acta* **719**, 363–369.
Tyagi, V. V. S., Henke, R. R. and Farkas, W. R. (1983). *Plant Physiol.* **73**, 687–691.

Vogel, H. J. and Hirvonen, A. P. (1971). *In* "Methods of Enzymology" (H. Tabor and C. W. Tabor, eds), Vol. XVII, pp. 146–150. Academic Press, New York and London.
Vogel, H. J., Thompson, J. S. and Shockman, G. D. (1970). *In* "Organization and Control in Prokaryotic and Eukaryotic Cells" (H. P. Charles and B. C. J. G. Knight, eds), pp. 107–119. Cambridge University Press, London and New York.
Wallsgrove, R. M. and Mazelis, M. (1980). *FEBS Lett.* **116**, 189–192.
Wallsgrove, R. M. and Mazelis, M. (1981). *Phytochemistry* **20**, 2651–2655.
Wallsgrove, R. M., Lea, P. J. and Miflin, B. J. (1983). *Plant Physiol.* **71**, 780–784.
Wampler, D. E., Takahashi, M. and Westhead, E. W. (1970). *Biochemistry* **9**, 4210–4216.
Wenko, L. K., Treick, R. W. and Wilson, K. G. (1985). *Plant Mol. Biol.* **4**, 197–204.
Yamada, Y., Kumpaisal, R., Hashimoto, T., Sugimoto, Y. and Susuki, A. (1986). *Plant Cell Physiol.* **27**, 607–617.
Yugari, Y. and Gilvarg, C. (1965). *J. Biol. Chem.* **240**, 4710–4716.
Zlotnick, G. W. and Gottlieb, M. (1986). *Anal. Biochem.* **153**, 121–125.

19 Threonine Biosynthesis

SVEN ERIK ROGNES

Department of Biology, Botany Division, University of Oslo, P.O. Box 1045, Blindern, 0316 Oslo 3, Norway

I.	Introduction	315
II.	Homoserine dehydrogenase	317
	A. Preparation of ASA	317
	B. Assay methods for HSDH	317
	C. Purification methods for HSDH	318
	D. Properties of HSDH-S and HSDH-R	318
III.	Homoserine kinase	318
	A. Assay methods for HK	320
	B. Purification methods for HK	320
	C. Properties of HK	321
IV.	Threonine synthase	321
	A. Preparation of HSP	322
	B. Assay methods for TS	322
	C. Purification methods for TS	323
	D. Properties of TS	323
	References	323

I. INTRODUCTION

Threonine is synthesised from aspartate in five steps involving four intermediates: aspartyl phosphate, aspartate semialdehyde (ASA), homoserine and homoserine O-phosphate (HSP) (Fig. 19.1). The first two steps, catalysed by aspartate kinase (AK; EC 2.7.2.4) and aspartate semialdehyde dehydrogenase (ASADH; EC 1.2.1.11) (see Bonner

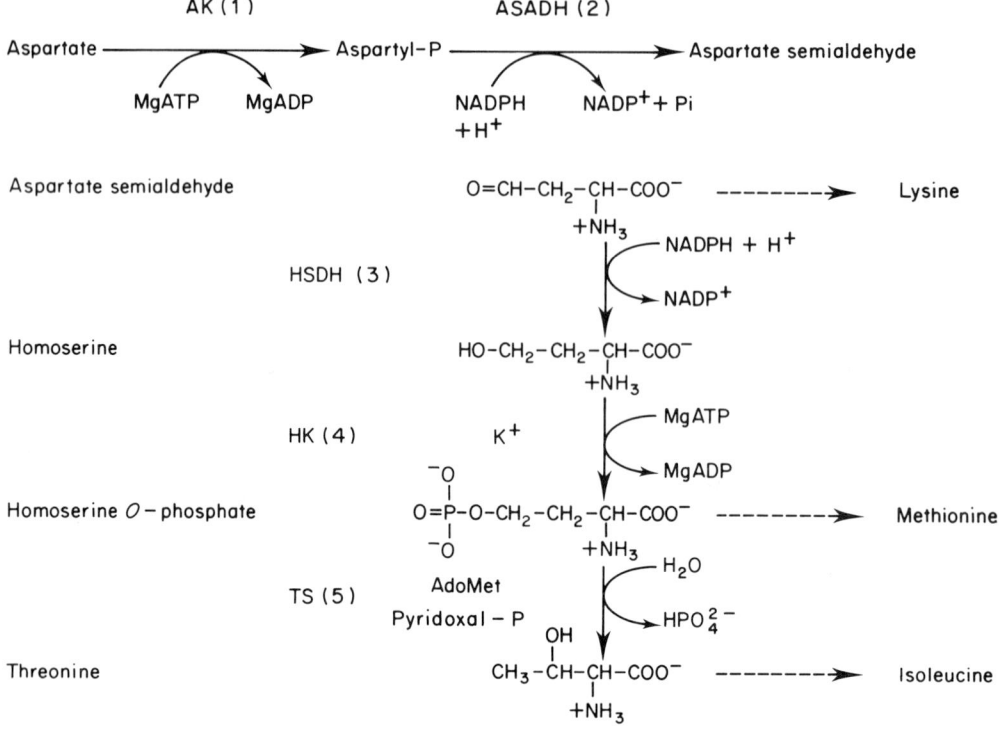

FIG. 19.1 The threonine biosynthetic pathway and its relation to lysine, methionine and isoleucine biosynthesis. L-forms of all amino acids are implied.

and Lea, Chapter 18, this volume), are common to lysine, threonine, isoleucine and methionine biosynthesis. ASA reduction to homoserine and homoserine phosphorylation are also parts of plant methionine biosynthesis, branching off at HSP (Madison, Chapter 21F, this volume). The final threonine synthase (TS; EC 4.2.99.2) step is a necessary reaction in isoleucine biosynthesis (Wallsgrove, Chapter 20, this volume).

In leaves, there is strong evidence for light-driven threonine synthesis confined to the chloroplasts (Miflin and Lea, 1982; Wallsgrove et al., 1983). Reviews and discussions of the threonine pathway and its regulation are found in Bryan (1980), Miflin and Lea (1982), Lea and Joy (1983), Rognes et al. (1983), Wallsgrove et al. (1983), Giovanelli et al. (1986, 1988) (plants) and Cohen and Saint-Girons (1987) (bacteria). For in vitro reconstitution of threonine synthesis and flux studies, see Szczesiul and Wampler (1976) and Shames et al. (1984). Due to the number of phosphorylated substrates, intermediates and products in the pathway, proton-coupled ^{31}P NMR spectra have proven useful for monitoring flux of substrates and inhibitor action.

The molecular genetics of the prokaryotic threonine pathway has been worked out in detail, but as yet no gene involved in threonine biosynthesis has been cloned from a higher plant. In yeast, the genes for AK, homoserine kinase (HK; EC 2.7.1.39) and TS have recently been cloned and partially characterised (Rafalski and Falco, 1988; Andresen and Rognes, 1988; Aas et al., 1988). Work with higher plants is complicated by the very common duplication of genes for enzymes of primary metabolism. Many steps in amino acid biosynthesis are catalysed by two or more isoenzymes.

II. HOMOSERINE DEHYDROGENASE

Homoserine dehydrogenase (HSDH; EC 1.1.1.3) catalyses the reduction of ASA to homoserine by NADPH; NADH is also a substrate. Assays in both directions are described in Section II.B below. The equilibrium constant is in the range 10^3 to 2×10^3 (Black and Wright, 1955). Due to its easy assay and high activity, HSDH has been the most studied enzyme of the three.

A. Preparation of ASA

This unstable intermediate is prepared by ozonolysis of DL-allylglycine (Black and Wright, 1955). The crude product contains, in addition to DL-ASA, aspartic acid, formaldehyde, performic acid and H_2O_2. It is strongly recommended to purify ASA over a Dowex 50W-H^+ column as described by Westerik and Wolfenden (1974). Two slightly overlapping ninhydrin-positive peaks are eluted by 0.275 N HCl, the first one being ASA, which is recognised by its reducing properties (aldehyde test) and reaction with 2,4-dinitrophenyl hydrazine. The H_2O_2 content of ASA solutions can be reduced by air bubbling or catalase addition to a working solution (pH 6). A high H_2O_2 content can lead to oxidation of NADPH by enzymes unrelated to HSDH (e.g. peroxidases: Aarnes, 1977) and cause misinterpretations.

ASA is stable for a year or more in 1 N HCl at $-20°C$. At pH 6 or below the half-life is several days at $0°C$. At pH 7.5 there is rapid loss of substrate activity (half-life < 20 min). A frozen stock solution in 1 N HCl should be freshly diluted, brought to pH 6 with solid $KHCO_3$ and kept in ice during the working period.

B. Assay Methods for HSDH

In the forward direction, the reaction mixtures contain in 1.00 ml (quartz cuvette; 1 cm light path): 200 mM potassium phosphate (pH 7.2); 0.5 mM EDTA; 1 mM dithiothreitol; 0.16 mM NADPH; 2 mM DL-ASA; H_2O and enzyme extract (0.05–0.3 nkat). Absorbance at 340 nm against a water blank is recorded (25°C). Following addition of enzyme, NADPH oxidation in the absence of ASA is followed for 1 min to establish the background rate to be subtracted. Then ASA is added and the decrease in A_{340} is monitored for 2–3 min. A decrease in A_{340} of 0.100 min^{-1} corresponds to 0.268 nkat (1 nkat (nanokatal) is the amount of enzyme that catalyses the ASA-dependent oxidation of 1 nmol NADPH s^{-1} = 60 nmol NADPH min^{-1}).

In the reverse direction, the reaction mixtures contain in 1.00 ml: 100 mM Tris-Cl (pH 9.0); 150 mM KCl; 0.5 mM EDTA; 1 mM dithiothreitol; 5 mM NAD$^+$ (or 0.5 mM NADP$^+$); 20 mM homoserine; H_2O and enzyme extract. The rate of NAD(P)$^+$ reduction in the absence of homoserine is first recorded at 340 nm (25°C). After addition of homoserine, the increase in A_{340} is monitored for 3–10 min. Rates are lower than in the forward direction. An increase in A_{340} of 0.100 min^{-1} corresponds to 0.268 nkat.

C. Purification Methods for HSDH

HSDH occurs as at least two distinct isoenzymes (Aarnes and Rognes, 1974; Walter et al., 1979; Grego et al., 1980; Sainis et al., 1981); a threonine-sensitive enzyme (HSDH-S), and a threonine-resistant enzyme (HSDH-R).* Both activities have been purified to apparent homogeneity from maize cell suspensions (Walter et al., 1979), and HSDH-S to homogeneity from maize seedlings by means of monoclonal antibodies and immunoaffinity chromatography (Krishnaswamy and Bryan, 1986). A critical factor in the isolation and purification of HSDH is the composition of the buffers used and the extraction conditions. HSDH-S is a sensitive allosteric enzyme which can easily be altered and desensitised. In this context, the history of maize HSDH is particularly instructive (Bryan, 1969; Matthews et al., 1975; DiCamelli and Bryan, 1975; Walter et al., 1979; Bryan and Lochner, 1981a,b). It is recommended to use a buffer containing 50 mM potassium phosphate pH 7.5; 15–20% (v/v) glycerol or ethylene glycol; 5–10 mM dithiothreitol; 2–5 mM threonine and 1 mM EDTA.

Some useful methods are summarised in Table 19.1. It is important to realise that the ratio of HSDH-S to HSDH-R varies with age, tissue and growth conditions. Old light-grown leaf material contains substances that inhibit recovery of native HSDH-S.

D. Properties of HSDH-S and HSDH-R

HSDH-S is a 150–190 kDa dimer of identical (Krishnaswamy and Bryan, 1986) or different (Walter et al., 1979) subunits with a chloroplast localisation in leaves (Bryan et al., 1977; Sainis et al., 1981). It is a complex allosteric enzyme displaying cooperative kinetics and ligand-induced interconversions among different dimeric or tetrameric states (Krishnaswamy and Bryan, 1983a,b). Threonine inhibits HSDH-S completely at 2 mM; the K_i is 50–100 µM. Threonine protects against thermal inactivation. HSDH-S is strongly activated by K^+—sometimes a nearly complete dependence is found (Grego et al., 1980; Di Marco and Grego, 1975; DiCamelli and Bryan, 1975). Maize HSDH-S shows higher activity with NAD^+ than $NADP^+$ in the reverse direction. Apparent K_ms for ASA and NADPH are in the range 66–660 µM and 30–50 µM, respectively, for HSDH-S from different plants.

HSDH-R is a 70–80 kDa dimer of identical subunits (Walter et al., 1979) located in the cytoplasm (Sainis et al., 1981). It shows Michaelis–Menten kinetics and is quite stable under conditions (e.g. 4 M urea, 24 h, 5°C) that completely inactivate HSDH-S (Bryan and Lochner, 1981b). Cysteine is a strong inhibitor, causing almost complete inhibition at 1 mM (Sainis et al., 1981). There is no K^+ effect as seen with HSDH-S. In the reverse direction, $NADP^+$ gives, typically, a higher rate than NAD^+ (opposite of HSDH-S) (Bryan and Lochner, 1981b). Apparent K_ms for ASA and NADPH have been reported to be 40–400 µM and 13–27 µM.

III. HOMOSERINE KINASE

Homoserine kinase catalyses the formation of HSP and ADP from homoserine and MgATP. The equilibrium constant is about 10^4 (Shames et al., 1984).

---

* Different authors have used different designations (e.g. HSDH I and II; Class II/III and I). To avoid confusion, HSDH-S and HSDH-R are adopted here.

TABLE 19.1. Chromatographic methods used for purification of HSDH-S and HSDH-R.

Matrix used	Plant source	Enzyme purified	References
Sephadex G-200, Sephacryl S-300 (Pharmacia)	Pea, barley, wheat	Both; separation obtained	1,2,3,6
Agarose A-0.5m, A-1.5m (Bio-Rad)	Maize seedling	HSDH-S	4
DEAE-Sephadex, DEAE-Sephacel (Pharmacia)	Pea, barley, maize cells, castor bean	Both; or HSDH-R	1,2,3,5
Octyl-Sepharose (Pharmacia)	Maize cells	Both	5
	Maize seedling	Not applicable	4
ADP-Agarose, type 2 (P-L Biochemicals)	Maize cells	Both	5
	Maize seedling	HSDH-S	4
Blue Sepharose (Pharmacia)	Maize cells	HSDH-S bound, HSDH-R unbound	5
	Pea leaf	HSDH-S unbound, HSDH-R bound	6
Matrex Gel Red A (Amicon)	Barley leaf	HSDH-S bound, HSDH-R unbound	6
MC-11 Agarose; monoclonal antibody to HSDH-S (immunoaffinity column)	Maize seedling	HSDH-S	7

1 Aarnes and Rognes (1974); 2 Aarnes (1977); 3 Grego et al. (1980); 4 Krishnaswamy and Bryan (1983a); 5 Walter et al. (1979); 6 Sainis et al. (1981); 7 Krishnaswamy and Bryan (1986).

A. Assay Methods for HK

The most sensitive and precise assay method involves the use of [^{14}C]homoserine as substrate and separation and measurement of [^{14}C]HSP after termination of the reaction. With purified enzyme lacking ATPase activity a coupled spectrophotometric assay of ADP formation can be useful.

1. Radioactive assay

The reaction mixture contains in 100 µl: 100 mM potassium phosphate (pH 7.8); 10 mM ATP; 12 mM MgCl$_2$; 1 mM dithiothreitol; 5 mM [^{14}C]homoserine (18.5 kBq µmol^{-1}); H$_2$O and enzyme extract. Incubate at 30°C for 15–60 min. Add 25 µl of 10% (w/v) trichloroacetic acid and spin briefly. Two methods can be used to separate [^{14}C]HSP from unreacted [^{14}C]homoserine:

(a) Apply 100µl (= 4/5) of the supernatant on a 0.5 ml bed volume Dowex 50W-H$^+$ column (2.5 × 0.5 cm) and collect the effluent. Wash with 4.9 ml H$_2$O. (HSP is very acidic and is recovered in the combined unbound solutions while homoserine is bound to the resin.) Determine the total radioactivity of the unbound fraction by scintillation counting. Determine the specific radioactivity of the substrate under the same conditions of measurement and calculate the enzyme activity. Corrections are made for acidic radioactive impurities in the substrate by running controls (minus ATP, zero time, minus enzyme).

(b) Paper electrophoresis in 45 mM sodium acetate, pH 5.2 (3 kV, 40 min) will separate HSP and homoserine. Eight 25 µl samples can conveniently be analysed in one run on a 26 cm × 57 cm Whatman No. 3 paper. HSP overlaps approximately the ATP area which is located with the aid of a 254 nm UV lamp. After drying, a paper square with the HSP spot is excised and its radioactivity determined.

2. Spectrophotometric assay

In a quartz cuvette at 25°C, the reaction mixture contains in 1.00 ml: 100 mM K-Hepes (pH 7.8); 100 mM KCl; 1 mM dithiothreitol; 5 mM ATP; 10 mM MgCl$_2$; 3 mM phosphoenolpyruvate; 0.2 mM NADH; 20 units each of pyruvate kinase and lactate dehydrogenase; 5 mM homoserine; H$_2$O and enzyme extract. Homoserine is added last. The rate of homoserine-dependent decrease in A_{340} against a water blank is taken as HK activity.

B. Purification Methods for HK

HK is located in the chloroplast (Wallsgrove et al., 1983; Muhitch and Wilson, 1983). It is a relatively labile enzyme, and the composition of the extraction buffer is important for some, but not all, plant tissues (Giovanelli et al., 1974). In several cases a substantial proportion of the activity was associated with the pellet fraction after centrifugation (10 000 × g) of a crude extract. In some plants this activity could be solubilised by washing with extraction buffer, but in others it remained firmly bound. A 'protective' buffer is recommended, e.g. 50 mM potassium phosphate (pH 7.5) containing 15% (v/v) glycerol, 0.5 mM EDTA, 5 mM dithiothreitol (high concentrations of 2-mercaptoethanol are not always equivalent) and, eventually, 0.05% Triton X-100 to facilitate solubilisation of bound forms. Another version is 'Medium A' (Giovanelli et al., 1974).

HK in a soluble form has been partially purified from green pea leaves and etiolated barley seedlings (Thoen et al., 1978a; Aarnes, 1976, 1978). Chromatography on CM-Sephadex C-50 was very effective with the pea enzyme; for barley HK results reported were strangely inconsistent. DEAE-Sephadex A-50 chromatography is effective. The purified enzyme from both of these sources was heterogeneous and gave two peaks during Sephadex G-200 gel filtration. A heat denaturation step (Aarnes, 1976, 1978) cannot be recommended, since an enzyme with altered properties may be isolated (e.g. desensitised to feedback inhibition). The properties of barley HK purified by a procedure involving an early heat treatment differ strikingly from the properties of pea and radish HK (Thoen et al., 1978a; Muhitch and Wilson, 1983; Baum et al., 1983).

Affi-Gel Blue A affinity chromatography was used to purify *Escherichia coli* HK (Shames et al., 1984) and could be worth a try with plant HK.

C. Properties of HK

There are several indications that HK may exist in at least two forms; whether these are separate gene products or not is unknown. In pea chloroplasts two HK forms with similar properties have been described; one soluble (presumably stroma-localised) and one associated with the thylakoid membranes (Muhitch and Wilson, 1983).

Barley, pea and radish HK are strongly dependent on K^+, with an optimal concentration at 100–200 mM. Strangely enough, the thylakoid-associated pea HK was stated not to be K^+-dependent. A 1–2 mM excess of Mg^{2+} over ATP is optimal.

The regulatory properties of the soluble pea and radish enzymes and the thylakoid-associated pea HK were similar. S-Adenosylmethionine (AdoMet), isoleucine and valine are effective feedback inhibitors. Threonine inhibits the radish HK and ornithine the pea HK. For radish HK, synergistic inhibition by AdoMet and isoleucine was shown at physiological concentrations (Baum et al., 1983). Isoleucine, valine and threonine may act at the same site, separate from the AdoMet site. Contrasting properties were found for a barley HK (Aarnes, 1978) which was neither inhibited by any of these nor by other amino acids tested at 10 mM.

The K_m for homoserine of the soluble pea enzyme is unusually high (6.7 mM) and this property may be a major factor in causing elevated concentrations of homoserine in germinating peas (see Lea and Joy, 1983).

IV. THREONINE SYNTHASE

The mechanism of HSP conversion to threonine was studied in detail with the *Neurospora* enzyme (Flavin and Slaughter, 1960a,b; Flavin and Kono, 1960; Kaplan and Flavin, 1965). There are few studies of TS in plants. The reaction is essentially irreversible, no [^{14}C]HSP formation was observed when *Neurospora* TS was incubated with [^{14}C]threonine and Pi, nor could enzymic exchange of ^{32}Pi into HSP be detected.

A. Preparation of HSP

Partially purified yeast (or *E. coli*) HK is used for enzymic synthesis of HSP or [^{14}C]HSP (Flavin and Slaughter, 1960a; Schildkraut and Greer, 1973; Thoen et al., 1978b; Giovanelli et al., 1984; Shames et al., 1984).

A crude protein extract is obtained by autolysis of frozen yeast in the presence of toluene or by mechanical or sonic disruption of the cells. An ammonium sulphate fraction between 30 and 45% saturation is prepared from the extract. The fraction (400 mg protein) is incubated for 24 h at 25°C with 100 mM Tris-Cl (pH 7.5); 1 mM dithiothreitol; 20 mM $MgCl_2$; 20 mM ATP; 50 mM KCl and 40 mM homoserine in a volume of 200 ml. The mixture is heated for 6 min at 100°C and then centrifuged. The supernatant is diluted to 300 ml and adjusted to pH 3 with HCl. The solution is applied on a Dowex 50W-H$^+$ column (150 ml bed) which is washed with 0.8 l H_2O. HSP is recovered in the unbound fractions and determined with ninhydrin. HSP is further purified over a Dowex 1-Cl$^-$ column by elution with formic acid (0–0.3M). After concentration and neutralisation the product is freeze-dried and dissolved in H_2O. The HSP should be free of Pi, adenine nucleotides and homoserine. The yield is 30–40% of added homoserine. [^{14}C]HSP is synthesised similarly in a smaller scale using [^{14}C]homoserine of desired specific radioactivity. Paper electrophoresis and chromatography can be used for rapid purification (Giovanelli et al., 1984).

B. Assay Methods for TS

TS can be assayed by (a) threonine formation, preferably using [U-^{14}C]HSP as substrate, or (b) by Pi formation. In many plants, the presence of phosphatase activity hydrolysing HSP to homoserine and Pi will disturb method (b) and demand a separation of threonine from HSP and homoserine (or measurement of a threonine-specific degradation product) if method (a) is used. This complicates the assay procedures. Two relatively simple methods are described below. For detailed procedures involving measurement of [^{14}C]glyoxylate and [^{14}C]acetaldehyde derived from periodate degradation of [^{14}C]threonine (see Madison and Thompson, 1976 and Giovanelli et al. 1984). If non-radioactive HSP is used, the products of threonine degradation can be estimated by means of lactate dehydrogenase (glyoxylate) or alcohol dehydrogenase (acetaldehyde) (Flavin and Slaughter, 1960a), but sensitivity is decreased.

1. *[^{14}C]Threonine formation from [^{14}C]HSP*

The reaction mixture contains in 100 µl: 50 mM K-Hepes (pH 8.0); 0.2 mM pyridoxal phosphate; 0.5 mM AdoMet; 2 mM L-[U-^{14}C]HSP (14 kBq µmol^{-1}); H_2O and enzyme extract. For very dilute extracts also add 0.2 mg bovine serum albumin. Incubate at 30°C for 30 min. Add 25 µl of 10% (w/v) trichloroacetic acid containing 5 mM threonine and 5 mM homoserine as carriers. Spin briefly and take out aliquots for paper chromatography in the solvent 1-butanol–acetone–diethylamine–H_2O (10 : 10 : 2 : 5). Threonine (R_f 0.63) is separated from homoserine (R_f 0.44) and HSP (R_f 0.13). Narrow guide strips are treated with ninhydrin to locate the amino acids. Cut out the threonine area and determine its radioactivity. Controls include tubes minus AdoMet and minus enzyme. For estimation of phosphatase activity the radioactivity in the homoserine spot is determined. If negligible phosphatase is present, threonine may be more rapidly separated from HSP (but not from homoserine) by paper electrophoresis at pH 5 (Thoen et al., 1978b).

2. Colorimetric assay of Pi formation

This rapid method is convenient for TS preparations with little or no HSP phosphatase activity. The reaction mixture is the same as in Section IV.B.1 above, except that the volume is increased to 0.30 ml and non-radioactive HSP is used. After incubation for the desired period, 0.10 ml of the assay solution is removed and mixed with 0.8 ml of the colour reagent (pH 1) of Lanzetta et al. (1979). After 1 min, 0.1 ml of the 34% sodium citrate solution is added. The 660 nm absorbance is read after 30 min at 20°C and compared to a Pi reference curve. Values are corrected for controls minus enzyme and minus AdoMet. TS activity is taken as AdoMet-dependent Pi formation.

C. Purification Methods for TS

Plant TS appears to be relatively stable. Hepes buffer is recommended for extraction solutions since Pi inhibits the reaction. TS was first detected in sugar beet and radish leaves (Madison and Thompson, 1976). It has been purified 14- to 38-fold to a specific activity in the range 0.12–0.44 nkat mg^{-1} protein from pea (Thoen et al., 1978b), barley (Aarnes, 1978) and *Lemna* (Giovanelli et al., 1984). Conventional methods were used and included ammonium sulphate fractionation, Sephadex/Sephacryl gel filtration and DEAE-Sephadex/DEAE-Sephacel chromatography. Yields ranged from 39 to 88%.

D. Properties of TS

TS is located in the chloroplast (Wallsgrove et al., 1983). In all plants examined, TS has a virtually absolute requirement for AdoMet as an allosteric activator. The Hill coefficient is 2.0–2.5. Half-maximal activation is obtained at 40–200 μM AdoMet. AdoMet increases V_m 5- to 25-fold and appears to decrease the K_m of HSP (Thoen et al., 1978b; Giovanelli et al., 1984). TS has a pH optimum around 8.0. The K_m for HSP of *Lemna* TS is very low (3–7 μM) compared to barley (50 μM), pea (0.67 mM) and sugar beet (2.2 mM). Pyridoxal phosphate is required for the activity of partially purified enzyme, but not for crude extracts.

TS is not feedback-inhibited by amino acids, but the level of TS activity is repressed by methionine supplementation to the growth medium (Giovanelli et al., 1984; Rognes et al., 1986). Isoenzymes of TS have not been detected.

REFERENCES

Aarnes, H. (1976). *Plant Sci. Lett.* **7**, 187–194.
Aarnes, H. (1977). *Plant Sci. Lett.* **9**, 137–145.
Aarnes, H. (1978). *Planta* **140**, 185–192.
Aarnes, H. and Rognes, S. E. (1974). *Phytochemistry* **13**, 2717–2724.

Aas, S. F., Brurberg, M. B., Brynestad, S., Andresen, A.-J. and Rognes, S. E. (1988). *Yeast* **4**, S313.
Andresen, A.-J. and Rognes, S. E. (1988). *Yeast* **4**, S315.
Baum, H. J., Madison, J. T. and Thompson, J. F. (1983). *Phytochemistry* **22**, 2409–2412.
Black, S. and Wright, N. G. (1955). *J. Biol. Chem.* **213**, 39–50.
Bryan, J. K. (1969). *Biochim. Biophys. Acta* **171**, 205–216.
Bryan, J. K. (1980). In "The Biochemistry of Plants" (B. J. Miflin, ed.), Vol. 5, pp. 403–452. Academic Press, New York.
Bryan, J. K. and Lochner, N. R. (1981a). *Plant Physiol.* **68**, 1395–1399.
Bryan, J. K. and Lochner, N. R. (1981b). *Plant Physiol.* **68**, 1400–1405.
Bryan, J. K., Lissik, E. A. and Matthews, B. F. (1977). *Plant Physiol.* **59**, 673–679.
Cohen, G. N. and Saint-Girons, I. (1987). In "*Escherichia coli* and *Salmonella typhimurium*. Cellular and Molecular Biology" (F. C. Neidhardt, ed.), Vol. 1, pp. 429–444. American Society for Microbiology, Washington, DC.
DiCamelli, C. A. and Bryan, J. K. (1975). *Plant Physiol.* **55**, 999–1005.
Di Marco, G. and Grego, S. (1975). *Phytochemistry* **14**, 943–947.
Flavin, M. and Kono, T. (1960). *J. Biol. Chem.* **235**, 1109–1111.
Flavin, M. and Slaughter, C. (1960a). *J. Biol. Chem.* **235**, 1103–1108.
Flavin, M. and Slaughter, C. (1960b). *J. Biol. Chem.* **235**, 1112–1118.
Giovanelli, J., Mudd, S. H. and Datko, A. H. (1974). *Plant Physiol.* **54**, 725–736.
Giovanelli, J., Veluthambi, K., Thompson, G. A., Mudd, S. H. and Datko, A. H. (1984). *Plant Physiol.* **76**, 285–292.
Giovanelli, J., Mudd, S. H., Datko, A. H. and Thompson, G. A. (1986). *Plant Physiol.* **81**, 577–583.
Giovanelli, J., Mudd, S. H. and Datko, A. H. (1988). *Plant Physiol.* **86**, 369–377.
Grego, S., Tricoli, D. and Di Marco, G. (1980). *Phytochemistry* **19**, 1619–1623.
Kaplan, M. M. and Flavin, M. (1965). *J. Biol. Chem.* **240**, 3928–3933.
Krishnaswamy, S. and Bryan, J. K. (1983a). *Arch. Biochem. Biophys.* **222**, 449–463.
Krishnaswamy, S. and Bryan, J. K. (1983b). *Arch. Biochem. Biophys.* **227**, 210–224.
Krishnaswamy, S. and Bryan, J. K. (1986). *Arch. Biochem. Biophys.* **246**, 250–262.
Lanzetta, P. A., Alvarez, L. J., Reinach, P. S. and Candia, O. A. (1979). *Anal. Biochem.* **100**, 95–97.
Lea, P. J. and Joy, K. W. (1983). In "Recent Advances in Phytochemistry" (C. Nozzolillo, P. J. Lea and F. A. Loewus, eds), Vol. 17, pp. 77–109. Plenum Press, New York.
Madison, J. T. and Thompson, J. F. (1976). *Biochem. Biophys. Res. Commun.* **71**, 684–691.
Matthews, B. F., Gurman, A. W. and Bryan, J. K. (1975). *Plant Physiol.* **55**, 991–998.
Miflin, B. J. and Lea, P. J. (1982). In "Encyclopedia of Plant Physiology, New Series" (D. Boulter and B. Parthier, eds), Vol. 14A, pp. 5–64. Springer, Berlin, Heidelberg and New York.
Muhitch, M. J. and Wilson, K. G. (1983). *Z. Pflanzenphysiol.* **110**, 39–46.
Rafalski, J. A. and Falco, S. C. (1988). *J. Biol. Chem.* **263**, 2146–2151.
Rognes, S. E., Bright, S. W. J. and Miflin, B. J. (1983). *Planta* **157**, 32–38.
Rognes, S. E., Wallsgrove, R. M., Kueh, J. S. H. and Bright, S. W. J. (1986). *Plant Sci.* **43**, 45–50.
Sainis, J. K., Mayne, R. G., Wallsgrove, R. M., Lea, P. J. and Miflin, B. J. (1981). *Planta* **152**, 491–496.
Schildkraut, I. and Greer, S. (1973). *J. Bacteriol.* **115**, 777–785.
Shames, S. L., Ash, D. E., Wedler, F. C. and Villafranca, J. J. (1984). *J. Biol. Chem.* **259**, 15331–15339.
Szczesiul, M. and Wampler, D. E. (1976). *Biochemistry* **15**, 2236–2244.
Thoen, Å., Rognes, S. E. and Aarnes, H. (1978a). *Plant Sci. Lett.* **13**, 103–112.
Thoen, Å., Rognes, S. E. and Aarnes, H. (1978b). *Plant Sci. Lett.* **13**, 113–119.
Wallsgrove, R. M., Lea, P. J. and Miflin, B. J. (1983). *Plant Physiol.* **71**, 780–784.
Walter, T. J., Connelly, J. A., Gengenbach, B. G. and Wold, F. (1979). *J. Biol. Chem.* **254**, 1349–1355.
Westerik, J. O'C. and Wolfenden, R. (1974). *J. Biol. Chem.* **249**, 6351–6353.

20 Enzymes of Leucine, Valine and Isoleucine Biosynthesis

ROGER M. WALLSGROVE

Biochemistry Department, AFRC Institute of Arable Crops Research, Rothamsted Experimental Station, Harpenden AL5 2JQ, UK

I.	Introduction	325
II.	Threonine dehydratase	327
	A. Extraction	327
	B. Assay	327
III.	Acetohydroxyacid synthase	328
	A. Extraction	328
	B. Assay	329
IV.	Acetohydroxyacid reductoisomerase	329
	A. Extraction	330
	B. Assay	330
V.	Dihydroxyacid dehydratase	330
	A. Extraction	330
	B. Assay	330
VI.	Branched-chain amino acid aminotransferase	331
	A. Extraction	331
	B. Assay	331
VII.	Enzymes of leucine biosynthesis	332
	References	333

I. INTRODUCTION

The branched-chain amino acids leucine, valine, and isoleucine are synthesised via a common pathway from pyruvate or threonine (Fig. 20.1.). All the enzymes of the

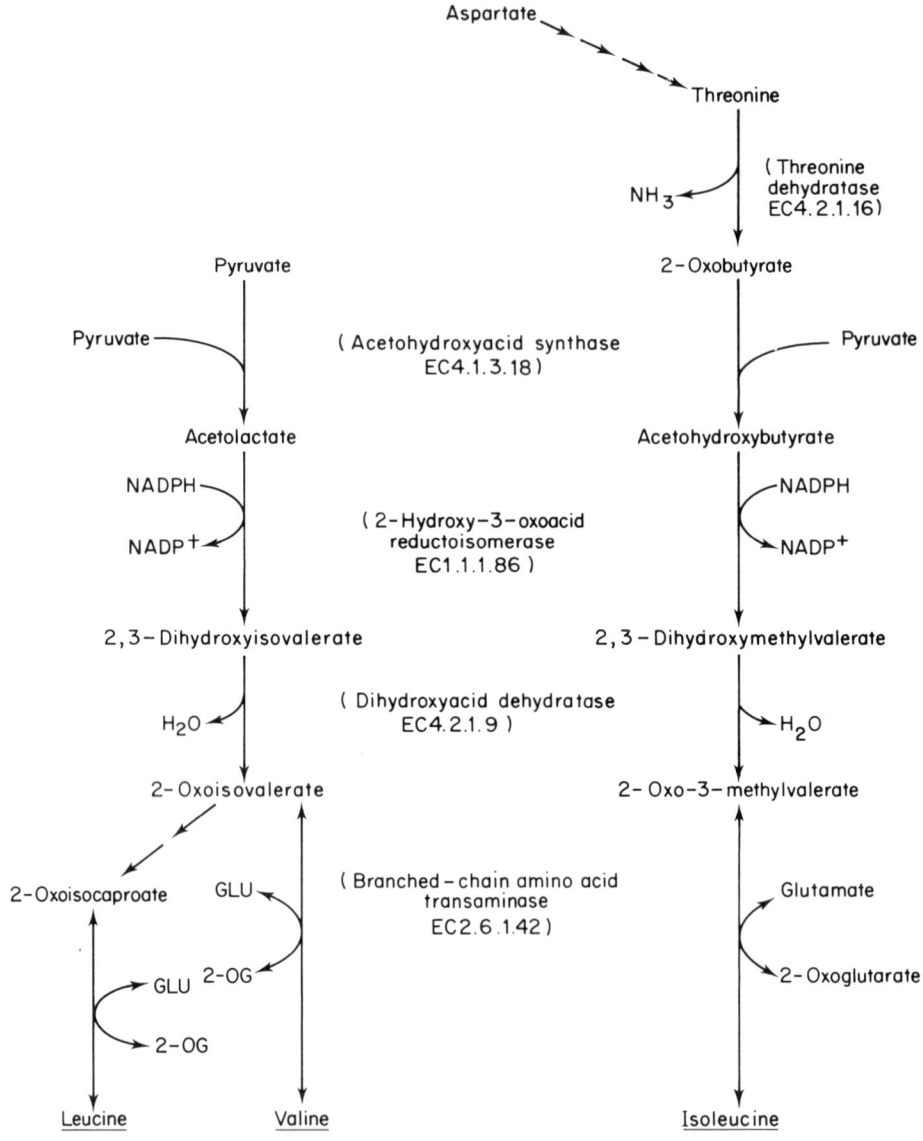

FIG. 20.1. The biosynthesis of leucine, valine and isoleucine.

pathway appear to be located in the plastids of plant cells (Shulze-Siebert et al., 1984). The biosynthesis of these amino acids is of interest because they are nutritionally essential for non-ruminant animals, and the pathway is the target of several classes of potent herbicides (Chaleff and Mauvais, 1984; Shaner et al., 1984). With few exceptions, these enzymes have not been intensively studied in plants, and the details given here for extraction and assay should be taken as no more than a starting point. Conditions may have to be altered significantly for optimal extraction and assay with other species and tissues.

Before discussing the enzymes individually, I would like to make some general points. Extraction conditions must be effective in breaking plastids if the full *in vivo* activity is to be measured. In some cases high concentrations of glycerol or ethanediol are necessary to stabilise the enzymes, and gentle homogenisation in such buffers may not break all plastids. The addition of Triton X-100 or other non-ionic detergent will overcome these problems, but care must be taken that the stability and activity of the enzyme are not adversely affected by the surfactant.

The activities of all these enzymes are low in plant tissues, and some are unstable, so great care should be taken to optimise the extraction conditions for the tissue being investigated. It is highly unlikely that a single extraction medium and set of extraction conditions will be equally effective for a wide variety of different species and tissues. In particular, the problems posed by phenolics and other secondary products will vary enormously. The addition of polyvinylpolypyrrolidone (PVPP) (0.2–0.5 g per g fresh weight) is recommended if the starting material contains significant amounts of polyphenols (if an extract turns brown or black on storage, repeat using PVPP!).

Whatever the extraction conditions, crude extracts should be routinely desalted prior to assay. For several of these enzymes, endogenous organic or amino acids will interfere with the assay. In addition, it is vital to remove the feedback inhibitors of threonine dehydratase and acetohydroxyacid synthase, which must be included in the extraction buffers to stabilise the enzymes. Small (2–5 ml) columns of Sephadex G-25 will rapidly desalt sufficient extract for several assays. Concentration of enzyme protein by $(NH_4)_2SO_4$ precipitation, followed by desalting, is often very useful, but note that the optimum salt concentration needed to precipitate any given enzyme may vary with the species.

II. THREONINE DEHYDRATASE (EC 4.2.1.16)

The extraction and assay described here are based on those optimised for the maize enzyme (J. K. Bryan, pers. comm.). The pH optimum and saturating substrate concentration may differ in other species, and should be determined for the material being used. Threonine dehydratase activity is regulated *in vivo* by isoleucine, which is a potent inhibitor (0.1–1 mM inhibiting completely). In maize, two apparent isozymes have been reported, separable by ion-exchange chromatography, but with very similar properties (Kirchner and Bryan, 1985).

A. Extraction

Extraction is carried out with 50 mM potassium phosphate (pH 7.5), containing 1 mM EDTA, 2 mM dithioerythritol (DTE) or DTT, 1 mM L-isoleucine, 20% (v/v) glycerol. Triton X-100, 0.05% (v/v), may be included. Fresh, not frozen, tissue should be used, and the extract must not be frozen at any stage. Partially purified enzyme is moderately stable at 4°C, particularly in low ionic strength buffers.

B. Assay

The product, 2-oxobutyrate, may be assayed (a) directly, as the dinitrophenylhydrazone, in fixed time assays, or (b) indirectly, via a continuous coupled spectrophotometric assay. For either assay the basic conditions are the same: 100 mM N-(2-hydroxyethyl)piperazine-N'-(3-propane sulphonic acid) (EPPS) (pH 8.7); 200 mM KCl; 100 mM L-threonine; 0.4 mM pyridoxal 5'-phosphate; and enzyme, in a total volume of 1.0 ml.

Method (a). Start the reaction by the addition of enzyme, and incubate at 30°C for up to 30 min. Stop the reaction by adding 0.2 ml of a 1:1 mixture of 30% (w/v) TCA and 0.2% (w/v) 2,4-dinitrophenylhydrazine in 2 M HCl. After 10 min add 1 ml 2.5 M NaOH, mix well, centrifuge, and read the absorbance at 540 nm. Compare with a standard curve obtained using 2-oxobutyrate in the assay medium.

Method (b). Add 100 units rabbit muscle LDH (Type II, Sigma), and 0.25 mM NADH. Record the decrease in absorbance at 340 nm with time.

For both assays run blanks lacking either enzyme or threonine. The pH optimum of the maize enzyme is very sharp, activity decreasing rapidly at lower pH. The enzyme activity requires a monovalent cation, and the apparent K_m (threonine) may be high (>20 mM). In my experience some batches of pyridoxal phosphate give unacceptably high backgrounds with method (a). The coupled assay is *not* recommended for crude extracts.

III. ACETOHYDROXYACID SYNTHASE (AHAS) (EC 4.1.3.18)

This is by far the best characterised and most studied of the enzymes in the pathway. It is the target for sulphonylurea and imidazolinone herbicides, and is regulated *in vivo* by leucine and valine, which cooperatively inhibit the enzyme. Several assay methods have been described, but the simplest and most commonly used is the colorimetric determination of acetolactate. This assay does not measure the other activity, acetohydroxybutyrate synthesis.

A. Extraction

Extraction is carried out with 50 mM Tris-HCl (pH 7.5) containing 10 mM $MgSO_4$, 20% (v/v) ethanediol or glycerol (basic buffer), plus 10 mM pyruvate, 0.1 mM thiamine pyrophosphate, 50 µM FAD, 1 mM L-leucine and 1 mM L-valine. The latter components should be added just before use, and the complete extraction medium kept cold and (ideally) in the dark. Grinding tissue in liquid nitrogen prior to extraction may be helpful. Ammonium sulphate precipitation is a useful concentration/purification step, though the high ethanediol or glycerol concentration may make it difficult to pellet the precipitated protein. Alternatively the crude extract may be applied to a column of Q-Sepharose (Pharmacia) equilibrated in the basic buffer, and AHAS eluted with 0.2 M KCl. Some workers include thiol reagents (dithiothreitol, mercaptoethanol) in the extraction buffer, but this has not proved necessary for either the barley or tobacco enzymes. Extraction of AHAS in the absence of Leu + Val significantly reduces the sensitivity of the enzyme to these feedback inhibitors, and the extractable activity. Purification of the barley enzyme has recently been reported (Durner and Böger, 1988).

B. Assay

Acetolactate formed in the reaction is decarboxylated in acid to acetoin, which is determined colorimetrically. The basic assay given here may need to be modified, as considerable variation in optimum pH and pyruvate concentration has been reported for AHAS from different sources.

The basic assay conditions are: 50 mM Tris-HCl (pH 7.5); 10 mM $MgSO_4$; 50 mM pyruvate; 1 mM thiamine pyrophosphate; 10 μM FAD; plus enzyme, in a total volume of 0.5 ml. Start the reaction by adding enzyme, and run zero time and minus enzyme blanks. Incubate at 30°C for 30–90 min, and stop the reaction by adding 20 μl 20% (v/v) H_2SO_4. Heat at 60°C for 15 min, and then rapidly add 0.15 ml 0.5% (w/v) creatine, and 0.15 ml 5% (w/v) α-naphthol in 1 M NaOH (both these solutions should be freshly prepared). Mix well and incubate for 15 min at 60°C. After a further 15 min at room temperature, centrifuge the tubes and measure the absorbance at 530 nm of the supernatant.

Whilst plant AHAS activity is not absolutely dependent on added FAD, this cofactor may enhance the activity of purified enzyme preparations. For colour development, the highest purity α-naphthol should be used to reduce background absorbance. If necessary, recrystallise commercial α-naphthol from chloroform, and always store the compound and its solutions in the dark. Reported pH optima for plant AHAS vary from pH 6–8, depending on species and tissue. A pyruvate concentration of 0.1 M may be used, but for the barley enzyme at least there is evidence of substrate inhibition when pyruvate exceeds 50 mM. Both these factors should be examined for the enzyme you are working with. Higher assay temperatures have been used, up to 50°C (Singh et al., 1988), but this may affect both stability and the kinetic properties of the enzyme. Plant AHAS is unstable, even at −80°C, and particularly at low protein concentrations. It is stabilised to some degree by glycerol/ethanediol (use up to 40%; v/v), and by its substrates, cofactors, and feedback inhibitors (Leu and Val).

If it is desired to assay acetohydroxybutyrate synthesis, some other assay method is needed. Whilst in theory the decarboxylation of this product should lead to a compound detectable by the acetoin method, in practice the yield is extremely poor. A gas chromatography method for the detection of both products has been described (Gollop et al., 1987). Alternatively, both substrates and both products give clear proton NMR signals, and bacterial and plant AHAS have been assayed by carrying out the reaction(s) in an NMR spectrometer (D. H. G. Crout and E. Lee, pers. comm.).

IV. ACETOHYDROXYACID REDUCTOISOMERASE (EC 1.1.1.86)

Only one detailed study of this enzyme in plants has been reported, Satyanarayana and Radhakrishnan (1965) describing a 70-fold purification of the enzyme from *Phaseolus radiatus*. Activity has been found in leaves of spinach (Kanamori and Wixom, 1963) and barley (R. M. Wallsgrove, unpubl. res.). The substrates are not commercially available and must be synthesised (see Krampitz, 1948; and Crout and Rathbone, 1989).

A. Extraction

Both Tris and phosphate buffers have been used. For initial studies I recommend extraction in 50 mM Tris-HCl (pH 8), 2 mM EDTA, 2 mM DTT. Ammonium sulphate precipitation (50% saturation) is effective in concentrating the barley enzyme, and helps to reduce the background rate of NAD(P)H oxidation.

B. Assay

NAD(P)H oxidation is monitored in a spectrophotometer at 340 nm. Tris buffer at pH 8–9 is recommended, the *Phaseolus* enzyme having a reported pH optimum of 8.6. The assay conditions are: 50 mM Tris-HCl (at an appropriate pH); 0.1 mM NAD(P)H; 20 mM 2-acetolactate or 2-acetohydroxybutyrate; plus enzyme, in a total volume of 1.0 or 3.0 ml (depending on available cuvettes). The cuvette is placed in a temperature-controlled holder (25 or 30°C) in the spectrophotometer. After measuring the background change in A_{340}, the reaction is started by the addition of substrate. Where the substrates have been prepared as the esters, they may be pre-treated with an esterase. Alternatively, crude plant extracts often contain sufficient esterase activity to rapidly hydrolyse the substrate esters in the reaction medium. The *Phaseolus* enzyme is reportedly more active with 2-acetohydroxybutyrate than with 2-acetolactate, and has a lower K_m for the former (Satyanarayana and Radhakrishnan, 1965). Crude barley leaf extracts are, however, more active with acetolactate, though the kinetics have not been compared in detail (R. M. Wallsgrove, unpubl. res.).

V. DIHYDROXYACID DEHYDRATASE (EC 4.2.1.9)

This is the only enzyme of the pathway to have been purified to homogeneity, from spinach leaves (Flint and Emptage, 1988). It is a dimeric iron–sulphur protein with a subunit molecular weight of 63 kDa. The enzyme is unstable at pH values below 8, and in the absence of Mg^{2+}. The substrates are not commercially available (see Kiritani and Wagner, 1970; and Armstrong *et al.*, 1977, for details of their synthesis).

A. Extraction

Homogenise fresh or frozen tissue in 0.1 M Tris-HCl (pH 8.2) containing 10 mM $MgSO_4$ and 0.05% (v/v) Triton X-100. Alternatively, an acetone powder may be prepared—homogenise tissue in acetone at $-15°C$, filter, and remove traces of acetone in a freeze-dryer. Extract the resulting powder as for fresh tissue.

B. Assay

The oxo-acid products may be assayed colorimetrically as their dinitrophenylhydrazones (Kiritani and Wagner, 1970), or directly by ion-exclusion HPLC (Wallsgrove *et al.*, 1986a,b). For either method, the reaction conditions are the same: 50 mM Tris-HCl (pH 8.2); 10 mM MgSO$_4$; 10 mM dihydroxymethylvalerate or dihydroxyisovalerate; plus enzyme, in a total volume of 0.5 ml. Start the reaction by adding the substrate or enzyme (run blanks lacking these), and incubate at 30°C for 30–90 min.

Method (a): Colorimetric assay. Stop the reaction by adding 50 µl 20% (w/v) TCA, followed by 0.25 ml saturated 2,4-dinitrophenylhydrazine in 2 M HCl. After 30 min at room temperature, add 0.5 ml 5 M NaOH, and vortex mix or sonicate the tubes. Centrifuge to remove any precipitate and measure the absorbance at 550 nm. Compare to standard curves prepared with the appropriate oxo-acid.

Method (b): HPLC. Either remove 100 µl aliquots and add to 10 µl 0.1 M H$_2$SO$_4$, or add 50 µl acid to the whole reaction mix. Centrifuge, and apply an aliquot to an ion-exclusion column (e.g. Aminex HPX-87H). The substrates and products are resolved by isocratic elution with 0.005 M H$_2$SO$_4$ at 60°C, and detected by absorbance at 230 nm. This method is less sensitive than method (a), but allows simultaneous assay of both reactions catalysed by the enzyme. The reported K_m' (dihydroxyisovalerate) for the spinach enzyme is 1.5 mM (Flint and Emptage, 1988), and the rate with this substrate is apparently two-fold greater than with dihydroxymethylvalerate (Kanamori and Wixom, 1963).

VI. BRANCHED-CHAIN AMINO ACID AMINOTRANSFERASE
(EC 2.6.1.42)

Two forms of this enzyme have been reported in extracts from barley (Aarnes, 1981) and soybean (Pathre *et al.*, 1987), separable by ion-exchange chromatography or isoelectric focusing. It is possible that the two forms represent distinct catabolic and anabolic enzymes, but this has yet to be confirmed.

A. Extraction

The extraction medium consists of 50 mM potassium phosphate (pH 7.5), containing 10% (v/v) glycerol, 14 mM β-mercaptoethanol and 0.01 mM pyridoxal 5'-phosphate. For soybean, the addition of 5 mM 2-oxoglutarate reportedly improved the stability of the enzyme (Pathre *et al.*, 1987).

B. Assay

Extracts *must* be desalted prior to assay, as any endogenous amino or oxo-acids will interfere with the assay. The reaction is fully reversible, and may be measured in either the forward (Ile/Val/Leu formation) or reverse (Glu formation) directions. Several assay methods have been used, and two basic procedures are described here. The choice will depend on the equipment and facilities available, and the purpose of the investigation.

Method (a): Colorimetric determination of the reverse reaction (Pathre *et al.*, 1987): 100 mM Tris-HCl (pH 8–9); 40 mM L-leucine; 10 mM 2-oxoglutarate; 0.2 mM pyridoxal phosphate; plus enzyme, in a total volume of 1.0 ml. Start the reaction by adding enzyme, and after 30 min at 30°C stop by adding 1 ml 0.3% (w/v) 2,4-dinitrophenylhydrazine in 2 M HCl. After a further 5 min at 30°C, add 5 ml cyclohexane, vortex mix for 1 min and centrifuge for 5 min. Remove 4 ml of the upper layer, add to 1.5 ml 10% (w/v) Na_2CO_3, vortex, and stand for 5 min. Take 1 ml of the upper layer, add to 2 ml 1 M NaOH, and measure the absorbance at 440 nm after 5 min. This method relies on the differential extraction of the dinitrophenylhydrazone of oxoisocaproate compared to that of oxoglutarate. It can only be recommended where more sophisticated equipment is not available.

Method (b): Radiochemical determination of forward or reverse reactions (Aarnes, 1981): 150 mM Tris-HCl (pH 8–9) plus either

(i) 4 mM L-[U-^{14}C]glutamate, 1 mM oxoisocaproate or oxomethylvalerate or oxoisovalerate (forward reaction); or

(ii) 4 mM L-[U-^{14}C]leucine or valine or isoleucine, 5 mM 2-oxoglutarate (reverse reaction);

plus enzyme, in a total volume of 0.1 ml.

The reaction is started by adding enzyme, and after 30 min at 30°C, stopped by addition of 0.2 ml 0.3% (w/v) 2,4-dinitrophenylhydrazine in 2 M HCl. After 30 min the dinitrophenylhydrazones are extracted by adding 0.3 ml ethylacetate. Mix thoroughly, and take 25 µl of the ethylacetate for determination of radioactivity in a liquid scintillation counter. This method is much more sensitive than method (a).

Useful variations of method (b) include the use of [^{14}C]oxo-acid substrates (all three branched-chain oxo-acids are obtainable from Amersham International) when measuring the forward reaction. This has been combined with the separation of substrates and products by high-voltage paper electrophoresis on Whatman 3MM paper (50 mM sodium acetate (pH 5), 30–40 min at 3 kV (Wallsgrove *et al.*, 1986a), also applicable to the standard assay (Aarnes, 1981).

Alternatively, ion-exclusion HPLC efficiently resolves all the oxo-acids, and this has been used as the basis for a 'cold' assay, in which all three reactions may be determined simultaneously. For the barley enzyme, assayed in this way, the relative rates of oxo-acid utilisation were oxoisocaproate > oxomethylvalerate > oxoisovalerate (R. M. Wallsgrove, unpubl. res.). These transaminases appear to be unstable in relatively crude extracts, but may be stabilised by the addition of glycerol, and by storage in the presence of phosphate rather than Tris buffers. The presence of any of the amino acid substrates significantly reduces the heat stability of the barley enzyme (Aarnes, 1981).

VII. ENZYMES OF LEUCINE BIOSYNTHESIS

The only study of the leucine biosynthetic enzymes in plant tissues appears to be that of Oaks (1965), but extraction and assay conditions were not optimised. Assays for the yeast and bacterial enzymes have been developed, and these are the best starting point for any study of the plant enzymes. The reader is advised to consult *Methods in Enzymology*, Vol. 17A (Tabor and Tabor, 1970). In addition, an improved assay for isopropylmalate synthase is described by Calvo *et al.* (1969).

REFERENCES

Aarnes, H. (1981). *Z. Pflanzenphysiol.* **102**, 81–89.
Armstrong, F. B., Muller, U. S., Reary, J. B., Whitehouse, D. and Crout, D. H. G. (1977). *Biochim. Biophys. Acta* **498**, 282–293.
Calvo, J. M., Bartholomew, J. C. and Stieglitz, B. I. (1969). *Anal. Biochem.* **28**, 164–181.
Chaleff, R. S. and Mauvais, C. J. (1984). *Science* **224**, 1443–1445.
Crout, D. H. G. and Rathbone, D. L. (1989). *Synthesis* 40–41.
Durner, J. and Böger, P. (1988). *Z. Naturforsch.* **43**, 850–856.
Flint, D. H. and Emptage, M. H. (1988). *J. Biol. Chem.* **263**, 3558–3564.
Gollop, N., Barak, Z. and Chipman, D. M. (1987). *Anal. Biochem.* **160**, 323–331.
Kanamori, M. and Wixom, R. L. (1963). *J. Biol. Chem.* **238**, 998–1005.
Kirchner, S. C. and Bryan, J. K. (1985). *Plant Physiol.* **77**, (Suppl.), 597.
Kiritani, K. and Wagner, R. P. (1970). In "Methods in Enzymology" (H. Tabor and C. W. Tabor, eds), Vol. 17A, pp. 755–764. Academic Press, New York and London.
Krampitz, L. O. (1948). *Arch. Biochem.* **17**, 81–85.
Oaks, A. (1965). *Biochim. Biophys. Acta* **111**, 79–89.
Pathre, V., Singh, A. K., Viswanathan, P. N. and Sane, P. V. (1987). *Phytochemistry* **26**, 2913–2917.
Satyanarayana, I. and Radhakrishnan, A. N. (1965). *Biochim. Biophys. Acta* **110**, 380–388.
Shaner, D. L., Anderson, P. C. and Stidham, M. A. (1984). *Plant Physiol.* **76**, 545–546.
Shulze-Siebert, D., Heineke, D., Scharf, H. and Schultz, G. (1984). *Plant Physiol.* **76**, 465–471.
Singh, B. K., Stidham, M. A. and Shaner, D. L. (1988). *Anal. Biochem.* **171**, 173–179.
Tabor, H. and Tabor, C. W. (1970). "Methods in Enzymology", Vol. 17A. Academic Press, New York and London.
Wallsgrove, R. M., Risiott, R., King, J. and Bright, S. W. J. (1986a). *Plant Sci.* **43**, 109–114.
Wallsgrove, R. M., Risiott, R., Negrutiu, I. and Bright, S. W. J. (1986b). *Plant Cell Rep.* **3**, 223–226.

21 Sulphur Metabolism
A. ATP-Sulphurylase

DANIEL SCHMUTZ

Pflanzenphysiologisches Institut der Universität Bern, Altenbergrain 21, CH-3013 Bern, Switzerland

I.	Introduction	335
II.	Methods for determination of enzyme activity	336
III.	Determination of enzyme activity with the luciferin–luciferase system	336
	A. Extraction	337
	B. Preparation of the luciferin–luciferase reagent	337
	C. Enzyme assay	337
	D. Calibration of the ATP meter	337
	References	337

I. INTRODUCTION

ATP-sulphurylase (EC 2.7.7.4) catalyses the first step in assimilatory sulphate reduction, forming adenosine 5'-phosphosulphate and pyrophosphate from adenosine triphosphate and sulphate.

$$SO_4^{2-} + ATP \rightleftarrows APS^{2-} + PPi$$

ATP-sulphurylase is widely distributed in nature. The presence of the enzyme has been demonstrated in animals, microorganisms and plants (De Meio, 1975). ATP-sulphurylase has been purified and characterised from several organisms, e.g. from spinach leaves (Balharry and Nicholas, 1970; Shaw and Anderson, 1972, 1974), as well as from leaves of green cabbage (Osslund et al., 1982). The latter enzyme seems to be an asymmetric dimer composed of 57 kDa subunits.

II. METHODS FOR DETERMINATION OF ENZYME ACTIVITY

Because of the extremely unfavourable equilibrium constant for APS formation, which seems to be about 10^{-8} (Robbins and Lipmann, 1958), the assay of ATP-sulphurylase is difficult due to the lack of accumulation of significant amounts of product. The forward and reverse reactions have both been used for the determination of the enzyme activity. Methods in which the formation of AP^{35}S from ^{35}SO$_4^{2-}$ and ATP is determined suffer from the fact that they necessitate the use of high amounts of radioactive SO$_4^{2-}$ and are extremely time-consuming because of the necessary separation of AP^{35}S from ^{35}SO$_4^{2-}$ by either high voltage paper electrophoresis (Ellis, 1969; Reuveny and Filner, 1976) or HPLC (Hommes and Moss, 1986).

Similar disadvantages are found for the assay of ATP-sulphurylase by PPi exchange, where [^{32}P]ATP is formed from PPi and ATP.

The molybdate method of Wilson and Bandurski (1958) is based on the fact that molybdate used as an analogue of sulphate forms an unstable adenosylphosphomolybdate and PPi. The PPi produced is hydrolysed by endogenous or added pyrophosphatase and the amount of phosphate produced is taken as a measure of ATP-sulphurylase. The disadvantages of this method are: (1) the assay is conducted with the wrong substrate: (2) ATPases can cause a high background; and (3) the hydrolysis of pyrophosphate may be incomplete.

A spectrophotometric assay for measuring ATP-sulphurylase was published by Burnell (1984). The reverse reaction catalysed by ATP-sulphurylase is coupled with hexokinase and glucose 6-phosphate dehydrogenase and the APS- or PPi-dependent reduction of NADP is measured.

Balharry and Nicholas (1970) have developed an assay for ATP-sulphurylase from plant tissues using the luciferin–luciferase system to measure the ATP produced in the reverse reaction.

This method has been further developed in our laboratory and the scintillation spectrometer has been replaced by an ATP meter (Lumac/3M Biocounter M 2010A, Fakola, Basel, Switzerland). This method is fast since one assay can be performed per minute and no radioactive wastes are produced. Additionally, the method contains corrections for ATPases, ATP, light-quenching substances or chemoluminescence present in crude extracts. The rates of ATP-sulphurylase activity determined by this method were 25 times higher than those measured in the forward reaction as AP[^{35}S] formed from ATP and SO$_4^{2-}$ (Schmutz and Brunold, 1982).

III. DETERMINATION OF ENZYME ACTIVITY WITH THE LUCIFERIN–LUCIFERASE SYSTEM

The method is modified from that originally described by Schmutz and Brunold (1982).

A. Extraction

Extracts are prepared by grinding 1 g of plant material in 10 ml of 0.1 M Tris-HCl (pH 8.0) containing 2 mM MgCl, 100 mM KCl and 10 mM DTE in a glass homogeniser cooled with ice. The homogenate is centrifuged for 10 min at 10 000 × g at 4°C.

B. Preparation of the Luciferin–Luciferase Reagent

One vial of a highly purified luciferin–luciferase mixture (Lumit HS) is dissolved in 60 ml of Lumit buffer (Fakola, Basel, Switzerland). The reagent may be stored in aliquots at −20°C for up to two months. To improve the stability of this system, the reagent is melted 1 h before the measurement and kept at room temperature in a light-protected glass vial.

C. Enzyme Assay

The assay contains 20 μl of extract, 20 μl of 0.1 mM APS, 100 μl of the luciferin–luciferase reagent and 100 μl of 165 μM PPi. The assay mixture minus PPi is pipetted into disposable polystyrol cuvettes and pre-incubated at room temperature for 1 min. The reaction is started by the addition of PPi. The signal produced at an interval between 30 and 60 s is used as a measure of the production of ATP in the reverse reaction of ATP-sulphurylase.

D. Calibration of the ATP Meter

For the calibration the following assay system was used: 100 μl of 165 μM PPi mixed with 20 μl of extract and 20 μl of an ATP solution containing 1 nmol of ATP. The reaction is started by the addition of 100 μl of luciferin–luciferase reagent.

REFERENCES

Balharry, G. J. E. and Nicholas, D. J. D. (1970). *Biochim. Biophys. Acta* **220**, 513–524.
Burnell, J. N. (1984). *Plant Physiol.* **75**, 873–875.
De Meio, R. H. (1975). *In* "Metabolism of Sulfur Compounds", (D. M. Greenberg, ed.), Vol. 7, pp. 287–358. Academic Press, New York.
Ellis, R. J. (1969). *Planta* **88**, 34–42.
Hommes, F. A. and Moss, L. (1986). *Anal. Biochem.* **154**, 100–103.
Osslund, T., Chandler, C. and Segel, I. H. (1982). *Plant Physiol.* **70**, 39–45.
Reuveny, Z. and Filner, P. (1976). *J. Biol. Chem.* **255**, 1858–1864.
Robbins, P. W. and Lipmann, F. J. (1958). *J. Biol. Chem.* **233**, 686–690.
Schmutz, D. and Brunold, Ch. (1982). *Anal. Biochem.* **121**, 151–155.
Shaw, W. H. and Anderson, J. W. (1972). *Biochem. J.* **127**, 237–247.
Shaw, W. H. and Anderson, J. W. (1974). *Biochem. J.* **139**, 27–35.
Wilson, L. G. and Bandurski, R. S. (1958). *J. Biochem. Chem.* **233**, 975–981.

21 Sulphur Metabolism
B. Adenosine 5'-Phosphosulphate Sulphotransferase

CHRISTIAN BRUNOLD and MARIANNE SUTER

Pflanzenphysiologisches Institut der Universität Bern, Altenbergrain 21, CH-3013 Bern, Switzerland

I.	Introduction	339
II.	Measurement of APS sulphotransferase activity	340
	A. Principles of measurement	340
	B. Preparation of [^{35}S]APS	340
	C. Assay system for the determination of APS sulphotransferase activity	341
III.	Extraction and properties of APS sulphotransferase	341
IV.	Occurrence and localisation of APS sulphotransferase	342
	References	342

I. INTRODUCTION

Adenosine 5'-phosphosulphate sulphotransferase (APS sulphotransferase; adenylylsulphate: thiol sulphotransferase) is the second enzyme in the pathway of assimilatory sulphate reduction (Schmidt, 1972; Goldschmidt *et al.*, 1975). It catalyses the transfer of the sulpho group of APS, formed by ATP-sulphurylase (EC 2.7.7.4), to a carrier thiol (CarSH):

$$APS^{2-} + CarSH \xrightarrow{\text{APS sulphotransferase}} CarS\text{-}SO_3^- + AMP^{2-} + H^+$$

The carrier thiol may be glutathione in *Chlorella* (Tsang and Schiff, 1978) or a somewhat larger molecule in spinach (Schmidt and Schwenn, 1971). Recently, phytochelatins ((γ-glutamyl-cysteine)$_n$-glycine, $n = 3$–7) have been proposed as carriers (Steffens et al., 1986), but APS sulphotransferase will transfer the sulpho group to other added thiols (Tsang and Schiff, 1976) in both the presence or the absence of the carrier.

Thiols forming rings on oxidation such as dithioerythritol (DTE) or dithiothreitol (DTT) produce sulphite (SO_3^{2-}) in this reaction (a). With monothiols such as glutathione (GSH) the Bunte salt (R—S—SO_3^-) of the thiol is synthesised (b). In the presence of excess thiol, sulphite is formed (c):

(a) $APS^{2-} + DTE_{red} \longrightarrow AMP^{2-} + DTE_{ox} + SO_3^{2-} + 2H^+$
(b) $APS^{2-} + GSH \longrightarrow AMP^{2-} + GS\text{-}SO_3^- + H^+$
(c) $GS\text{-}SO_3^- + GSH \longrightarrow GSSG + SO_3^{2-} + H^+$

II. MEASUREMENT OF APS SULPHOTRANSFERASE ACTIVITY

A. Principles of Measurement

The activity of APS sulphotransferase can be determined qualitatively with APS as a substrate in the presence of DTE, by measuring the SO_2 evolved from the acidified incubation mixture using pararosanilin as a reagent (Brunold and Schiff, 1976). The quantitative determination of APS sulphotransferase activity includes the use of radioactive [^{35}S]APS as substrate, from which [^{35}S]SO_3^{2-} is produced by the enzyme in the presence of DTE or DTT. Addition of carrier SO_3^{2-} to the incubation mixture causes the production of large amounts of radioactive SO_2 after acidification, which may be trapped in a base. The trapped radioactivity is used as a measure of APS sulphotransferase activity (Schiff and Levinthal, 1968).

B. Preparation of [^{35}S]APS

[^{35}S]APS is produced according to Tsang et al. (1976), using extracts from *Chlorella*. With this method [^{35}S]PAPS (adenosine 3′-phosphate 5′-phosphosulphate) is synthesised in a first step from carrier-free $^{35}SO_4^{2-}$ and ATP. [^{35}S]PAPS is then dephosphorylated by 3′ nucleotidase. Using the same procedure non-radioactive APS can be synthesised, but this substrate can also be purchased.

We found that the use of a micro-dismembrator (Braun; Bender + Hobein, Zürich, Switzerland) for breaking the *Chlorella* cells was more convenient than the French pressure cell in the original method and resulted in extracts with higher efficiency for PAPS production. Portions of about 0.5 g of wet-packed cells are transferred into the Teflon containers of the micro-dismembrator and are frozen using liquid nitrogen. Homogenisation to a fine powder is performed for 3 min at maximal speed. For the separation of [^{35}S]PAPS and [^{35}S]APS from the other compounds in the incubation mixtures we routinely use a DEAE-Sephacel column (5 × 30 cm, Pharmacia) and apply a triethanolamin-CO_2 buffer (Tsang et al., 1976) in a 2 l linear gradient from 0.25 to 0.8 M for elution.

C. Assay System for the Determination of APS Sulphotransferase Activity

The assay system routinely used for the determination of APS sulfotransferase activity consists of: 1 M Tris-HCl (pH 9.0), 50 µl; 0.2 M dithioerythritol (DTE), 10 µl; 2 M MgSO$_4$, 200 µl; plant extract, 10–130 µl; 3.75 µM [^{35}S]APS, 10 µl; 1 M Na$_2$SO$_3$, 100 µl; H$_2$O, add, 500 µl.

The incubation is for 30 min in a 1.5 ml Eppendorf tube at 37°C. The Eppendorf tube is then transferred into a vial for scintillation counting, containing 1 ml 1 M triethanolamine solution. The enzyme reaction is stopped by addition of 200 µl 2 M H$_2$SO$_4$ to the assay system. The cover is screwed immediately on the vial. The distillation of the ^{35}SO$_2$ formed into the triethanolamine solution is for 12–24 h at room temperature. The Eppendorf tube is taken out of the scintillation vial, scintillation fluid is added and the radioactivity is determined. The specific activity of the [^{35}S]APS used is determined under identical quenching conditions as that of the assay system. The radioactivity of [^{35}S]APS varies according to the activity of APS sulphotransferase and the amount of extract added.

We found APS sulphotransferase activity in root material was about 0.1 nmol min^{-1} (mg protein)$^{-1}$, whereas in leaves, activities from 1 up to 20 nmol min^{-1} (mg protein)$^{-1}$ were measured.

Lower APS sulphotransferase activities are measured below pH 9. DTE may be used in lower concentrations and may be replaced by DTT. High ionic strength causes an increase in APS sulphotransferase activity; MgSO$_4$ can therefore be replaced by various other salts, such as Na$_2$SO$_4$ or NaCl. The concentration of APS applied should be checked for the plant material examined, since moderate concentrations can cause substrate inhibition. Mixing of the plant extract and APS separately from the complete assay system should be avoided, because this can result in rapid degradation of APS.

III. EXTRACTION AND PROPERTIES OF APS SULPHOTRANSFERASE

The buffer routinely used for extraction of APS sulphotransferase activity from roots or shoots is 0.1 M Tris-HCl (pH 8.0), containing 0.1 M KCl, 0.02 M MgCl$_2$, and 0.01 M DTE or DTT. Other buffer systems adopted successfully include Hepes-KOH or potassium phosphate. KCl and MgCl$_2$ may be omitted, but may be useful when required for the measurement of other enzyme activities in the crude extracts. DTE or DTT may also be omitted for the extraction from spruce needles (C. Brunold, unpubl. res.), so the necessity of its use should be examined for each tissue. For the extraction of APS sulphotransferase activity from tissue containing high amounts of phenols addition of 1% (v/v) Tween 80 to the extraction buffer was found to be essential (Brunold et al., 1983).

The highest activities of APS sulphotransferase were obtained when the plant tissues were homogenised using a glass homogeniser at 0–4°C, but a Polytron (Type PT 10/35, Kinematica, Kriens, Switzerland) was used sucessfully for this purpose for spruce needles (Tschanz et al., 1986).

To our knowledge the conditions for stabilising APS sulphotransferase activity from tissues of higher plants have not yet been found. This explains why the enzyme has not yet been purified to homogeneity.

APS sulphotransferase has been partially purified from *Chlorella* (Tsang and Schiff, 1976) and from spinach leaves (Schmidt, 1976). The enzyme from spinach has a molecular weight of 110 kDa. The K_m for APS is 10 µM. AMP is a competitive inhibitor. Cysteine causes a 50% inhibition at 68 µM with the APS sulphotransferase from *Chlorella* (Schmidt, 1973) but has no effect on the enzyme from *Lemna* (Brunold and Schmidt, 1976).

Changes in enzyme level seem to be more important in controlling the APS sulphotransferase step than variation of enzyme activity. Cysteine and H_2S decrease the enzyme level in various systems (Brunold and Schmidt, 1976, 1978; Wyss and Brunold, 1979; Jenni *et al.*, 1980; Brunold *et al.*, 1981), while lack of SO_4^{2-} or high demand for reduced sulphur cause an appreciable increase in extractable activity (Brunold and Suter, 1984; Brunold *et al.*, 1987; Nussbaum *et al.*, 1988).

IV. OCCURRENCE AND LOCALISATION OF APS SULPHOTRANSFERASE

APS sulphotransferase activity has been demonstrated in many higher plants (Schmidt, 1975), in algae and in cyanobacteria (Tsang and Schiff, 1975; Schmidt, 1977a,b). Within the leaves of spinach APS sulphotransferase activity is localised predominantly, or even exclusively, in the chloroplasts (Schwenn *et al.*, 1976; Fankhauser and Brunold, 1978). In the roots, APS sulphotransferase activity is similarly distributed intracellularly as nitrite reductase, indicating a predominant proplastid localisation (C. Brunold and M. Suter, unpubl. res.). In *Euglena*, the enzyme was detected in the mitochondria (Brunold and Schiff, 1976; Saida *et al.*, 1988).

REFERENCES

Brunold, C. and Schiff, J. A. (1976). *Plant Physiol.* **57**, 430–436.
Brunold, C. and Schmidt, A. (1976). *Planta* **133**, 85–88.
Brunold, C. and Schmidt, A. (1978). *Plant Physiol.* **61**, 342–347.
Brunold, C. and Suter, M. (1984). *Plant Physiol.* **76**, 579–583.
Brunold, C., Zryd, J. P. and Lavanchy, P. (1981). *Plant Sci. Lett.* **21**, 167–174.
Brunold, C., Landolt, W. and Lavanchy, P. (1983). *Physiol. Plant* **59**, 313–318.
Brunold, C., Suter, M. and Lavanchy, P. (1987). *Physiol. Plant.* **70**, 168–174.
Fankhauser, H. and Brunold, C. (1978). *Planta* **143**, 285–289.
Goldschmidt, E. E., Tsang, M. L.-S. and Schiff, J. A. (1975). *Plant Sci. Lett.* **4**, 293–299.
Jenni, B. E., Brunold, C., Zryd, J. P. and Lavanchy, P. (1980). *Planta* **150**, 140–143.
Nussbaum, S., Schmutz, D. and Brunold, C. (1988). *Plant Physiol.* **88**, 1407–1410.
Saida, T., Na, S.-Q., Li, J. and Schiff, J. A. (1988) *Biochem. J.* **25**, 533–539.
Schiff, J. A. and Levinthal, M. (1968). *Plant Physiol.* **43**, 547–554.
Schmidt, A. (1972). *Arch. Mikrobiol.* **84**, 77–86.
Schmidt, A. (1973). *Arch. Mikrobiol.* **93**, 29–52.
Schmidt, A. (1975). *Plant Sci. Lett.* **5**, 407–415.
Schmidt, A. (1976). *Planta* **130**, 257–263.
Schmidt, A. (1977a). *FEMS Microbiol. Lett.* **1**, 137–140.
Schmidt, A. (1977b). *Arch. Mikrobiol.* **112**, 263–270.

Schmidt, A. and Schwenn, J. D. (1971). *Proc. 2nd Int. Congr. Photosynthesis.* Junk, The Hague.
Schwenn, J. D., Depka, B. and Hennies, H. H. (1976). *Plant Cell Physiol.* **17**, 165–176.
Steffens, J. D., Hunt, D. F. and Williams, B. G. (1986). *J. Biol. Chem.* **261**, 13879–13882.
Tsang, M. L.-S. and Schiff, J. A. (1975). *Plant Sci. Lett.* **4**, 301–307.
Tsang, M. L.-S. and Schiff, J. A. (1976). *Plant Cell Physiol.* **17**, 1209–1220.
Tsang, M. L.-S. and Schiff, J. A. (1978). *Plant Sci. Lett.* **11**, 177–183.
Tsang, M. L.-S., Lemieux, J., Schiff, J. A. and Bojarski, T. B. (1976). *Anal. Biochem.* **74**. 623–626.
Tschanz, A., Landolt, W. and Brunold, C. (1986). *Physiol. Plant.* **67**, 235–241.
Wyss, H.-R. and Brunold, C. (1979). *Planta* **147**, 37–42.

21 Sulphur Metabolism
C. Sulphite Reductase

CHRISTOPH VON ARB

Pflanzenphysiologisches Institut der Universität Bern, Altenbergrain 21, CH-3013 Bern, Switzerland

I.	Introduction	345
II.	Measurement of sulphite reductase	345
	A. Principal possibilities	345
	B. Assay system for sulphite reductase	346
III.	Properties of sulphite reductase	347
IV.	Occurrence and localisation of sulphite reductase	348
	References	348

I. INTRODUCTION

Sulphite reductase (SiR) catalyses the reduction of free sulphite to free sulphide in a step requiring six electrons. While NADPH serves as the electron donor in microorganisms (EC 1.8.1.2; Siegel, 1975), ferredoxin (Fd) is the donor in higher plants (EC 1.8.7.1; Hennies, 1975; Aketagawa and Tamura, 1980; Krueger and Siegel, 1982; Hirasawa *et al.*, 1987):

$$SO_3^{2-} + 6\,Fd_{red} \xrightarrow{Fd\text{-}SiR} S^{2-} + 6\,Fd_{ox} + 3H_2O$$

II. MEASUREMENT OF SULPHITE REDUCTASE

A. Principal Possibilities

The various possible assay systems to measure SiR activity involve: (a) formation of methylene blue from the sulphide produced directly in the reaction mixture (Siegel, 1965) or from H_2S after acidification (Tamura et al., 1978); (b) estimation of the radioactivity of $H_2^{35}S$ formed from $^{35}SO_3^{2-}$ using Cd^{2+} solution as a distillation trap (Hennies, 1975); (c) using S^{2-}-electrode (Ng and Anderson, 1979); (d) determination of the absorbance change of Fd at 422 nm (Krueger and Siegel, 1982); and (e) measurement of cysteine in a coupled assay system using O-acetyl-L-serine sulphydrylase (OASSase, EC 4.2.99.8) (von Arb and Brunold, 1983). A general disadvantage of the methods (a)–(d) is the fact that sulphide which accumulates in the assay system inhibits SiR (50% at 18 μM) (von Arb and Brunold, 1985). Individual disadvantages of the different methods have been summarised by Siegel (1965) and von Arb and Brunold (1983).

B. Assay System for Sulphite Reductase

Although the assay system which is described here appears to be rather complicated, it seems to be well suited for most purposes because it takes the disadvantages mentioned above into consideration. The complete assay contains in a total volume of 1 ml (von Arb and Brunold, 1983):

Hepes-NaOH (pH 7.8)	25 μmol
Na_2SO_3	1 μmol
Ferredoxin from spinach	7 μmol
O-acetyl-L-serine	5 μmol
O-acetyl-L-serine sulphydrylase	0.3 μmol
Dithiothreitol	10 μmol
DCMU	10 μmol

Extract from plant tissue
A ferredoxin-reducing system including: Na-ascorbate, 20 μmol; dichlorophenol indophenol, 10 μmol; chloroplast thylakoids heated for 5 min at 55°C, containing 50 μg chlorophyll.

The assay is performed in 25 ml Erlenmeyer flasks, which are sealed with rubber stoppers, evacuated and flushed five times with Ar or N_2. Incubation is for 5–40 min at 30°C and a quantum flux density of about 500 μE m^{-2} s^{-1} on a shaking device (160 rpm). Incubation is terminated by adding 0.2 ml of 1.5 M trichloroacetic acid to the assay. After centrifugation (1 min at $2200 \times g$) at room temperature, the supernatant is mixed with 1 ml ninhydrin reagent (Gaitonde, 1967) (250 mg ninhydrin + 4 ml conc. HCl + 16 ml acetic acid). The assays are kept in a boiling water bath for 5 min, cooled rapidly in iced-water and mixed with 2 ml 96% ethanol. The absorbance at 546 nm is determined. Calibration curves are established by adding known amounts of cysteine to the assay. Serine, O-acetyl-L-serine and cystine do not give a colour reaction.

Inhibition of the enzyme activity by the sulphide formed is prevented since OASSase eliminates sulphide beyond detection limits. The measured amount of cysteine is equivalent to the decrease of sulphite in the assay (von Arb and Brunold, 1983). The reaction is linear for at least 40 min incubation time and 0.8 mg protein.

OASSase activity is endogenously present in many plant extracts with rates 10^2–10^3 higher than SiR. It has to be tested in every new system to ensure that the OASSase present in the extract is not rate-limiting for the measurement of SiR.

OASSase is a stable enzyme. A partial purification useful for the assay has been described by von Arb and Brunold (1983) as follows:

Destemmed spinach leaves (3.3 kg) are homogenised in a Waring blender using 3.3 l 0.02 M Tris-HCl (pH 8.0) at 4°C. Temperature and buffer remain the same throughout the purification. After a 45–70% $(NH_4)_2SO_4$ cut of the filtered crude extract and subsequent centrifugation, the clear enzyme preparation is desalted on a Sephadex G-25 column (3.5 × 11 cm). The active fractions are applied to a DEAE-Sephadex column (1.8 × 20 cm) and eluted in steps with buffer containing 0.11, 0.15 and 0.19 M NaCl (300 ml per step).

The fractions containing OASSase activity are pooled and applied to a Sephadex G-100 column (2.5 × 59 cm) after reduction to 15 ml by ultrafiltration. For the elution on the second DEAE-Sephadex column (1.8 × 20 cm) a linear gradient of 0.1–0.2 M NaCl (200 ml) is used. The active fractions are concentrated to the desired volume by ultrafiltration. The enzyme is stored at −20°C. Fd can be purchased or is purified from spinach according to Buchanan and Arnon (1971). Chloroplast thylakoids are prepared as described by Schürmann and Brunold (1980). The preparation starts with the isolation of broken chloroplasts according to Elias and Givan (1978). The 1000 g pellet is resuspended in 0.1 M Hepes-NaOH (pH 7.8), centrifuged at 30 000 × g and washed twice in the same buffer. Heating at 55°C for 5 min destroys the oxygen-evolving capacity and any remaining SiR activity. Photosystem I remains active for 10 days at 4°C.

Reduction of Fd in the dark by electrons derived from glucose 6-phosphate via glucose 6-phosphate dehydrogenase, NADPH and Fd-$NADP^+$ reductase (EC 1.18.1.2) reduces SiR activity to about 10% of its normal rate in the light.

An assay in which Fd and its reducing system are replaced by NADPH and glutathione-reduced (GSH) (Ng and Anderson, 1979) is probably limited to chloroplast preparations with a high percentage of intact chloroplasts. 3-(3,4-Dichlorophenyl)-1,1-dimethyl urea (DCMU) can be omitted in routine measurements after its negligible effect has been proven. Incubation in air instead of N_2 or Ar reduces SiR activity by 40%, indicating that the system is susceptible to O_2. A simplified but unphysiological modification of the assay consists of the replacement of Fd and its reducing system by $Na_2S_2O_4$ (15 mM), $NaHCO_3$, (30 mM) and methyl viologen (5 mM). The enzyme reaction is run under aerobic conditions and does not require light. However, this assay has the disadvantage that free sulphite is formed from dithionite and therefore does not allow the establishment of exactly defined substrate concentrations.

An alternative to dithionite is to reduce methyl viologen prior to incubation by H_2 and Pt asbestos (Krueger and Siegel, 1982). However, both preparation and incubation have to be performed under absolute anaerobic conditions. It does therefore not seem to be suitable for crude extracts or incubation times longer than a few minutes.

III. PROPERTIES OF SULPHITE REDUCTASE

This sirohaem-containing enzyme, which has great similarities to Fd-dependent nitrite reductase (EC 1.7.7.1), has been purified to homogeneity from spinach by Krueger and Siegel (1982). It consists of 2 subunits of 63 kDa and 69 kDa with 1 sirohaem and 1 Fe_4S_4 centre per subunit. The K_m for the substrate sulphite was estimated to be 20–25 μM.

IV. OCCURRENCE AND LOCALISATION OF SULPHITE REDUCTASE

SiR activity has been measured in *Chlorella* (Abrams and Schiff, 1973) and in higher plants such as spinach, peas, wheat and maize (Tamura *et al.*, 1978; von Arb and Brunold, 1986; Sawhney and Nicholas, 1975; Schmutz and Brunold, 1985). The enzyme activity which has been localised in wheat chloroplasts (Sawhney and Nicholas, 1975) appears to be tightly particulate bound in pea leaves (von Arb and Brunold, 1983). Mayer (1967) measured SiR in roots and shoots of barley with an activity ratio of 1:10.

REFERENCES

Abrams, W. A. and Schiff, J. A. (1973). *Arch. Microbiol.* **94**, 1–10.
Aketagawa, J. and Tamura, G. (1980). *Agric. Biol. Chem.* **44**, 2371–2378.
Buchanan, B. B. and Arnon, D. T. (1971). *In* "Methods in Enzymology" (A. San Pietro, ed.), Vol. 23, Part A, pp. 410–440. Academic Press, New York and London.
Elias, B. A. and Givan, C. V. (1978). *Planta* **142**, 317–320.
Gaitonde, M. C. (1967). *Biochem. J.* **104**, 627–633.
Hennies, H. H. (1975). *Z. Naturforsch.* **30c**, 359–362.
Hirasawa, M., Boyer, J. M., Gray, K. A., Davis, D. J. and Knaff, D. B. (1987). *FEBS Lett.* **221**, 343–348.
Krueger, R. J. and Siegel, L. M. (1982). *Biochemistry* **21**, 2905–2909.
Mayer, A. M. (1967). *Plant Physiol.* **42**, 324–326.
Ng, B. H. and Anderson, J. W. (1979). *Phytochemistry* **18**, 573–580.
Sawhney, S. K. and Nicholas, D. J. D. (1975). *Phytochemistry* **14**, 1499–1503.
Schürmann, P. and Brunold, C. (1980). *Z. Pflanzenphysiol.* **100**, 257–268.
Schmutz, D. and Brunold, C. (1985). *Physiol. Plant.* **64**, 523–528.
Siegel, L. M. (1965). *Anal. Biochem.* **11**, 126–132.
Siegel, L. M. (1975). *In* "Metabolism of Sulfur Compounds" (D. M. Greenberg, ed.), Vol. 7, pp. 287–358. Academic Press, New York and London.
Tamura, G., Hosoi, T. and Aketagawa, J. (1978). *Agric. Biol. Chem.* **42**, 2165–2167.
von Arb, Ch. and Brunold, C. (1983). *Anal. Biochem.* **131**, 198–204.
von Arb, Ch. and Brunold, C. (1985). *Physiol. Plant.* **64**, 290–294.
von Arb, Ch. and Brunold, C. (1986). *Physiol. Plant.* **67**, 81–86.

21 Sulphur Metabolism
D. Cysteine Synthase

AHLERT SCHMIDT

Botanisches Institut, Tierärztliche Hochschule Hannover, Bunteweg 17d, D-3000 Hannover 71, FRG

I.	Introduction	349
II.	Measurement of cysteine synthase activity	350
III.	Specificity of the assay	350
IV.	Extraction and purification	351
V.	Specificity of cysteine synthase	351
VI.	Occurrence and localisation of cysteine synthase	352
VII.	Regulatory properties	352
	References	353

I. INTRODUCTION

L-Cysteine is the first compound with a carbon–sulphur bond formed during sulphur assimilation. Incorporation of sulphur for amino acid synthesis in plants is catalysed by the conversion of activated serine and sulphide forming cysteine (Ellis, 1963; Greenberg, 1975). The biosynthesis of coenzymes and iron–sulphur clusters within the cell is dependent on the availability of cysteine. The pathways leading to sulphide have been discussed previously, therefore the main focus here will be on the enzymes forming L-cysteine from *O*-acetyl-CoA, L-serine and sulphide (Brüggemann *et al.*, 1961; Smith and Thompson, 1971; Brunold and Suter, 1982) according to the following reactions:

$$\text{L-Serine} + \text{acetyl-S-CoA} \xrightarrow{\text{Serine acetyltransferase}} O\text{-Acetyl-L-serine} + \text{CoA-SH} \qquad (1)$$

$$O\text{-Acetyl-L-serine} + H_2S \xrightarrow[\text{synthase}]{\text{Cysteine}} \text{L-Cysteine} + \text{acetate} \qquad (2)$$

Cysteine synthase (O-acetyl-L-serine sulphydrylase, EC 4.2.99.8 [equivalent to: O-acetyl-L-serine(thiol)-lyase or O-acetyl-L-serine-sulphydrase]) is responsible for sulphur assimilation at the reduced level (Greenberg, 1975; Anderson, 1980; Schmidt, 1979, 1982a,b, 1986; Giovanelli, 1987; Soda, 1987). Kinetic experiments have shown that sulphide influx into organic material has only this one entry point in plants, although cell-free experiments demonstrated the possible sulphydration of activated homoserine derivatives (Anderson, 1980; Giovanelli et al., 1980).

II. MEASUREMENT OF CYSTEINE SYNTHASE ACTIVITY

Different techniques have been used for the measurement of this enzyme, including: (a) the measurement of cysteine as a red ninhydrin complex (Gaitonde, 1967; Krauss, 1984); (b) its determination after derivatisation of the thiol group and identification by HPLC methods (Fahey and Newton, 1987; Cooper and Turnell, 1987); (c) its determination as pyruvic acid after coupling with cysteine desulphydrase (Wedding, 1987); and (d) using a binding protein assay (Smith et al., 1987).

The acid ninhydrin test has been used in my laboratory and this test system will be described as a reference. The main assay system contains: 1 M Tris-HCl (pH 7.5), 100 µl; 0.1 M DTE, 30 µl; 50 mM O-acetyl-L-serine, 100 µl; 0.1 M sulphide, 50 µl; enzyme, 10–200 µl; H_2O to 1000 µl.

The reaction should be started by the addition of sulphide. Normal incubation is carried out at 37°C for 15 min, the vials being covered by glass marbles to prevent loss of sulphide (pH 7.5). The reaction is stopped by addition of 1 ml of Gaitonde reagent (see below) and heated afterwards for 10 min at 95°C. The red ninhydrin complex formed is determined at 560 nm and evaluated against a calibrated cysteine standard (cysteine reference) made on the same day with the same ninhydrin reagent. The measurement should be made within 15 min after the heating procedure.

The acidic ninhydrin reagent is set up by dissolving 1.25 g ninhydrin in 20 ml conc. HCl and 80 ml conc. acteic acid (Krauss, 1984).

III. SPECIFICITY OF THE ASSAY

The acidic ninhydrin reagent is not specific for cysteine; it will detect cysteine derivatives such as glutathione and S-sulphocysteine, which can be hydrolysed to cysteine under acidic conditions (Krauss, 1984). Addition of DTE (or DTT) improves this assay since it is specific for the reduced cysteine and will not detect cystine.

IV. EXTRACTION AND PURIFICATION

Cysteine synthase can be purified by conventional methods. An ammonium sulphate precipitation, between 30 and 80% followed by dialysis gives yields of 90%. For further purification the following steps have been found useful: (1) DEAE-cellulose chromatography, separation on hydrophobic columns, especially phenyl-Sepharose and separation according to molecular weights on gels. Similar principles have been used for purification of cysteine synthases from spinach (Murakoshi et al., 1985) and barley (Rosichan et al., 1983).

Native gels can be run for this enzyme with about 40% yield. Active staining on native gels is possible using L-cysteine as a substrate and lead acetate for determination of free sulphide as a brownish band of PbS. The sensitivity can be enhanced by addition of cyanide if β-cyanoalanine synthase is absent (see specificity of the cysteine synthase). A combination of the methods stated above allowed purification of the cysteine synthase to homogeneity and the raising of specific antisera (A. Schmidt, unpubl. res.).

V. SPECIFICITY OF CYSTEINE SYNTHASE

Cysteine synthase is a pyridoxal phosphate-containing enzyme with two subunits of about 34 kDa (Greenberg, 1975; Diessner and Schmidt, 1981; Leon et al., 1987). The enzyme catalyses the conversion of O-acetyl-L-serine to L-cysteine. It is specific for O-acetyl-L-serine; other derivatives including L-serine, O-phospho-L-serine, L-cysteine, O-acetyl-D-serine, DL-homocysteine and O-phospho-L-homoserine have activities below 1% (Schmidt, 1977a,b). Cysteine synthase from barley accepts sodium azide to form β-azido-alanine (Rosichan et al., 1983); furthermore, at a rate of about 1%, cyanide can be added instead of sulphide to form β-cyanoalanine (A. Schmidt, unpubl. res.). Cysteine synthases do not discriminate between sulphide and selenide, thus formation of selenocysteine is catalysed by this enzyme (Ng and Anderson, 1978).

The catalysis of L-cysteine formation requires an aminoacrylate intermediate, which is the sulphide-accepting species. This aminoacrylate can be formed from L-cysteine as well (although with less efficiency), allowing principally a back reaction. Such a back reaction will lead to free sulphide from cysteine, allowing an isotopic exchange reaction (Schmidt, 1977a,b; Diessner and Schmidt, 1981). This can also be demonstrated for sulphide liberation by methanethiol or ethanethiol or cysteine formation from S-methyl- and S-ethyl-cysteine in the presence of sulphide (Schmidt, 1977a,b; Murakoshi et al., 1985). Synthesis of substituted cysteines as summarised by Greenberg (1975) might not be catalysed by cysteine synthase itself, as demonstrated for a tryptophan synthase from Escherichia coli (Esaki et al., 1983). Cysteine synthases from higher plants will not accept thiosulphate to form S-sulphocysteine.

The back reactions discussed above are favoured at pHs above 9, thus they are not of physiological significance. However, they might be useful for assay principles such as staining of the cysteine synthase on native gels. The 'real' pH optimum for cysteine synthase cannot be measured due to a chemical conversion of O-acetyl-L-serine to N-acetyl-L-serine at pH values above 7.8.

Since cysteine synthases are pyridoxal phosphate-containing enzymes, they can be inhibited by compounds such as hydroxylamine, aminooxyacetic acid or cycloserine. The K_i for inhibition of hydroxylamine is high (3 mM), allowing a selective inhibition of D-cysteine desulphydrase and β-cyanoalanine synthase at lower hydroxylamine concentrations (A. Schmidt, unpubl. res.).

Cysteine synthase and β-cyanoalanine synthase are different enzymes in higher plants: (1) they can be separated by column chromatography using hydrophobic techniques; (2) antibodies raised against purified cysteine synthase do not inhibit β-cyanoalanine synthase and antibodies raised against purified β-cyanoalanine synthase do not inhibit cysteine synthase (A. Schmidt, unpubl. res.).

VI. OCCURRENCE AND LOCALISATION OF CYSTEINE SYNTHASE

Cysteine synthases have been partly purified from different sources ranging from microorganisms to higher plants (see Schmidt, 1979, 1986). Isoenzymes have been detected in phototrophic bacteria (Hensel and Trüper, 1976) and cyanobacteria (Diessner and Schmidt, 1981). In eukaryotes, cysteine synthase has been detected in the cytoplasm. The chloroplast of higher plants is the specific organelle for sulphate reduction, and all enzymes for sulphate reduction, including cysteine synthase, are also found within the chloroplast (Schmidt and Trebst, 1969; Trebst and Schmidt, 1969). Careful studies, using marker enzymes, correcting for differences in the yields of organelles, have shown that about 20% of the total cysteine synthase activity is located within the chloroplasts (Fankhauser et al., 1976) and similar data have been obtained for proplastides isolated from spinach roots (Fankhauser and Brunold, 1978). Multiple forms of cysteine synthase from spinach can be demonstrated by biochemical methods such as gel electrophoresis (Fankhauser and Brunold, 1978) or during purification of the enzyme on hydrophobic gels (see above).

Cysteine synthase in C_4 plants is detected in mesophyll and bundle sheath cells, in contrast to the sulphate-activating enzymes, which are found only in the bundle sheath cells (Schmutz and Brunold, 1982). So far, evidence for cysteine synthase activity within mitochondria in plants is missing, although the presence of cysteine synthase activity within mitochondria has been reported for *Euglena* (Brunold and Schiff, 1976) and the mould *Aspergillus* (Bal et al., 1975).

VII. REGULATORY PROPERTIES

Cysteine synthase activity in higher plant systems reacts only slowly to changing environmental factors such as sulphate or nitrogen availability (Brunold and Schmidt, 1976, 1978), however a response during plant development is evident (Schmutz and Brunold, 1982; von Arb and Brunold, 1986).

Regulation of cysteine synthase activity in green algae can be observed in response to sulphur starvation (Biedlingmaier and Schmidt, 1983; Krauss, 1984; Krauss and Schmidt, 1987; Leon et al., 1987, 1988). However, regulation of this enzyme due to environmental changes is slow compared to the induction of sulphate uptake or arylsulphatase activity (Niedermeyer et al., 1986). This suggests that the regulation of cysteine synthase is controlled by the availability of an unknown compound derived from cysteine, as discussed by Schmidt (1986).

REFERENCES

Anderson, J. W. (1980). In "The Biochemistry of Plants" (B. J. Miflin, ed.), Vol. 5, pp. 203–225. Academic Press, New York.
Bal, J., Maleszka, R., Stephien, P. and Cybis, J. (1975). FEBS Lett. **58**, 164–166.
Biedlingmaier, S. and Schmidt, A. (1983). Arch. Microbiol. **136**, 124–130.
Brüggemann, J., Schlossmann, K., Merkenschlager, M. and Waldschmidt, M. (1961). Biochem. Z. **335**, 392–399.
Brunold, C. and Schiff, J. A. (1976). Plant Physiol. **57**, 430–436.
Brunold, C. and Schmidt, A. (1976). Planta **133**, 85–88.
Brunold, C. and Schmidt, A. (1978). Plant Physiol. **62**, 343–347.
Brunold, C. and Suter, M. (1982). Planta **155**, 321–327.
Cooper, J. D. H. and Turnell, D. C. (1987). In "Methods in Enzymology" (W. B. Jakoby and O. W. Griffith, eds), Vol. 143, pp. 141–143. Academic Press, New York.
Diessner, W. and Schmidt, A. (1981). Z. Pflanzenphysiol. **102**, 57–68.
Ellis, R. J. (1963). Phytochemistry **2**, 129–136.
Esaki, N., Tanaka, H., Miles, E. W. and Soda, K. (1983). Agric. Biol. Chem. **47(12)**, 2861–2864.
Fahey, R. C. and Newton, G. L. (1987). In "Methods in Enzymology" (W. B. Jakoby and O. W. Griffith, eds), Vol. 143, pp. 85–96. Academic Press, New York.
Fankhauser, H. and Brunold, C. (1978). Plant Sci. Lett. **14**, 185–192.
Fankhauser, H., Brunold, C. and Erismann, K. H. (1976). Experientia **32**, 1494–1497.
Gaitonde, M. C. (1967). Biochem. J. **104**, 627–633.
Giovanelli, G. (1987). In "Methods in Enzymology" (W. B. Jakoby and O. W. Griffith, eds), Vol. 143, pp. 419–426. Academic Press, New York.
Giovanelli, J., Mudd, S. H. and Datko, A. H. (1980). In "The Biochemistry of Plants" (B. J. Miflin, ed.), Vol. 5, pp. 453–505. Academic Press, New York.
Greenberg, D. M. (1975). In "Metabolic Pathways" (D. M. Greenberg, ed.), Vol. VII, pp. 505–528. Academic Press, New York.
Hensel, G. and Trüper, H. G. (1976). Arch. Microbiol. **109**, 101–112.
Krauss, K. F. (1984). "Funktion, Eigenschaften und Regulation der multifunktionalen Cysteinsynthase und einiger am Cysteinstoffwechsel beteiligter Enzyme beim Wachstum von Chlorella fusca auf verschiedenen schwefelhaltigen Verbindungen". Thesis, München.
Krauss, K. F. and Schmidt, A. (1987). J. Gen. Microbiol. **133**, 1209–1219.
Leon, J., Romero, L. C., Galvan, F. and Vega, J. (1987). Plant Sci. **53**, 93–99.
Leon, J., Romero, L. C. and Galvan, F. (1988). J. Plant Physiol. **132**, 618–622.
Murakoshi, I., Ikegami, F. and Kaneko, M. (1985). Phytochemistry **24**, 1907–1911.
Ng, B. H. and Anderson, J. W. (1978). Phytochemistry **17**, 2069–2074.
Niedermeyer, I., Biedlingmaier, S. and Schmidt, A. (1987). Z. Naturforsch. **42c**, 530–536.
Rosichan, J. L., Blake, N., Stallard, R., Owais, W. M., Kleinhofs, A. and Nilan, R. A. (1983). Biochim. Biophys. Acta **748**, 367–373.
Schmidt, A. (1977a). Z. Naturforsch. **32c**, 219–225.
Schmidt, A. (1977b). Z. Pflanzenphysiol. **84**, 435–446.
Schmidt, A. (1979). In "Encyclopedia of Plant Physiology New Series" (M. Gibbs and E. Latzko, eds), Vol. 6, pp. 481–496. Springer, Berlin and New York.
Schmidt, A. (1982a). Z. Pflanzenphysiol. **107**, 301–312.
Schmidt, A. (1982b). In "On the Origins of Chloroplasts" (J. A. Schiff, ed.), pp. 179–197. Elsevier/North Holland, New York and Amsterdam.
Schmidt, A. (1986). Progress in Botany **48**, 133–150.
Schmidt, A. and Trebst, A. (1969). Biochim. Biophys. Acta **180**, 529–535.
Schmutz, D. and Brunold, C. (1982). Plant Physiol. **70**, 524–527.

Smith, I. K. and Thompson, J. F. (1971). *Biochim. Biophys. Acta* **227**, 288–295.
Smith, M., Furlong, C. E., Greene, A. A. and Schneider, J. A. (1987). *In* "Methods in Enzymology" (W. B. Jakoby and O. W. Griffith, eds), Vol. 143, pp. 144–148. Academic Press, New York.
Soda, K. (1987). *In* "Methods in Enzymology" (W. B. Jakoby and O. W. Griffith, eds), Vol. 143, pp. 453–459. Academic Press, New York.
Trebst, A. and Schmidt, A. (1969). *In* "Progress in Photosynthesis Research" (H. Metzner, ed.), Vol. 3, pp. 1510–1516. Tubingen, Liechenstein.
Wedding, R. T. (1987). *In* "Methods in Enzymology" (W. B. Jakoby and O. W. Griffith, eds), Vol. 143, pp. 29–31. Academic Press, New York.
von Arb, Ch. and Brunold, V. (1986). *Physiol. Plant.* **67**, 81–86.

21 Sulphur Metabolism
E. Synthesis of Glutathione

SIGRID KLAPHECK[1] and HEINZ RENNENBERG[2]

[1]*Botanisches Institut der Universität zu Köln, Gyrhofstrasse 15, D-5000 Köln 41, FRG*

[2]*Fraunhofer Institut für Atmosphärische Umweltforschung, Kreuzeckbahnstrasse 19, D-8100 Garmisch-Partenkirchen, FRG*

I.	Introduction	355
II.	Glutathione synthetase	356
	A. Methods of determination	356
	B. Enzyme extraction and purification	357
	C. Properties	358
	D. Distribution of the enzyme	359
	References	359

I. INTRODUCTION

The tripeptide glutathione (γ-glutamyl-cysteinyl-glycine; GSH) is the most abundant low molecular weight thiol in higher plants. It plays an important role in numerous physiological processes, e.g. the storage and long distance transport of reduced sulphur, redox activation/inactivation of enzymes, detoxification of harmful oxygen species, pesticides and metal ions (Rennenberg, 1982, 1987). Until recently, the synthesis of glutathione has only been intensively studied in animals and microorganisms (Meister and Anderson, 1983). These studies showed that glutathione is synthesised enzymatically in a two-step procedure. In the first reaction, catalysed by a γ-glutamylcysteine synthetase (γ-GC-synthetase, EC 6.3.2.2), the dipeptide γ-glutamylcysteine is produced from glutamate and cysteine in an ATP-dependent reaction:

$$\text{Glu} + \text{Cys} + \text{ATP} \longrightarrow \gamma\text{-Glu-Cys} + \text{ADP} + \text{Pi} \quad (1)$$

In the second step glycine is added to the C-terminal site of the dipeptide by a glutathione synthetase (GSH-synthetase, EC 6.3.2.3):

$$\gamma\text{-Glu-Cys} + \text{Gly} + \text{ATP} \longrightarrow \gamma\text{-Glu-Cys-Gly} + \text{ADP} + \text{Pi} \quad (2)$$

Both enzymes have been purified to apparent homogeneity and have been characterised for their catalytic and physical properties (Seelig and Meister, 1985; Meister, 1985).

Several pieces of evidence suggest that glutathione is synthesised in plant cells via the same pathway. γ-Glutamylcysteine was found to be present in many plant species (Kasai and Larsen, 1980). In spinach its concentration declines upon illumination to about the same extent as the concentration of glutathione increases (Buwalda et al., 1988). In crude homogenates of maize roots and tobacco suspension cultures, glutathione is synthesised from its constituent amino acids in the presence of ATP (Carringer et al., 1978; Rennenberg et al., 1982). Recently the enzymes of glutathione synthesis have been studied in plant cells. A preliminary report by Steffens and Williams (1987) indicates the presence of a γ-GC-synthetase in cultured tomato cells. A reliable, reproducible method for the determination of γ-GC-synthetase has, to our knowledge, not been published. GSH-synthetases have, however, been found and characterised in several plant species (Law and Halliwell, 1986; Macnicol, 1987; Hell and Bergmann, 1988; Klapheck et al., 1988).

II. GLUTATHIONE SYNTHETASE

A. Methods of Determination

Several methods have been applied for the determination of GSH-synthetases in higher plants. Enzyme activity has been analysed by measuring the formation of the ^{14}C-labelled tripeptide using [^{14}C]glycine as a substrate. The radioactively labelled product was separated from the [^{14}C]glycine by paper electrophoresis (Law and Halliwell, 1986) or anion exchange chromatography (Macnicol, 1987). Determination of enzyme activity by measuring the formation of inorganic phosphate or ADP, as used for enzyme preparations from animal cells (Meister, 1985), is complicated by the fact that plant extracts contain high activities of ATPase. This method may, therefore, only be applied to highly purified enzyme preparations.

A sensitive and specific method for the determination of GSH-synthetase activity is accomplished by the derivatisation of GSH and other thiols present in the assay mixture with monobromobimane (mBBr), followed by reverse phase HPLC and fluorescence detection. The use of mBBr for HPLC determination of thiols was introduced by Newton et al. (1981) and adapted for the quantification of GSH-synthetase activity by Hell and Bergmann (1988). This method is superior to others since the stability of γ-glutamylcysteine and glutathione in the assay mixture can be controlled. Such control is essential, when the natural substrate of GSH-synthetase, γ-glutamylcysteine, is used, as this compound can undergo significant oxidation. The method of Hell and Bergmann (1988) makes use of γ-glutamyl-α-aminobutyrate, a substrate invented to avoid difficulties in the determination of GSH-synthetase activity by oxidation of γ-glutamylcysteine (Meister, 1985; Law and Halliwell, 1986), unnecessary. In addition, this method allows the determination of GSH-synthetase activity by measuring simultaneously the formation of GSH and the consumption of γ-glutamylcysteine.

1. Assay procedure

In a total volume of 200 µl the assay mixture contained: 100 mM Tris-HCl (pH 8.5), 10 mM glycine, 1 mM γ-glutamylcysteine, 4 mM Na_2-ATP, 4 mM phosphoenolpyruvate, 2 U pyruvate kinase, 50 mM $MgCl_2$, 10 mM KCl, and 0–100 µl enzyme extract. γ-Glutamylcysteine was prepared as described by Strumeyer and Bloch (1962); mBBr was obtained as 'Thiolyte MB' from Calbiochem, Frankfurt. All other chemicals were purchased from Boehringer, Mannheim; Sigma, St. Louis or Merck, Darmstadt. The enzyme reaction was started by the addition of glycine. After 0–120 min incubation at 30°C, the reaction was terminated by addition of the derivatisation reagent. For this purpose an aliquot of 10 µl of the assay mixture was transferred into a vial containing 200 µl 0.1 M Tris-HCl (pH 8.0) and 12 µl 10 mM mBBr in acetonitrile. After 15 min, 5% acetic acid was added to a total volume of 1 ml. The derivatised samples may be stored at 4°C in the dark for several days.

2. HPLC analysis

Separation of thiol derivatives was achieved on an ODS column (Hypersil 5 µm corn size, 4.6 mm × 250 mm) with an HPLC system consisting of two pumps, a controller, an autosampler, a fluorescence detector and an integrator. Aliquots of 25 µl were injected. Solvent A was 0.1 M K-acetate (pH 5.5), solvent B was 100% methanol. The column was equilibrated for 6 min with 10% methanol in solvent A at a flow rate of 1 ml min^{-1}. Thiol derivatives were eluted with an exponential gradient of 10–20% methanol in solvent A in 12 min, followed by 3 min isocratic elution at 20% methanol in solvent A. Subsequently, the column was washed with 100% methanol for 4 min and re-equilibrated as described above. Fluorescence was determined at 480 nm with excitation at 380 nm. Quantification of the thiols was based on internal standards added to the assay mixtures. The procedure given allows the determination of 1.3 pmol GSH per injection volume (25 µl), or 1 nmol GSH per assay mixture (0.2 ml).

B. Enzyme Extraction and Purification

GSH-synthetase was extracted from plant material (10 g) by homogenisation with 0.1 M Tris-HCl (pH 8.5), 20 mM $MgCl_2$ (20 ml) and insoluble polyvinylpyrrolidone (1 g). After centrifugation (40 000 × g; 20 min) the supernatant was subjected to gel filtration on Sephadex G-50 or dialysis overnight, to remove low molecular weight substances interfering in the enzyme assay. Protein extracts of several plant species obtained by this or similar methods yielded GSH-synthetase activities of 0.2–2.0 nmol (mg protein)$^{-1}$ min^{-1} (Hell and Bergmann, 1988; Klapheck *et al.*, 1988).

There are two reports on the purification of GSH-synthetase from plant sources. Using a modification of the procedure described by Oppenheimer et al. (1979), Law and Halliwell (1986) achieved a 103-fold purification of the GSH-synthetase from spinach leaves. Macnicol (1987) partially purified the enzyme from shoots of *Pisum sativum* (170-fold) and *Vigna radiata* (320-fold). The procedure used by this author included polyethyleneglycol fractionation, chromatography on DEAE-cellulose, calcium phosphate cellulose, and Blue Sepharose. In none of the purification procedures reported was GSH-synthetase purified to homogeneity.

C. Properties

GSH-synthetases from plants were found to have a molecular weight of 85 kDa (Macnicol, 1987). This molecular weight is slightly lower than that observed for the enzyme from animal cells and microorganisms (118–123 kDa; Mooz and Meister, 1967; Oppenheimer et al., 1979; Dennda and Kula, 1986). Whereas the subunit composition of the plant enzyme has not been investigated, several reports on GSH-synthetase from other sources indicate a composition of two, possibly identical, subunits (Oppenheimer et al., 1979; Dennda and Kula, 1986). The plant enzyme is inhibited by thiol reagents, indicating the presence of a thiol group at the catalytic centre of the enzyme (Law and Halliwell, 1986).

The catalytic properties of plant GSH-synthetases have been analysed either in purified extracts (Law and Halliwell, 1986; Macnicol, 1987) or in protein fractions obtained by ammonium sulphate fractionation (Hell and Bergmann, 1988; Klapheck et al., 1988). The enzymes exhibit a broad pH optimum between 7.5 and 9.5. The temperature dependency and hence the activation energy of the enzyme have so far not been investigated. GSH-synthetase activity is highly dependent on Mg^{2+} with optimum concentrations of 2–30 mM; Mn^{2+} can only partially substitute for Mg^{2+}. The presence of K^+ in the assay mixture stimulates the enzyme activity up to 50%. Plant GSH-synthetases apparently obey monophasic Michaelis–Menten kinetics, if the affinity for each substrate is measured in the presence of saturating concentrations of the two others (Hell and Bergmann, 1988; Klapheck et al., 1988). K_m values for ATP are in the range of 0.17–0.45 mM; the affinity of the enzyme for other nucleotide triphosphates has not been investigated. Plant GSH-synthetases exhibit activities with several γ-glutamyl-dipeptides, e.g. γ-glutamylcysteine, γ-glutamyl-α-aminobutyrate, and γ-glutamyl-S-methylcysteine. As observed with the animal enzyme, K_m values reported for γ-glutamylcysteine by different authors differ by over one order of magnitude (0.02–0.75 mM). In pea and tobacco, GSH-synthetases have apparently the same affinity for γ-glutamyl-α-aminobutyrate as they do for γ-glutamylcysteine (Macnicol, 1987; Hell and Bergmann, 1988). Considerable differences have been observed in the specificity of plant GSH-synthetases with respect to the amino acid added at the C-terminal site of γ-glutamylcysteine. Besides glycine, β-alanine can be used by the enzyme as a substrate (Macnicol, 1987; Klapheck et al., 1988). This observation is consistent with the occurrence of homoglutathione (γ-glutamyl-cysteinyl-β-alanine) in many legumes (Klapheck, 1988). In those plants containing high amounts of homoglutathione as compared to glutathione (Klapheck, 1988), the affinity of GSH-synthetase for β-alanine was found to be high as compared to glycine (Macnicol, 1987; Klapheck et al., 1988). This observation led to the designation of the enzyme as homo-GSH-synthetase (Macnicol, 1987). With homo-GSH-synthetases from two legumes, K_m values of 0.33 and 1.34 mM for β-alanine and of 7.5 and 98 mM for glycine have been determined (Macnicol, 1987; Klapheck et al., 1988). GSH-synthetases of several plant species exhibit K_m values of 0.17–1.4 mM for glycine (Law and Halliwell, 1986; Macnicol, 1987; Hell and Bergmann, 1988); the affinity for β-alanine is low ($K_m = 14$ mM; Macnicol, 1987) or β-alanine is not accepted as a substrate (Hell and Bergmann, 1988). Both enzymes, homo-GSH- and GSH-synthetase, are strongly inhibited by 5 mM ADP and are only slightly affected by 5 mM GSH (Law and Halliwell, 1986; Macnicol, 1987).

D. Distribution of the Enzyme

GSH- or homo-GSH-synthetase activity has so far been reported in eight different species (Law and Halliwell, 1986; Macnicol, 1987; Hell and Bergmann, 1988; Klapheck et al., 1988). In three of these species the subcellular distribution of the enzyme has been investigated. Between 17 and 56% of the enzyme activity was found in the chloroplasts. The rest of the enzyme activity seems to be localised in the cytoplasm (Klapheck et al., 1987; Hell and Bergmann, 1988).

REFERENCES

Buwalda, F., de Kok, L. J., Stulen, J. and Kuiper, P. J. C. (1988). *Physiol. Plant.* **74**, 663–668.
Carringer, R. D., Rieck, C. E. and Bush, L. P. (1978). *Weed Sci.* **26**, 167–171.
Dennda, G. and Kula, M.-R. (1986). *J. Biotechnol.* **4**, 143–158.
Hell, R. and Bergmann, L. (1988). *Physiol. Plant.* **72**, 70–76.
Kasai, T. and Larsen, P. O. (1980). In "Progress in the Chemistry of Organic Natural Products". (W. Herz, H. Grisebach and W. Kirby, eds), Vol. 39, pp. 173–285. Springer, Wien.
Klapheck, S. (1988). *Physiol. Plant.* **74**, 727–732.
Klapheck, S., Latus, C. and Bergmann, L. (1987). *J. Plant Physiol.* **131**, 123–131.
Klapheck, S., Zopes, H., Levels, H.-G. and Bergmann, L. (1988). *Physiol. Plant.* **74**, 733–739.
Law, M. Y. and Halliwell, B. (1986). *Plant Sci.* **43**, 185–191.
Macnicol, P. K. (1987). *Plant Sci.* **53**, 229–235.
Meister, A. (1985). In "Methods in Enzymology" (A. Meister, ed.), Vol. 113, pp. 393–399. Academic Press, New York.
Meister, A. and Anderson, M. E. (1983). *Ann. Rev. Biochem.* **52**, 711–760.
Mooz, E. D. and Meister, A. (1967). *Biochemistry* **6**, 1722–1734.
Newton, G. L., Dorian, R. and Fahey, R. C. (1981). *Anal. Biochem.* **114**, 383–387.
Oppenheimer, L., Wellner, V. P., Griffith, O. W. and Meister, A. (1979). *J. Biol. Chem.* **254**, 5184–5190.
Rennenberg, H. (1982). *Phytochemistry* **21**, 2771–2781.
Rennenberg, H. (1987). In "Plant Molecular Biology" (D. von Wettstein and N.-H. Chua, eds), NATO ASI Ser. A, Life Sci., Vol. 140, pp. 279–292. Plenum Press, New York.
Rennenberg, H., Birk, Ch. and Schaer, B. (1982). *Phytochemistry* **21**, 5–8.
Seelig, G. F. and Meister, A. (1985). In "Methods in Enzymology" (A. Meister, ed.), Vol. 113, pp. 379–392. Academic Press, New York.
Steffens, J. C. and Williams, B. (1987). *Plant Physiol.* **83**, (Suppl.), 666.
Strumeyer, D. and Bloch, K. (1962). *Biochem. Prep.* **9**, 52–55.

21 Sulphur Metabolism
F. Enzymes Involved in the Synthesis of Methionine

JAMES T. MADISON

Agricultural Research Service, United States Department of Agriculture, United States Plant, Soil and Nutrition Laboratory, Tower Road, Ithaca, New York 14853, USA

I.	Introduction	361
II.	Cystathionine γ-synthase	362
	A. Assay	362
	B. Purification and properties	363
III.	Cystathionine β-lyase	364
	A. Assay	364
	B. Purification of β-cystathionase from spinach	365
	C. Properties	366
IV.	Homocysteine methylase	366
	A. Assay	367
	B. Purification	368
	C. Properties	368
	References	369

I. INTRODUCTION

The enzymes that are responsible for the pathway that is unique to methionine synthesis are cystathionine γ-synthase, β-cystathionase and 5-methyltetrahydropteroylglutamate-homocysteine methyltransferase. For a recent overview of the synthesis of the sulphur amino acids by plants, see Giovanelli (1987a). The best recent evidence suggests that the direct sulphydration of homoserine (Equation 1) makes an insignificant contribution to the synthesis of homocysteine in plants (Giovanelli, 1987a). The production of homocysteine by direct sulphydration that can be detected in extracts from various plants is a result of H_2S substituting for cysteine in the reaction catalysed by cystathionine-γ-synthase.

$$O\text{-Phosphohomoserine} + H_2S \longrightarrow \text{Homocysteine} + Pi \qquad (1)$$

The pathway in plants and bacteria where the S atom is transferred from cysteine to methionine is the opposite of what happens in animals, fungi and some bacteria where the S of homocysteine is transferred to cysteine.

In the discussion that follows all the amino acids and derivatives are in the L-configuration unless stated otherwise.

II. CYSTATHIONINE γ-SYNTHASE (EC 4.2.99.-)

The first step unique to methionine synthesis in plants is the condensation of O-phosphohomoserine (OPH) with cysteine to form cystathionine (Equation 2).

$$\text{OPH} + \text{Cysteine} \longrightarrow \text{Cystathionine} + Pi \qquad (2)$$

A. Assay

Cystathionine-γ-synthase can be assayed by measuring the disappearance of cysteine (Kaplan and Guggenheim, 1971), by measuring the production of cystathionine, or by measuring the production of α-ketobutyrate from an O-acyl derivative of homoserine. Cystathionine can be determined as described below. In bacteria, cystathionine γ-synthase can be assayed by measuring the production of α-ketobutyrate from O-succinylhomoserine (Equation 3).

$$O\text{-Succinylhomoserine} + H_2O \longrightarrow \alpha\text{-Ketobutyrate} + NH_3 + \text{succinate} \qquad (3)$$

But the most useful and sensitive reaction (particularly with relatively crude extracts) is probably the incorporation of labelled substrates into cystathionine; [^{14}C] or [^{35}S]cysteine has been used by Nagai and Flavin (1967). In this case the cystathionine had to be separated from the precursor by paper electrophoresis. A more convenient assay uses labelled O-phosphohomoserine (Aarnes, 1980; Datko et al., 1974; Madison and Thompson, 1976), since in this case the precursor can be separated from cystathionine in one step, by chromatography on Dowex 50-H$^+$. Carbon-14- or tritium-labelled OPH can be produced by the use of homoserine kinase in crude extracts of radish leaves (Madison and Thompson, 1976). Recently carbon-14-labelled homoserine has not been commercially available, but homoserine can presumably be labelled with tritium using a tritium-labelling service (DuPont; New England Nuclear).

Labelled homoserine could also be made from [^{14}C]aspartate by reduction of the β-ester of aspartate by [^{3}H]NaBH$_4$ (Murphy and Gottschalk, 1961).*

---

* Mention of company names or commercial products does not imply recommendation or endorsement by the United States Department of Agriculture over others not mentioned.

The assay conditions that we have used are: 0.07 µCi labelled OPH, 0.2 µmol cysteine, 50 µmol 3-[N-Morpholino]propanesulphonic acid (MOPS) (pH 7.3), and 2 µmol dithiothreitol (DTT) in 0.2 ml (Madison and Thompson, 1976). A control without cysteine is included. After incubation for 30 min at 30°C, the reaction is stopped by the addition of 0.8 ml 5% trichloroacetic acid. The precipitated protein was removed by centrifugation (10 000 × g for 10 min) and the clear supernatant was applied to a 1.5 × 0.5 cm column of Dowex 50X8-H$^+$. The labelled OPH is rinsed through with 4 ml of H_2O. Cystathionine is eluted with 1 ml of 2 M NH_4OH followed by 4 ml of H_2O. An aliquot of the combined eluents is neutralised with HCl and counted in a liquid scintillation spectrometer. Cystathionine was the only product that was detected.

If labelled OPH is not readily available the assay can be carried out using labelled cysteine (Datko et al., 1974; Thompson et al., 1982). Cystathionine (0.14 mM), DTT (0.64 mM), [^{14}C]cysteine (0.3 mM, 0.1–1 µCi), MOPS (pH 7.7) (0.6 M), OPH (20 mM) and crude extract (0.25–0.15 mg protein) in a total volume of 50 µl are incubated at 30°C for 1 h under N_2. The reaction is stopped by the addition of 0.5 ml of an ice-cold solution containing trichloroacetic acid (6.8%), cysteine (252 mM) and cystathionine (0.68 mM). Insoluble material is removed by centrifugation and cystathionine isolated by chromatography on Dowex 50-H$^+$ as described above. A suitable aliquot of the NH_4OH eluate from the column is taken to dryness and dissolved in 200 µl of performic acid. Each tube is stoppered and incubated at 30°C for 1 h. The oxidation is terminated by the addition of 10 µl of 48% HBr. After bubbling stops and the colour changes from orange to yellow, each sample is diluted with 10 ml of H_2O and reapplied to a Dowex 50-H$^+$ column. Each column is washed with 12 ml of H_2O to remove the labelled cysteic acid and the cystathionine sulphone is eluted with 3.9 ml of 2 N NH_4OH and counted.

B. Purification and Properties

Cystathionine γ-synthase has apparently not been extensively purified from a higher plant. Aarnes (1980) achieved about a 12-fold purification from barley seedlings, after ammonium sulphate precipitation and gel filtration. The 12-fold purified enzyme was stable to repeated freezing and thawing and to prolonged storage at −25°C. The pH optimum of the barley enzyme is about 6.8 with 50% activity measured when the pH is either one pH unit higher or lower. 2-[N-Morpholino]ethanesulphonic acid (Mes), MOPS and Tricine buffers give significantly more activity than Hepes, Tris or phosphate buffers. The addition of EDTA or sulphydryl compounds had no effect. The molecular weight of the barley enzyme is about 180 kDa (Aarnes, 1980). The barley enzyme has a K_m for OPH of 0.13 mM and 0.4 mM for cysteine (Aarnes, 1980), while the enzyme from sugar beet leaves had a higher K_m for OPH (6.6 mM) and a lower K_m for cysteine (<0.1 mM) (Madison and Thompson, 1976).

Since cystathionine γ-synthase catalyses the first step after the branch point, where OPH can be used as a precursor for either threonine or methionine, it would be expected to be under metabolic control. However, there is no evidence that inhibition of this enzyme is important in the regulation of methionine synthesis. Thompson et al. (1982) have found that the specific activity of cystathionine γ-synthase in Lemna is inversely correlated with the level of intracellular methionine. Exogenous methionine (2 µM) decreased the activity to 15% of the control. Conversely, lowering the concentration of methionine by the addition of aminoethoxyvinylglycine, or by the addition of lysine

(36 μM) plus threonine (4 μM) increased the activity of cystathionine γ-synthase two- to three-fold. From these data it appears likely that methionine (or one of its metabolites, such as AdoMet) is involved in the regulation of the synthesis of cystathionine γ-synthase.

III. CYSTATHIONINE β-LYASE (β-Cystathionase EC 4.4.1.8)

The next step in the pathway to methionine is carried out by β-cystathionase (Equation 4).

$$\text{Cystathionine} + H_2O \longrightarrow \text{Homocysteine} + \text{Pyruvate} + NH_4^+ \quad (4)$$

Cleavage of cystathionine to cysteine has not been found in any of the plant tissues that have been examined (Giovanelli and Mudd, 1971).

A. Assay

β-Cystathionase can be assayed by measuring the homocysteine or the pyruvate produced. Kase and Nakayama (1974) measured the homocysteine produced by using the reaction with 5,5'-dithiobis-(2-nitrobenzoic acid) (DTNB) (Ellman, 1959). Kase and Nakayama (1974) also measured the pyruvate produced, by following the decrease in absorbance at 340 nm after the addition of NADH and lactate dehydrogenase.

1. Preparation of labelled cystathionine

In crude extracts the preferred assay uses cystathionine labelled in the 3-carbon part of the molecule. [^{14}C]Cystathionine can be made from [^{14}C]L-serine and homocysteine using partially purified cystathionine β-synthase (Mudd *et al.*, 1965; Selim and Greenberg, 1959) as described by Giovanelli (1987b). [^{14}C]Serine (0.6 mCi, 78 μmol) and homocysteine (150 μmol) (freshly prepared from the thiolactone) is incubated with a rat liver extract (Mudd *et al.*, 1965) along with Tris-Cl (pH 8.3) (60 μmol), EDTA (2 μmol), and pyridoxal phosphate (0.025 μmol) in 1.3 ml at 37°C. After 135 min, the protein is precipitated with 5% trichloroacetic acid and the labelled cystathionine is purified from the supernatant by chromatography over Dowex 50-H$^+$ columns (0.9 × 2.9 cm). The remaining [^{14}C]serine is eluted with 18 ml of H$_2$O followed by 35.5 ml of 0.4 M HCl. The column is then rinsed with 12 ml of H$_2$O and the [^{14}C]cystathionine eluted with 3.9 ml of 2 M NH$_4$OH. After lyophilisation the residue is dissolved in 1.5 ml of H$_2$O and clarified by the addition of 2 M HCl until the solution is red-orange to thymol blue. The entire solution is subjected to paper chromatography on Whatman No. 1 using 2-propanol–88% formic acid–H$_2$O (70:10:20). The spot containing the radioactivity is eluted with 0.1 M HCl and lyophilised.

2. Radioactive assay

The β-cystathionase assay is performed in 0.5 ml containing: Tris-Cl, (pH 8.65) (100 μmol), [^{14}C]cystathionine (9.8 nCi, 47.5 nmol), Na pyruvate (5 μmol), and enzyme. The reaction is started by addition of enzyme. The incubation is carried out under N$_2$

for 30 min at 30°C, when the reaction is stopped by the addition of 1 ml of 5% trichloroacetic acid and the precipitate removed by centrifugation. The supernatant is applied to a 0.9×2.9 cm column of Dowex 50-H$^+$ followed by 3 ml of H$_2$O. The radioactivity in the combined effluent and wash is counted to determine the amount of pyruvate formed.

3. Spectrophotometric assay

More purified enzyme preparations can be assayed by measuring the pyruvate produced using the change in absorbance at 340 nm with lactate dehydrogenase and NADH (Giovanelli, 1987b). The incubations are carried out in a 1 cm cuvette in a volume of 1 ml which contains: Tris-Cl (pH 8.65) (200 µmol), L-cystathionine (10 µmol), NADH (0.175 µmol), lactate dehydrogenase (30 µg) and β-cystathionase. The reaction is started by the addition of the enzyme extract and the change in absorbance at 340 nm monitored at 30°C. The spectrophotometric assay is linear with up to four times the amount of enzyme and gives 2.2 times more pyruvate per unit of enzyme as compared to the radioactive assay, which is to be expected since the cystathionine concentration is much higher.

B. Purification of β-Cystathionase from Spinach

Giovanelli (1987b; Giovanelli and Mudd, 1971) has purified β-cystathionase approximately 400-fold from spinach leaves. Fresh spinach leaves (160 g) are deveined and homogenised in a Waring blender with 2.4 l of acetone at −40°C. The homogenate is filtered on a Buchner funnel, the protein suspended in 2 l of diethylether, refiltered, and the residual ether removed in a vacuum desiccator containing paraffin flakes and MgClO$_4$. The dry powder (9 g) is stable for at least 1 year at 4°C.

The acetone powder is ground in a mortar and pestle with an equal weight of aluminium oxide and 20 times the weight of Buffer A (0.1 M KPO$_4$, (pH 7.25), 0.1 mM EDTA, 0.14 mM 2-mercaptoethanol) and centrifuged for 40 min at $9000 \times g$ and 4°C. The supernatant is filtered through a loosely-woven cloth and recentrifuged for an additional 2 h to give a clear solution. The supernatant solution is treated with solid ammonium sulphate and the fraction precipitating between 24 and 34% saturation is collected by centrifugation. The precipitate is dissolved in a minimal volume of Buffer A and desalted on a column of Sephadex G-25 in Buffer B (5 mM KPO$_4$ (pH 7.25), 0.1 mM EDTA, 0.1 mM pyridoxal phosphate, 0.14 mM mercaptoethanol).

The protein peak from the gel filtration step is diluted to a concentration of 7.9 mg protein ml^{-1}. To each ml, 0.4 ml of a slurry containing 24 mg of bentonite in 0.1 M acetic acid is added and stirred for 15 min. The bentonite is removed by centrifugation at $20\,000 \times g$. The clear supernatant (30 ml) is immediately acidified with 6 ml of 25 mM acetic acid and 12 ml aliquots applied to columns (0.9×2.2 cm) of hydroxyapatite. Several small columns are preferred because a single large column gives a much slower flow rate. Each column is washed with 2 ml of 0.1 M potassium acetate (pH 5.5) and the enzyme eluted with 4 ml of 0.5 M KPO$_4$ (pH 6.0). The eluates are combined, concentrated 10-fold by dialysis against Aquacide II (sodium salt of carboxymethyl cellulose, molecular weight 500 kDa; Calbiochem), and finally gel filtered through Sephadex G-25 equilibrated with Buffer B (but without pyridoxal phosphate). The purification is about 420-fold with 17% recovery of β-cystathionase activity (Giovanelli, 1987b).

C. Properties

1. Stability

Crude extracts and the $(NH_4)_2SO_4$ fractions are stable at $-80°C$ for several months. The purified fractions lose up to 70% of their activity in 5 days either at 4° or at $-80°C$.

2. Cofactor

Addition of pyridoxal phosphate did not increase the activity of the enzyme at any stage of purification. However, the finding that treatment with hydroxylamine abolished the activity and that activity could be restored by the addition of pyridoxal phosphate suggests that the holoenzyme contains pyridoxal phosphate that is tightly bound (Giovanelli, 1987b).

3. pH Optimum

The pH optimum is 8.8 in Tris-Cl buffer.

4. Substrate specificity

The enzyme is most active with the L-isomers of cystathionine and djenkolate and the mixed disulphide of cysteine and homocysteine. The K_m for cystathionine is 0.13 mM and that for djenkolate is 0.25 mM. The V_{max} value for cystathionine is about 90% of that for djenkolate (Giovanelli and Mudd, 1971).

5. Inhibitors

β-Cyanoalanine is a strong competitive inhibitor ($K_i = 40$ μM). The enzyme is also inhibited by the sulphydryl reacting compounds DTNB and N-ethylmaleimide. The cystathionine analogue, rhizobitozine, is a potent inhibitor both *in vitro* (Giovanelli *et al.*, 1971) and *in vivo* (Giovanelli *et al.*, 1973).

IV. HOMOCYSTEINE METHYLASE (Methionine synthase; EC 2.1.1.14)

The final step in the synthesis of methionine is the methylation of homocysteine by 5-methyltetrahydropteroyltriglutamate-homocysteine methyltransferase (Equation 5).

$$N^5\text{-Me-H}_4\text{PteGlu} + \text{Homocysteine} \longrightarrow \text{Methionine} + \text{H}_4\text{PteGlu} \tag{5}$$

The enzyme from plants is the vitamin-B12 independent form. The cobalamin-independent enzyme can be distinguished from the cobalamin-dependent form because it does not require S-adenosylmethionine (AdoMet) or a reducing system and because it requires Pi and is stimulated by Mg^{2+} and is most active with the polyglutamate forms of N^5-methyltetrahydropteroylglutamate (5-Me-H$_4$PteGlu). Detectable rates of reaction are, however, found with the more available monoglutamate form.

The enzyme has been detected in beans, barley seedlings, and spinach (Burton and Sakami, 1969); in pea leaf chloroplasts (Shah and Cossins, 1970) and in carrot roots (Fedec and Cossins, 1976).

Methionine can also be formed by the transfer of methyl groups from AdoMet and S-methylmethionine (SMM) to homocysteine (Dodd and Cossins, 1970) (Equations 6 and 7; SAHC = S-adenosylhomocysteine). Still another source of methionine is the conversion of the methylthioadenosine that is produced when AdoMet is used to form 1-aminocyclopropane-1-carboxylic acid (ACC), the precursor of ethylene (Miyazaki and Yang, 1987) (Equation 8). In all these cases, however, there is no net synthesis of methionine since methionine was used to form AdoMet, S-methylmethionine or methylthioadenosine.

$$\text{AdoMet} + \text{Homocysteine} \longrightarrow \text{Methionine} + \text{SAHC} \qquad (6)$$

$$\text{SMM} + \text{Homocysteine} \longrightarrow 2 \text{ Methionine} \qquad (7)$$

$$\text{Methylthioadenosine} \longrightarrow \longrightarrow \longrightarrow \text{Methionine} \qquad (8)$$

A. Assay

The enzyme can be assayed by measuring the amount of methionine formed using methionine requiring microbial mutants (Burton and Sakami, 1969) or by measuring the radioactivity transferred from N^5-[^{14}C]Me-H_4PteGlu to homocysteine. With more active preparations of the enzyme, the activity can be followed by measuring the formation of H_4PteGlu (tetrahydropteroylglutamate) by converting the H_4PteGlu to 5,10-methylene-H_4PteGlu, which can be quantitated by its absorbance at 350 nm (Burton and Sakami, 1969).

1. Carbon-14 method

As described by Dodd and Cossins (1970), the reaction mixture contains: enzyme extract, homocysteine (2 μmol), 5-[^{14}C]Me-H_4PteGlu (1.6 nmol, 0.1 μCi; Amersham) and KPO$_4$ buffer (pH 6.9) (50 μmol) in 0.5 ml. After incubation at 30°C for 1 h the reaction is stopped by rapid cooling in an ice bath. An aliquot (0.1 ml) is placed on a 0.5 × 2.5 cm column of Dowex 1X8-Cl$^-$. Labelled methionine is rinsed through the column with six 0.2 ml portions of H_2O. The combined effluents are counted in a liquid scintillation spectrometer.

To increase sensitivity it seems likely that the entire incubation mixture can be put through the Dowex 1 column (Taylor and Weissbach, 1971).

2. Spectrophotometric method

As described by Burton and Sakami (1971), the enzyme is incubated in a volume of 0.3 ml containing 0.1 M KPO$_4$ (pH 7.0), 0.2 M mercaptoethanol, 5 mM homocysteine, and 1.5 mM 5-Me-H_4PteGlu for up to 1 h. Burton and Sakami (1969) recommended 37°C, but 30°C may be more appropriate for the plant enzyme.

At the end of the incubation, 1.2 ml of 97% formic acid containing 0.1 M mercaptoethanol is added and the tubes heated at 100°C for 5 min, which converts H_4PteGlu to the 5,10-methylene form. After cooling to room temperature the absorbance at 350 nm is determined. The amount of $H_4PteGlu$ that was produced is calculated from the extinction coefficient of 5,10-methylene-$H_4PteGlu$ ($E_m = 26\,000$) (Rosenthal et al., 1965). If between 5 and 10 nmol of $H_4PteGlu$ is produced, the rate is a linear function of both enzyme concentration and time.

The above assays were both performed using the monoglutamate form of tetrahydrofolate. Using extracts from *Phaseolus* fruits, spinach leaves and barley sprouts, Burton and Sakami (1969) found that the triglutamate derivative was from 5 to 10 times more active as a methyl donor than the monoglutamate form.

B. Purification

Dodd and Cossins (1970) partially purified two forms of methionine synthase from germinating pea seeds. Pea seeds were germinated for 1 day in the dark. All subsequent procedures were carried out at 4°C. The seed coats were removed and 50 g of the cotyledons were homogenised in a Waring blender for 1 min with 100 ml of 50 mM KPO_4 (pH 6.9), 5 mM mercaptoethanol (Buffer C). The homogenate was squeezed through cheesecloth and the liquid centrifuged for 20 min at $10\,000 \times g$. The supernatant was brought to a 20% saturation with $(NH_4)_2SO_4$, stirred for 20 min, centrifuged as above, and the precipitate discarded. The precipitate that was collected after adjustment of the supernatant to 60% saturation with $(NH_4)_2SO_4$ was desalted on Sephadex G-50 using Buffer C. The protein fractions were combined and precipitated from 60% saturated $(NH_4)_2SO_4$. The precipitate (25 mg of protein in 15 ml) was applied to a 50×2.8 cm column of Sephadex G-100 in Buffer C.

The enzyme active with 5-Me-$H_4PteGlu$ as the methyl donor was purified 27-fold with a recovery of about 3% of the original activity. The form that utilises AdoMet or SMM was purified about 135-fold with recovery of 5% of the original activity.

C. Properties

Both forms of the pea seed enzyme were inhibited with fairly high levels of methionine. The 5-Me-$H_4PteGlu$-specific enzyme was 50% inhibited by about 4 mM methionine, while about 20 mM methionine was required to inhibit the SMM-specific enzyme by 50% (Dodd and Cossins, 1970).

TABLE 21.1. Some properties of seed methionine synthase.

		K_m (μM)	Optimum pH
Pea seed (Dodd and Cossins, 1970)			
5-Me-$H_4PteGlu$ form	5-Me-$H_4PteGlu$	0.026	6.5
	Homocysteine	600	
SMM form	SMM	2.2	7.3
	AdoMet	4.0	
	Homocysteine	460	
Jack bean seed (Abrahamson and Shapiro, 1965)			
SMM form	SMM	55	7.9
	Homocysteine	41	

The transmethylase that utilises SMM was more stable than the 5-Me-H$_4$PteGlu-specific form. After 10 days at $-15°C$ the SMM form lost 6% of its initial activity, while the 5-Me-H$_4$PteGlu form lost 30% of its original activity.

Abrahamson and Shapiro (1965) had partially purified a methionine synthase from jack bean meal that would use SMM as the methyl donor. These workers achieved a 250-fold purification and a 9% yield of activity. The jack bean meal enzyme was inhibited less than 50% by 20 mM methionine. Dimethyl-β-propiothetin was as active a methyl donor as SMM, while AdoMet was only 10% as active. The properties of the enzymes from the two sources are shown in Table 21.1.

REFERENCES

Aarnes, H. (1980). *Plant Sci. Lett.* **19**, 81–89.
Abrahamson, L. and Shapiro, S. K. (1965). *Arch. Biochem. Biophys.* **109**, 376–382.
Burton, E. G. and Sakami, W. (1969). *Biochem. Biophys. Res. Commun.* **36**, 228–234.
Datko, A. H., Giovanelli, J. and Mudd, S. H. (1974). *J. Biol. Chem.* **249**, 1139–1155.
Dodd, W. A. and Cossins, E. A. (1970). *Biochim. Biophys. Acta* **201**, 461–470.
Ellman, G. L. (1959). *Arch. Biochem. Biophys.* **82**, 70–77.
Fedec, P. and Cossins, E. A. (1976). *Phytochemistry* **15**, 1819–1823.
Giovanelli, J. (1987a). *Meth. Enzymol.* **143**, 419–426.
Giovanelli, J. (1987b). *Meth. Enzymol.* **143**, 443–449.
Giovanelli, J. and Mudd, S. H. (1971). *Biochim. Biophys. Acta* **227**, 654–670.
Giovanelli, J., Owens, L. D. and Mudd, S. H. (1971). *Biochim. Biophys. Acta* **227**, 671–684.
Giovanelli, J., Owens, L. D. and Mudd, S. H. (1973). *Plant Physiol.* **51**, 492–503.
Kaplan, M. and Guggenheim, S. (1971). *Meth. Enzymol.* **17B**, 425–433.
Kase, H. and Nakayama, K. (1974). *Agric. Biol. Chem.* **38**, 2235–2242.
Madison, J. T. and Thompson, J. F. (1976). *Biochem. Biophys. Res. Commun.* **71**, 684–691.
Miyazaki, J. H. and Yang, S. F. (1987). *Plant Physiol.* **84**, 277–281.
Mudd, S. H., Finkelstein, J. D., Irreverre, F. and Laster, L. (1965). *J. Biol. Chem.* **240**, 4382–4392.
Murphy, W. H. and Gottschalk, A. (1961). *Biochim. Biophys. Acta* **52**, 349–360.
Nagai, S. and Flavin, M. (1967). *J. Biol. Chem.* **242**, 3884–3895.
Rosenthal, S., Smith, L. C. and Buchanan, J. M. (1965). *J. Biol. Chem.* **240**, 836–843.
Selim, A. S. M. and Greenberg, D. M. (1959). *J. Biol. Chem.* **234**, 1474–1480.
Shah, S. P. J. and Cossins, E. A. (1970). *FEBS Lett.* **7**, 267–270.
Taylor, R. T. and Weissbach, H. (1971). *Meth. Enzymol.* **17B**, 379–388.
Thompson, G. A., Datko, A. H., Mudd, S. H. and Giovanelli, J. (1982). *Plant Physiol.* **69**, 1077–1083.

22 Protein Kinase

ALISTAIR M. HETHERINGTON[1], NICHOLAS H. BATTEY[2] and PAUL A. MILLNER[3]

[1]*Department of Biological Sciences, University of Lancaster, Lancaster LA1 4YQ, UK*

[2]*Department of Horticulture, University of Reading, Reading RG6 2AS, UK*

[3]*Department of Biochemistry, University of Leeds, Leeds LS2 9JT, UK*

I.	Introduction	371
II.	Assay	372
	A. *In vivo* assays	372
	B. *In vitro* assay	373
	C. Precautions	373
	D. Future developments in kinase identification and assay	374
III.	Purification	376
	A. Organellar kinases	376
	B. Soluble kinases	377
	C. Membrane kinases	377
	D. Novel methods	379
IV.	Properties	380
	References	381

I. INTRODUCTION

The protein kinases are a group of pleiotropic enzymes which catalyse the reversible transfer of the γ-phosphoryl group from nucleoside triphosphates to proteins. This post-translational covalent modification has a major role in the regulation of both enzymic and non-enzymic proteins (Boyer and Krebs, 1986, 1987; Trewavas, 1976; Bennett, 1984; Ranjeva and Boudet, 1987). The equation below describes the activity of protein kinase:

$$\text{Protein} + n\text{ATP-Mg}^{2+} \xrightarrow{\text{Protein kinase}} \text{Protein-P}_n + n\text{ADP}$$

In physiological terms ATP-Mg^{2+} is probably the most important substrate; however, the metal ion requirement can often be met by Mn^{2+} in *in vitro* studies. Similarly, although ATP appears to be the most significant phosphoryl donor *in vivo*, other nucleoside triphosphates can substitute in experimental manipulations. A protein substrate can be phosphorylated at single or multiple sites, which are usually serine, threonine and tyrosine residues. Protein kinases are regulated by a diverse array of molecules, some of which fall into the category of second messengers. This chapter will selectively survey the biochemistry of these enzymes in higher plants.

II. ASSAY

A. *In vivo* Assays

Protein phosphorylation in isolated single cells, tissue slices or whole organs can be investigated following the addition of [^{32}P]orthophosphate to the tissue bathing medium. During the incubation period the phosphate will become incorporated into the cell's ATP pool. Once equilibrium has been reached the remaining extracellular [^{32}P]orthophosphate can be washed away and the system challenged with putative agonists and antagonists. [^{32}P]Orthophosphate has been used in studies of protein phosphorylation in animals (Garrison, 1983) and higher plants (Trewavas, 1976; Ranjeva and Boudet, 1987).

In plants Poovaiah and coworkers have studied *in vivo* protein phosphorylation extensively (Veluthambi and Poovaiah, 1986; McFadden and Poovaiah, 1988). A typical protocol comes from their study of protein phosphorylation in root tips (Raghothama *et al.*, 1987). Root tips were incubated for 2 h at room temperature in 10 mM sodium citrate (pH 6.3), 1.5% sucrose and 5 mM MgCl$_2$, and carrier free [^{32}P]orthophosphoric acid (HCl free) was added. Phosphorylation was terminated by removing the incubation buffer and washing the roots three times with homogenisation buffer (50 mM Mes-NaOH [pH 7.0], 10 mM KH$_2$PO$_4$, 1 mM EDTA, 10 mM NaF, 0.5 mM PMSF and 1 mM DTT). The root tips were then frozen in liquid nitrogen and homogenised in homogenisation buffer containing RNase 1. After centrifugation (27 000 × *g* for 20 min) the supernatant was treated with DNase 1 for 20 min and then 20% TCA was added and proteins allowed to precipitate for 15 min at 4°C. The protein pellet obtained by centrifugation (27 000 × *g* for 20 min) was washed twice using ice-cold 30% acetone with 2% sucrose followed by 100% acetone. The samples were then freeze-dried and analysed by two-dimensional gel electrophoresis and the phosphopeptides detected by autoradiography. A useful practical description of [^{32}P]phosphopeptide detection by autoradiography has been published by Garrison (1983).

The major disadvantages with this technique are that it does not identify the kinases responsible and it is usually difficult to assign identities to substrates.

B. *In vitro* Assay

The assay relies on the protein kinase-mediated transfer of the γ-phosphoryl group of [γ-^{32}P]ATP to an acceptor protein substrate in the presence of Mg^{2+}. [γ-^{32}P]ATP is available at high specific activities from a number of manufacturers, or alternatively can be synthesised (Maxam and Gilbert, 1980). The reaction is terminated by precipitating the phosphoproteins on to filter disks with TCA and the radiolabelled phosphoproteins are quantified by liquid scintillation counting.

Since the description of the assay by Corbin and Reimann (1974) a number of modifications have been reported. These chiefly concern the filter disk material. In addition to the use of Whatman 3MM disks (Corbin and Reimann, 1974), other materials have been used including Whatman P81 phosphocellulose (Roskoski, 1983), 0.45 µm nitrocellulose filters (Kikkawa *et al.*, 1983), and Whatman 3MM coated with ATP (McPherson and Ramachandran, 1980). A recent evaluation by Sahal and Fujita-Yamaguchi (1987) concluded that a flat sheet matrix of Whatman P81 phosphocellulose paper is the material of choice when combined with TCA washes. A spectrophotometric assay has also been described (Roskoski, 1983). However, these methods have not yet been fully evaluated by plant biochemists.

Plant protein kinase may be conveniently assayed using the following procedure (Hetherington and Trewavas, 1982). The purified or partially purified preparation (see Section III) is resuspended in an appropriate buffer such as 50 mM Hepes (pH 7) containing 10 mM $MgCl_2$, and assayed at 25°C. Exogenous protein substrates, lipid cofactors, other divalent cations and inhibitors such as PMSF (proteases) and NaF (phosphataes) may also be added. The reaction is initiated by the addition of [γ-^{32}P]ATP (specific activity of stock solution, 110 TBq mmol^{-1}). The final concentration of ATP is 2×10^{-5} M. Samples are removed at intervals and pipetted on to Whatman 3MM filters which have been pre-treated with 10% TCA, 20 mM sodium pyrophosphate, 10 mM EDTA. The filters are then dropped into the TCA mixture and left overnight at 4°C. In the morning the filters are washed once in 5% TCA, heated to 90°C for 15 min in 10% TCA, 20 mM sodium pyrophosphate, 10 mM EDTA and after a further 5% TCA wash treated with chloroform–methanol (3:1; v/v) before drying. Radioactivity is determined by Cerenkov counting. It is important to note that if histone H1 is used as an exogenous substrate 20% TCA must be used throughout the procedure. Blank tubes contain the reagents described above and also 25 mM EDTA. After the addition of [γ-^{32}P]ATP a sample is immediately pipetted onto a filter disk and blanks are processed as above. Protein kinases may also be detected in gels (Geahlen *et al.*, 1986; Harmon *et al.*, 1987) and nitrocellulose filters (Allis *et al.*, 1986; Blowers *et al.*, 1985).

C. Precautions

There are a number of potential pitfalls in the measurement of protein kinase activity. These especially relate to studies which use crude extracts. In such extracts the presence of ATPases, phosphoprotein phosphatases, other protein kinases, endogenous inhibitors and proteolytic enzymes will all contribute to the apparent activity of the kinase under study. It is possible to inhibit phosphoprotein phosphatase activity by including 50 mM NaF in the incubation medium (Garrison, 1983); additionally micromolar levels of vanadate are reported to inhibit phosphoprotein phosphatase (Swarup et al., 1982). The proteolytic activity can be minimised by inclusion of 200 μM PMSF, 1 μM pepstatin A, 1 μM leupeptin or other inhibitors. The activity determinations will also be influenced by the level of endogenous protein phosphorylation of the substrates. To minimise this, in vitro studies often use a dephosphorylated exogenous substrate.

A further complication is that in crude extracts or in vivo studies not all the bound phosphate will be associated with protein. Nucleic acids, lipids and sugars can all act as phosphate acceptors. Contamination from these sources can be removed by DNase 1 and RNase 1 treatment (Raghothama et al., 1987) or washing the preparations with $CHCl_3$–MeOH (2:1; v/v) or acid (Buss and Stull, 1983). Furthermore, protein-bound phosphate is itself present in three forms. It can be found as protein-bound acyl phosphate attached to the carboxyl group of either aspartic or glutamic acids. Acylphosphates are specifically hydrolysed by hydroxylamine (Buss and Stull, 1983) but are also labile to alkali or hot acid. The acyl phosphates are transiently formed (within milliseconds of the addition of ATP) and are found as reaction intermediates of several enzymes such as ATPases. Protein-bound phosphate is also found as phosphoamidate with either of the two imidazole nitrogen atoms of histidine and the ε-amino group of lysine. These may also be formed as reaction intermediates and are stable to alkali but labile to acid. Protein kinase activity results in the formation of an O-monoester with the hydroxyl group of serine, threonine or tyrosine. These phosphoamino acids are stable to acid, but only phosphotyrosine and phosphothreonine are relatively stable to alkali. Through a knowledge of the stability of these different forms of bound phosphate it is possible to devise filter and gel washing protocols which will distinguish one form from another (Weller, 1979; Groppi and Browning, 1980; Buss and Stull, 1983; Cooper et al., 1983).

In in vivo studies of protein phosphorylation which require the presence of an untreated control it is also essential to take certain precautions. It is important to ensure that the specific activity of the $[^{32}P]ATP$ in the cell is not changed by the treatment. Protocols for the determination of the specific activity of cellular $[^{32}P]ATP$ have been described (Garrison, 1983). Finally, it must be recognised that under the correct conditions (the use of high specific activity $[\gamma-^{32}P]ATP$ and abundance of a particular substrate protein), detectable but physiologically irrelevant phosphorylation of just about any protein can be obtained. In order to test whether the observed phosphorylation of a particular protein is meaningful it may be necessary to determine the stoichiometry of phosphate to protein binding and whether the properties of the protein are altered by phosphorylation/dephosphorylation (Budde et al., 1985).

D. Future Developments in Kinase Identification and Assay

1. Affinity methods

The provision of a radiolabelled kinase often facilitates the development of isolation procedures since it is then easier to track relatively small quantities through the various purification steps. In many cases protein kinases are able to autophosphorylate. This feature has been made use of in the isolation of the thylakoid membrane-bound kinase that is responsible for the phosphorylation of the light-harvesting complex associated with photosystem II (LHC II) (Coughlan and Hind, 1986a,b, 1987a,b). In the absence of autophosphorylation, there are a number of strategies that can be pursued in order to tag the protein kinase of interest. Such approaches usually employ affinity ligands that are adenine nucleoside analogues. For example, Farchaus *et al.* (1985) used the adenosine analogue 5′-*p*-fluorosulphonylbenzoyl adenosine (FSBA) in order to study the kinetics of spinach thylakoid membrane protein kinases. [^{14}C]FSBA covalently labelled a 50 kDa candidate protein kinase. Another affinity labelling methodology relies on the fact that periodate-oxidised adenine nucleotides are often competent substrates. Here the ribose ring of ATP is opened between carbons C′2 and C′3 by cleavage with periodic acid. The 2′,3′ dialdehyde produced can form Schiff-bases with nearby amine side chains in the nucleotide binding site. Borohydride or cyanoborohydride is then used to reduce the Schiff-base and to attach covalently the oxidised nucleotide. This technique has led to useful findings concerning the nucleotide binding sites of pyruvate carboxylase (Easterbrook-Smith *et al.*, 1976), alkaline phosphatase (Chang *et al.*, 1981) and NAD-dependent isocitrate dehydrogenase (King and Colman, 1983). Radiolabelling is accomplished either by using [^{3}H]borohydride as reductant or by generating the oxidised compound from [^{32}P]ATP. The former approach has the advantage that unlabelled periodate-oxidised nucleotides are commercially available, whilst the latter strategy allows the inherently greater energy and specific activity of ^{32}P to be employed.

Photoaffinity labelling methodologies are essentially identical to the affinity methods described above except that the chemically active reagent is generated *in situ*, photochemically. The most commonly used photoactivateable group is probably the azido group ($-N\!\!=\!\!N^+\!\!=\!\!N^-$), which on activation by UV light rapidly loses N_2 leaving a highly reactive nitrene which covalently attaches by nucleophilic addition. Since the active species is in theory liberated at the nucleotide binding site itself, labelling should be more specific than with non-photoaffinity reagents that are intrinsically chemically active. In practice, the extreme reactivity of the nitrene group and the incompleteness of azido ATP binding to the target molecule always leads to labelling of a number of components, as with other affinity methods. With radiolabelled reagents this background can be minimised by initially carrying out the modification with non-radioactive affinity reagents, either under conditions where the target protein kinase is quiescent and/or with ATP present. This serves to block any non-specific reactive sites. The experiment is then repeated using radiolabelled affinity reagent. The ATP analogues, 2-azido- and 8-azidoadenosine-5′-triphosphate have been used in a number of studies to label the nucleotide-binding sites of ATP-synthetases (Czarnecki *et al.*, 1983; Wagenvoord *et al.*, 1981). However, they could also be used for labelling protein kinases. Synthetic routes for 8-azido-ATP and related compounds have been developed (Czarnecki *et al.*, 1979). However, they are complex and in many cases dangerous; fortunately 8-azido-ATP itself is commercially available in the ^{32}P-labelled form.

2. Anti-phosphoaminoacid antibodies

To date these have been used primarily for phosphotyrosine residues (Wang, 1988).

Phosphotyrosine (Ek and Heldin, 1984) or an analogue (Comoglio et al., 1984) was coupled to a carrier protein, such as immunoglobulin or keyhole limpet haemocyanin, and the phosphotyrosine–carrier complex used as immunogen. The anti-phosphotyrosine antibodies raised did not appear to cross-react with phosphothreonine or phosphoserine. This strategy is almost certainly applicable to phosphothreonine and phosphoserine and may be useful for the rapid screening of a large number of polypeptides, e.g. components of a two-dimensional electrophoretic separation. Since there are a number of amplification systems that can be applied to immune-labelled material on protein blots, such as the use of biotinylated secondary antibody followed by streptavidin coupled to a visualisation system (enzymic or radiolabelled), this approach should be at least as sensitive as using ^{32}P.

3. Determination of phosphorylation sites

At present protein sequencing of [^{32}P]polypeptides or peptides is most often used to determine phosphorylation site structure (Findlay et al., 1989; Geisow and Aitken, 1989). Of the sequencing methodologies available, solid phase sequencing (Findlay et al., 1989) is the most appropriate since the phosphoproducts of the cleavage reactions are hydrophilic and are poorly extracted by the hydrophobic solvents used in gas-phase sequencing (Geisow and Aitken, 1989). However, since ^{32}P-labelling of a polypeptide only gives an idea of the relative turnover of the phosphate groups on that protein and no estimate of the basal level of phosphorylation, the stoichiometry of phosphate attachment cannot be determined. Recently mass spectrometric methods have been developed for the sequencing of peptides (Biemann, 1989). In addition to the extreme sensitivity of these methods, the material need not be radiolabelled and the absolute level of phosphorylation can be determined with precision. Using the technique of tandem mass-spectrometry, Michel et al. (1988) were able to determine unequivocally the structures of phosphopeptides derived from photosystem II proteins. Since information of this level of detail cannot be aquired by other means it seems likely that use of this technology will increase despite the high cost and complexity of the equipment required.

III. PURIFICATION

Protein kinases have been isolated from organellar, soluble and membrane fractions. We shall describe in turn the methods used to purify kinases from these different fractions.

A. Organellar Kinases

The first plant nuclear casein kinase to be extensively purified and characterised was from soybean (Murray et al., 1978a). High salt was used to dissociate the kinase from chromatin; purification was by protamine sulphate precipitation, gel filtration and ion-exchange chromatography. Other nuclear cAMP-independent kinases, phosphorylating either casein or histone, have been isolated with high salt and purified by various procedures (Lin and Key, 1976, 1980; Murray et al., 1978b; Erdmann et al., 1982, 1985).

Ca^{2+}-calmodulin-activated protein kinase has been extracted from wheat germ chromatin and purified by ion-exchange chromatography, ammonium sulphate fractionation and gel filtration (Polya and Davies, 1982; Polya et al., 1983).

Ribosomal protein kinase was demonstrated in both peas and *Lemna*, and after sucrose density gradient centrifugation most of the activity could be eluted with 0.3 M KCl (Keates and Trewavas, 1974). Mitochondrial fractions containing protein kinase activity have been prepared (Rao and Randall, 1980; Danko and Markwell, 1985), but plant mitochondrial protein kinases have not been extensively characterised and purified. Chloroplast fractions are best prepared by the method of Nakatini and Barber (1977); methods of further purification have been discussed by Soll (1985). A thylakoid protein kinase that phosphorylates the light-harvesting chlorophyll a/b protein complex (LHC) has been characterised and purified (Coughlan and Hind, 1986a,b, 1987a,b); other thylakoid kinases have been prepared that were unable to phosphorylate the LHC *in vitro* (Lin et al., 1982; Lucero et al., 1982). Thylakoids have been shown to contain non-LHC phosphoproteins in photosystem II (Millner et al., 1986; Ikeuchi et al., 1987). Chloroplasts also contain an outer membrane kinase, a ribosomal kinase and a soluble kinase that have been characterised but not purified (Soll, 1985; Guitton et al., 1984; Foyer, 1984, 1985).

B. Soluble Kinases

Soluble protein kinases from a variety of sources have been extensively purified and characterised (see Table 22.1). Ca^{2+}-activated soluble kinase from wheat germ has been purified by Polya and coworkers by procedures similar to those used for the nuclear kinase (see above) but including a phenyl Sepharose step (Polya and Micucci, 1984). A cAMP-independent kinase from barley was purified using affinity chromatography on casein-Sepharose and ion-exchange chromatography and the purified kinase (linked to Sepharose) was used to extract a protein substrate (Reddy et al., 1987). Using a similar principle, purified cytokinin binding protein linked to Sepharose has been used to purify a kinase that phosphorylates this protein (Polya and Davies, 1983). Soluble Ca^{2+}-dependent but calmodulin-independent protein kinase was purified 1000-fold from soybean cells using ion-exchange chromatography, phenothiazine affinity chromatography, gel filtration and Cibacron Blue-Sepharose (Harmon et al., 1987).

C. Membrane Kinases

Standard differential centrifugation techniques provide a preliminary separation of organellar and membrane-bound kinases. Membranes containing kinase activity have been further fractionated by isopycnic separation on sucrose density gradients (Hetherington and Trewavas, 1984; Kato and Fujii, 1985) or by phase partitioning using polyethylene glycol and dextran (Blowers et al., 1985). Solubilisation of membrane kinases is usually achieved with detergents or by acetone precipitation. Detergents are used most commonly, but their presence during subsequent purification can be a problem. Acetone precipitation has been found to solubilise protein kinase activity from membranes of pea buds (Hetherington and Trewavas, 1984) and apple fruit (Battey and Venis, 1988a,b). Spinach membrane kinase has also been solubilised with acetone (Aducci et al., 1987). Sonication has been used successfully to solubilise kinase from

Lemna endoplasmic reticulum probably because it is located in the lumen rather than in the membrane (Kato and Fujii, 1985). Some assays for protein kinase activity involve pre-treatment for 15 min in 0.1% Triton X-100 (Salimath and Marmé, 1983; Paliyath and Poovaiah, 1985). This treatment probably solubilises significant amounts of protein kinase and/or substrate (Hetherington and Trewavas, 1984), and so may give different results from those obtained with membrane-bound activity.

TABLE 22.1. Distribution, characteristics and regulation of plant protein kinases that have been partially or fully purified.

Species and tissue	Mol. wt. (kDa)	Substrate Exogenous	Substrate Endogenous	Regulators	Reference
A. Nuclear kinases					
Soybean hypocotyls	55	Casein Phosphvitin		nd	Murray *et al.* (1978a,b)
Cauliflower inflorescence	39	Histone		nd	Murray *et al.* (1978a,b)
Soybean	48.5	Histone		nd	Lin and Key (1980)
Tobacco cell suspension	85, 23	Histone Casein	Auto-P	nd	Erdmann *et al.* (1985)
Wheat germ	90, 86	Casein		Ca^{2+} Polyamines	Polya and Micucci (1984)
B. Membrane-bound kinases					
Lemna ER Lumen	220	Histone		nd	Kato and Fujii (1985)
Oat root PM	nd	nd	ATPase	Ca^{2+}	Schaller and Sussman (1988)
Pea bud PM	18	Histone	Auto-P	Ca^{2+} CaM	Blowers and Trewavas (1987)
Apple microsomes	45–56	Histone	Auto-P	Ca^{2+}	Battey and Venis (1988b)
Spinach thylakoid	64	Histone	Auto-P LHCP	Redox poise	Coughlan and Hind (1986a)
C. Soluble					
Wheat germ	20	Casein	Ribosomes	nd	Rychlik and Zagorski (1980)
Wheat germ	nd		38 kDa subunit of eIF2	nd	Ranu (1980)
Wheat germ	36	Casein Phosphvitin	T-Substrate	nd	Yan and Tao (1982b)
Wheat germ	30	Casein	Cytokinin binding protein		Polya and Davies (1983)
Tobacco cell suspension	60	Histone	Auto-P	nd	Erdmann *et al.* (1985)

Continued

TABLE 22.1. (*continued*)

Species and tissue	Mol. wt. (kDa)	Substrate Exogenous	Endogenous	Regulators	Reference
Soybean cotyledons	39	Casein	Auto-P	nd	Gowda and Pillay (1982)
Soybean cotyledons	52 37	Casein	Auto-P 40s ribosome	nd	Göwda and Pillay (1982) Gowda and
	35		subunit		Pillay (1982)
Maize seedling	10	Casein		nd	Muszynska et al. (1983)
Barley embryo	95	Phosphvitin Casein	52 kDa protein	nd	Reddy et al. (1987)
Cicer embryo	94	Casein	nd	nd	Bansal et al. (1987)
Soybean cell suspension	46, 51	Histone	nd	Ca^{2+}	Harmon et al. (1987)
Amaranthus seedling	85	Histone	Auto-P	Ca^{2+} phospholipid	Elliott and Kokke (1987b)
Zucchini hypocotyls		Histone	Microsomal H^+ ATPase	nd	Scherer et al. (1988)
Alfalfa cell suspension	50, 56	Histone	Auto-P	Ca^{2+}	Bogre et al. (1988)

There are also two reports of tyrosine phosphorylation (Elliott and Geytenbeek, 1985; Torruella et al., 1986).

Abbreviations: nd, not determined; Auto-P, autophosphorylation.

Effective detergent solubilisation has been obtained using octylglucoside and sodium cholate (Coughlan and Hind, 1986a,b). $(NH_4)_2SO_4$ precipitation and sucrose density gradient centrifugation, followed by chromatography on Affi-Gel blue in the presence of Triton X-100 (yielding pure but denatured kinase) or on histone Sepharose in the presence of 10 mM 3-[(3-cholamidopropyl)-dimethylammonio]-1-propane sulphonate (CHAPS) (yielding active kinase) were used in this purification schedule. Acetone-solubilised membrane protein kinase has been further purified using Western blotting under non-denaturing conditions (Blowers et al., 1985). This allows enzyme activity to be detected on blots; unblotted samples run in parallel can then be electroeluted and used for studies of kinase properties (Blowers and Trewavas, 1987).

D. Novel Methods

Recently, immobilised metal-ion affinity chromatography (Porath and Olin, 1983) has been developed for the purification of phosphopeptides (Andersson and Porath, 1986). This technique relies on the strong affinity displayed by Fe^{3+}-iminodiacetyl sepharose towards phosphopeptides. Initially, the iminodiacetyl sepharose resin is pre-equilibrated with $FeCl_3$ at low pH (pH 3.0) in order to allow Fe^{3+} ions to be chelated by the iminodiacetyl moieties. The phosphopeptides are then applied to the resin and, after washing at low pH to remove non-specifically bound material, are eluted by a buffer of near neutral pH. The requirement for such a low pH in the phosphopeptide binding step probably precludes the use of this method for purification of phosphopolypeptides. However, it may be possible to carry out binding under slightly less acid conditions. Using this methodology, Bennett and coworkers (Michel and Bennett, 1987; Michel et al., 1988) successfully purified phosphopeptides derived from the N-terminal domains of phosphorylated thylakoid polypeptides.

The development of solid-phase peptide synthesis (Kent, 1988) has meant that relatively short peptides, i.e. up to 20 amino acids, can now be routinely prepared. This, in turn, has enabled the systematic probing of the phosphorylation site primary structure required for several animal protein kinases (Pinna et al., 1984; Meggio et al., 1984; Marin et al., 1986; Woodgett et al., 1986; Glass et al., 1986). Furthermore, peptides that are either effective substrates (low K_m) or potent inhibitors (low K_i) should clearly find use as affinity ligands for the preparation of specific kinases. Already the efficacy of affinity procedures has been demonstrated for the isolation of several protein kinases (see earlier). The use of specifically designed peptides or affinity ligands represents the logical development of this approach.

IV. PROPERTIES

Summarised in Table 22.1 is information on the distribution, characteristics and regulation of plant protein kinases which have been partially or completely purified. There is little information on the tissue and organ distribution of the different classes of kinase. It would be interesting to determine whether kinases with common regulatory characteristics predominate in certain tissue types. It is apparent from Table 22.1 that the status of both cAMP and calmodulin as regulators is not as certain as in animals. Although there are occasional reports of cAMP stimulating kinase activity (Kato et al., 1983, 1984; Janistyn, 1986, 1988) the majority of reports have failed to show any effect. In contrast, there are several reports of calmodulin-regulated kinases, but there are also Ca^{2+}-dependent, calmodulin-independent kinases (see Table 22.1). An initial problem in identifying calmodulin dependence is that endogenous calmodulin must be separated from the protein kinase. Salimath and Marmé (1983), reported that 1 μM calmodulin stimulated kinase activity in zucchini only after repeatedly washing the membranes with EDTA. However, in pea and apple membranes, significant quantities of calmodulin remain even after EDTA washing (Hetherington and Trewavas, 1984; N. H. Battey, unpubl. obs.). These differences may arise from genuine variation in the cellular distribution of calmodulin, or may be caused by the different extraction protocols. However, calmodulin can be removed by Ca^{2+}-dependent hydrophobic interaction chromatography on phenyl Sepharose (Battey and Venis, 1988a), Cibacron Blue-Sepharose (Harmon et al., 1987), and DEAE-cellulose plus CM-cellulose (Polya and Davies, 1982). Calmodulin-agarose is potentially a powerful means of purifying calmodulin-regulated enzymes, and has been successfully used for both NAD kinase (Cormier et al., 1981), and Ca^{2+}-ATPase (Dieter and Marmé, 1980). However, there is only one report of its use to purify Ca^{2+}-activated kinase (Polya and Davies, 1982). Ca^{2+} and phospholipid regulated protein kinases have been reported in plants (Schäfer et al., 1985; Muto and Schimogawara, 1985; Elliott and Skinner, 1986; Elliott et al., 1988). However, they may not be identical to protein kinase C as phorbol ester sensitivity has only been found in one case (Oláh and Kiss, 1986). Recent papers have shown that the Ca^{2+} requirement for soluble kinase activity can be reduced or abolished by fatty acids (Lucantoni and Polya, 1987; Klucis and Polya, 1987). It seems clear that the regulatory mechanisms which control plant protein kinase activity are similar only in broadest outline to those found in animal systems.

The most outstanding question in plant protein kinase research concerns the identity and function of the phosphorylated protein substrates. There are a few clearly defined substrates: the light-harvesting chlorophyll a/b protein complex of chloroplasts, pyruvate-dehydrogenase, quinate: NAD^+ oxidoreductase, pyruvate-Pi dikinase, and PEP carboxylase (reviewed by Ranjeva and Boudet, 1987). Recent work strongly suggests that the H^+ ATPase of plant membranes is regulated by phosphorylation (Schaller and Sussman, 1988; Scherer et al., 1988), and there are indications that changes in the rate of protein synthesis may be regulated by phosphorylation (see Schröder-Lorenz and Rensing, 1987). However, only the kinase that phosphorylates the light-harvesting chlorophyll a/b protein complex has been purified and fully characterised (Coughlan and Hind, 1986a,b, 1987a,b). Paradoxically, there are now a number of relatively well characterised kinases whose function remains obscure (see Table 22.1). We suggest that this situation can be remedied by studying specific cellular control processes first, and then moving on to phosphorylation if it suggests itself as a means of regulation (as has been the case with H^+ ATPase).

Note added in proof. A useful account of protein phosphorylation which complements this chapter has appeared since the manuscript was submitted. Budde, J. A. and Randall, D. D. (1989). *In* "Second Messengers in Plant Growth and Development" (W. F. Boss and D. S. Morré, eds). Alan R. Liss, New York (in press).

REFERENCES

Aducci, P., Ballio, A., Marra, M. and Walton J. D. (1987). *J. Plant Physiol.* **128**, 327–335.
Allis, C. D., Chicoine, G., Clover, C. V. C., White, E. M. and Gorovsky, M. A. (1986). *Anal. Biochem.* **159**, 58–66.
Andersson, L. and Porath, J. (1986). *Anal. Biochem.* **154**, 250–254.
Bansal, A., Saluja, D. and Sachar, R. C. (1987). *Phytochemistry* **26**, 1877–1881.
Battey, N. H. and Venis, M. A. (1988a). *Anal. Biochem.* **170**, 116–122.
Battey, N. H. and Venis, M. A. (1988b). *Planta* **176**, 91–97.
Bennett, J. (1984). *Physiol. Plant.* **60**, 583–590.
Biemann, K. (1989) *In* "Protein Sequencing—A Practical Approach" (J. B. C. Findlay and M. J. Geisow, eds) pp. 99–118. IRL Press, Oxford.
Blowers, D. P. and Trewavas, A. J. (1987). *Biochem. Biophys. Res. Commun.* **143**, 691–696.
Blowers, D. P., Hetherington, A. M. and Trewavas, A. J. (1985). *Planta* **166**, 208–215.
Bogre, L., Oláh, Z. and Dudits, D. (1988). *Plant Sci.* **58**, 135–144.
Boyer, P. D. and Krebs, E. G (1986). "The Enzymes", Vol. XVII, Part A, 3rd edn. Academic Press, New York.
Boyer, P. D. and Krebs, E. G. (1987). "The Enzymes", Vol. XVIII, Part B, 3rd edn. Academic Press, New York.
Budde, R. J. A., Holbrook, G. P. and Chollet, R. (1985). *Arch. Biochem. Biophys.* **242**, 283–290.
Buss, J. E. and Stull, J. T. (1983). *In* "Methods in Enzymology" (J. D. Corbin and J. G. Hardman, eds), Vol. 99, pp. 7–14. Academic Press, New York.
Chang, G. G., Wang, S. C. and Pan, F. (1981). *Biochem. J.* **199**, 281–287.
Comoglio, P. M., Di Renzo, M. F., Tarone, G., Giancotti, F. G., Naldini, L. and Marchisio, P. C. (1984). *Embo. J.* **3**, 483–489.

Cooper, J. A., Sefton, B. M. and Hunter, T. (1983). *In* "Methods in Enzymology" (J. D. Corbin and J. G. Hardman, eds), Vol. 99, pp. 387–402. Academic Press, New York.
Corbin, J. D. and Reimann, E. M. (1974) *In* "Methods in Enzymology" (J. G. Hardman and B. W. Omally, eds), Vol. 38, pp. 287–290. Academic Press, New York.
Cormier, M. J., Charbonneau, H. and Jarrett, H. W. (1981). *Cell Calcium* **2**, 313–331.
Coughlan, S. J. and Hind, G. (1986a). *J. Biol. Chem.* **261**, 11378–11385.
Coughlan, S. J. and Hind, G. (1986b). *J. Biol. Chem.* **261**, 14062–14068.
Coughlan, S. J. and Hind, G. (1987a). *J. Biol. Chem.* **262**, 8402–8408.
Coughlan, S. J. and Hind, G. (1987b). *Biochemistry* **26**, 6515–6521.
Czarnecki, J., Geahlen, R. and Haley, B. (1979). *In* "Methods in Enzymology" (S. Fleischer and L. Packer, eds), Vol. LVI, pp. 642–653. Academic Press, New York.
Czarnecki, J., Abbott, M. S. and Selman, B. R. (1983). *Eur. J. Biochem.* **136**, 19–24.
Danko, S. J. and Markwell, J. P. (1985). *Plant Physiol.* **79**, 311–314.
Dieter, P. and Marmé, D. (1980). *Proc. Natl. Acad. Sci. USA* **77**, 7311–7314.
Easterbrook-Smith, S. B., Wallace, J. C. and Keech, D. B. (1976). *Eur. J. Biochem.* **62**, 125–130.
Ek, B. and Heldin, C-H. (1984). *J. Biol. Chem.* **259**, 11145–11152.
Elliott, D. C. and Geytenbeek, M. (1985). *Biochim. Biophys. Acta.* **345**, 317–323.
Elliott, D. C. and Kokke, Y. S. (1987a). *Phytochemistry* **26**, 2929–2935.
Elliott, D. C. and Kokke, Y. S. (1987b). *Biochem. Biophys. Res. Commun.* **145**, 1043–1047.
Elliott, D. C. and Skinner, J. D. (1986). *Phytochemistry* **25**, 39–44.
Elliott, D. C, Fournier, A. and Kokke, Y. S. (1988). *Phytochemistry* **27**, 3725–3730.
Erdmann, H., Bocher, M. and Wagner, K. G. (1982). *FEBS Lett.* **137**, 245–248.
Erdmann, H., Bocher, M. and Wagner, K. G. (1985). *Plant Sci.* **41**, 81–89.
Farchaus, J., Dilley, R. A. and Cramer, W. A. (1985). *Biochim. Biophys. Acta.* **809**, 17–26.
Findlay, J. B. C., Pappin, D. J. C. and Keen, J. N. (1989). *In* "Protein Sequencing—A Practical Approach" (J. B. C. Findlay and M. J. Geisow, eds), pp. 69–84. IRL Press, Oxford.
Foyer, C. H. (1984). *Biochem. J.* **222**, 247–253.
Foyer, C. H. (1985). *Biochem. J.* **231**, 97–103.
Garrison, J. C. (1983). *In* "Methods in Enzymology" (J. D. Corbin and J. G. Hardman, eds), Vol. 99, pp. 20–36. Academic Press, New York.
Geahlen, R. L., Anostano, M., Low, P. S. and Harrison, M. L. (1986). *Anal. Biochem.* **153**, 151–158.
Geisow, M. J. and Aitken, A. (1989). *In* "Protein Sequencing—A Practical Approach" (J. B. C. Findlay and M. J. Geisow, eds), pp. 84–98. IRL Press, Oxford.
Glass, D. B., Cheng, H.-C., Kemp, B. and Walsh, D. A. (1986). *J. Biol. Chem.* **261**, 12166–12171.
Gowda, S. and Pillay, D. T. N. (1982). *Plant Sci. Lett.* **25**, 49–59.
Groppi, V. E. and Browning, E. T. (1980). *Molec. Pharmacol.* **18**, 427–437.
Guitton, C., Dorne, A-M. and Mache, R. (1984). *Biochem. Biophys. Res. Commun.* **121**, 297–303.
Harmon, A. C., Putnam-Evans, C. L. and Cormier, M. J. (1987). *Plant Physiol.* **83**, 830–837.
Hetherington, A. M. and Trewavas, A. J. (1982). *FEBS Lett.* **145**, 67–71.
Hetherington, A. M. and Trewavas, A. J. (1984). *Planta* **161**, 409–417.
Ikeuchi, M., Plumley, G. F., Inoue, Y. and Schmidt, G. W. (1987). *Plant Physiol.* **85**, 638–642.
Janistyn, B. (1986). *Z. Naturforsch.* **41c**, 579–584.
Janistyn, B. (1988). *Phytochemistry* **27**, 2735–2736.
Kato, R. and Fujii, T. (1985). *Plant Cell Physiol.* **26**, 1379–1386.
Kato, R., Uno, I., Ishikawa, T. and Fujii, T. (1983). *Plant Cell Physiol.* **24**, 841–848.
Kato, R., Uno, I., Ishikawa, T. and Fujii, T. (1984). *Plant Cell Physiol.* **25**, 691–696.
Keates, R. A. B. and Trewavas, A. J. (1974). *Plant Physiol.* **54**, 95–99.
Kent, S. B. (1988). *Ann. Rev. Biochem.* **57**, 957–959.
Kikkawa, U., Minakuchi, R., Takai, Y. and Nishizuka, Y. (1983). *In* "Methods in Enzymology" (J. D. Corbin and J. G. Hardman, eds), Vol. 99, pp. 288–298 Academic Press, New York.
King, M. M. and Colman, R. F. (1983). *Biochemistry* **22**, 1621–1630.
Klucis, E. and Polya, G. M. (1987). *Biochem. Biophys. Res. Commun.* **147**, 1041–1047.
Lin, P. P.-C. and Key, J. L. (1976). *Biochim. Biophys. Res. Commun.* **73**, 396–403.
Lin, P. P.-C. and Key, J. L. (1980). *Plant Physiol.* **66**, 360–367.
Lin, Z.-F., Lucero, H. A. and Racker, E. (1982). *J. Biol. Chem.* **257**, 12153–12156.

Lucantoni, A. and Polya, G. M. (1987). *FEBS Lett.* **221**, 33–36.
Lucero, H. A., Lin, Z.-F. and Racker, E. (1982). *J. Biol. Chem.* **257**, 12157–12160.
Marin, O., Meggio, F., Marchiori, F., Borin, G. and Pinna, L. A. (1986). *Eur. J. Biochem.* **160**, 239–244.
Maxam, A. M. and Gilbert, W. (1980). *In* "Methods in Enzymology" (L. Grossman and K. Moldave, eds), Vol. 65, pp. 499–560. Academic Press, New York.
McFadden, J. J. and Poovaiah, B. W. (1988). *Plant Physiol.* **86**, 332–334.
McPherson, M. A. and Ramachandran, J. (1980). *Biochem. Biophys. Res. Commun.* **94**, 1057–1065.
Meggio, F., Marchiori, F., Borin, G., Chessas, G. and Pinna, L. A. (1984). *J. Biol. Chem.* **259**, 14567–14579.
Michel, H.-P. and Bennett, J. (1987). *FEBS Lett.* **212**, 103–108.
Michel, H.-P., Hunt, D. F., Shabanowitz, J. and Bennett, J. (1988). *J. Biol. Chem.* **263**, 1123–1130.
Millner, P. A., Marder, J. B., Gounaris, K. and Barber, J. (1986). *Biochim. Biophys. Acta.* **852**, 30–37.
Murray, M. G., Guilfoyle, T. J. and Key, J. L. (1978a). *Plant Physiol.* **61**, 1023–1030.
Murray, M. G., Guilfoyle, T. J. and Key, J. L. (1978b). *Plant Physiol.* **62**, 434–437.
Muszynska, G., Dobrowolska, G. and Ber, E. (1983). *Biochim. Biophys. Acta.* **757**, 316–323.
Muto, S. and Shimogawara, K. (1985). *FEBS Lett.* **193**, 88–92.
Nakatini, H. Y. and Barber, J. (1977). *Biochim. Biophys. Acta.* **461**, 510–512.
Oláh, Z. and Kiss, Z., (1986). *FEBS Lett.* **195**, 33–37.
Paliyath, G. and Poovaiah, B. W. (1985). *Plant Cell Physiol.* **6**, 977–986.
Pinna, L. A., Meggio, F., Marchiori, F. and Borin, G. (1984). *FEBS Lett.* **171**, 211–214.
Polya, G. M. and Davies, J. R. (1982). *FEBS Lett.* **150**, 167–171.
Polya, G. M. and Davies, J. R. (1983). *Plant Physiol.* **71**, 482–488.
Polya, G. M. and Micucci, V. (1984). *Biochim. Biophys. Acta.* **785**, 68–74.
Polya, G. M., Davies, J. R. and Micucci, V. (1983). *Biochim. Biophys. Acta.* **761**, 1–12.
Porath, J. and Olin, B. (1983). *Biochemistry* **22**, 1621–1630.
Raghothama, K. G., Reddy, A. S. N., Friedmann, M. and Poovaiah, B. W. (1987). *Plant Physiol.* **83**, 1008–1013.
Ranjeva, R. and Boudet, A. M. (1987). *Ann. Rev. Plant Physiol.* **38**, 73–93.
Ranu, R. S. (1980). *Biochem. Biophys. Res. Commun.* **97**, 1124–1132.
Rao, K. P. and Randall, D. D. (1980). *Arch. Biochem. Biophys.* **200**, 461–466.
Reddy, A. S., Raina, A., Gunnery, S. and Datta, A. (1987). *Plant Physiol.* **83**, 988–993.
Roskoski, R. (1983). *In* "Methods in Enzymology" (J. D. Corbin and J. G. Hardman, eds), Vol. 99, pp. 3–7. Academic Press, New York.
Rychlik, W. and Zagorski, W. (1980). *Eur. J. Biochem.* **106**, 653–659.
Sahal, D. and Fujita-Yamaguchi, Y. (1987). *Anal. Biochem.* **167**, 23–30.
Salimath, B. P. and Marmé, D. (1983). *Planta* **158**, 560–568.
Schäfer, A., Bygrave, F., Matzenauer, S. and Marmé, D. (1985). *FEBS Lett.* **187**, 25–28.
Schaller, G. E. and Sussman, M. R. (1988). *Planta* **173**, 509–518.
Scherer, G. F. E., Martiny-Baron, G. and Stoffel, B. (1988). *Planta* **175**, 241–253.
Schröder-Lorenz, A. and Rensing, L. (1987). *Planta* **170**, 7–13.
Soll, J. (1985). *Planta* **166**, 394–400.
Swarup, G., Cohen, S. and Garbers, D. L. (1982). *Biochem. Biophys. Res. Commun.* **107**, 1104–1109.
Trewavas, A. (1976). *Ann. Rev. Plant Physiol.* **27**, 349–374.
Torruella, M., Casano, L. M. and Vallejos, R. H. (1986). *J. Biol. Chem.* **261**, 6651–6653.
Veluthambi, K. and Poovaiah, B. W. (1986). *Plant Physiol.* **81**, 836–841.
Wagenvoord, R. J., Vershoor, G. J. and Kemp, A. (1981). *Biochim. Biophys. Acta.* **634**, 229–236.
Wang, J. Y. J. (1988). *Anal. Biochem.* **172**, 1–7.
Weller, M., (1979). "Protein Phosphorylation". Pion, London.
Woodgett, J. R., Gould, K. L. and Hunter, T. (1986). *Eur. J. Biochem.* **161**, 177–184.
Yan, T.-F. and Tao, M. (1982a). *J. Biol. Chem.* **257**, 7037–7043.
Yan, T.-F. and Tao, M. (1982b). *J. Biol.. Chem.* **257**, 7044–7049.

23 Tonoplast Adenosine Triphosphatase and Inorganic Pyrophosphatase

PHILIP A. REA and JANICE C. TURNER

Department of Biochemistry, AFRC-IACR, Rothamsted Experimental Station, Harpenden, Herts AL5 2JQ, UK

I.	Introduction	386
	A. Primary and secondary transport	386
	B. V- and P-type H^+-ATPases	387
	C. H^+-Translocating inorganic pyrophosphatase (PPase)	388
	D. Membrane vesicles	389
II.	Preparation of tonoplast vesicles	389
	A. Reagents	390
	B. Procedure	390
	C. Further comments	391
	D. Identification of tonoplast	392
III.	Measurement of phosphohydrolase activity	393
	A. Standard Ames method	393
	B. Modified Ames method	394
	C. Coupled assay	394
	D. Bencini method	396
	E. Radiometric method	396
	F. Uncouplers and latency	397
IV.	Measurement of H^+-translocation	398
	A. Fluorimetric methods	399
	B. Radiometric method	400
V.	Summary	403
	References	403

I. INTRODUCTION

A. Primary and Secondary Transport

Essentially all transmembrane transport is energised by membrane-bound adenosine triphosphatases (ATPases; EC 3.6.1.3), whether energy transduction is at the level of the reaction chemistry of the ATPase itself (primary transport) or through the action of the ionic gradients established by the pump on other porters (secondary transport) (Harold, 1986). In plants, most transport processes are energised by the primary translocation of protons (Fig. 23.1).

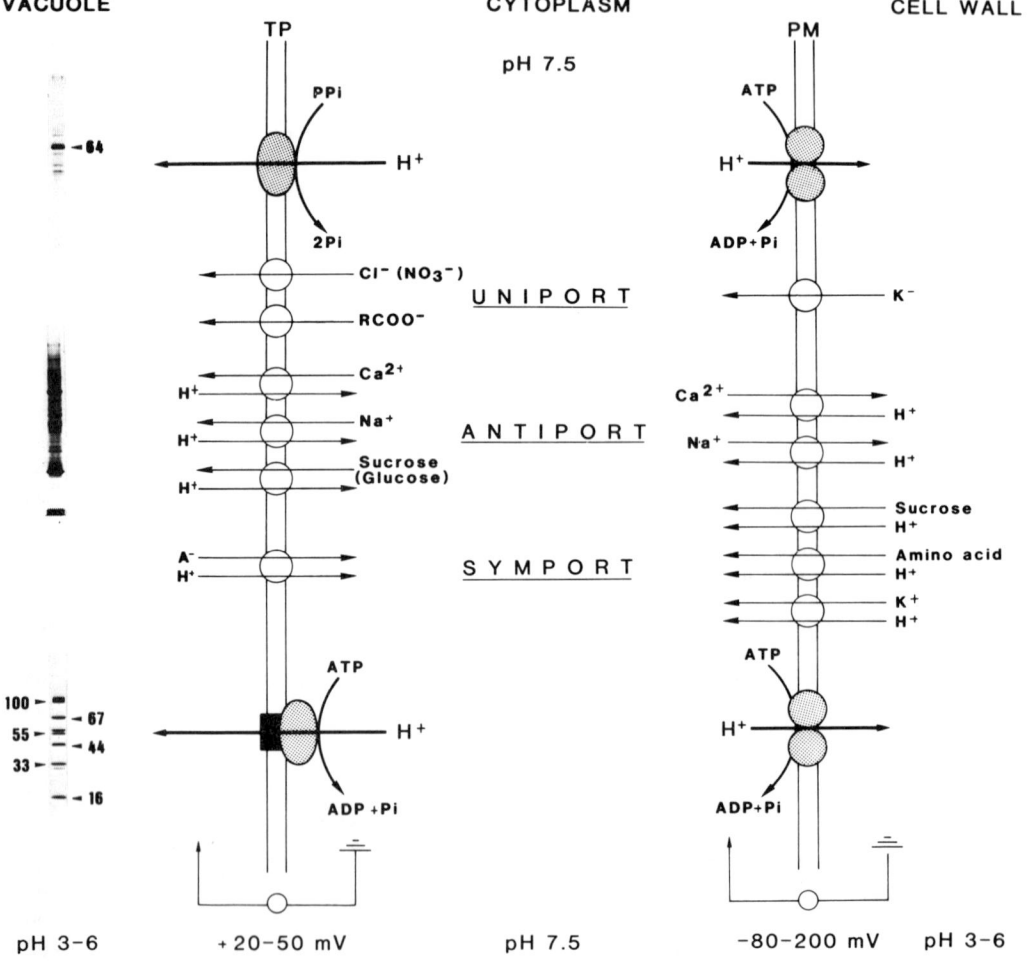

FIG. 23.1. Schematic diagram showing examples of primary H^+-translocation and secondary, $\Delta\bar{\mu}_{H^+}$-energised, solute transport at the tonoplast and plasma membrane. Note that the transport pathways shown were selected to illustrate the directions of the gradients and how $\Delta\bar{\mu}_{H^+}$-energised transport might be achieved: the list is not exhaustive.

H^+-translocating ATPases located at the plasma membrane and tonoplast translocate H^+ from the cytosol to extracellular and vacuolar compartments, respectively, to establish a H^+-electrochemical potential difference ($\Delta\bar{\mu}_{H^+}$), consisting of the sum of the chemical and electrical potential difference for H^+. Formally:

$$\Delta\bar{\mu}_{H^+} = \Delta\Psi - (2.3\ RT/F).\Delta pH \tag{1}$$

or in mV at 25°C,

$$\Delta\bar{\mu}_{H^+} = \Delta\Psi - 59\Delta pH \tag{2}$$

where $\Delta\Psi$ is the transmembrane electrical potential difference, ΔpH is the transmembrane pH difference, F is the Faraday constant and R and T have their usual meanings. Energy is liberated when H^+ or any cation moves in the opposite direction or when an anion moves in the same direction as H^+ pumping. The controlled (coupled) liberation of the potential energy contained in an electrochemical potential difference forms the basis of all secondary transport mechanisms, and the transport of any solute (charged or not; protonated or not) can be coupled to the flow of H^+, or electrical charge, by an appropriate secondary transport system (Fig. 23.1). Thus, a transport system will be driven by ΔpH if the substrate crosses the membrane with, or in exchange, for H^+. Similarly, a transport system will be driven by $\Delta\Psi$, if it carries a net charge.

B. V- and P-Type H^+-ATPases

The H^+-ATPases responsible for energisation at the plasma membrane and tonoplast are functionally and structurally distinct. The plasma membrane pump belongs to a broad category of P-type ('plasma membrane-type') enzymes whereas the tonoplast ATPase is but one member of the ubiquitous V-type ('vacuolar membrane-type') category of H^+-translocases (Carafoli and Pederson, 1987; Rea and Sanders, 1987).

The two categories are readily distinguished. P-type H^+-ATPases consist of a dimer of one 100 kDa polypeptide which forms a phosphorylated (acyl-phosphate) intermediate during catalysis. Members of this category are subject to inhibition by the transition state analogue orthovanadate and operate at low thermodynamic efficiency ($H^+/ATP = 1$). Other P-type enzymes are the ($Na^+ + K^+$)-ATPase of animal cell plasma membranes, the Ca^{2+}-ATPase of sarcoplasmic and endoplasmic reticulum (ER), the ($H^+ + K^+$)-ATPase of gastric mucosa, and the K^+-ATPase of prokaryotic plasma membranes, all of which have sequence homologies with the plasma membrane H^+-ATPase (Hager et al., 1986; Serrano et al., 1986).

V-type H^+-ATPases, by contrast, have an apparent functional mass of 400–600 kDa and comprise 3–8 different subunits of which the nucleotide-binding (67–72 kDa and 55–62 kDa) polypeptides and a 16 kDa proteolipid are universal components (Rea and Sanders. 1987). The best characterised members of this category—the H^+-ATPases of plant and fungal vacuolar membranes, and of (animal) clathrin-coated vesicles and chromaffin granules—are immunologically cross-reactive (Manolson et al., 1987) and all are inhibited by nitrate (e.g. Carafoli and Pederson, 1987; Rea and Sanders, 1987).

The degree of structural conservation within the V-type category is pronounced.

Genomic and cDNA clones of the genes encoding the $M_r = 67$–72 kDa subunits of the enzymes from *Neurospora* and *Daucus* are more than 60% homologous (Bowman et al., 1988a; Zimniak et al., 1988); the corresponding genes for the $M_r = 55$–62 kDa polypeptides from *Arabidopsis* (Manolson et al., 1988) and *Neurospora* vacuoles (Bowman et al., 1988b), and bovine chromaffin granules (Nelson et al., 1989) exhibit greater than 70% amino acid sequence identity. Although it has been proposed that the V-type ATPases from animal and plant sources may fall into different subgroups on the basis of subunit composition (Nelson, 1988), this no longer appears to be tenable. Recent improvements in the purification of the enzyme from higher plant vacuolar membranes yield near-homogeneous enzyme preparations containing stoichiometric proportions of six subunits of 100, 67, 55, 44, 32 and 16 kDa, respectively (Fig. 23.1; Parry et al., 1989); a polypeptide composition highly reminiscent of the '100–116, 72, 58, 41, 33, 16 kDa' size distribution of the V-type ATPases from animal sources (Nelson, 1988).

The functions of the 100, 44 and 33 kDa subunits of the enzyme are unknown. The 67–72, 55–62 and 16 kDa polypeptides, however, have the characteristics of catalytic (e.g. Bowman et al., 1986), regulatory (Manolson et al., 1985) and H^+-channel-forming subunits (Rea et al., 1987a; Sun et al., 1987), respectively. The nucleotide-binding 67–72 and 55–62 kDa subunits are peripherally associated with the membrane (Arai et al., 1988; Rea et al., 1987b), while the 16 kDa subunit is deeply embedded in the bilayer (Rea et al., 1987a; Sun et al, 1987).

C. H^+-Translocating Inorganic Pyrophosphatase (PPase)

The primacy of ATP for the energisation of transport across the membranes of plant cells is not in dispute, but it is now clear that the tonoplast of higher plant cells contains, in addition to the V-type ATPase, an inorganic pyrophosphatase (PPase; EC 3.6.1.1) which catalyses electrogenic H^+-translocation (Fig. 23.1; Rea and Poole, 1985).

The PPase is a K^+-stimulated phosphohydrolase: both PPi hydrolysis and PPi-dependent H^+-translocation are maximally stimulated by K^+ and relatively independent of the nature of the counter-anion (Table 23.1; Rea and Poole, 1985). ATP hydrolysis and ATP-dependent H^+-translocation by the tonoplast ATPase are, on the other hand, most stimulated by Cl^-, relatively unaffected by the counter-cation, and inhibited by nitrate (Table 23.1). While the initial studies on tonoplast vesicles left some room for doubt concerning the existence of the H^+-PPase as an entity distinct from the ATPase, chromatographic resolution of the two activities eliminates this possibility (Rea and Poole, 1986). The separated PPase is specific for PPi as substrate whereas the ATPase shows essentially no activity towards this substrate.

A reversible H^+-translocating PPase on the energy coupling membranes of mitochondria, chloroplasts and photosynthetic bacteria (notably *Rhodospirillum*) has been identified (Baltscheffsky and Nyren, 1984). However, the tonoplast H^+-PPase is the first example of a PPi-dependent H^+ pump at a 'non-energy-coupling' membrane.

Our understanding of the structure of the PPase is still in its infancy, by comparison with the ATPase, but the enzyme appears to comprise one major polypeptide of 64 kDa (Fig. 23.1; Britten et al., 1989). The enzyme seems to be a ubiquitous component of the vacuolar membrane of higher plants and capable of generating a H^+-electrochemical potential difference equivalent to that established by the ATPase (Rea and Sanders, 1987). There are indications that fungi also contain the enzyme (Ohsumi et al., 1985;

Rea et al., 1987c) but it remains to be determined if animal cells also contain endomembrane H^+-PPases.

TABLE 23.1. Inhibitor-sensitivities of tonoplast-enriched fraction from *Beta vulgaris* storage root. The tonoplast-enriched fraction was prepared as described in Section II. Fifty mM KCl was included in all of the assay media except when substituted by 50 mM NaCl for the PPase assays. Potassium nitrate (100 mM), Na_2MoO_4 (200 µM), sodium orthovanadate (100 µM) or NaN_3 (2 mM) were added as indicated. Five µM gramicidin-D was present throughout and ATPase and PPase activity were assayed at ATP and PPi concentrations of 3 mM and 0.6 mM, respectively.

Effector	Activity (µmol mg^{-1} h^{-1})	
	ATPase	PPase
None	69.6	21.6
KNO_3	17.4	21.6
Na_2MoO_4	66.8	21.2
Orthovanadate	72.6	21.2
NaN_3	71.3	21.5
$KNO_3 + NaN_3$	17.1	21.6
NaCl	—	5.1

D. Membrane Vesicles

Investigations of H^+ pumping and H^+-coupled transport are contingent on an experimental system in which the substrate concentrations on both sides of the membrane, ΔpH and $\Delta \Psi$ can be measured and manipulated. Membrane vesicles—membrane fragments which have spontaneously sealed on themselves to form hollow spheres—satisfy most of these requirements. Vesicles are essentially devoid of endogenous solutes, can be made to equilibrate rapidly with the bulk medium and are mechanically robust (Blumwald et al., 1987). In the discussion that follows we deal exclusively with methods for the preparation of transport-competent tonoplast vesicles and procedures for the assay of membrane-bound phosphohydrolase activity and H^+-translocation.

II. PREPARATION OF TONOPLAST VESICLES

Tonoplast vesicles can be prepared from most tissue homogenates by a combination of differential and density gradient centrifugation. Tonoplast is a low density membrane fraction (c. 1.10 g cm^{-3}) and can usually be resolved from other membranes on this basis. It should, however, be appreciated that the method described below has been optimised for beet (*Beta vulgaris* L.) storage root. Those readers interested in preparing tonoplast vesicles from other tissues should refer to Churchill et al. (1983) (*Avena* roots), Mettler et al. (1982) (*Zea* roots), Dupont (1987) (*Hordeum* roots), Smith et al. (1984) (*Kalanchoe* leaves) and Blumwald and Poole (1987) (suspension cultured cells) for examples of modifications which might be appropriate to their system.

A. Reagents

1. Homogenisation medium

A suitable medium consists of 10 mM glycerophosphate, 0.65 M ethanolamine (adjusted to pH 7.2–8.0 with conc. H_2SO_4 before addition), 0.28 M choline chloride, 2 mM salicylhydroxamic acid, 0.2% (w/v) BSA (fraction V, substantially fatty acid free), 10% (w/v) insoluble PVP (polyvinyl-polypyrrolidone), 0.5 mM butylated hydroxytoluene, 3 mM Tris-EDTA, 26 mM $K_2S_2O_5$ and 5 mM DTT buffered to pH 7.2–8.0 with 70 mM Tris-Mes buffer. This solution can be stored for several months at $-20°C$ provided that the $K_2S_2O_5$ and DTT are not added until immediately before use. The pH of the medium should be adjusted before the addition of insoluble PVP since it strongly binds glass and impairs electrode performance. Choline and ethanolamine are included in the homogenisation medium to diminish membrane degradation by phospholipase D, while glycerophosphate is added to inhibit phosphatidic acid phosphatase activity (Scherer and Morré, 1978). Butylated hydroxytoluene is added to scavenge free radicals and diminish lipid peroxidation.

2. Suspension medium

The suspension medium contains 1.1 M glycerol, 1 mM Tris-EDTA, 0.45 mM butylated hydroxytoluene and 1 mM DTT buffered to pH 7.2–8.0 with 5 mM Tris-Mes. A stock solution of 230 mM butylated hydroxytoluene is prepared in 2-propanol and 2 ml are added to 1 litre of suspension medium with stirring. The solution is then filtered through Whatman No. 1 filter paper to remove any undissolved butylated hydroxytoluene.

3. KI and sucrose solutions

KI (0.24 M) and 10 and 23% (w/w) sucrose are prepared in fresh suspension medium.

B. Procedure

1. Plant material

Both fresh and dormant storage root from red beet (*Beta vulgaris* L.) or sugar beet (*Beta vulgaris* L.) yield transport-competent tonoplast vesicles.

2. Homogenisation and differential centrifugation

All solutions and equipment are pre-cooled to 4°C. Three hundred and thirty grams of beets are thoroughly cooled in ice-cold distilled water, peeled and cut into 1 cm × 1 cm × 3 cm segments and left in ice-cold distilled water until homogenisation. The tissue segments are homogenised at high speed in a Waring blender for 30 s in 330 ml homogenisation medium and the homogenate is first filtered through two layers and then four layers of cheesecloth to remove large cell debris and some of the insoluble PVP. The filtrate is immediately centrifuged at 7000–10 000 × g for 15 min at 4°C. The pellet, containing cell debris and mitochondria, is discarded and the supernatant is centrifuged at $80\,000 \times g$ for 30 min at 4°C. After centrifugation the supernatant is aspirated and the pellet is resuspended with a Teflon pestle homogeniser in suspension medium. The volume of the resuspended pellet is adjusted to 10 ml with suspension medium and 25 ml of 0.24 M KI in suspension medium are added with stirring to give a final concentration of 0.17 M KI. The membranes are again centrifuged at $80\,000 \times g$ for 30 min at 4°C.

3. Density gradient centrifugation

For the routine preparation of tonoplast vesicles, the KI-treated microsomal pellet is resuspended in 18 ml suspension medium and 3 ml volumes are layered on top of six step gradients of 5 ml 10% (w/w) and 5 ml 23% (w/w) sucrose. The gradients are centrifuged in a swing-out rotor at $80\,000 \times g$ for 2 h, after which time the membranes at the 10/23% sucrose interface are carefully removed with a Pasteur pipette. The membranes are diluted 10- to 20-fold with the appropriate experimental solution, sedimented at $80\,000 \times g$ for 30 min and resuspended in 1–2 ml of the same medium to a protein concentration of 1–4 mg ml^{-1}.

If plasma membrane, as well as tonoplast, is required, the step gradients consist of 2.5 ml each of 10% (w/w), 23% (w/w), 34% (w/w) and 40% (w/w) sucrose. Plasma membrane is collected from the 34/40% sucrose interface.

The membranes may be used immediately or frozen in liquid N_2 and stored at $-70°C$. There is no discernible effect of freezing and storage on ATPase or PPase activity but there may be a slight (10–15%) diminution of the steady state ΔpH established by ATP-dependent H^+-translocation.

C. Further Comments

Additional points worthy of mention concerning the above method are outlined below:
   (1) The relative yield of tonoplast vesicles diminishes with increase in the ratio tissue fresh wt:volume homogenisation medium. The ratio given is near-optimal for red beet storage root.
   (2) The necessity of the first low-speed (7000–10 000 $\times g$) mitochondrial spin depends on the use to which the membranes are to be put. The low density of tonoplast vesicles enables their efficient separation from mitochondria and submitochondrial particles by sucrose density centrifugation alone. If, however, plasma membrane as well as tonoplast are to be prepared, a low-speed centrifugation is necessary to minimise contamination of the plasma membrane fraction with mitochondrial fragments
   (3) Caution is required when using KI and other chaotropic anions for the removal of loosely associated proteins from tonoplast because the tonoplast ATPase itself is sensitive to irreversible chaotropic effects (Rea *et al.*, 1987b). Although KI at a concentration of 0.17 M has no effect on the specific activity of the tonoplast ATPase of *Beta*, higher concentrations are irreversibly inhibitory (Rea *et al.*, 1987b). It is therefore advisable that when pre-treatment with KI is employed, the concentrations used are optimised for the preparation concerned.

D. Identification of Tonoplast

The tonoplast vesicles collected from the 10/23% interface are identified according to three primary criteria:

(1) They have two predominant phosphohydrolase activities, the anion-sensitive V-type ATPase (Section I.B) and the K^+-stimulated PPase (Section I.C), which are quantitatively and qualitatively similar to those of the membranes of isolated vacuoles (Walker and Leigh, 1981).
(2) If isolated intact vacuoles are homogenised and centrifuged on a sucrose gradient, they form vesicles of the same density with similar hydrolytic and H^+-translocase activities (Bennett et al., 1984).
(3) Only low activities of marker enzymes for other identifiable membranes are found at this density.

The most commonly employed criterion by which tonoplast vesicles are identified is reversible inhibition of the associated V-type ATPase by nitrate. This criterion should, however, be used judiciously as a means of identifying tonoplast fractions and tonoplast ATPase-mediated H^+-translocation:

(a) The F_1F_0-ATPases of both animal and plant mitochondria (Wang and Sze, 1985) are also inhibited by nitrate at concentrations similar to those necessary to inhibit the tonoplast ATPase. Nitrate inhibitability only has significance for the identification of tonoplast if insensitivity to azide can also be demonstrated (Table 23.1).
(b) Nitrate inhibits the tonoplast ATPase with pseudo-competitive kinetics with respect to Mg.ATP and yields K_i values of about 39 mM and 22 mM in the absence and presence of chloride, respectively (Griffith et al., 1986). Consequently, it is important not only to state the nitrate concentration employed but also the Mg.ATP and chloride concentrations when assessing the degree of inhibition of a particular membrane preparation.
(c) Nitrate diminishes the ΔpH-dependent quench of some fluorescent probes, notably acridine orange (Pope and Leigh, 1988; also see Section IV.B). Diminution of ΔpH formation by nitrate is only a sound means of identifying tonoplast ATPase-mediated H^+-translocation when alternative methods for the measurement of ATP-dependent H^+-translocation are used (Section IV.A.3).

Where possible, the H^+-PPase should be used as an independent tonoplast marker. The apparent ubiquity of the enzyme in higher plant vacuolar membranes and its strict dependence on K^+ for maximal activity (Table 23.1) make this enzyme particularly well-suited for this purpose. Evidence for a tight association of the H^+-PPase with the tonoplast comes from a number of investigations. Intact vacuoles of EGTA-permeabilised cells of Characean algae are capable of both PPi- and ATP-dependent intravacuolar acidification (Shimmen and MacRobbie, 1987), single patches of mechanically isolated vacuoles of Beta mediate both PPi- and ATP-dependent H^+-translocation (Hedrich and Kurkdjian, 1988) and there is a strictly linear relationship between tonoplast ATPase-catalysed and PPi-dependent H^+-translocation on any linear sucrose of dextran gradient (P. A. Rea, unpubl. res.). H^+-PPase activity may be associated with the Golgi of some preparations (Chanson et al., 1985) but the enzyme is essentially absent from endoplasmic reticulum and plasma membrane (Rea and Poole, 1985).

Markers that should be absent or minimal in the tonoplast-enriched fraction are glucan synthetase II and orthovanadate-sensitive ATPase (plasma membrane), latent IDPase (Golgi) and cytochrome oxidase and azide-sensitive ATPase (mitochondria). NADH cytochrome c reductase (an 'ER marker') is often present but this activity is also found in intact vacuole preparations (Walker and Leigh, 1981; Poole et al., 1984). Hodges and Leonard (1972) describe methods for the assay of these and other markers.

III. MEASUREMENT OF PHOSPHOHYDROLASE ACTIVITY

Factors which influence the method chosen for the measurement of substrate hydrolysis include the quantities of membrane material available, the substrate concentration range over which the measurements are to be made, and the nature and amounts of potential interferents in the preparation. Five methods which cover most applications are outlined below.

A. Standard Ames Method

1. Principle

The method of Ames (1966) is employed for the routine determination of ATPase and PPase activity at saturating substrate concentrations when neither continuous measurements nor high sensitivity are necessary. Activity is measured as the rate of liberation of Pi from ATP or PPi. Orthophosphate is quantitated as the molybdenum blue complex formed upon reduction of ammonium phosphate in strong acid (Ames, 1966).

2. Reagents

(a) 10% (w/v) Ascorbic acid. This should be stored at 4°C and is stable for about a month.
(b) 0.42% (w/v) Ammonium molybdate.$4H_2O$ in 1 N H_2SO_4. This solution is stable at room temperature.
(c) Ames reagent. Mix 1 part of solution (a) with 6 parts of solution (b). This reagent should be prepared fresh daily.

3. Procedure

Pi liberation is measured in a 0.5 ml reaction volume containing 50 mM KCl, 1–3 mM Tris-ATP (or 50–300 µM Tris-PPi), 3 mM $MgSO_4$ and 30 mM Tris-Mes adjusted to the desired pH. The reaction is initiated by the addition of 5–15 µg membrane protein and allowed to proceed for 10–30 min at 20–37°C. One and a half millilitres of Ames reagent are added to the reaction mixture at the end of the incubation period. Colour development is allowed to continue for 15–45 min at room temperature after which time the absorbance at 820 nm (A_{820}) is measured against reagent blanks. Calibration should be in the range 0–120 nmol Pi using reaction conditions identical to those employed for the assays.

4. Comments

All glassware should be scrupulously clean or disposable plastic-ware be used. Since ATP and PPi are subject to acid hydrolysis after the addition of Ames reagent, it is essential that a strict time-schedule be followed. Commonly encountered agents which interfere with the Standard Ames Method are mannitol, sorbitol and other sugar alcohols (glycerol excepted), detergents and added phospholipid.

B. Modified Ames Method

1. Principle

Both detergent and exogenous phospholipid often have to be incorporated into the assay medium when the activity of detergent-solubilised, delipidated, purified membrane protein is to be maximised (e.g. Manolson et al., 1985, for tonoplast ATPase; Rea and Poole, 1986, for tonoplast PPase). Since both detergent and phospholipid interfere with the Standard Ames Method, it is necessary to remove these components before the estimation of Pi. One method employing the precipitation of detergent-lipid micelles with 10% TCA:4% perchloric acid is outlined below.

2. Reagent

10% (w/v) TCA:4% (w/v) perchloric acid. This is the only reagent required in addition to those listed in Section III.A, above.

3. Procedure

Pi liberation is measured in a 0.3 ml reaction volume as detailed in Section III.A. However, at the end of the incubation period, 0.3 ml ice-cold 10% (w/v) TCA:4% (w/v) perchloric acid is added to the reaction mixture. The samples are thoroughly mixed, left on ice for 2 min and centrifuged for 3 min in an Eppendorf microfuge. Half millilitre volumes of the supernatant are then assayed for Pi as described in Section III.A.

4. Comments

The samples may be left on ice for up to 2 h after the addition of 10% TCA:4% perchloric acid without detectable non-enzymic hydrolysis of either ATP or PPi. Calibration should be performed using reaction conditions identical to those used for the samples.

C. Coupled Assay

1. Principle

Continuous measurements of ATPase activity necessitate the coupled method. Here, ATP hydrolysis is measured as the rate of ADP-dependent NADH oxidation in a reaction system containing phosphoenolpyruvate (PEP), pyruvate kinase and lactate dehydrogenase:

$$ATP \longrightarrow ADP + Pi \qquad (A)$$

$$ADP + PEP \longrightarrow ATP + \text{pyruvate} \qquad (B)$$

$$\text{Pyruvate} + NADH \longrightarrow \text{Lactate} + NAD^+ \qquad (C)$$

Pyruvate kinase catalyses transfer of the phosphate group of PEP to reform ATP [Reaction (B)] from the ADP formed in Reaction (A). The pyruvate so-formed from PEP is reduced by one mole of NADH to form lactate and NAD^+. Hence, for every mole of ATP hyrolysed, one mole of NAD^+ is generated.

2. *Reagents*

(a) Coupling enzymes. Both pyruvate kinase (EC 2.7.1.40) and lactate dehydrogenase (EC 1.1.1.27) should be purchased as suspensions in glycerol. Suspensisons in $(NH_4)_2SO_4$ are unsuitable if coupling between ATP hydrolysis and H^+-translocation needs to be evaluated as NH_4^+ collapses transmembrane pH gradients. Recommended preparations are pyruvate kinase, Type VII from rabbit muscle and lactate dehydrogenase, Type XXXV from porcine heart, both of which are suspensions in 50% glycerol and available from Sigma Chemical Co. Ltd, St. Louis, MO.

(b) NADH. Solutions of NADH are unstable and should therefore be prepared fresh daily.

3. *Procedure*

The assays are performed with a reaction volume of 1 ml containing 50 mM KCl, 30 mM Tris-Mes, 3 mM $MgSO_4$, 3 mM phosphoenolpyruvate (PEP), 0.2 mM NADH, 2 U pyruvate kinase, 2 U lactate dehydrogenase and 5–50 µg membrane protein in a recording spectrophotometer. The reaction is initiated by addition of the desired concentration of Tris-ATP and the rate of ADP-dependent NADH oxidation is measured as the linear decrease in A_{340} ($\varepsilon_{\text{red-ox}} = 6.22$ mM^{-1} cm^{-1}) against time.

4. *Comments*

Endogenous NADH oxidase activity in some membrane preparations can yield anomalously high rates at low ATP concentrations, but this component is readily enumerated by the measurement of NADH oxidation in the absence of ATP. Pyruvate kinase requires K^+ as cofactor; whenever KCl is omitted from the reaction system, it must therefore be replaced by an alternative K^+ salt.

D. Bencini Method

1. Principle

Whenever problems are encountered with the Ames method (Section III.A)—for instance, high rates of acid-catalysed hydrolysis of PPi or insufficient sensitivity—the method of Bencini (Bencini et al., 1983) is recommended as a first alternative. The Bencini method does not require the addition of strong acid for colour development and is approximately twice as sensitive as the Ames method.

2. Reagents

Bencini Reagent: this solution consists of 5 mM ammonium molybdate, 5 mM zinc acetate and 250 mM sodium acetate adjusted to pH 4.0 with acetic acid. For assay media containing more than 40 µg ml^{-1} membrane protein, 2% (w/v) sodium-SDS can be added without loss of sensitivity. Both solutions are stable for more than one year at room temperature.

3. Procedure

The assay procedure is identical to the standard Ames method (Section III.A) except that at the end of the 10–30 min incubation at 20–37°C, 0.5 ml Bencini reagent is added. The solution is thoroughly mixed, incubated for 2–60 min at room temperature and the A_{335} or A_{350} measured against reagent blank.

4. Comments

The Bencini method is subject to the same interferences as the Ames method with additional complications from chelators such as EGTA and citrate, which complex Zn^{2+}. However, unlike the Ames method, the absorption complex generated on the addition of Pi to Bencini reagent forms within seconds at room temperature ($t_{98\%} = 69$ s) and is stable for at least 48 h.

E. Radiometric Method

1. Principle

Radiometric methods find application when high sensitivity without ATP regeneration is required, for instance when the effects of nucleoside diphosphates on activity are to be investigated (e.g. Griffith et al., 1986). The method lends itself to increased sensitivity because of the high specific activities of commercially available [γ-^{32}P]ATP preparations (> 5000 Ci mmol^{-1}) and the comparative ease of measuring ^{32}Pi by liquid scintillation counting.

2. Reagents

(a) 15% (w/v) Norit A activated charcoal in 5% (w/v) TCA. Norit A (250–350 mesh;

from BDH Chemicals) is prepared as a fine slurry in 5% TCA. The slurry is thoroughly mixed immediately before use to ensure uniformity between samples.
(b) [γ-^{32}P]ATP. A specific activity of 1–4 Ci mmol^{-1} is adequate for most applications.

3. Procedure

Cleavage of the terminal phosphate of ATP is measured in a 200 µl reaction volume containing 50 mM KCl, 3 mM MgSO$_4$, 30 mM Tris-Mes and the desired concentration of [γ-^{32}P]ATP. The reaction is initiated by the addition of membrane protein (1–10 µg) and allowed to proceed for 1–10 min at 20–37°C, after which time 775 µl of ice-cold Reagent (a) are added. The samples are thoroughly mixed, left on ice for 1 h and centrifuged at 12 000 $\times$ g (13 000 rpm) for 3 min in an Eppendorf microfuge. Five hundred microlitres of the supernatant are removed without disturbing the pellet, added to 5 ml Aquasol liquid scintillation cocktail (New England Nuclear) and the ^{32}Pi is estimated by liquid scintillation counting. Reagent blanks without added membranes are used to correct for acid-catalysed ATP breakdown after the addition of Reagent (a).

4. Comments

To ensure that the estimates of ATPase activity are valid, it is important to determine that the amount of ^{32}Pi liberated from [γ-^{32}P]ATP increases linearly with time and with increase in membrane protein concentration.

F. Uncouplers and Latency

Two factors should be recognised when estimating the H$^+$-phosphohydrolase activity of membrane vesicles: (1) the effects of the $\Delta\mu_{H^+}$ established by the pump on substrate hydrolysis; (2) membrane orientation and therefore the accessibility of the hydrolase to substrate. As indicated by the stimulatory effects of uncouplers, substrate hydrolysis by both the tonoplast ATPase and PPase is subject to stalling by the $\Delta\bar{\mu}_{H^+}$ developed during the course of H$^+$-translocation (Poole *et al.*, 1984; Bennett *et al.*, 1984; Rea and Poole, 1985). The tighter the sealing (or coupling), the greater the potential for $\Delta\bar{\mu}_{H^+}$-limited substrate hydrolysis. It is therefore advisable to include ionophores in the assay media when assaying H$^+$-phosphohydrolase activity so as to eliminate stalling. Examples of suitable ionophores, or ionophore combinations, are gramicidin-D (5 µM), carbonyl cyanide *m*-chlorophenylhydrazone (CCCP) or carbonyl cyanide *p*-trifluoromethoxyphenylhydrazone (FCCP) (2 µM) + valinomycin (2 µM) or nigericin (5 µM). CCCP or FCCP, alone, are not completely uncoupling unless K$^+$-valinomycin is also included to ensure charge compensation upon the dissipation of ΔpH (e.g. Poole *et al.*, 1984).

The effects of membrane orientation can be dramatic. For instance, plasma membrane vesicles prepared by the dextran–polyethylene glycol two-phase method (Larsson, 1985) can exhibit greater than 10-fold stimulations of ATPase activity upon permeabilisation with low detergent concentrations. Since the ATP-binding site is located on the cytoplasmic face of the membrane *in vivo* (Fig. 23.1), the latency observed is attributed to a 10:1 prevalence of rightside-out over inside-out vesicles. Most tonoplast vesicle preparations isolated by sucrose density centrifugation, on the other hand, appear to be randomly oriented and consist of equal proportions of rightside-out (catalytically active) and inside-out (catalytically silent) vesicles; low detergent concentrations stimulate activity by only about two-fold (Rea et al., 1987b).

Enumeration of the latency of a membrane preparation requires the titration of phosphohydrolase activity with detergent over a concentration range equivalent to 0.5–2.0 times the critical micelle concentration. Since supraoptimal detergent concentrations are frequently inhibitory while suboptimal concentrations underestimate the latent component, the detergent concentration necessary for maximal activity must be determined rigorously within relatively narrow limits. Triton X-100 has found wide application as a permeabilising agent but recent investigations suggest that lysophosphatidylcholine (lyso-PC) (Gronzis et al., 1987) might be more suitable. Lyso-PC relieves latency at low concentrations but, unlike most detergents, exerts no inhibitory action on the plasma membrane Mg-ATPase of Zea at supraoptimal concentrations (Gronzis et al., 1987), possibly because its interaction with the membrane is limited to the phospholipid bilayer (Tokumura et al., 1985).

IV. MEASUREMENT OF H^+-TRANSLOCATION

The study of H^+-translocation by membrane vesicles requires simple and effective methods for the estimation of ΔpH and $\Delta \Psi$. Conventional electrophysiological and chemical methods are, however, inappropriate. Estimates of the intravesicular volume of tonoplast vesicles yield a value of only approximately 10 µg mg^{-1} membrane protein (Poole et al., 1985). Fluorimetric and radiometric methods of high sensitivity have therefore had to be developed.

Since the instrumentation and techniques required for the estimation of ΔpH and $\Delta \Psi$ are very similar, we restrict our discussion to the former: Bashford and Smith (1979) excellently review methods for the measurement of $\Delta \Psi$ in membrane vesicles.

1. Principle

The inside-acid ΔpH of membrane vesicles—established by Mg.ATP- or Mg.PPi-dependent H^+-translocation—is most commonly determined from the equilibrium distribution of a weak base, usually a primary amine, whether the distribution is measured fluorimetrically or radiometrically.

At any given pH, the free (B) and protonated (BH$^+$) forms of a weak base are in equilibrium with the H^+ activity of the medium. If B is lipophilic and BH$^+$ is not, B will freely traverse the membrane and be trapped as BH$^+$ in the acidic interior of the vesicle. Thus, since the base is protonated on both sides of the membrane and $B_i = B_o$ at equilibrium:

$$K = \frac{[H^+]_i \cdot [B]_i}{[BH^+]_i} = \frac{[H^+]_o \cdot [B]_o}{[BH^+]_o} \tag{3}$$

Thence:

$$pH_o - pH_i = \Delta pH = \log([BH^+]_i/[BH^+]_o) \tag{4}$$

where subscripts i and o indicate inside (intravesicular) and outside (extravesicular), respectively.

A. Fluorimetric Methods

Fluorimetric methods for the measurement of ΔpH make use of the altered spectral properties of the weak base when it is accumulated. For the fluorescent probes discussed below, the estimation of ΔpH depends on fluorescence quenching by self-interaction at high internal probe concentrations (Lee et al., 1982) such that:

$$\log([BH^+]_i/[BH^+]_o) = \log[Q/(100-Q)] \tag{5}$$

where Q is the percentage fluorescence quench.

The choice of probe is dictated by the system being investigated. The need for high sensitivity determines that quinacrine or acridine orange be used for most higher plant vesicle preparations.

1. Procedure

The quenching of the fluorescence of quinacrine (2 µM) or acridine orange (5 µM) is measured at 25°C in a stirred cuvette. The changes in fluorescence are monitored with a recording spectrofluorimeter. Recommended instrumental settings for quinacrine are excitation and emission wavelengths of 420 and 495 nm, respectively, and a slit width of 5 nm. The corresponding values for acridine orange are 495 and 540 nm, and 5 nm, respectively. A typical assay medium consists of 250 mM sorbitol (as asmoticum), 50 mM KCl and 3 mM Tris-ATP (or 100–300 µM Tris-PPi) buffered to the desired pH with 3 mM Tris-Mes. Vesicles (20–50 µg membrane protein) are pre-incubated in assay medium and the reaction is started by the addition of MgSO$_4$ to a final concentration of 3 mM. In order to correct for the binding of probe to the membranes, which can cause an attenuation of the fluorescence in the absence of a pH gradient, the gradient should be collapsed by adding gramicidin-D to a final concentration of 2 µM when a steady-state ΔpH has been achieved. Pump-dependent ΔpH formation is then expressed as $\Delta F/F$ where F is the protonophore-reversible quench (Q) and ΔF is the change in fluorescence.

2. Data analysis

Conventional methods for estimating the initial rate of fluorescence quenching from a hand-drawn tangent to a non-linear analogue pen recording are both time-consuming and prone to error. Appropriate software should therefore be used when quantitative data are required.

Jennings et al. (1988) describe software specifically designed for the logging and editing of ΔpH-related changes in fluorescence. The software enables the fluorescence changes to be fitted to a wide range of functions by a non-linear least-squares method (Marquardt, 1963), but in practice most data are best fitted by an exponential of the form:

$$F = P_1 - \frac{P_2}{\exp[P_3(t - P_4)]} \tag{6}$$

where F is fluorescence intensity (%), t is time and P_1 to P_4 are constants derived by curve fitting. Baseline shifts in the fluorescence trace are accommodated in a finite value of P_4. Thus, providing these are negligible, Equation (6) reduces to:

$$F = P_1 - P_2 \exp(-P_3 t) \qquad (7)$$

in which P_1 is the final steady-state fluorescence, P_2 is the overall change in fluorescence intensity and P_3 is the rate constant of the change (min^{-1}).

A typical fluorescence trace for ATP-dependent H$^+$-translocation is shown in Fig. 23.2A. The time period over which least-squares fitting is performed is shown in Fig. 23.2B. Fig. 23.2C summarises the results of the data analysis in which the fluorescence trace has been fitted to Equation (7) above. The initial rate of ATP-dependent fluorescence quenching is calculated as the first derivative [$dF/dt = P_2 P_3 \exp(-P_3 t) = 91.7\%\ F\ \text{min}^{-1}$].

3. Probe calibration and anion effects

For calibration of the fluorescence quench against ΔpH, artificial inside-acid pH gradients may be imposed by the addition of small volumes of concentrated base (e.g. NaOH) to the medium (Blumwald and Poole, 1985) or by pre-incubation of the vesicles in buffer of a predetermined pH (Blumwald et al., 1987). The validity of probe calibrations of this type for the determination of pump-generated pH gradients is, however, limited. Usually it is not known what proportion of the sealed vesicles in any given preparation actually participate in substrate-dependent H$^+$-translocation. Note that:

$$\Delta\text{pH} = \log[Q/(100 - Q)] + \log(V_o/V_i) \qquad (8)$$

Thus, a 10-fold difference in the extravesicular to ('active') intravesicular volume ratio (V_o/V_i) is equivalent to a ΔpH of 1 unit. The significance of the volume term in Equation (8) is evident when account is taken of the fact that electron spin resonance measurements on tonoplast vesicles indicate that less than 10% of the vesicles which are sealed to H$^+$ undergo ATP-dependent intravesicular acidification (Poole et al., 1985).

Caution should also be exercised in the choice of fluorescent probe. Acridine orange is especially prone to artifact. In the presence of the lipophilic anions I$^-$, ClO$_3^-$, NO$_3^-$, Br$^-$ or SCN$^-$, acridine orange reports smaller pH gradients than either quinacrine or [^{14}C]methylamine (Pope and Leigh, 1988; Table 23.2A). Although the mode of interaction of acridine orange with lipophilic anions has not been elucidated, this probe appears to elicit a genuine diminution of ΔpH. Addition of 5 µM acridine orange, for instance, causes a 73% reduction of net [^{14}C]methylamine uptake in the presence of 50 mM nitrate; 2 µM quinacrine, on the other hand, has no effect (Table 23.2B). It is not known if these effects are general, but it would seem prudent to assume that the pH gradients in any membrane preparation might be dissipated by acridine orange and, where possible, to use quinacrine instead. Control experiments employing the [^{14}C]methylamine uptake assay method (Section IV.B) are also recommended whenever a fluorescent indicator of ΔpH is used for the first time.

B. Radiometric Method

Most radiometric estimations of ΔpH employ [^{14}C]methylamine as the primary amine:

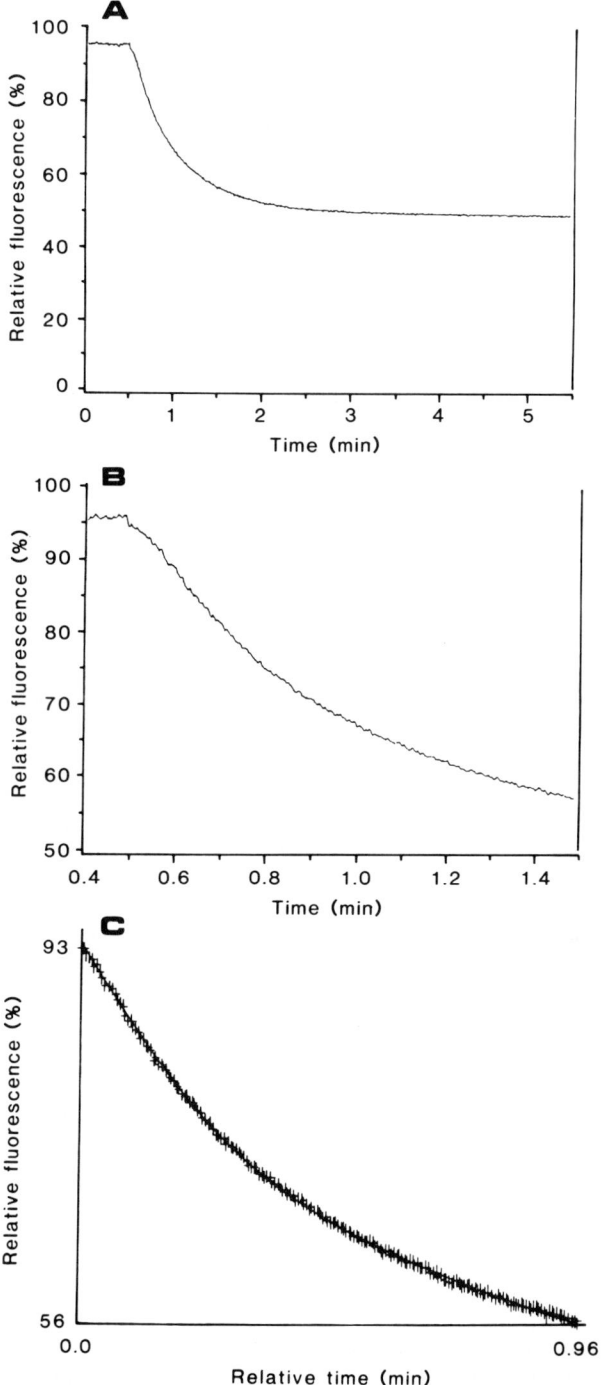

FIG. 23.2. Analysis of pump-dependent fluorescence quench data by method of Jennings *et al.* (1988). (A) ATPase-dependent quenching of acridine orange fluorescence by addition of 1 mM $MgSO_4$ to a suspension of tonoplast vesicles at 0.48 min. (B) Scale expansion of selected data from Panel A. (C) Non-linear least-squares fit of Mg.ATP-dependent fluorescence quenching with time redefined to 0 at the time of addition of $MgSO_4$. Equation (6) was used for fitting with P_4 constrained to 0. The least-squares estimates ($\pm$SE) were: $P_1 = 51.8 \pm 0.1\%$; $P_2 = -41.5 \pm 0.1\%$; $P_3 = 2.21 \pm 0.01$ min^{-1}. The initial rate of Mg.ATP-dependent fluorescence quenching was calculated as the first derivative of Equation (7) [$= P_2 P_3 \exp(-P_3 t)$], taken at $t = 0$, resulting in a value of 91.7% F min^{-1}.

it equilibrates rapidly, does not bind membranes or proteins and is not metabolised. Longer chain aliphatic amines may be used but both the permeability of the protonated species and accumulation within the membrane bilayer increase with chain length (Rottenberg, 1979).

TABLE 23.2.
A. Comparison of effects of various anions on PPi-dependent ΔpH formation by *Avena* tonoplast vesicles measured with [^{14}C]methylamine, quinacrine (2 μM) or acridine orange (5 μM).

Salt added	ΔpH		
	Methylamine	Quinacrine	Acridine orange
KCl	100	100	100
KBr	99	102	75
KNO$_3$	93	104	21
KI	75	103	0
K$_2$SO$_4$	21	21	29
KSCN	19	15	4

B. Effect of 2 μM quinacrine or 5 μM acridine orange on PPi-dependent [^{14}C]methylamine accumulation in the presence of 50 mM KCl or 50 mM KNO$_3$.

	[^{14}C]Methylamine accumulation		
	Control	+Quinacrine	+Acridine orange
KCl	100	96	66
KNO$_3$	110	88	27

Reproduced from Pope and Leigh (1988), with permission.

1. Materials

(a) [^{14}C]-Methylamine: this is purchased as its hydrochloride at a specific activity of 50–60 mCi mmol^{-1} and used without dilution with carrier.
(b) Membrane filters: Whatman WCN cellulose nitrate or Millipore HATF filters (0.45 μm pore size) are suitable.

2. Procedure

To initiate the reaction, membrane vesicles (25–100 μg membrane protein) are added to a 0.5 ml reaction volume containing 250 mM sorbitol (as osmoticum), 3 mM MgSO$_4$, 1–3 mM Tris-ATP (or 100–300 μM Tris-PPi) and 15–20 μM [^{14}C]methylamine (*c.* 1.2 μCi ml^{-1}) buffered to the desired pH with 20 mM Tris-Mes buffer. H$^+$-Translocation and [^{14}C]methylamine accumulation are allowed to proceed for 5–10 min at 20–25°C, after which time uptake is terminated by vacuum filtration through pre-wetted Whatman WCN or Millipore HATF filters.

The volume of the reaction mixture to be filtered must be optimised for the membrane preparation concerned: overloading plugs the membrane filters, increases the wash time and thereby increases the loss of accumulated probe; underloading decreases the counts and hence increases error. Volumes of 50–150 µl (i.e. 2.5–30 µg membrane protein) are suggested for preliminary investigations.

Either a direct or dilution filtration procedure may be adopted. In the dilution method, the 50–150 µl aliquot is diluted into 3 ml ice-cold wash medium (250 mM sorbitol, 3 mM Tris-Mes), mixed briefly on a vortex mixer, filtered and rinsed with 0.5 ml ice-cold wash medium. The direct method, on the other hand, involves wetting the filter with 1 ml ice-cold wash medium, directly filtering an aliquot of the reaction medium and quickly rinsing with 1 ml of ice-cold wash medium. The filters are air-dried and radioactivity determined by liquid scintillation counting.

To correct for non-energised [^{14}C]methylamine uptake and extravesicular solution trapped on the filter, samples without added ATP (or PPi) are assayed. Amine binding to the filters may be determined by filtering an aliquot of reaction mixture without vesicles.

3. Comments

The concentration of [^{14}C]methylamine in the assay medium should be kept low (less than 25 µM) since high amine concentrations dissipate ΔpH (e.g. Churchill and Sze, 1983). In order to minimise the losses of intravesicular [^{14}C]methylamine during filtration it is important to use as short a wash time as possible; the entire stop and wash times should not exceed a total of 10 s. For the reasons given in Section IV.B.1 above, fluorescence estimates of ΔpH should always be tested independently by the [^{14}C]methylamine distribution method.

V. SUMMARY

Methods for the preparation and assay of tonoplast vesicles from tissue homogenates are described. While much progress has been made in understanding transport at the tonoplast by the use of isolated intact vacuoles, tonoplast vesicles are more suitable for most membrane biochemical applications. Isolated vacuoles suffer from three major disadvantages: (1) they contain high concentrations of soluble proteins, inorganic and organic salts, and neutral compounds; (2) they are extremely fragile mechanically; and (3) their membranes contain many loosely associated proteins which can interfere with enzymic assays but which cannot be removed without membrane disruption. Tonoplast vesicles, on the other hand, obviate most of these difficulties (Section I.D). In addition, their high surface area:volume ratios and ease of preparation with standard equipment make tonoplast vesicles a rich source of membrane proteins for biochemical and molecular biological investigations.

REFERENCES

Ames, B. N. (1966). *Methods Enzymol.* **8**, 115–118.

Arai, H., Terres, G., Pink, S. and Forgac, M. (1988). *J. Biol. Chem.* **263**, 8796–8802.
Baltscheffsky, M. and Nyren, P. (1984). In "Bioenergetics" (L. Ernster, ed.), pp. 187–206. Elsevier, Amsterdam.
Bashford, C. L. and Smith, J. C. (1979). *Methods Enzymol.* **55**, 569–586.
Bencini, D. A., Wild, J. R. and O'Donovan, G. A. (1983). *Anal. Biochem.* **132**, 254–258.
Bennett, A. B., O'Neill, S. D. and Spanswick, R. M. (1984). *Plant Physiol.* **74**, 538–544.
Blumwald, E. and Poole, R. J. (1985). *Proc. Natl. Acad. Sci. USA* **82**, 3683–3687.
Blumwald, E., Rea, P. A. and Poole, R. J. (1987). *Methods Enzymol.* **148**, 115–123.
Bowman, E. J., Mandala, S. M., Taiz, L. and Bowman, B. J. (1986). *Proc. Natl. Acad. Sci. USA* **83**, 48–52.
Bowman, E. J., Tenny, K. and Bowman, B. J. (1988a). *J. Biol. Chem.* **263**, 13994–14001.
Bowman, E. J., Allen, R., Wechser, M. and Bowman, B. J. (1988b). *J. Biol. Chem.* **263**, 14002–14007.
Britten, C. J., Turner, J. C. and Rea, P. A. (1989). *FEBS Lett.* **256**, 200–206.
Carafoli, E. and Pederson, P. L. (1987). *Trends Biochem. Sci.* **12**, 186–189.
Chanson, A., Fichmann, J., Spear, D. and Taiz, L. (1985). *Plant Physiol.* **79**, 159–164.
Churchill, K. A. and Sze, H. (1983). *Plant Physiol.* **76**, 490–497.
Churchill, K. A., Holaway, B. and Sze, H. (1983). *Plant Physiol.* **73**, 921–928.
Dupont, F. M. (1987). *Plant Physiol.* **87**, 526–534.
Griffith, C. J., Rea, P. A., Blumwald, E. and Poole, R. J. (1986). *Plant Physiol.* **81**, 120–125.
Gronzis, J-P, Gibrat, R., Rigaud, J. and Grignon, C. (1987). *Biochim. Biophys. Acta* **903**, 449–464.
Hager, K. M., Mandala, S. M., Davenport, J. W., Speicher, D. W., Benz, E. J. and Slayman, C. W. (1986). *Proc. Natl. Acad. Sci. USA* **83**, 7693–7697.
Harold, F. M. (1986). "The Vital Force: A Study of Bioenergetics". Freeman, New York.
Hedrich, R. and Kurkdjian, A. (1988). *EMBO J.* **7**, 3661–3666.
Hodges, T. K. and Leonard, R. T. (1972). *Methods Enzymol.* **32**, 392–406.
Jennings, I. R., Rea, P. A., Leigh, R. A. and Sanders, D. (1988). *Plant Physiol.* **86**, 1257–1263.
Larsson, C. (1985). In "Modern Methods of Plant Analysis—New Series Cell Components" (H. F. Linskens and J. F. Jackson, eds), Vol. 1, pp. 85–104. Springer, New York.
Lee, H. C., Forte, J. G. and Epel, D. (1982). In "Intracellular pH: Its Measurement, Regulation and Utilization in Cellular Functions" (R. Nuccitelli and D. W. Deamer, eds), pp. 135–160. Alan R. Liss Inc., New York.
Manolson, M. F., Rea, P. A. and Poole, R. J. (1985). *J. Biol. Chem.* **260**, 12273–12279.
Manolson, M. F., Percy, J. M., Apps, D. K., Xie, X.-S., Stone, D. K. and Poole, R. J. (1987). In "Membrane Proteins" (S. C. Goheen, L. Hjelmeland, M. McNamee and R. Gennis, eds), pp. 427–434. Bio-Rad, New York.
Manolson, M. F., Quellette, B. F. F., Filion, M. and Poole, R. J. (1988). *J. Biol. Chem.* **263**, 17987–17994.
Marquadt, D. (1963). *J. Soc. Industr. Appl. Math.* **11**, 431–441.
Mettler, I. J., Mandala, S. and Taiz, L. (1982). *Plant Physiol.* **70**, 1738–1742.
Nelson, H., Mandiyan, S. and Nelson, N. (1989). *J. Biol. Chem.* **264**, 1775–1778.
Nelson, N. (1988). *Plant Physiol.* **86**, 1–3.
Ohsumi, Y., Uchida, E. and Anraku, Y. (1985). In "Biochemistry and Function of Vacuolar Adenosine-triphosphatase in Fungi and Plants", (B. Marin, ed.), pp. 141–150. Springer, Berlin.
Parry, R. V., Turner, J. C. and Rea, P. A. (1989). *J. Biol. Chem.*, in press.
Poole, R. J., Briskin, D. P., Kratky, Z. and Johnstone, R. M. (1984). *Plant Physiol.* **74**, 549–556.
Poole, R. J., Mehlhorn, R. J. and Packer, L. (1985). In "Biochemistry and Function of Vacuolar Adenosine-Triphosphatase in Fungi and Plants" (B. Marin, ed.), pp. 114–118. Springer, Berlin.
Pope, A. J. and Leigh, R. A. (1988). *Plant Physiol.* **86**, 1315–1322.
Rea, P. A. and Poole, R. J. (1985). *Plant Physiol.* **77**, 46–52.
Rea, P. A. and Poole, R. J. (1986). *Plant Physiol.* **81**, 126–129.
Rea, P. A. and Sanders, D. (1987). *Physiol. Plant.* **71**, 131–141.
Rea, P. A., Griffith, C. J. and Sanders, D. (1987a). *J. Biol. Chem.* **262**, 14745–14752.

Rea, P. A., Griffith, C. J. and Sanders, D. (1987b). *In* "Plant Vacuoles: Their Importance in Plant Cell Compartmentation and Their Applications in Biotechnology" (B. Marin, ed.), pp. 157–172. Plenum, New York and London.

Rea, P. A., Griffith, C. J., Manolson, M. F. and Sanders, D. (1987c). *Biochim. Biophys. Acta* **904**, 1–12.

Rottenberg, H. (1979). *Methods Enzymol.* **55**, 547–569.

Scherer, G. F. E. and Morré, D. J. (1978). *Plant Physiol.* **62**, 933–937.

Serrano, R., Kielland-Brandt, M. C. and Fink, G. R. (1986). *Nature* **319**, 689–693.

Shimmen, T. and MacRobbie, E. A. C. (1987). *Protoplasma* **136**, 205–207.

Smith, J. A. C., Uribe, E. G., Ball, E., Heuer, S. and Luttge, U. (1984). *Eur. J. Biochem.* **141**, 415–420.

Sun, S.-Z., Xie, X.-S. and Stone, D. K. (1987). *J. Biol. Chem.* **262**, 14790–14794.

Tokumura, A., Mostafa, M. H., Nelson, D. R. and Hanahan, D. J. (1985). *Biochim. Biophys. Acta* **812**, 568–574.

Walker, R. R. and Leigh, R. A. (1981). *Planta* **153**, 150–155.

Wang, Y. and Sze, H. (1985). *J. Biol. Chem.* **260**, 10434–10443.

Zimniak, L., Dittrich, P., Gogarten, J. P., Kibak, H. and Taiz, L. (1988). *J. Biol. Chem.* **263**, 9102–9112.

Index

A

Acetohydroxyacid reductoisomerase
  assay, 330
  extraction, 330
Acetohydroxyacid synthase
  assay, 329
  extraction, 328
Acetohydroxybutyrate, 328
Acetoin, 329
Acetolactate, 329
Acetyl-CoA, 195, 203, 298
Acetyl-CoA carboxylase, 195–197
  assay, 197
  localisation, 197
  purification, 197
Acyl carrier protein, 197, 208
  measurement, 198
  properties, 200
  purification, 199
Acyl carrier protein thioesterase (Oleoyl), 210
Acyl carrier protein transacylase(Acetyl-CoA)
  assay, 201
Acyl carrier protein transacylase(Malonyl-CoA), 202
  assay, 203
  properties, 203
Acyl diaminopimelate aminotransferase, 298
Acyl diaminopimelate deacylase, 298
Adenosine diphosphate(Glucose), 96
Adenosine diphosphate glucose pyrophosphatase, 94–97
  assay, 95
  extraction, 95
Adenosine 5′-phosphosulphate 335
Adenosine 5′-phosphosulphate transferase, 339–342
  assay, 341
  extraction, 341
  properties, 341
Adenosine triphosphatases H^+ translocating, 374, 381, 387, 392

Adenosine triphosphatases (*cont.*)
  binding site, 397
  P-type, properties, 387
  H-type, properties, 387
Adenosine triphosphate, 46, 96, 146, 195, 335, 372, 386, 388
S-Adenosylmethionine, 298
Adenylate kinase, 41, 60–62
  assay, 61
  properties, 62
  purification, 61
Albizziine, 291
Alanine, 59, 137, 282–283
Alanine aminotransferase, 59
  assay, 60
  properties, 60
  purification, 60
Alanine:glyoxylate aminotransferase, 136
Aldolase, 23, 147, 153–154
  assay, 153
  extraction, 153
  properties, 154
  purification, 153
Amino acid synthesis, 258, 278, 316
Aminoadipic acid, 297
Aminotransferases, 41, 57–59, 71, 136–137, 277–286, 294, 298
  assay, 136, 279, 280
  extraction, 58
  purification, 137
  regulation, 137
Ammonia, 138, 242, 258, 289, 291
Amylopectin, 94
Amylose, 94
Asparaginase, 291
  assay, 292
  extraction, 293
  properties, 294
  purification, 293
Asparagine, 287, 288, 291, 295
Asparagine:glyoxylate aminotransferase, 136
Asparagine aminotransferases, 294

INDEX

Asparagine synthetase, 288
  assay, 288–289
  extraction, 290
  properties, 291
  purification, 290
Aspartate, 57, 280, 289, 291, 315
Aspartate aminotransferase, 41, 57–59, 71, 278
  assay, 58, 280
  extraction, 281
  properties, 59, 282
  purification, 58, 281
  regulation, 59
Aspartate kinase, 298, 315, 316
Aspartate kinase, lysine sensitive, 299
  assay, 299
  purification, 302, 304
Aspartate kinase, threonine sensitive, 300
Aspartate semialdehyde, 298, 306, 315–316
Aspartate semialdehyde dehydrogenase, 298, 305, 315
Aspartyl phosphate, 315
Azaserine, 291
Azide, 13

B

Bicarbonate, 138, 195
Branched chain amino acid aminotransferase
  assay, 331
  extraction, 331

C

C_3 plants
  growth in elevated levels of CO_2, 130
C_4 plants, 40
C_4 photosynthesis, 39–43
  activity of enzymes, 42
  extraction of enzymes, 17, 19
  location of enzymes, 42
Calvin cycle, 15–37
  activity measurement, 18
  light activation of enzymes, 17, 19
Carbon dioxide, 2–5, 138
  as activator (cofactor), 4
  assimilation in C_4 plants, 41
  concentration in solutions, 2
  dark evolution, 130
  hydration, 2, 11–12
Carbonic anhydrase, 2, 11–13
  activity, 12
  in C_4 plants, 2
  inhibition, 13
  extraction, 12
  properties, 13
  purification, 12–13
2-Carboxy-D-arabinitol-1-phosphate, 5–6

Cardiolipin, 234, 238
Calcium, 291
Casein kinase, 376
Catalase, 134
  assay, 135
  purification, 135
  regulation, 135
Choline, 230
Choline phosphotransferase, 232
Citric acid cycle, 176
Coenzyme A, 176
Coenzyme A synthetase(Oleoyl), 210
Condensing enzymes of fatty acid synthesis, 203
3-Cyanoalanine hydrolase, 295
3-Cyanoalanine synthase, 295
β-Cystathionase
  assay, 364
  properties, 366
  purification, 365
Cystathionine, 362, 364
Cystathionine β-lyase, β-cystathionase, 364–366
Cystathionine γ-synthase, 361
  assay, 362
  properties, 363
  purification, 363
Cysteine, 295, 349, 350–351, 356, 363
Cysteine synthase
  activity, 350
  assay, 350
  localisation, 352
  properties, 352
  purification, 351
  specificity, 351
Cytidinediphosphocholine, 231, 233
Cytidinephosphodiacylglycerol, 231, 234–235, 237
Cytidinephosphodiacylglycerol ethanolamine, 231
Cytochrome c reductase (NADH), 242, 247

D

Dehydrogenase (NADH), 242
D-enzyme, 109–110
  assay, 109
  extraction, 109
Diacylglycerol, 233
2,6,Diaminopimelate, 298
Diaminopimelate decarboxylase, 298, 310
  assay, 310
  purification, 311
Diaminopimelate epimerase, 298, 309
  assay, 309
  isolation, 309

Dichlorophenol indophenol reductase (NADH), 242
Dihydrodipicolinate reductase, 298, 309
　assay, 309
　purification, 308
Dihydrolipoamide dehydrogenase
　measurement, 179
　purification, 184
Dihydrolipoyl acetyltransferase, 176
Dihydrolipoyl transacetylase
　measurement, 179
　purification, 185
Dihydropicolinate synthase, 298, 306
　assay, 306
　purification, spinach, 306
　purification, wheat, 307
Dihydroxyacetone-phosphate, 88, 153
Dihydroxyacid dehydratase
　assay, 350
　extraction, 330

E

Elongases, purification, 213
Enolase
　assay, 163
　extraction, 163
　properties, 165
　purification, 163
Enoyl-acyl carrier protein reductase
　type 1, 207
　type 2, 207
Ethanolamine phosphotransferase, 233

F

Fatty acids, 194, 230
　elongation, 212
　systems for studies, 212
Fatty acid synthetase, 197
Ferricyanide reductase (NADH), 242, 247
Ferredoxin–thioredoxin system, 17, 24–28
Flavin adenine dinucleotide, 176
Fluoride, 79
Fructans, synthesis, 76
Fructose, 75
Fructose 1,6-bisphosphatase, 17–18, 23–25, 67, 74
　assay, 25
　chloroplastic enzyme, 24
　cytosolic enzyme, 24
　properties, 24
　purification, 27
　regulation, 24
Fructose 1,6-bisphosphate, 4, 74, 146, 150, 153

Fructose 1,6-bisphosphate aldolase, 67
Fructose 2,6-bisphosphatase, 75, 87–88, 147
　purification, 91
Fructose 2,6-bisphosphate, 87–92
　assay, 89
　degradation assay, 90
　extraction, 88
Fructose 6-phosphate, 77, 87–88, 146–147
Fructose 6-phosphate,2-kinase, 87
　activation, 88
　assay, 89
　extraction, 89
　inhibition, 88
　purification, 91

G

Glucan hydrolases, 110
Glucan phosphorylases, 104, 108, 112–116
　activity assay, 109
　compartmental specific immunological studies, 122
　extraction, 108
　type 1, 112
　　extraction, 118
　　purification, 118
　type 2, 112
　　extraction, 120
　　purification, 120
Glucan polymerising reactions, 108
Glucose, 75, 116
Glucose 1-phosphate, 74, 94, 96
Glucose 6-phosphate, 74
Glucose 6-phosphate dehydrogenase, 116
Glutamate, 57, 59, 242, 256, 258, 277, 280, 282–283, 288
Glutamate dehydrogenase, 258, 277, 281
Glutamate:glyoxylate aminotransferase, 136
Glutamate:pyruvate aminotransferase, 136
Glutamate synthase (GOGAT), 132, 258
　ferredoxin dependent, 258–260
　　assay, 268
　　extraction, 267
　　properties, 271
　　purification, 269
　　structure, 270
　NADH dependent, 258–260
　　assay, 270
　　extraction, 267
　　purification, 273
　　structure, 273
　NADPH dependent, 259
Glutamate synthase cycle, 258–277
Glutamine, 258
Glutamine synthetase(GS), 132, 258–260
　assay, biosynthetic, 262

Glutamine synthetase (*cont.*)
  assay, synthetase, 261
  assay, transferase, 262
  extraction, 260
  properties, 265
  purification, 264
γ-Glutamylcysteine, 356
γ-Glutamylcysteine synthetase, 355
Glutathione, 340, 355–356
Glutathione synthetase, 356–359
  assay, 357
  distribution, 359
  extraction, 357
  properties, 358
  purification, 357
Glyceraldehyde P-dehydrogenase, 17
Glyceraldehyde 3-phosphate, 153–154
Glyceraldehyde 3-phosphate dehydrogenase (NADP), 17, 67
  assay, 22
  properties, 21
  purification, 22
Glycerate kinase, 141–142
  assay, 141
  purification, 141
  regulation, 141
3-Glycerate 3-phosphate, 88
Glycerol 3-phosphate dehydrogenase, 147, 156–159
  assay, 157
  extraction, 157
  properties, 158
  purification, 157
Glycerol P-phosphatidyl transferase, 235
Glycine, 136, 138, 283
Glycine decarboxylase, 138
Glycolate, 134
Glycolate oxidase, 134–136
  assay, 134
  purification, 135
  regulation, 135
Glycolysis, 146
Glycosyl acceptor, 104
Glycosyl donor, 104
Glyoxylate, 134, 138, 283
Glyoxylate aminotransferases, 283–285
  assay, 284
  extraction, 284
  properties, 285
  purification, 284

H

Heavy metals, 83
Hexokinase, 116
Hexose 6-phosphate, 75

Homocysteine, 362, 364, 366
Homocysteine methylase, 366–368
Homoserine, 315–316, 318
Homoserine dehydrogenase, 317
  assay, 320
  properties, 321
  purification, 318
  threonine resistant isoenzyme
    properties, 318
    purification, 318
  threonine sensitive isoenzyme
    properties, 318
    purification, 318
Homoserine kinase, 316, 318
  assay, 320
  properties, 321
  purification, 320
Homoserine O-phosphate, 315–316, 318, 322
Homoserine phosphorylation, 316
Hydrogen peroxide, 134
Hydrolases, 104, 105–107, 110–112
  debranching enzyme activity, 106
  detection procedure, 111
  endoamylase activity, 105
  hydrolytic activity, 105
Hydroxamate, 299
β-Hydroxyacyl-acyl carrier protein dehydrogenase, 206
Hydroxypyruvate, 137
Hydroxypyruvate reductase, 137, 140
  assay, 140
  NADH as cofactor, 140
  NADPH as cofactor, 140
  purification, 141
Hysteric behaviour, 19

I

Imidazolinone herbicide, 328
Inositol, 230–231
Inositol phosphatidyltransferase, 235
Invertase, 75, 82–83
  assay, 82
  inhibition, 83
  properties, 82
  purification, 82
Invertase, acid, 75
Invertase, alkaline or neutral, 75
Isoleucine, 298, 316, 325
Isoleucine biosynthesis, 316
Isopropylmalate synthase, 332

K

β-Ketoacyl-acyl carrier protein reductase
  assay, 206

INDEX

β-Ketoacyl-acyl carrier protein reductase
  assay (cont.)
  properties, 205
  purification, 205
β-Ketoacyl-acyl carrier protein, substrate specificities in spinach leaves, 204
β-Ketoacyl-acyl carrier protein synthase 1, 203
  assay, 204–205
  properties, 203
β-Ketoacyl-acyl carrier protein synthase 2, 203
  assay, 204–205
  properties, 203
α-Ketobutyrate, 362
Kinases, membrane, 337
Kinases, organellar, 376
Kinases, soluble, 377

L

Leucine, 325
Leucine biosynthesis, 332
Linoleate, 210
Linoleate desaturation *in vivo*, 211
γ-Linolenate formation, 213
Lipases, 220–224
  activity, 221
  assay, 222
  assay difficulties, 225
Lipid acid degrading enzymes
  group 1, 220
  group 2, 220
Lipoic acid, 176
Lipoyl dehydrogenase, 176
Luciferin–Luciferase system, 336
Lysine, 297–298, 316

M

Magnesium ion, 4–6, 19, 25, 28, 51, 79, 88, 150, 176, 372
Malate, 51, 53
Malate dehydrogenase(NADP), 17–18, 41
  assay, 49
  extraction, 49
  properties, 50
  purification, 49
  regulation, 50
Malic enzyme(NAD)
  assay, 49
  extraction, 49
  properties, 55
  purification, 54
  regulation, 55
Malic enzyme(NADP), 71
  assay, 51

Malic enzyme(NADP) (*cont.*)
  extraction, 51
  inhibition by CO_2, 53
  inhibition by malate, 53
  properties, 52
  purification, 51
  regulation, 53
Malonyl-CoA, 195, 208, 212
Maltose phosphorylase, 109
Manganese ion, 54, 79, 163, 372
Methionine, 298, 305, 316, 366
Methionine biosynthesis, 216, 261
5-10-Methylene tetrahydrofolate, 138
5-Methyltetrahydropteroylglutamate-homocysteine methyltransferase, 361, 366
Mitochondria, purification, 182
Monogalactosyldiacylglycerol, 210
Mutations, 130

N

Nicotinamide adenine dinucleotide, 21, 279
Nicotinamide adenine dinucleotide reduced, 137, 177, 205, 208, 212, 242
Nicotinamide adenine dinucleotide phosphate, 21–22, 279
Nicotinamide adenine dinucleotide phosphate, reduced, 9, 21, 205, 208, 212, 242
Nitrate assimilation, 242
Nitrate reductase, 241
  immunochemistry, 247–249, 253
Nitrate reductase($FADH_2$), 242
Nitrate reductase($FMNH_2$), 242, 246
Nitrate reductase(NADH), 242–249
  assay, 245
  function, 242
  properties, 244
  purification, 244
  structure, 242
Nitrate reductase(NADPH), 247
Nitrate reduction, 242
Nitrite reductase, 249–253
  assay, 252
  bromophenol blue specific, 242, 246
  function, 250
  methyl viologen specific, 242, 246
  properties, 251
  purification, 251
  structure, 250
Nitrite reduction, 242, 249

O

Oleate, 208, 210
Oligosaccharides, synthesis, 76

Oxaloacetate, 44, 49, 55, 57, 280
2-Oxoglutarate, 57, 59, 257, 258, 280, 282–283
2-Oxoglutarate aminotransferase, 282
 assay, 283
 extraction, 283
 properties, 283
 purification, 282
Oxyanions, 4

P

Phosphate, inorganic, 26–27, 64, 78–79, 150, 231
 measurement, 133
Phosphatidate phosphatase, 236
Phosphatidic acid, 231
Phosphatidylcholine, 210, 231–233
Phosphatidylethanolamine, 210, 231–232
Phosphatidylglycerol, 210, 231, 234
Phosphatidylglycerol(bis), 231
Phosphatidylglycerol phosphate phosphatase, 236
Phosphatidylinositol, 231, 234
Phosphatidylserine, 231, 234
Phosphoenolpyruvate, 46, 48, 163, 165, 169
Phosphoenolpyruvate carboxykinase, 41, 55–57, 71
 assay, 56
 inhibition, 57
 properties, 56
 purification, 56
 regulation, 57
Phosphoenolpyruvate carboxylase, 41, 46–48, 71
 extraction, 44
 properties, 45
 purification, 44
 regulation, 46
Phosphoenolpyruvate phosphatase, 169
 assay, 170
 extraction, 170
 properties, 171
 purification, 170
Phosphofructo-2-kinase, 75
Phosphofructokinase, ATP dependent (PFK), 87, 146
 extraction, 147
 properties, 149
 purification, 147
Phosphofructokinase, PPi dependent (PFP), 147, 150
 assay, 151
 extraction, 151
 properties, 152
 purification, 151

6-Phosphogluconate, 4
3-Phosphoglycerate(PGA), 2, 22, 130
3-Phosphoglycerate kinase, 17, 67
 assay, 159
 chloroplastic enzyme, 20
 cytosolic enzyme, 20
 extraction, 160
 inhibition by ADP and ATP, 20
 properties, 161
 purification, 21, 160
Phosphoglycerate mutase
 assay, 161
 extraction, 162
 properties, 163
 purification, 162
Phosphoglycollate, 2, 130, 132
Phosphoglycollate phosphatase, 132
 assay, 133
 isoenzymes, 133
 purification, 133
 regulation, 134
O-Phosphohomoserine, 362
Phosphohydrolase activity, latency, 397
Phosphohydrolase activity, measurement
 Ames method, 393–394
 Bencini method, 396
 coupled assay, 394
 radiometric method, 396
Phosphohydrolase activity uncouplers, 397
Phospholipase D
 assay, 225
 properties, 225
Phospholipids, 230
 inositol exchange, 237
 ethanolamine exchange, 236
 serine exchange, 236
Phosphopyruvate dehydrogenase phosphatase
 measurement, 182
 properties, 190
 structure, 186
Phosphoribulokinase, 17
Phosphorylases, 107–110
 glucan activity, 107
 maltose activity, 107
Photorespiration, 130, 278
Photorespiratory cycle, 130
Δ'-Piperidine dicarboxylate acylase, 298
Phytochelatins, 340
Protein kinases
 assay, affinity method, 373
 assay, anti-phosphoamino acid antibodies, 375
 assay, in vitro, 373
 assay, in vivo, 372
 distribution, 378–379
 properties, 380

Protein kinases (*cont.*)
  purification, 376
  regulation, 378–379
Protons, 386
Pyridoxil 5-phosphate, 278
Pyrophosphatase, 41
  assay, 63
  inhibition by Pi, 64
  properties, 63
  purification, 63
Pyrophosphatase, inorganic H^+ translocating, 338, 392
Pyrophosphate, 46, 79, 88, 96, 335, 388
Pyruvate, 46, 53, 59, 146, 165, 169, 282–283, 298, 325, 364
Pyruvate dehydrogenase, 176
  measurement, 179
  purification, 184
Pyruvate dehydrogenase complex (mitochondrial), 176
  activity, 177
  measurement *in situ*, 190
  purification, 183
Pyruvate dehydrogenase kinase, 176
  measurement, 179
  properties, 189
  purification, 184
Pyruvate dehydrogenase phosphatase, 176
Pyruvate Pi dikinase, 71
  assay, 47
  extraction, 46
  properties, 48
  purification, 47
  regulation, 8
Pyruvate 6-i dikinase regulatory protein
  assay, 64
  extraction, 64
  properties, 65
  purification, 65
  regulation, 66
Pyruvate kinase
  assay, 165
  extraction, 166
  properties, 168
  purification, 167

R

Raffinose, 76
Reductive pentose phosphate pathway, 15–37
Ribose 5-phosphate isomerase, 17, 30–34
  activation by RuBP, 30
  assay, 31
  inhibition by AMP, 30
  properties, 31
  purification, 33

Ribose 5-phosphate isomerase (*cont.*)
  tobacco enzyme, 30
Ribulose 1,5-bisphosphate carboxylase, 16–17
  activity, 8–9
  initial and total activity, 5–6
Ribulose 1,5-bisphosphate carboxylase/oxygenase(Rubisco), 1–11, 41, 66, 130
  activation, 4
  extraction, 5
  nocturnal inhibitor, 5–6
  purification, 6–7
  specificity factor, 10
  storage, 11
  structure, 3
Ribulose 1,5-bisphosphate oxygenase
  activity, 9–10
Ribulose 5-phosphate isomerase, 68
Ribulose 5-phosphate kinase, 16–17, 68
Ricinoleic acid formation, 214
Rubisco activase, 4

S

Sedoheptulose-1,7-bisphosphatase, 17, 23, 28–29, 67
  assay, 28
  properties, 28
  purification, 29
Serine, 137–138, 230, 349
Serine acetyltransferase, 350
Serine:glyoxylate aminotransferase, 136–137, 294
Serine phosphatidyltransferase, 234
Serine:pyruvate aminotransferase, 136
Serine transhydroxymethylase, 138–139
  assay, 139
  purification, 139
Stachyose, 76
Starch branching enzyme, 94
  assay, 100
  extraction, 99
  purification, 101
Starch degrading enzymes, 110
Starch metabolising enzymes, 104
Starch synthase, 94
  assay, 98
  bound enzyme, 98–99
  purification, 99
  soluble enzyme, 98–99
O-Succinylhomoserine, 362
Sucrose, 73
  biosynthesis, 74–75
  breakdown, 75
  enzymes of metabolism, 76
  synthesis, 88
Sucrose phosphatase, 74

Sucrose phosphate
  assay, 79
  inhibition, 79
  properties, 79
  purification, 79
Sucrose phosphate synthase
  assay, 77
  inhibition, 78
  properties, 77
  purification, 76
Sucrose synthase
  assay, 80
  properties, 81
  purification, 80
Sulphate, 9, 335
Sulphate reduction, 335, 339
Sulphide, 349
Sulphite reductase, 345
  assay, 346
  location, 348
  properties, 347
Sulphonylurea herbicides, 328
Sulphur assimilation, 349
Sulphurylase(ATP), 335
  assay, 336
  assay, using Luciferin–Luciferase system, 336

T

Thiamine pyrophosphate, 28, 176
Threonine, 298–300, 315–316, 325
Threonine dehydratase
  assay, 327
  extraction, 327

Threonine synthase, 316, 321
  assay, 322
  properties, 323
  purification, 323
Transketolase, 28
Translocation, H^+, 394, 398
  measurement, fluorimetric, 399
  measurement, radiometric, 400
Transport, primary, 386
Transport, secondary, 386–387
Transport, transmembrane, 386
Triacylglycerols, 220
Triose phosphate isomerase, 23, 147
  activity in chloroplasts, 23
  assay, 155
  extraction, 155
  properties, 156
  purification, 156
Triphosphates, 371–372

U

UDP-glucose, 74, 77

V

Valine, 325
Vesicles, membrane, 389
Vesicles, tonoplast, 389
  identification, 392
  preparation, 389